VOLUME THREE

Developments in Marine Geology

GLOBAL SEDIMENTOLOGY OF THE OCEAN: AN INTERPLAY BETWEEN GEODYNAMICS AND PALEOENVIRONMENT

VOLUME THREE

DEVELOPMENTS IN MARINE GEOLOGY

GLOBAL SEDIMENTOLOGY OF THE OCEAN: AN INTERPLAY BETWEEN GEODYNAMICS AND PALEOENVIRONMENT

CHRISTIAN M. ROBERT

Aix-Marseille Université

ELSEVIER

Amsterdam • Boston • Heidelberg • London • New York • Oxford
Paris • San Diego • San Francisco • Singapore • Sydney • Tokyo

Elsevier
Linacre House, Jordan Hill, Oxford OX2 8DP, UK
Radarweg 29, PO Box 211, 1000 AE Amsterdam, The Netherlands

First edition 2009

British Library Cataloguing in Publication Data
A catalogue record for this book is available from the British Library

Library of Congress Cataloging-in-Publication Data
A catalog record for this book is available from the Library of Congress

ISBN: 978-0-444-51817-0
ISSN: 1572-5480

Transferred to Digital Printing in 2012

CONTENTS

Preface

A great wealth of information on oceanic sediments has been gained during the past 30 years, principally from a variety of international programs aimed at investigating the deep ocean. These programs led to a number of thematic and regional syntheses, but more general syntheses would be desirable.

Global Sedimentology of the Ocean was designed as an effort to make such a synthesis available. It is aimed at describing the way oceanic sediments are being formed, their variability in relation to the history of oceanic systems, their diagenetic alteration and their potential as archives of past environments. New information on oceanic sediments is integrated within a fundamental context of plate tectonics, global circulation and climate principally. The book is organized in three parts:

A general presentation of the geodynamic framework, sedimentation processes and major characteristics of oceanic sediments.

A description of the relationships between plate tectonics and sedimentation, following the evolution of ocean systems as described in the cycle of Wilson from initial break-up and rift formation to the stage of collision. For each major stage of ocean evolution, generalities are associated to case studies.

A description of the formation and evolution of oceanic sediment series. For each major type of sediment, the origin, transport and preservation of sediment particles, as well as the accumulation and diagenetic alteration of sediment series are detailed, using examples.

The objective of this approach, which combines basics and generalities together with detailed examples from research and associates closely related topics, is to make information available from undergraduate to young researcher level, and to help preparing lectures.

This book benefited from the encouragements of Hervé Chamley and Jim Kennett, the support of many colleagues who helped gathering documents, the aid of the Administrative Editors and Project Manager at Elsevier and the remarkable patience of Mireille and Caroline. To all of them, I express my sincere thanks.

Part 1: Generalities

The formation and transformation of oceanic sediments involve geological, biological, physical and chemical processes. The knowledge of oceanic sediments and other objects of the Earth and Ocean Sciences therefore requires a multidisciplinary approach. This knowledge considerably increased during the past 50 years, closely following significant progress in the methods of investigation at sea and in the laboratory. In addition, ocean exploration is deeply rooted in History. Our understanding of the Ocean (including oceanic sediments and related processes) progressed step-by-step, following the evolution of techniques and ideas. Chapters 1 and 2 summarize the historical aspects of Oceanography (focusing on Marine Geology), along with the variety of processes that drive the formation and transport of sediment particles as well as their accumulation and transformation in oceanic sediments.

CHAPTER ONE

Introduction

1.1. Historical Aspects: Milestones of Oceanography

1.1.1. The Visible Ocean: Exploration, Maps and Hydrology

More than 2/3 of Planet Earth is covered by seawater, and this is no surprise that humans have been attracted by oceans early in history. As early as 40,000 years ago, some of them may have already used watercrafts to settle Greater Australia from mainland Asia, via island hopping through a restricted Indonesian Seaway. By that time, expansion of polar ice-sheets and related drop of sea level had resulted in the emersion of shelves and islands, making the journey easier. However, early sailors still had to cross deep-sea channels, up to 80 km wide. Later in history, improved boat design allowed more distant settlements and the development of communication. For example, Polynesians settled Pacific islands from about 5,000 years BP, and Egyptians had established trading routes in the Eastern Mediterranean by about 3,500 years BP.

Concurrently, humans became interested in oceanic surface phenomena and processes. Early maps used by Polynesian sailors featured sticks and shells for dominant wave direction and island position. In Europe, the oldest comments on wave dynamics and their relationships to wind activity have been attributed to Aristotle, about 2,300 years ago. By the same time, Greek and Massilian expeditions to the Atlantic had reached cold and icy areas to the North, and crossed the torrid regions of Africa to the South. During his journey up North to the British islands, Scandinavia and probably Iceland, Pytheas observed the tides and suggested that they are caused by the moon. As early as 2,200 years ago, Erathostenes described parts of the world already known from Mediterranean cultures on a map, which extended from the British islands to Ceylon and Ethiopia. Then, more and more geographical information on coastal areas was made available as the Roman Empire expanded. About 1,850 years ago (AD 150) in Alexandria, Claudius Ptolemy compiled and synthesized all existing knowledge in his "Geography", which includes a set of maps of Europe, Africa and Asia. None of his maps has survived, but he provided an index where data are expressed in coordinates and discussed, and instructions on how to create the maps (Figure 1.1).

Further progress in science and techniques (rudder, sails, compas, astrolabe, quadrant, etc.) progressively facilitated travel and allowed better knowledge of nearshore and distant oceanic areas. During the 14th century, Chinese ships reached South Africa and in Europe new maps were created from Ptolemy's Geography. With the revival of Greek and Roman concepts, the European belief of a flat Earth was replaced by the geocentric model, stating that the stars and planets orbit Earth.

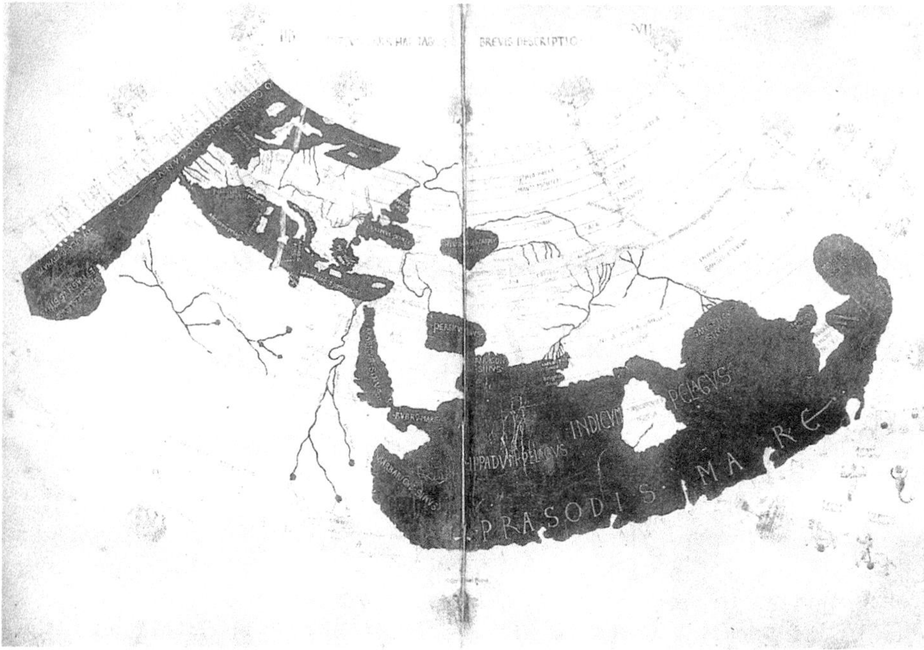

Figure 1.1 Ptolemy's world map. Reproduced from the 15th century. From http://fr.wikipedia.org/wiki/Ptolémée

In 1410, Pierre d'Ailly published "Imago mundi", based on the work of ancient Greeks and Romans: together with maps, it contains a critical assessment of Ptolemy's coordinates and distances, and the idea that India is easily reachable from West Africa across a small Western Ocean. Pierre d'Ailly's work inspired Christopher Columbus for his journey across the Atlantic. In the beginning of 16th century, intense exploration and search for new trading routes resulted in more precise geographical knowledge. Besides numerous Spanish, British, Portuguese, French and Dutch expeditions to the Americas, the voyages of Vasco de Gama and Albuquerque to the Indian Ocean, the first circum-navigation of Magellan, and the journey of Tasman to New Zealand were of special interest. This period of intense exploration was accompanied by a great wealth of geographical and hydrological information, and stimulated scientific and technological activities related to the ocean. They concerned physical geography and surface waters (tides, waves and currents) principally, the deep ocean being still considered as calm and motionless. Only a few milestones, of major importance, are indicated here.

In 1569, the introduction of the Mercator projection, making the meridians parallel on a map, allowed navigators to easily plot locations, routes and distances (Figure 1.2). Two centuries later the invention and improvement of the sextant from 1731 to 1750, followed by the invention of the chronometer by John Harrison in 1757, allowed navigators to calculate positions more accurately and favored a renewal of exploration and geographical knowledge (among others, the voyages of Bougainville, James Cook, La Pérouse to the Pacific Ocean). The degree of

Figure 1.2 Mercator's world map. From http://fr.wikipedia.org/wiki/Gerardus.Mercator

precision introduced by these instruments was unsurpassed until the invention of the radar in the late thirties, and the development of Global Positioning Systems in the seventies. During the 19th century, systematic hydrological measurements (including temperature, direction and velocity at different depths) by the first research vessels and their management by newly created national agencies, allowed publication of the first chart of the Gulf Stream in 1845, and the first map of Atlantic surface currents in 1848, under direction of M.F. Maury, superintendent of the United States Naval Observatory.

In 1687, Isaac Newton published the first mathematical explanation of tidal forces, based on combined gravitational attraction of the Sun and Moon (static theory). During the 18th century, scientists adapted Newton's theory to the variability of ocean's depth, presence and morphology of landmasses, consequences of the Earth's rotation, etc. In his "Hydrodynamica" published in 1738, Daniel Bernoulli assimilated the tides to long waves, but also used observation to predict the tides. The dynamic theory of the tides was established by Pierre-Simon Laplace in 1775, with applications to shallow coastal areas. Harmonic analyses of tides were developed by Ferrel and Kelvin during the 19th century, based on the assumption that the observed tide is the sum of partial tides (resulting from the relative movements of Earth, Moon and Sun), each partial tide being characterized by unique amplitude and phase at any given location. The harmonic theory allowed them to build the first tide-predicting machine in 1872.

In 1835, Coriolis demonstrated that Newton's laws could be used in a rotating frame of reference (such as the Earth's motion), providing that an acceleration parameter is added to the equations. When applied to a moving object this parameter, the Coriolis effect, is proportional to its speed, perpendicular to its direction, and increases with the latitude from equator to pole. Later in the 19th century, the polar explorer Fridtjov Nansen observed that floating objects including ships and icebergs were moving to the right of the wind direction, and Vag Walfrid Ekman tried to explain this offset between wind and water directions. His demonstration (1902) used the force exerted by the wind on the ocean surface, the viscosity of the fluid layers and the Coriolis effect, applied to the equations of movement. He proved that the direction of surface currents is deflected by 45° relative to the wind direction, the current speed decreases exponentially with depth, the current direction veers with depth, and the main flux of water is perpendicular to the wind direction. These results are illustrated by the Ekman spiral, and had deep implications for our understanding of geotropic circulation and surface currents and processes such as upwellings.

By comparison, only little information on deep-ocean circulation and processes was available. In 1814, von Humboldt estimated that cold deep waters in the low-latitude areas could only flow to the equator from high latitudes. Taking Mediterranean waters as an example and considering evidence for subtropical climate at polar latitudes in the geologic record, T.C. Chamberlin suggested in 1906 that warmer and more saline waters originating from evaporating tracts may have flown in the deep ocean during intervals of warm climate. In their synthesis published in 1942 (The Oceans: their Physics, Chemistry and general Biology), which is the first comprehensive textbook in oceanography, Harald Sverdrup, Martin Johnson and Richard Fleming provided detailed information on surface to intermediate water masses and circulation for each Great Ocean, but allotted only a small part to deep-water processes. Most knowledge was derived from observation of frontal zones and concerned their origin and distribution. Their circulation was still considered as sluggish. Although deep-water sampling started early in the 20th century and bathyscaph expeditions conducted by Auguste Piccard started exploring the deepest parts of the ocean during the 1930s, only limited evidence for deep circulation was available, like the demonstration by Georg Wüst and Albert Defant that sharp temperature and salinity gradients at close hydrological Meteor profiles in the Central Atlantic implied strong water flows driven by density. Extensive knowledge of deep-water circulation and processes followed the development of unmanned submarines and instrumented stations in the sixties.

For the past 50 years, extensive investigation of ocean water masses associating *in situ* measurements and new techniques like numerical modelling has led to more and more precise knowledge of water processes. However, the most remarkable breakthroughs came from the development of remote-sensing data obtained from satellites. During the 1970s and the 1980s, the Geos, Seasat and Geosat experiments proved the validity and precision of satellite altimetry. In 1992 Topex/Poseidon was launched, embarking dual-frequency radar altimeters. Key tools for international programs such as WOCE (World Ocean Circulation Experiment) and TOGA (Tropical Ocean and Global Atmosphere), satellite radars together with other devices have been especially used to explore the links between ocean and climate (El-Nino events, sea-level variations, etc.), making wide use of numerical modelling techniques. Satellite data are also being used to gain information on tides, ocean dynamics, processes at the ocean/atmosphere interface, and on the transport of heat, water mass, nutrients and salt within the ocean.

1.1.2. Within the Ocean: Physiography, Biology and Sediments

Until the 16th century, animal descriptions (bestiaries) included some incredible creatures based on uncertain reports by early explorers. During the 17th century, John Ray proposed the concept of species in an attempt to improve things. He provided the basis for the 18th century classification of Carl von Linnaeus, later improved by Georges Cuvier. By the same time, Nicolaus Stenon and Robert Hache established that fossils were petrified biological remains by demonstrating the link between petrified teeth and living sharks. However, other fossils appeared to represent animals that no longer existed.

The first attempt for explaining the presence of marine fossils on continents, up to mountain areas, has been formulated by Buffon in 1750. He theorized that much of the Earth's surface had been once beneath an ancient sea, which carved continental relief. His work raised interest for geology, and before the 18th century was over, three major theories had emerged. The neptunist theory, formulated by Abraham Werner, stipulated that sedimentary rocks result from the accumulation of debris on the floor of a silty sea which once covered the entire planet. Erosion carved the morphology of the continents and oceans as the body of water receded. The plutonist theory, expressed by James Hutton, acknowledged the marine origin of most sedimentary rocks and stressed the role of volcanic heat and pressure. Volcanic activity was involved in the formation of certain rocks and altered the surface of the Earth raising landscapes, which gradually receded because of weathering and erosion by running waters. The catastrophist theory of Georges Cuvier specified that periodic and catastrophic floodings transformed the Earth's surface, formed sedimentary rocks and fossil layers, giving way for further life development.

In 1830, Lyell suggested that the Earth's surface had changed only gradually over time, the same forces that had been active in the past being still effective in modern times. His uniformitarianist theory proposes the modern Earth as a model to understand its past. One year later in 1831, Charles Darwin embarked on the *Beagle* as a naturalist, for a science expedition around the world that lasted until 1836. He made geological and biological observations on Atlantic and Pacific islands and continental coastal areas, from Cape Verde to the Galapagos, New Zealand and Australia. Lyell's concepts were of interest to Darwin, whose theory of evolution published in 1859 is based on life adaptation to environmental change over long-time intervals.

There was some doubt on how living creatures could exist in the deep ocean until the middle of the 19th century (the azoic theory of Edward Forbes stipulated that life was impossible due to oxygen depletion), when soundings in the North Atlantic brought up shell and deep coral fragments. Charles Wyville Thomson,

Professor at Edinburgh University, and his assistant John Murray designed the Challenger expedition (1872–1876) to determine the conditions of the deep sea in all major oceans. The Challenger visited 362 stations where experiments included sampling of planktonic and benthic faunas and sediments, using nets and dredges. The expedition firmly established the basis of oceanography as a science and, among a huge quantity of new results, allowed the discovery of about 4,500 new species. The expedition also recognized the control of water temperature on the distribution of many species of microorganisms, and their gradient of distribution from the tropics to the high latitudes. In his general introduction to the Challenger Reports, Charles Wyville Thomson estimates that the most remarkable biological result of the expedition is that the distribution of living beings has no depth limit, and notices that the modern marine fauna (Figure 1.3) has close relations to the deep-water fauna of the oolite, chalk and tertiary formations. From the late 19th century, oceanographic studies mostly focussed on the emergent field of Marine Biology and also on shallower coastal areas of easier access. Marine laboratories were built in Europe and North America, among them the Institut Océanographique de Monaco and the Scripps Institution of Oceanography were built in California. Knowledge progressed rapidly and in 1919, D.W. Johnson published a synthesis on coastal processes, "Shore processes and shoreline development". Some marine laboratories progressively developed activities in all major domains of oceanography. This global and comprehensive approach ensured their lasting leadership.

Early descriptions of a shallow plain nearshore along the Mediterranean coast, down to 150 m water depth, were made by Marsigli in 1725. However, this is more

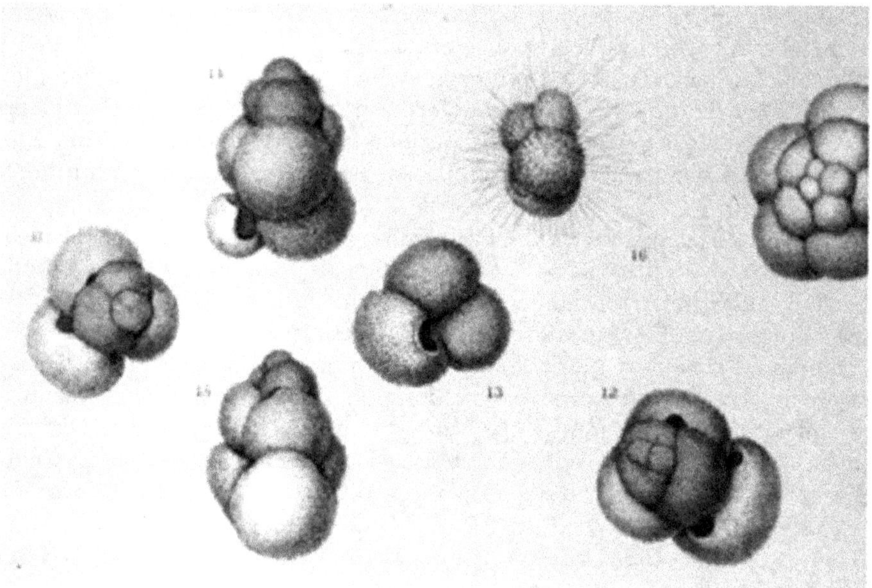

Figure 1.3 Specimens of *Globigerina*, from the H.M.S. Challenger reports. Courtesy of NOAA, http://oceanexplorer.noaa.gov/history/quotes/early/media/life.html

than a century later that the U.S. Coast Survey identified the shelf break and continental slope in the western North Atlantic in 1849, and a submarine canyon in the East Pacific off Monterey in 1857. By the same time, the success of the electric telegraph invented by Samuel Morse in 1839 and trans-Channel communication via the first submarine cable (1851) fuelled systematic studies of the ocean floor, as a preliminary for laying the first telegraph cables across the Atlantic ocean. Bathymetric charts were published by the U.S. Naval Observatory in 1855, under direction of M.F. Maury (Figure 1.4). One important feature was the discovery of an elevated submarine relief in the middle of the ocean, named the Telegraph Plateau. With the Challenger expedition (1872–1876) and the multiplication of oceanographic expeditions late in the 19th century and early in the 20th century aboard specifically designated research vessels, depth soundings allowed the creation of bathymetric maps for all oceans. The soundings being performed at irregularly spaced stations using ropes and weights, or mechanical sounding machines, errors were frequent and the maps rather imprecise. However, it was clear that the Telegraph Plateau extended to the south and that a similar elevation existed in the South Atlantic: the relief became the Mid-Atlantic Ridge. In 1917, Paul Langevin built a device for the detection of submarines, using the vibrations of a quartz crystal. The device was also able to pick up a signal from the seafloor and provided

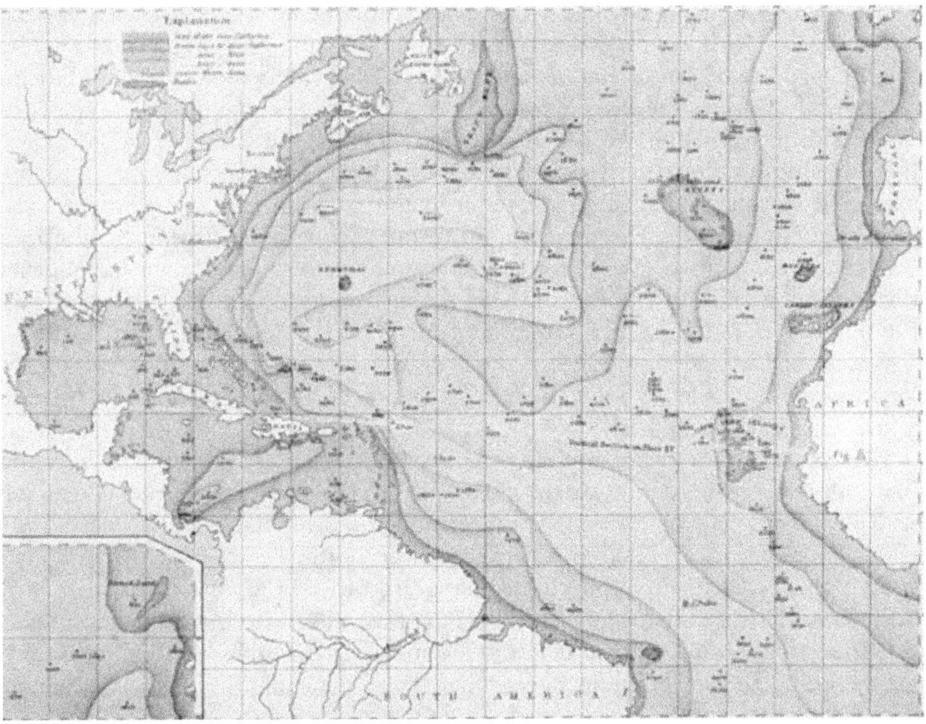

Figure 1.4 Bathymetry of the North and Central Atlantic Ocean by M.F. Maury, 1855. Courtesy of NOAA, http://oceanexplorer.noaa.gov/history/quotes/early/media/seafloor.html

the basis for the development of the echosounder, which allowed a continuous record of the water depth and significant progress in the knowledge of the ocean morphology. Using this technique, the Meteor expedition of 1925–1927 proved the continuity of the Mid-Atlantic Ridge. A further significant step in the knowledge of the seafloor was the introduction of the Precision Depth Recorder which accurately proved to be above 99% of the water depth. Our knowledge of the deep ocean greatly benefited from this tool, together with the continuous presence of oceanographic vessels (like the Vema) at sea. The scientific teams of the Lamont Geological Observatory (including Maurice Ewing, Bruce Heezen, Charles Hollister and Mary Tharpe) discovered the rift valley on top of the Mid-Atlantic Ridge in 1953, and later proved the continuity and relationships of all mid-oceanic ridge segments and related rift valleys across the oceans. The huge quantity of data accumulated during the 1950s and 1960s led to the publication of a precise bathymetric chart of the Atlantic by Elazar Uchupi in 1971, and of the famous "Physiographic maps of the Oceans" by Bruce Heezen and Mary Tharpe during the 1970s. The most recent outbreak in ocean morphology started when the first multi-beam sounding system was installed in 1963. Progressively improved with the evolution of electronics and computing systems, multi-beam sounders are now basic equipment of many research vessels and, associated to global positioning navigation systems, provide instant, detailed three-dimension maps of the seafloor.

Oldest information on deep-sea sediments was obtained when weights used for depth measurement, covered with grease, brought up a few particles. By the middle of the 19th century the construction of grabs and dredges, towed by ropes or piano wire, led to the discovery of a *Globigerina* ooze in the Gulf Stream area in 1853. During the next 20 years or so, extensive sampling led to charts of sediment patterns of the NW Atlantic margins by Delesse and later by Pourtales. However, information on deep-sea sediments remained very meager until the expedition of the Challenger that produced the basis for marine sedimentology. Largely descriptive, this extensive investigation of marine sediments synthesized by Murray and Renard (1891) for example defined the broad range of biogenic (calcareous and siliceous) oozes and terrigenous muds and allowed the discovery of polymetallic nodules. With the introduction of steel cables first used on R/V *Blake* in 1877, dredging deep sediment was made easier, and in 1888 the U.S. Coast Survey published a detailed map of surface sediments in the NW Atlantic (Figure 1.5).

While studies in Marine Biology and coastal processes progressed rapidly in the early 20th century, only little attention was paid to deep marine sediments. During the 1920s and 1930s in Europe, winter cruises of the polar vessel *Pourquoi Pas?* under direction of Jean Charcot investigated the bottom of the NE Atlantic and adjacent seas. Related studies by Louis Dangeard proved the continental nature of the floor of the English Channel in 1928. During this period, attempts were also made to retrieve sediment from below seafloor using steel tubes attached to weights or explosive devices. Sediment cores were short, generally less than two meters long. Such short cores were taken from the tropical South Atlantic during the Meteor cruise of 1925–1927. Subsequent investigations by Schott (1935)

Figure 1.5 Surface sediments of the western North Atlantic by the U.S. Coast Survey, 1888. Siliceous shore deposits and terrigenous clays and silts dominate nearshore, grading to pteropod and/or globigerina ooze and red clay to the deep Atlantic, with coral sands near some Caribbean islands. Courtesy of NOAA, http://oceanexplorer.noaa.gov/history/quotes/soundings/media/bottom.html

recognized the presence of the planktonic foraminifer *Globorotalia menardii* in the upper part of the cores, which disappeared a few tens of centimeters below seafloor. He attributed the absence of the foraminifer to the presence of colder waters during the last glacial interval, and deduced sedimentation rates for the Holocene. In the chapter on marine sediments of their synthesis on "The Oceans: their Physics, Chemistry and general Biology", Harald Sverdrup, Martin Johnson and Richard Fleming (1942) highlighted the importance of investigating marine sediments because they can bring important knowledge concerning the history of the Earth, phases of geochemistry and geology, and factors of the environment of deposition. Major sources of sediment particles (eolian, riverine, volcanic, biologic activity, etc.) and processes (settling, mud flows, transport by currents, etc.) had been identified. Also, the distribution of the major types of sediments had been recognized, but not explained yet. For instance, it was well known that biogenic carbonates decreased toward the shores, the high latitudes and the deepest areas, but not fully understood. Possible explanations included: (i) higher production of calcareous forms at low latitudes; (ii) conditions more favorable to dissolution at high latitudes; (iii) the role of oceanic morphology, the basins acting as traps for inorganic debris; (iv) longer exposure of planktonic carbonates to seawater in the deeper areas and (v) higher contents of carbon dioxide near bottom due to oxidation of organic matter. Obviously many questions were raised already, the answers waiting for the appropriate technology.

The next technical step forward comes with the Swedish Deep-Sea Expedition aboard R/V Albatross (from 1947 to 1949), which introduced the Küllenberg piston corer. The tube and weight are linked to a lever arm and counterweight. The system allows the corer to fall from a chosen height when the counterweight reaches seafloor. The tube contains a piston which regulates the penetration of the sediment. The Küllenberg piston corer commonly retrieves sediment cores 10 m to 20 m long and allowed the Swedish Deep-Sea Expedition to discover eastern Mediterranean sapropels. Together with technological progress in mineral chemistry (atomic absorption, X-ray fluorescence), mineralogy (X-ray diffraction) and especially isotope chemistry (mass spectrometer) beginning in the 1950s, the Küllenberg piston corer initiated a long period of intense investigation in Quaternary paleoclimatology and paleoceanography. For example oxygen isotope measurements, performed on Pacific sediments taken during the Swedish Deep-Sea Expedition using a Küllenberg piston corer, allowed identification by Cesare Emiliani (1955) of 15 cycles of Quaternary glaciation in place of the 4 glacial intervals previously known from continents. Recently in the late 1990s, the introduction of very light and strong aramide (kevlar) cables made possible the construction by the Institut Paul-Emile Victor of the Calypso giant piston corer, which is operated on the R/V Marion-Dufresne. A derivative of the Küllenberg system, the Calypso corer is equipped with a weight of 8–10 tons and a tube up to 75 m long, and routinely allows retrieval of sediment cores more than 50 m long, with a diameter of 10.5 cm.

Built from the experience of offshore oil exploration, the first scientific drilling ship (the Glomar Challenger) started operations in 1968. The Deep Sea Drilling Project (DSDP) was organized by a consortium of U.S. institutes and universities, the Joint Oceanographic Institutions for Deep Earth Sampling (JOIDES), and the scientific operations were managed by the Scripps Institution of Oceanography. Based on the capacities of the standard rotary system developed for oil industry, the initial phase of the project was designed to explore the deep sediments, and test the accuracy of the emerging plate tectonic theory through retrieval and datation of the oceanic crust and overlying sediment. Core recovery was poor, but for the first time investigation of deep sediments older than the Quaternary was made possible. Previous knowledge was from areas of very low sedimentation rates and from emerged series, already altered. The wealth of information that emerged from the first cruises raised interest for continuing the program. The DSDP turned international in the mid-1970s, when France, Germany, Japan, the United Kingdom and USSR joined for the International Phase of Ocean Drilling. Drilling activities were reorganized in 1984 with the addition of new members (Australia, Canada, European Science Foundation and later China). The newly created Ocean Drilling Program (ODP) started operations using a new ship, the JOIDES-Resolution, under scientific management of Texas A&M University. Since the early times of the DSDP, technical improvement has been continuous. Concerning the retrieval of sediments, the introduction of new coring devices, the extended core barrel and the advanced piston corer in the early 1980s allowed recovery of continuous series ranging from very soft sediment to hard rock, sometimes more than 1 km long. From the beginning of the DSDP in 1968 to the end of the ODP in 2003, 1,277 sites have been visited in all oceans (Figure 1.6).

Scientific drilling played a major role in our current understanding of the dynamics of the oceans at geological scale, from the processes of the lithosphere to the history of coral reefs. Among major advances in the specific domain of oceanic sediments, ocean drilling expeditions helped understanding the processes of ocean opening and creation of the margins, the formation of sediments and their diagenesis, the development of glaciation in both hemispheres and the relationship

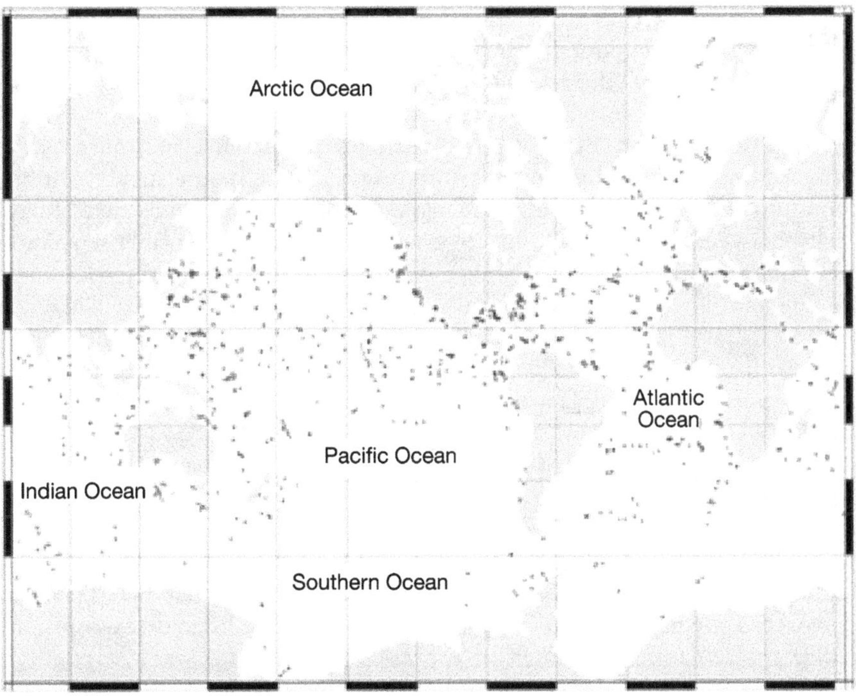

Figure 1.6 Distribution of DSDP and ODP sites in all oceans. Courtesy of ODP, http://www-odp.tamu.edu/sitemap/sitemap.html

between geophysics and sedimentology. Drilling expeditions provided the necessary scientific background and material for the full development of paleoceanography. Most of the information contained in this book is derived from the scientific knowledge generated from the ocean drilling cruises of the past 35 years. Next step, the implementation of the riser drilling ship *Chikyu* in 2007 allows the Integrated Ocean Drilling Program (IODP) to investigate oceanic areas and sediments still unexplored, for safety reasons. The main targets of the IODP include the deep biosphere and its relations to the subseafloor ocean, processes and effects of environmental change and solid Earth cycles and geodynamics.

1.1.3. The Groundwork of the Ocean: Geophysics, Lithosphere and Tectonics

None of the early theories on the dynamics of the Earth published in the late 18th century and early 19th century by Buffon, Werner, Hutton and Lyell (see Section 1.2) incorporated the idea that continents may have moved through time at the surface of the Earth. However, early cartographers from the late 16th century like Abraham Ortelius already suspected that landmasses might have not always been fixed in their present-day position, whereas Leonardo da Vinci and Francis Bacon had already noticed in 1620 the similarity of the coasts of Africa and South America. In 1782, Benjamin Franklin expressed the belief that solid superficial parts of the globe might swim upon fluid internal parts. The theory of the atolls, published by Charles Darwin in 1842 from his observations of Pacific islands during the expedition of the *Beagle*, contains the first suggestion of seafloor subsidence: fringing coral reefs develop at the periphery of young, active volcanoes, and grow continuously near surface when volcanoes cease activity and sink below sea level. A few years later in 1859, Edward Forbes noticed the presence of similar species of mollusks on both sides of the North Atlantic and suggested some ancient continuity or contiguity of both coastlines. The same year, Antonio Snider explained the similarity between African and South American coasts by a possible separation of both continents, material from within the Earth pushing the continents.

By the middle of the 19th century, observations and theories had evolved sufficiently to raise interest for further investigating the mobility of the crust, but this did not happen. One possible reason is that the necessary techniques were not available. Another reason is that despite increasing exploration of the oceans, most knowledge was obtained from marine series exposed on the continents. European basins of England, France and Germany as well as the Hercynian belt and the Alps, and the Appalachians in North America, had been intensely investigated in the search for ore deposits and coal to fulfill the needs of the early industrial era. The first unifying theory to explain the formation of mountains was proposed by James Hall in 1857, later modified by James Dwight Dana (1873), Emile Haug (1900) and other workers: the geosynclinal theory. Geosynclines were huge depressions on the borders (Appalachians) or the interior (Alps) of continental masses, where sediments deposited during episodic inflow from the ocean, forming epeiric seas. The weight of accumulating sediments caused subsidence and unstability of the structure, followed by thrusting and folding. Energy from the interior of the Earth made

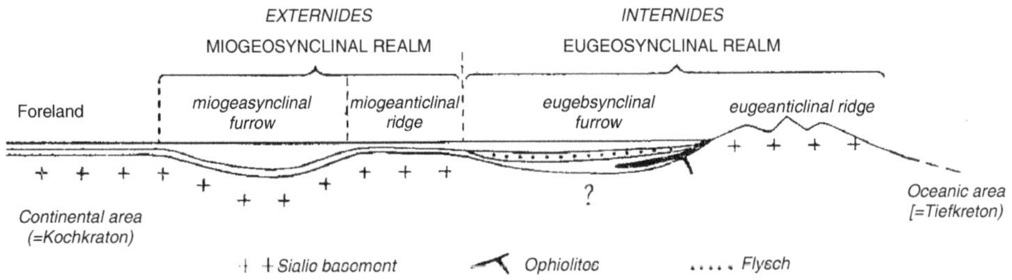

Figure 1.7 Summary of a geosynclinal structure. The miogeosynclinal realm is characterized by shallow water deposits and shows similarities with passive margins. The eugeosynclinal realm is characterized by ophiolites (ocean floor) and flyschs (alternances of pelagic and clastic proximal deposits) and shows similarities with active margins. Modified from Aubouin, J., 1965. Geosynclines, Elsevier, Amsterdam.

possible the exhumation and elevation of the structure as well as the metamorphose of sediments, followed by erosion. However, many observations were not compatible with the theory. This led to conflicting interpretations progressively complicating the principles. In 1940, Hans Stille introduced the concepts of eugeosynclinal, mostly filled with deep-sea sediments and volcaniclastics, and miogeosynclinal, mostly filled with shallow water sediments (Figure 1.7). In 1963, Robert Dietz highlighted the similarities between the sedimentary series of the geosynclines and those from the oceans. For example, faulted and thrusted miogeosynclinal sediments were comparable to those from the shelf, slope and rise, whereas the deformed and metamorphosed flyschs of the eugeosynclinal had similarities with abyssal deposits, the ophiolites being fragments of the ocean floor incorporated to the sediment during deformation. The geosynclinal theory was still the dominant paradigm in the early 1970s, but was then progressively abandoned as growing evidence from the ocean but also from mountain areas favored the theory of plate tectonics.

The geosynclinal theory focused attention on emerged mountain belts where deformation was concentrated, the oceans being considered as permanent (fixist theories). In the late 19th and early 20th century, Eduard Suess noticed that fern assemblages from India closely resembled those from Australia, Madagascar and South America, but were quite different from those in Europe and North America. He speculated that India and the southern continents were once connected by subaerial areas (the land bridges) to form a single landmass (the Gondwanaland), while they remained separated from Europe by open waters. He also hypothesized that cooling and contraction of the Earth (following Kelvin's theory) caused submergence of land bridges and the formation of mountains. Later in the 20th century, the land bridge theory was adjusted to the physiography of the oceans. For example, faunal similarities between South Africa and South America were explained by migrations across the South Atlantic, via subaerial parts of the Walvis Ridge, Mid-Atlantic Ridge and Rio Grande Rise.

Only little attention was paid to the mobilist theories. Building on the concept of Suess suggesting that the Earth's crust made of silica and alumina (Sial) principally covers a mantle made of silica and magnesia (Sima), Osmond Fischer proposed in

1881 that the solid crust may float over a viscous mantle. In the early 20th century, deformation as a result of a contracting Earth did not satisfy some scientists including Frank Taylor who proposed in 1910 that displacement of landmasses results in deformation and mountain building on their leading edge. He assumed an initial continuity between the old Caledonide and Appalachian belts to infer westward displacement of the Americas and formation of the Western Cordilleras. Continental drift was established as a theory by Alfred Wegener in 1915 and 1922 when he published a book on "The origin of continents and oceans". He used the principles of isostasy (first proposed by George Airy in 1854) to demonstrate that land bridges could not sink because of the buoyancy of the continental crust. He used the apparent fit of African and South American coastlines, and the similarity of rocks and fold belts on both sides of the Atlantic to justify initial coincidence of the continents. He used paleontological similarities to explain the necessity of dry land connections between the southern continents for part of the Mesozoic, Cenozoic differences being compatible with their separation. Wegener also noticed that Carboniferous and Permian tillites are now in tropical areas and widely dispersed, but closely match at higher latitudes when continents are put together: he hypothesized the existence of a unique continent by that time, the Pangea. He concluded that continents must have moved across the globe since then, but was aware that the driving forces were still missing although he already noticed that the Mid-Atlantic Ridge should be regarded as the place where hot Sima raises from depth.

In the late 1920s, Arthur Holmes provided the driving mechanism for continental drift. He demonstrated that radioactive decay within the globe interior keeps it hot and viscous, and that thermal exchanges occur through convection in such a medium. Convective cells in the mantle provide the energy to move the continents, and differential heating of the crust promotes deformation. By the same time, Alexander Du Toit accumulated geological evidence in favor of continental drift through detailed comparison of South American and South African geology. He demonstrated the identical nature of terranes in the Falkland islands and the Cape Town area, the continuity of the geological units and tectonic events of the Samfrau geosyncline that extended from Argentina to South Africa and Australia, as well as the coeval, continuous and uniform character of volcanism in Brazil and Angola. Du Toit also noticed the continuity of the Mozambique Trough in South Africa and Patagonia, and grouped the early Mesozoic landmasses of the northern hemisphere beyond the Tethys to form the Laurasia.

During the beginning of 1920s, methodological and technical development increased. Natural seismic waves were first used in 1909 by Andrija Mohorovicic, who inferred different physical properties of the Earth's crust and upper mantle from a delay in the reception of primary waves from an earthquake, and a first estimate of crust thickness. Artificial seismic waves and seismographs were used extensively during World War I to calculate the position of artillery batteries. This provided the basic tools for seismic prospection, and the first artificial reflection from a lithological contact (using the laws of Snell-Descartes) was obtained in 1921. Applications to oceanography started during the 1930s. The seismicity of the Mid-Atlantic Ridge was noticed by Nicholas Heck who published a world

Figure 1.8 Map of earthquake distribution by Nicholas Heck, 1932. Darker areas on oceans and continents indicate seismically active areas. Courtesy of NOAA, http://oceanexplorer.noaa.gov/history/quotes/early/media/eq_map.yell.html

seismicity map in 1932 (Figure 1.8), and the increasing depth of earthquakes below Japan with distance from the Pacific Rim was observed by Kyoo Wadati in 1935. The first offshore experiment of seismic reflexion was conducted in 1935 by Maurice Ewing, who demonstrated in the late 1940s that continental shelves are not permanent features but form progressively through accumulation of sediments. Explosives were used as a source until the 1950s, and then progressively replaced by airguns (air is compressed in a chamber to high pressure, and suddenly released in the water). In the early 1930s, Felix Vening-Meinesz started measuring gravity from submarines and demonstrated that isostatic equilibrium is not reached in oceanic trenches areas, due to tectonic activity. The dynamo theory explaining the Earth's magnetic field was proposed during the 1940s, at the time early magnetic detectors were being built for military purposes. The first towed magnetometer developed by the Scripps Institution of Oceanography was installed on the *Pioneer* and marine magnetic surveys started in 1955. The first cruises recognized frequent changes in the magnetic direction of the ocean floor.

The multiplication of oceanographic cruises implementing newly developed and continuously improved methodology including seismic reflection, bathymetry, magnetometry, gravimetry and sampling of deep sediments (Küllenberg coring, dredging) collected a considerable quantity of data. Their analysis brought puzzling information on the geology of the oceans: (i) the ocean floor was mafic in nature and very thin (about 5 km thickness); (ii) heat flow was very high below mid-oceanic ridges compared to continental areas; (iii) magnetic banding of the seafloor was duplicated on both sides of mid-oceanic ridges; (iv) the depth of guyots increased with distance from mid-oceanic ridges; (v) the sediment cover was generally thin while thicknesses of more than 20 km were expected from fixist theories; (vi) sediment thickness decreased from the margins to the crest of mid-oceanic ridges where young basalts outcropped and (vii) nothing very old was brought up from the ocean floor. A different picture of the Earth was coming out. Harry Hess put together all this new information in an article on "the history of ocean basins" published in 1962, prudently introduced as an "essay in geopoetry", which presented the seafloor spreading theory. The basic idea was that the ocean floors are moving like conveyor belts, carrying passively the continents along with them. Oceanic crust is constantly produced in the rift valleys of mid-oceanic ridges, old crust being pulled and destroyed in deep-ocean trenches near the edges of continents. Therefore, oceanic basins were not permanent features anymore, and first estimates suggested that all ocean floors should be younger than 200 Ma.

The seafloor spreading theory helped focussing attention on critical points of geodynamics. Building on this theory, Fred Vine and Drummond Matthews suggested in 1963 that the magnetic stripes of the seafloor represent the direction of the magnetic field at the time the crust was created, successive magnetic anomalies being preserved in the lava. In 1966, they compared seafloor anomalies with the new geomagnetic timescale established for the past 4 Ma by Allan Cox, Richard Doell and Brent Dalrymple (by K/Ar datation of lavas containing magnetic inversions), providing a chronology of seafloor spreading. A network of seismic stations was developed in 1963, to monitor new regulation of nuclear testing. This allowed a precise mapping of earthquake concentration by Lynn Sykes in 1965, yielding new information on activity and motion of the seafloor. The same year, John Tuzo Wilson discovered that mid-oceanic ridges are offset by perpendicular faults, and that only fault sections separating ridge segments, named transform faults, are seismically active. Rapidly, it appeared that movement was concentrated in narrow bands separating vast portions of oceans and continents where activity was comparatively weak. These vast portions of relatively passive lithosphere delimited by narrow, active areas were considered as rigid plates. A comprehensive model integrating continental drift, seafloor spreading and mountain building was progressively established via a set of articles published in 1967 and 1968. Brian Isacks, Xavier Le Pichon, Dan Mc Kenzie, Drummond Matthews, Jason Morgan, Lynn Sykes, John Tuzo Wilson, Fred Vine among others played key roles in its conception. Principles imply a rheologic decoupling of the lithosphere and astenosphere, the production of lithosphere along mid-oceanic ridges and its destruction in island arc areas along Wadati–Benioff sheer plans, horizontal movements induced by convection in the astenosphere and vertical movements due to isostasy. In 1968 a paper by Le Pichon presented the six major plates. Together with magnetic anomalies and timescale, the principles of plate tectonics were then used to reconstruct the absolute ages of continental breakup and history of the oceans. In 1970, John Dewey and John Bird demonstrated that the plate tectonic theory also helped understanding mountain building (collision belts) and the history of past oceans.

The 1970s confirmed plate tectonics as a paradigm for Earth science processes. Leg 3 of the DSDP under direction of Arthur Maxwell and Richard von Herzen recovered samples of basalt crust and overlying sediments of increasing age from the crest of the mid-oceanic ridge to the deep basins in the South Atlantic. Following the race for reaching record depth using bathyscaphs, small manned submersibles resistant to high pressure and easier to operate were developed in the late 1960s. In 1973 and 1974, the French submersible *Cyana* and the U.S. submersible *Alvin*, together with the bathyscaph *Archimede*, participated in the French American Mid Ocean Undersea Survey (FAMOUS) to explore the rift valley of a segment of the Mid-Atlantic Ridge West of the Azores. They observed open fissures of the oceanic crust, continuous flows of pillowed basalts, hydrothermal vents and concretions, and deep organisms drawing energy from hydrothermal activity. The theory of plate tectonics was supported by physical evidence.

1.2. Objectives

1.2.1. Oceanic Sediments in their Context

The basics for the classification of oceanic sediments have been provided by Murray and Renard in 1891, from the many observations made on surface sediments during and after the Challenger Expedition. They had identified components of biologic, clastic and authigenic origins. Most of the time one group of components was dominant, providing the name of the sediment (i.e. *Globigerina* ooze). They mapped the distribution of oceanic sediments, highlighting the presence of biogenic calcareous (foraminifer and nannofossil) and siliceous (radiolarian and diatom) oozes in most of the pelagic realm, the dominance of clastic muds near the continents, as well as the occurrence of red clays and manganese nodules in the deepest parts of the oceans. Further progress came from systematic survey, and from the development of coring and drilling tools for the recovery of sedimentary series. Progressively, the variability of oceanic sediments through space and time became evident.

- *Variability through space.* Besides the global distribution of oceanic sediments outlined by Murray and Renard, significant regional variations have been evidenced. For example, sediments are siliciclastic to hemipelagic in the Mediterranean and the Red Sea. Siliciclastic deposits are observed on and near the margins of the Atlantic Ocean with the exception of some upwelling areas where siliceous biogenic deposits locally dominate. Calcareous biogenic sediments dominate in central areas of the Atlantic, grading to hemipelagic sediments toward the basins. In the Pacific Ocean, siliciclastic and hemipelagic sediments dominate near landmasses and in backarc areas. Volcaniclastics are locally important near island arcs, and siliceous biogenic oozes in areas of cold, upwelled waters. Calcareous biogenic oozes dominate in shallower parts of the tropical East Pacific and near Australia, red clays and polymetallic nodules being dominant in the deep areas of the central and western basins. The nature and composition of oceanic sediments vary with latitude, distance from the shore, water depth and hydrology. Oceanic sediments are influenced by the morphology and geological structure of the ocean, as well as by the distribution of water masses and circulation that are closely related to global climate.

- *Variability through time.* Many DSDP and ODP sites drilled for reconstructing the history of the ocean show typical sequences where oldest deposits are essentially siliciclastic in nature, and frequently contain relatively high proportions of organic matter. They usually grade to hemipelagic and sometimes to pelagic biogenic sediments upwards. Transitions to hemipelagic and/or biogenic sediments are either abrupt or progressive, and recurrences of former sediments are sometimes observable. However, abrupt transitions sometimes reveal significant hiatuses. For example, siliciclastic deposits drilled off Tasmania during ODP Leg 189 grade progressively to biogenic oozes over 15 Myr or so, but transition looks progressive (300 m) west of Tasmania because of high sedimentation rates, and abrupt east of Tasmania (<10 m) because of several long hiatuses. Also, the Miocene transition from nannofossil to diatom ooze at ODP Site 689 on Maud Rise in the Weddell Sea contains recurrent intervals (several meters) of pure nannofossil ooze. Shore based studies have shown that transitions from siliciclastic to biogenic sediments often coincide with different stages of ocean evolution; for example, final separation of Australia from Antarctica and regional subsidence for the transition off Tasmania. Also, transitions from calcareous to siliceous biogenic sediments are often associated with changes of water masses and circulation; for example, the expansion of Antarctic waters and circumpolar circulation for the transition on Maud Rise in the Weddell Sea.

Oceanic sediments have something to teach, in relation to the history of the oceans, their tectonics, biology and hydrology: this is about ocean widening and deepening, opening of passageways for surface and deep-water circulation, succession of water masses, supply of nutrients, etc. This information is closely connected to major processes including plate tectonics, global climate and evolution of the biosphere.

Using oceanic sediments as a base, this book deals with their interaction with these processes, and addresses the following topics:

- formation (in the oceans and on the continents), transport and deposition of sediment components;
- early diagenetic evolution of oceanic sediments; and
- major characteristics of oceanic sediments and their variation in relation to the history of the oceans.

1.2.2. Specificity of the Book

This book is not about general or structural geology, oceanology or sedimentology. Many other books are more general, dealing with all aspects of sedimentology and sedimentary geology, or physical oceanography. Others focus either on regional aspects or thematic aspects of oceanography. Some books, published during the past

20 years, have a global approach of the ocean using one or several disciplines, and they inspired the organization and contents of this book.

Marine Geology (1982) by James P. Kennett, is a comprehensive synthesis including information on geophysics and structure, rocks and sediments, microfossils and stratigraphy. It also explains how information from oceanic deposits, together with modern and classical concepts of geology, can be used for understanding the history of the ocean basins and margins, as well as past water masses and climates. Deep Marine Environments (1989) by Kevin T. Pickering, Richard N. Hiscott and Frances J. Hein, deals with modern and ancient deep marine sedimentation, with focus on plate tectonic aspects, deep-sea mechanisms and environmental processes. Sedimentary Basins (1992) by Gerhard Einsele, is about qualitative and quantitative aspects of sedimentology and sedimentary geology including flux rates, diagenesis and fluid flow with focus on oceanic basins in a context of plate tectonics. The Sea Floor (1996) by Eugen Seibold and Wolfgang H. Berger provides information on ocean morphology and tectonics, summarizes geologic processes in the deep sea and shelf areas, and reviews the climatic record of deep-sea sediments. Géologie Sédimentaire (1999) by Bernard Biju-Duval, copes with the formation and evolution of sedimentary basins in a geodynamical context, mechanisms and environments of deposition, and diagenesis, with special interest in processes related to oil and gaz formation.

This book provides an overview of oceanic sedimentation, with focus on historical, evolutionary and synthetic aspects. General topics are illustrated by regional examples: the analysis of sediment series in diverse oceanic systems is used for understanding the history of oceans and ancient environments and their links to global processes. The substance is principally derived from deep sea drilling expeditions (DSDP and ODP) and other programs of the past 30 years. This information is placed in a framework provided by plate tectonics and history since the Jurassic, a time span encompassing most of the creation and evolution of modern oceans and the Tethys.

The book is divided in 13 chapters. The first chapter summarizes some historical aspects of oceanography, from surface to lithosphere, and provides the objectives. The second chapter is a general presentation of plate tectonics, physical oceanography and marine sedimentology, to be used as a framework for understanding the evolution of oceans and oceanic environments. Chapters three to nine deal with the broad characteristics of sedimentation during the evolution of the ocean, from early opening in rift systems to collision of continental margins. Chapters 10 to13 describe the origin of sediment particles, as well as the formation and transformation of major types of oceanic sediments.

FURTHER READING

Arcyana, 1978. Atlas FAMOUS. Bordas, Paris.

Aubouin, J., 1965. Geosynclines. Elsevier, Amsterdam.

Biju-Duval, B., 1999. Géologie Sédimentaire. Technip, Paris.

Bird, J.M., Isacks, B. (Editors), 1972. Plate tectonics. Selected papers from the journal of geophysical research. American Geophysical Union, Washington, DC.

Chamberlin, T.C., 1906. On a possible reversal of deep-sea circulation and its influence on geologic climates. Journal of Geology, 14: 363–373.
Du Toit, A.L., 1937. Our wandering continents. Oliver & Boyd, Edinburgh.
Einsele, G., 1992. Sedimentary basins. Springer, Berlin.
Exon, N.F., Kennett, J.P., Malone, M.J., 2001. Proceedings of the Ocean Drilling Program, Initial Reports, volume 189. Ocean Drilling Program, College Station, TX.
Heezen, B.C., Tharp, M., 1977. World ocean floor. U.S. Navy Office of Naval Research.
Hess, H., 1962. History of ocean basins. In J.L. Engle, H.L. James, B.F. Leonard (Editors), Petrologic studies: A volume to honor A.F. Buddington. Geol. Soc. Amer., Denver, Co.
http://oceanexplorer.noaa.gov/history.html
http://penelope.uchicago.edu/Thayer.html
http://rst.gsfc.nasa.gov
http://wrgis.wr.usgs.gov
http://www.cosmovisions.com
http://www.iodp.org
http://www.oceansonline.com
http://www-odp.tamu.edu/publications.html
http://www.pbs.org
Kennett, J.P., 1982. Marine geology. Prentice-Hall, Englewood Cliffs, NJ.
Kious, J., Tilling, R.I., 2001. This dynamic earth: The story of plate tectonics. U.S. Geological Survey, Denver, CO.
Maury, M.F., 1855. The physical geography of the seas. Harper, New York.
Murray, J., Renard, A.F., 1891. Report on deep-sea deposits, based on specimens collected during the voyage of H.M.S. Challenger in the years 1872 to 1876. Eyre & Spottiswoode, London.
Pickering, K.T., Hiscott, R.N., Hein, F.J., 1989. Deep-marine environments. Unwin-Hyman, London.
Seibold, E., Berger, W.H., 1996. The sea floor, an introduction to Marine Geology. Springer, Berlin.
Suplee, C., 2000. Milestones of science. National Geographic Society, Washington, DC.
Sverdrup, H., Johnson, M., Fleming, R., 1942. The Oceans: Their physics, chemistry and general biology. Prentice-Hall, Englewood Cliffs.
Wegener, A., 1929. The origin of continents and oceans. Dover Publishers Inc., New York.

CHAPTER TWO

Generalities: Geodynamics of the Ocean

Composition, fluxes and distribution of sediment components in the ocean are influenced by a variety of factors including the morphology of oceans and continents, conditions of weathering and erosion on the continents, volcanic activity, transport by runoff, winds and currents, availability of nutrients, pattern of oceanic circulation and chemistry of water masses. These conditions vary through time, but are closely related to two major domains: global tectonics and global climate. This chapter provides basic information on both domains and places the main features of oceanic sediments in their evolving global context.

2.1. The Geological Structure of the Ocean

2.1.1. The Lithosphere and Lithospheric Plates

In 1909, Andrija Mohorovicic observed a delay in the reception of primary waves emitted by a single earthquake in Croatia. He deduced that the waves travelled through terranes of different properties: a faster upper mantle and a slower crust (separated by the discontinuity of Mohorovicic). They both form the most rigid (very high viscosity) envelop of the Earth, the lithosphere.

The Earth's crust shows differences below continents and oceans (Figure 2.1). The continental crust has an average thickness of 30 km and average density of 2.8, and mostly consists of rocks (siliciclastics, shales, granites, gneiss, etc.) and minerals (quartz, feldspars, micas, clays, etc.) enriched in silica. The lower crust (below the discontinuity of Conrad) is slightly different in composition as it includes intrusions from the upper mantle in depressurized, faulted areas. The continental crust is derived from the mantle through a succession of geological processes (melting, crystallization, alteration and weathering, erosion, deposition, diagenesis, metamorphism, etc.). Due to low density, the continental crust stays in surface and is very old (Precambrian rocks) in some places. As a consequence, its structure is very complex and contains a record of successive events of geological history. The rocks and minerals of the continental crust have a brittle comportment (faults) near surface. As pressure and temperature increase with depth they become ductile (shear zones) in the lower crust where they are metamorphosed and melt at temperatures as low as 600–700°C into silica-dominated magmas (at the origin of granites, rhyolites, etc.). The oceanic crust has an average thickness of 7 km and average density of 2.9. The oceanic crust mostly consists of basalts, gabbros and sometimes peridotites, occasionally serpentinized. More than 1,700 m of oceanic crust have been drilled at DSDP Site 504 off the Galapagos Islands: below 571 m of basalt pillow-lavas, Hole 504 penetrated 209 m of breccias and pillow-lavas with intrusive basaltic dykes and 948 m of massive basaltic dykes. More homogeneous in

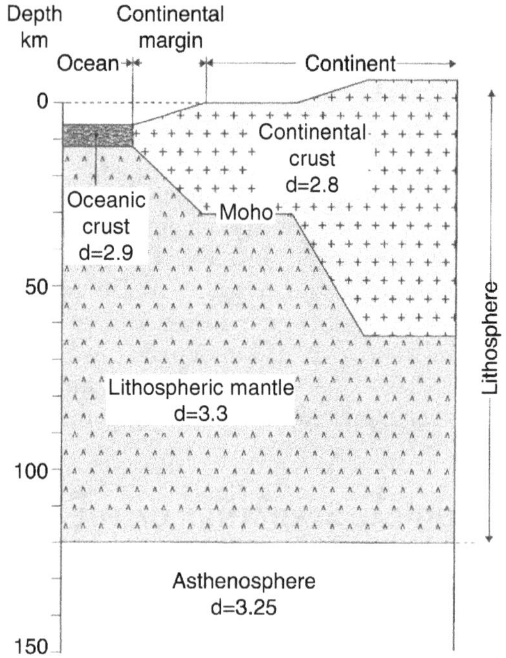

Figure 2.1 Main characteristics of the Earth's lithosphere. Modified from Boillot, G., Coulon, C., 1998. La déchirure continentale et l'ouverture océanique, Overseas Publishers Association, Amsterdam.

composition than the continental crust, the oceanic crust is essentially of Cenozoic and Mesozoic age and has a brittle comportment.

The upper mantle has an average density of 3.3 and consists of rocks (peridotites) and minerals (olivine, pyroxenes, amphiboles, oxides, etc.) enriched in iron and magnesium by comparison with the crust. Upper mantle rocks drilled on the Mid-Atlantic Ridge during ODP Leg 209 consist of peridotites with intrusive dykes of gabbros and traces of hydrothermal alteration. Differences in the pressure exerted by oceanic and continental crusts are compensated within the upper mantle. For isostatic equilibrium, its thickness varies below oceans and continents: low mass seawater ($d = 1$) and thin oceanic crust are compensated by a thicker upper mantle, whereas high mass and variable thickness of continental crust (up to 60 km below mountain areas) are compensated by a thinner upper mantle. Differences in conditions of isostatic equilibrium between both types of crust are illustrated by average elevations of continents (+1,000 m) and oceans (−4,000 m). The rocks and minerals of the upper mantle have a brittle comportment to pressures and temperatures much higher than those in the continental crust. Therefore, the upper mantle has a brittle comportment changing to ductile with depth below 40–60 km and for temperatures of 600–800°C.

Including both types of crust and the upper mantle, the lithosphere consists of a variety of rocks enriched in silica in surface and alternances of brittle and ductile terranes. The constraints applied to the lithosphere vary with direction (anisotropy).

In brittle terranes, elastic deformation is followed by rupture for higher stress. In ductile terranes, elastic deformation is followed by creep deformation and rupture as stress increases. Rupture occurs during earthquakes. Conduction processes ensure upward heat transfer within the lithosphere (geothermal gradient) and release in the atmosphere and ocean. Heat transfer affects some physical properties of the lithosphere, higher fluxes resulting in lower density and higher volume while low fluxes induce higher density and lower volume. The physical properties of the Earth's mantle change with higher pressure and temperature, lithological and chemical compositions being still the same: conduction is supplanted by convection for temperatures above 1,300°C, at an average depth of 120 km where density decreases to 3.25 asthenosphere and viscosity decreases. This transition marks the lower limit of the lithosphere and upper limit of the asthenosphere, and is illustrated by low velocities of seismic waves. The isotherm 1,300°C moves upwards above mantle plumes, resulting in a thinner lithosphere. The physical properties of the asthenosphere facilitate some geological processes: subsidence and subduction of the lithosphere, intrusion of the asthenosphere in fractured areas of the lithosphere. To summarize, the lithosphere is a rigid envelop made of an alternance of brittle and ductile layers where transfer of energy occurs through conduction, which drifts over a less viscous, convective asthenosphere. Generally the lithosphere experiences only rare and limited geological activity, as attested by the scarcity of earthquakes and volcanoes in most areas. Some vertical deformation is related to isostasy, for example, in areas of periodic ice-cap growth and of important accumulation of sediments.

Stress is transmitted within the lithosphere to specific geographic areas where geological activity is concentrated. These areas consist of long and relatively narrow belts where important deformation is associated to intense seismic and volcanic activity. They include mid-oceanic ridges, oceanic trenches and island arcs and young mountain belts such as the Andes, the Alps, the Caucasus, and the Himalayas. Areas of intense geological activity delimit portions of lithosphere where geological activity is by comparison negligible: the lithospheric plates (Figure 2.2). The lithosphere is separated into seven major plates including Africa, Antarctica, Australia, Eurasia, Pacific, North America and South America, associated to smaller plates such as Arabia, Caribbean, Cocos, India, Nazca, Philippines and Scotia. Most lithospheric plates include portions of oceanic and continental crusts, but some carry almost exclusively oceanic (Pacific plate) or continental (Turkish plate) crust.

2.1.2. The Motion of Lithospheric Plates

Lithospheric plates move at the surface of the Earth, considered as a sphere. According to Euler's theorem, any point of a sphere can be moved by a single rotation about an axis through the center of the sphere. The intersections of the axis of rotation (Eulerian axis) with the surface of the sphere are the rotation poles (Eulerian poles). As a consequence, the relative motion of lithospheric plates can be described using the position of rotation poles, angle of rotation and angular velocity which is the same for any point of the plate (Figure 2.3). At the surface, the velocity

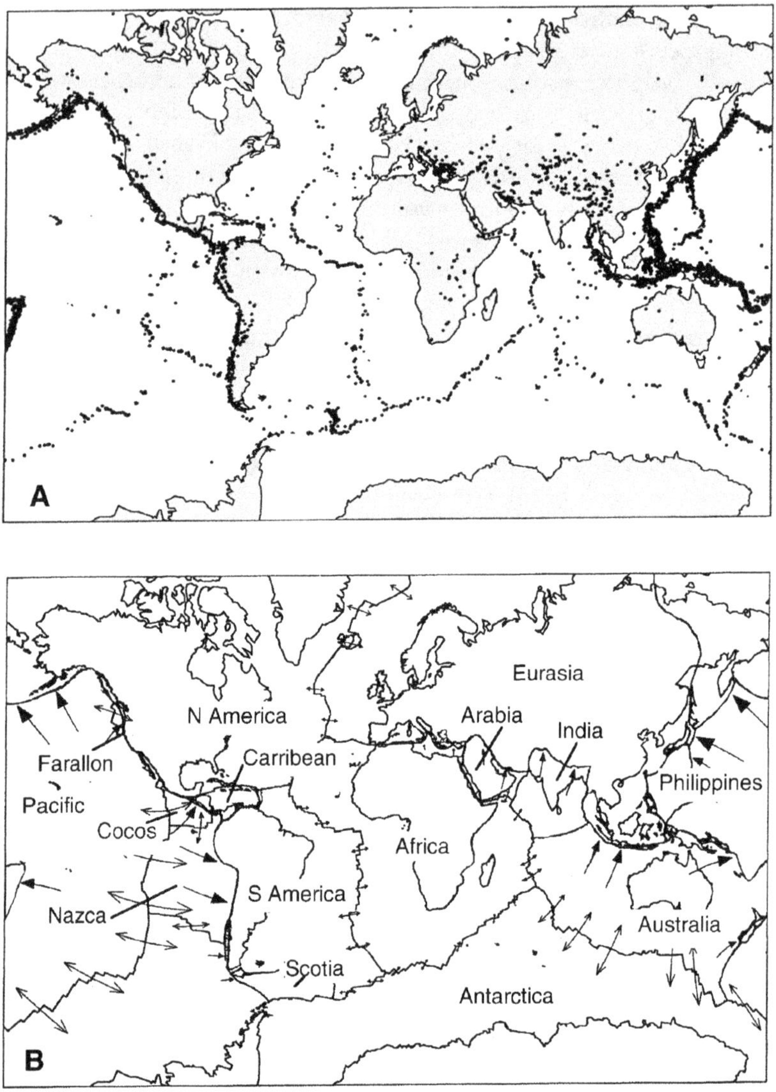

Figure 2.2 (A) Earthquakes of magnitude 5.5 and above, 1963–1987 and (B) boundaries of major lithospheric plates (arrows are proportional to angular velocity). Modified from Gordon, R.G., Stein, S., 1992. Science, 256, 333–342.

of any given point of the lithospheric plate (linear velocity, V) is deduced from the equation

$$V = \omega R \sin \theta$$

where ω is the angular velocity, R the radius of the Earth and θ the angular distance between any given point and the rotation pole. Therefore, the linear velocity of lithospheric plates is null at rotation poles and increases with distance to a maximum at the Eulerian equator (angular distance of 90°). Linear velocities between 2 and

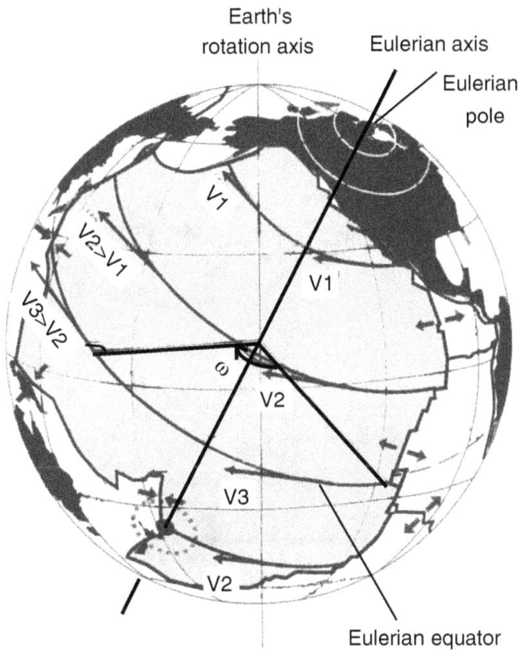

Figure 2.3 Eulerian parameters of the modern Pacific plate. ω: angular velocity; V_1, V_2, V_3: linear velocities. Modified from Renard, V., Pomerol, C., 2000. Eléments de Géologie, Colin, Paris.

20 cm/yr are commonly observed. The volume of the Earth being constant, the movements of lithospheric plates are (to some extent) interdependent: changes of rotation poles or angular velocity of any given plate have repercussions in the motion of other plates. Two major types of relative movements between lithospheric plates are observed: divergence and convergence.

The motion of lithospheric plates is not continuous but mostly occurs during earthquakes, when the stress accumulated for some time is suddenly released. Relative motion of lithospheric plates is being monitored along the San-Andreas Fault (California) where the Pacific plate slides northward along the North American plate (taken as a reference) at an average velocity of 5 cm/yr. An earthquake of magnitude 7.1 occurred in Loma Prieta (90 km south of San Francisco) in October 1989, the hypocenter being located at 18.5 km beneath the continental crust on an oblique fault plane adjacent to the main fault (Figure 2.4). Resulting motion of the Pacific plate was of about 1.30 m upward and 1.85 m northward. The main shock was followed by 4,760 aftershocks in three weeks, most of them of low magnitude (below 3). They were distributed over an 80-km long stretch from surface to about 19-km depth, indicating regional readjustment of multiple fault segments. The Loma Prieta earthquake and aftershocks filled a gap in the density of Central California earthquakes for the past 20 years, releasing excessive accumulation of stress in the area. Other areas of low seismicity are indicative of strain build-up and likely to be the place of further plate motion in the near future. The deficit of the Parkfield area was partly compensated by a 5.4 earthquake and aftershocks in September 2004.

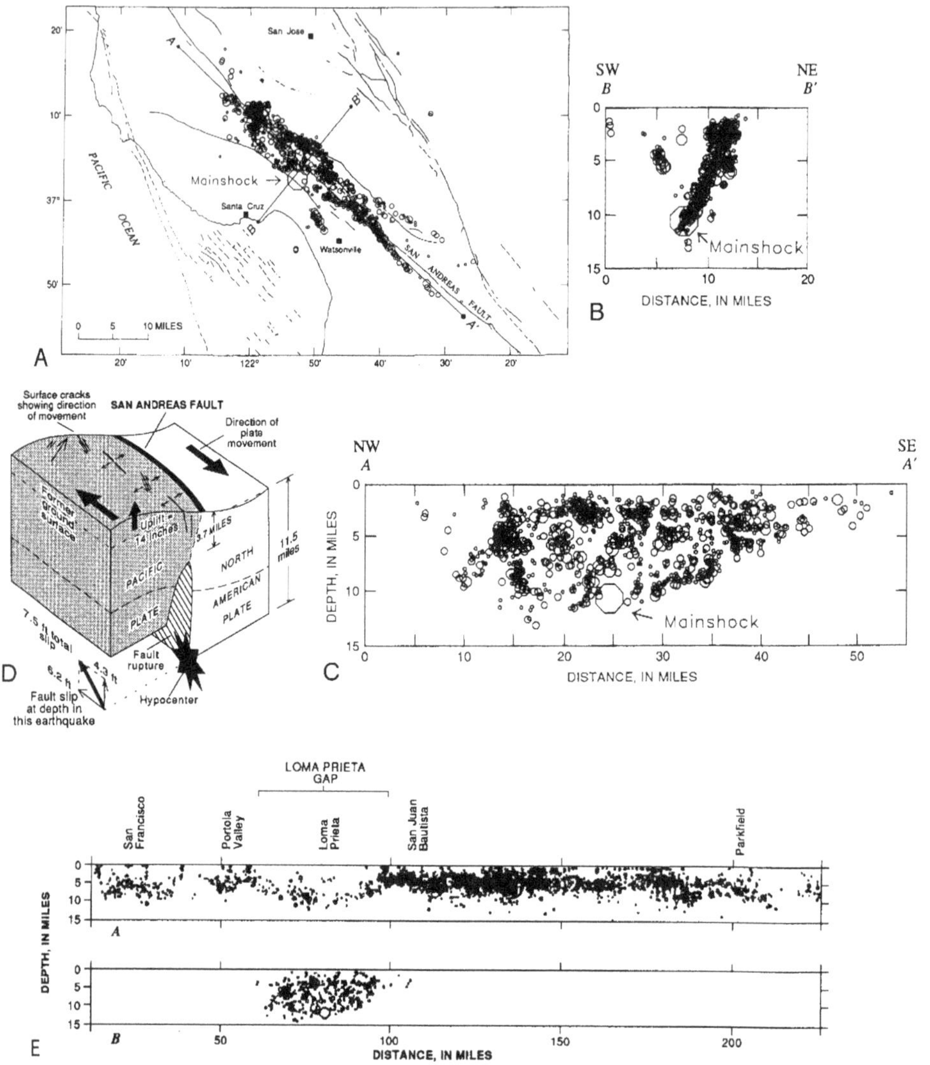

Figure 2.4 The Loma-Prieta earthquake (California, 1989). (A) Location map showing epicenters of the mainshock and aftershocks. (B and C) Vertical distribution of the mainshock and aftershocks along two cross-sections perpendicular and parallel to the San Andreas Fault. (D) Inferred motion of the Pacific plate relative to the North American plate. (E) Cross-section showing regional seismicity for the 1969–1989 interval and during the Loma Prieta earthquake. The Loma Prieta earthquake filled a gap in the regional seismic activity. Other gaps remain in the San Francisco and Parkfield areas. Courtesy of U.S. Geological Survey: Plafker, G., Galloway, J.P., 1989. U.S. Geological Survey Circular 1045 (http://www.usgs.gov).

2.1.3. The Divergence of Lithospheric Plates

Divergent boundaries are located at mid-oceanic ridges, where plates are moving away and new material is added. There, earthquakes occur at depths less than 30 km.

The asthenosphere raises in areas where the lithosphere is thinned and fractured, its decompression being associated with increased temperatures (adiabatic expansion). Thus, upper mantle rocks partially melt (anhydrous fusion). Silicates being more sensitive to high temperatures, the resulting magma is enriched in silica. Because of lower density and viscosity (a consequence of high temperature and chemical composition), the magma raises in fractured areas of the lithosphere where gabbros crystallize in magma chambers, the remaining fluid magma forming basaltic intrusions in the upper crust or welling in the axial rift valley as pillow-lavas. When magma cools and crystallizes it is magnetized in the direction of the Earth's magnetic field below the point of Curie. Newly created lithosphere moves away from the active area, driven by convection in the underlying asthenosphere. New oceanic crust being continuously created at mid-oceanic ridges, the ocean floor extends and preserves the imprint of successive magnetic reversals, providing a chronology of seafloor spreading: the age of the oceanic crust increases with distance from mid-oceanic ridges. As a consequence, the sediment cover is very thin and all very recent on top of mid-oceanic ridges. The thickness of the sediment column increases, as does the age of the oldest sediment in contact with the oceanic crust, with distance from the active area of mid-oceanic ridges. Leg 3 of the DSDP verified the theory by drilling seven holes along an east–west transect across the mid-oceanic ridge in the South Atlantic (Figure 2.5). K–Ar ages of the seafloor, paleomagnetic data from basalts and overlying sediments and biostratigraphic ages were all in accordance: ages of the seafloor and stratigraphy of overlying sediment increase from Late Miocene to Maestrichtian over a stretch of about 1,100 km, yielding an average spreading rate of about 2 cm/yr, and show a remarkable symmetry about the ridge axis.

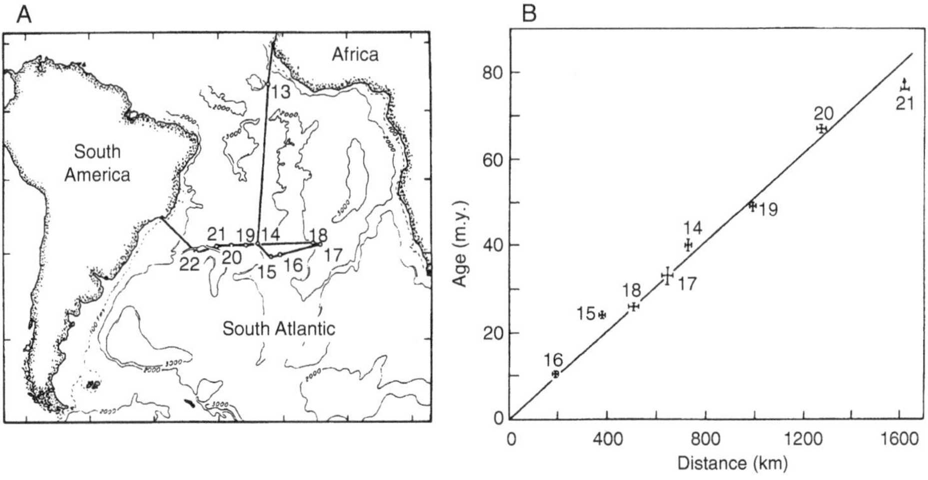

Figure 2.5 K–Ar ages of the South Atlantic seafloor. (A) Location of DSDP Leg 3 sites and (B) relationship between age of the seafloor and distance from the mid-oceanic ridge. Modified from Maxwell, A.E., von Herzen, R.P. et al., 1970. Initial Reports of the Deep Sea Drilling Project, volume 3. U.S. Gov. Print. Office, Washington, DC.

Most solids decrease in density and expand in volume when their temperature increases. Heat flow from abnormal asthenosphere and upper mantle at mid-oceanic ridges favors the expansion and decreased density of oceanic lithosphere. Isostatic conditions being modified, the active area of mid-oceanic ridges is elevated by 1,500–2,000 m above the deep oceanic basins for equilibrium. As newly created ocean floor moves away from active areas and cools, its density increases and volume decreases: for isostatic reasons, cool and older oceanic lithosphere forms the basement of deep oceanic basins. The depth of oceanic basins varies with the age of the oceanic lithosphere according to the empirical formula:

$$D = 2,500 + 350\sqrt{t}$$

where D (in meters) is the average depth of mid-oceanic ridges and t the time elapsed since creation of the considered portion of oceanic crust (Figure 2.6). Beyond 70 Myr the subsidence slows down and the ocean floor descends asymptotically to maximum possible depth of 5,500–6,500 m. Greater water depths are only found in tectonically active transform fault and oceanic trench areas. The activity of mid-oceanic ridges is variable. One consequence of variable activity is that some characteristics of mid-oceanic ridges may change according to time and/or location (Figure 2.7). High activity (as evidenced through high heat flow and volcanic activity and larger magnetic anomalies) results in spreading rates around 10 cm/yr and heterogenous oceanic crust made of basalts and gabbros. Fast ridges are higher and larger than average (higher volume), and their rift valley is narrow and sometimes absent. Low activity (as evidenced through low heat flow and volcanic activity and narrow magnetic anomalies) results in spreading rates of a few centimeters per year, and homogeneous oceanic crust made of basalts. In this case, gabbros mostly crystallize at depth in magma chambers. Slow ridges are lower and narrower than average (lower volume). Their rift valley is generally deep (around 1,000 m) and large (10–20 km), and includes lava fields and volcanoes. Important hydrothermal activity metamorphoses magmatic rocks in depth and resulting hot springs (around 350°C) of highly mineralized waters favor the accumulation of

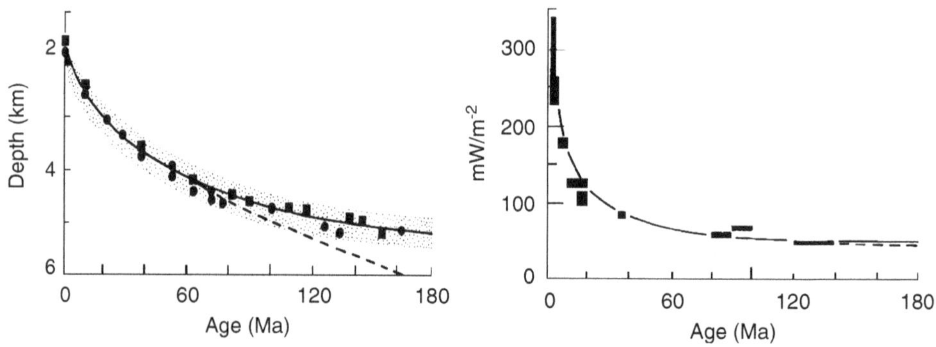

Figure 2.6 Relationship between heat flow, depth and age of the oceanic lithosphere. The thin solid lines represent the envelop of the data (dots). Thick lines represent the results from different plate models. Reprinted by permission from Macmillan Publishers Ltd.: Stein, C.A., Stein, S., Nature, 359, 123–129, copyright 1992.

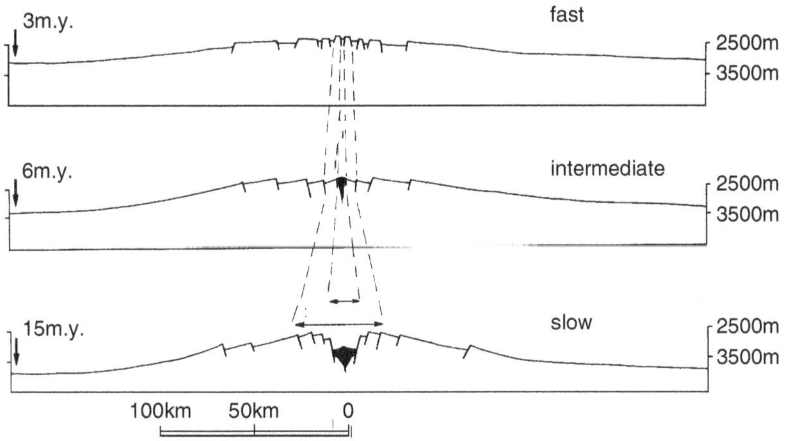

Figure 2.7 Changing morphologies of mid-oceanic ridges as a function of their activity. Modified from Choukroune, P. et al., 1984. Earth and Planetary Science Letters, 68, 115–127.

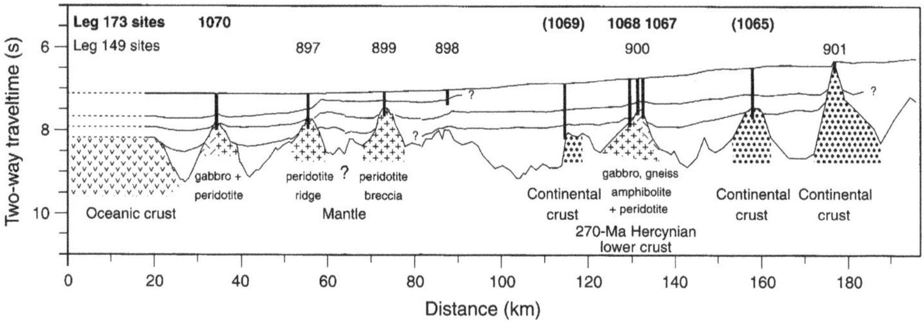

Figure 2.8 Transition from continental to oceanic lithosphere off the Atlantic coast of Spain. Modified from Whitmarsh, R.B., Beslier, M.-O., Wallace, P.J. et al., 1998. Proceedings of the Ocean Drilling Program, Initial Reports, volume 173. Ocean Drilling Program, College Station, TX.

metalliferous deposits near hydrothermal vents (black and white smokers). During intervals of increased spreading rates such as those observed in the Cretaceous, higher heat flow results in increased volume of mid-oceanic ridges. This in turn decreases the average depth of the oceans and facilitates the transgression of ocean waters over continental areas. The oldest and deepest oceanic crust is found near continental areas. However this is not always reflected by water depth, due to thicker sediment cover and progradation of slope and shelf areas.

The transition from oceanic to continental crust is gradual. An eastern North Atlantic area of relatively thin sediment cover has been surveyed and drilled during ODP Legs 149 and 173 off Galicia (Figure 2.8). Seismic data and basement cores evidenced blocks of continental crust thinning seaward, with local exhumation of lower crustal rocks. They shift to blocks of serpentinized peridotites and then to typical oceanic crust seaward. Both thinned crust and peridotites have been veined by intrusions of gabbros derived from melting of heterogenous mantle rocks during past intervals of extension. These areas of transitional crust overlain by shelf and slope sediments have not been geologically active since early extensional stages and represent passive continental margins.

2.1.4. Sliding Lithospheric Plates

Detailed bathymetric maps show that mid-oceanic ridges are not continuous, but segmented. As linear velocity changes with distance from the rotation poles, different portions of lithospheric plates move at different velocities. Lithospheric plates being rigid, these differences are accommodated along transform faults, where two plates slide along each other (conservative boundaries). Transform faults offset active plate boundaries, are bounded by spreading centers (or trenches) and are seismically active. Their activity ceases beyond the geological structures they offset, but traces of past activity are visible on the ocean floor of divergent plates (offset of magnetic anomalies, steep relief decreasing with distance from active transform fault, etc.), where they correspond to fracture zones. In the Central Atlantic, the Romanche transform fault shifts two segments of the Mid-Atlantic Ridge by about 900 km and consists of a deep (more than 5,000 m with a maximum at 7,600 m water depth) U-shaped axial valley bounded by two steep-sloped transverse ridges where basalts, gabbros and peridotites outcrop (Figure 2.9). The morphology of the transverse ridges looks controlled by tectonics (slight convergence associated to the main strike-slip motion) and subsidence (lithospheres of different ages and densities in contact), as suggested by their asymmetry: steep crustal highs are separated by a suspended valley on the transform fault side, while their slope declines smoothly on the basin side. The northern transverse ridge culminates up to 5,500 m above the axial valley of the Romanche transform fault. Its shallowest flat surface (Pillsbury Seamount) is currently at minimum water depth of 950 m but shows traces of subaerial weathering and erosion, and a succession of shallow water to deep deposits. This illustrates the complexity and variable activity of transform faults, where vertical displacement is associated to the transverse strike-slip component (but two orders of magnitude smaller).

2.1.5. The Convergence of Lithospheric Plates

Lithospheric plates approach each other along convergent boundaries, where one plate descends underneath the other into the asthenosphere and is destroyed. New oceanic lithosphere being continuously created at divergent boundaries, an equivalent quantity of lithosphere is destroyed at convergent boundaries. Seismic activity there is of higher magnitude than at divergent boundaries and includes both shallow and deep earthquakes. Highest magnitude of 9.5 was recorded in 1960 in Chili. Most hypocenters are concentrated along an oblique area dipping at variable angle from 30° to 80° (average 45°) to a depth of about 350 km (but earthquakes may occur as deep as 700 km), the Wadati-Benioff zone (Figure 2.10). Variations in velocity of seismic waves, gravity and heat flow illustrate the subduction of one lithospheric plate underneath the other along Wadati-Benioff zones. Earthquakes

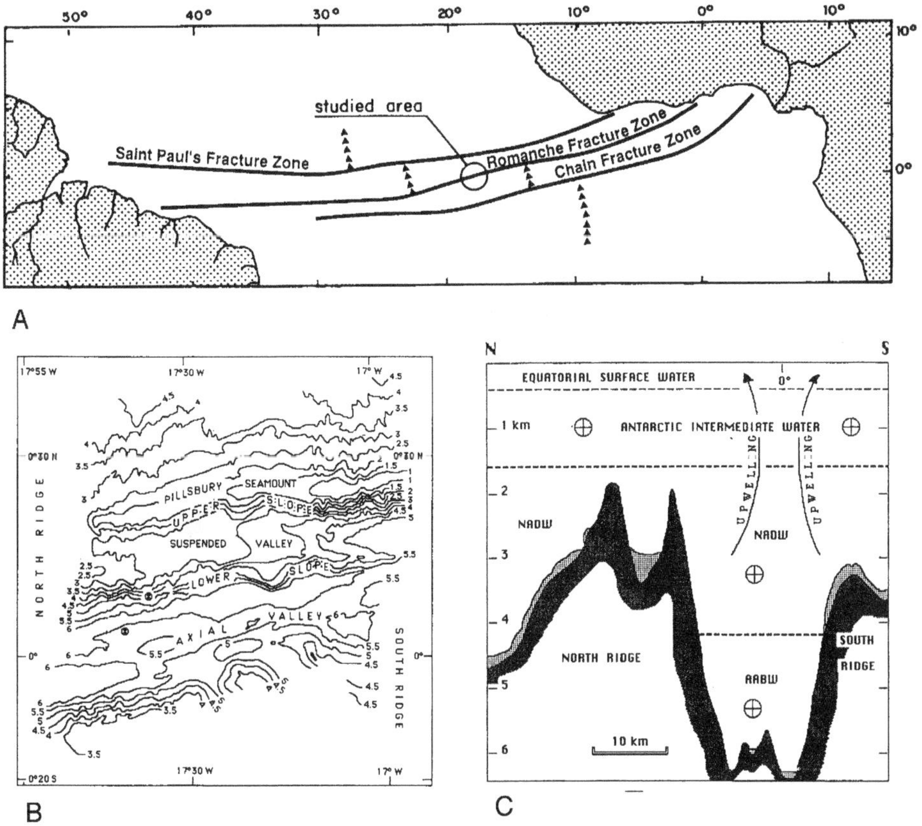

Figure 2.9 The Romanche transform fault in the South Atlantic. (A) Location map. (B) Detailed seabeam bathymetry showing U-shaped axial valley bounded by steep crustal highs between segments of the mid-oceanic ridge. (C) Schematic cross-section perpendicular to the transform fault showing sediment cover and water masses (the axial valley acts as a passageway for the deepest water masses). Reprinted from Westall, F., Rossi, S., Mascle, J., 1993. Sedimentary Geology, 82, 157–171.

result from compression and friction (reverse faulting) between converging plates. Earthquakes also occur in the overlapping plate where they result from alternating compressive and distensive movements.

The oceanic lithosphere mostly consists of upper mantle rocks ($d = 3.30$), the oceanic crust being very thin. The asthenospheric mantle is similar to the upper mantle in nature, but its density is slightly lower ($d = 2.25$ average) because of higher temperature, principally. This facilitates the subduction of old and dense oceanic lithosphere into the asthenosphere under lithospheres of lower density: mainly continental lithosphere, but also sometimes younger oceanic lithosphere. The progressive metamorphose of the subducted plate releases water, and the presence of water decreases the melting temperature of mantle rocks. Hydrated fusion of mantle rocks produces abundant, low-density, silica-rich magma that moves upwards across the fractures of the overlapping lithosphere. Some

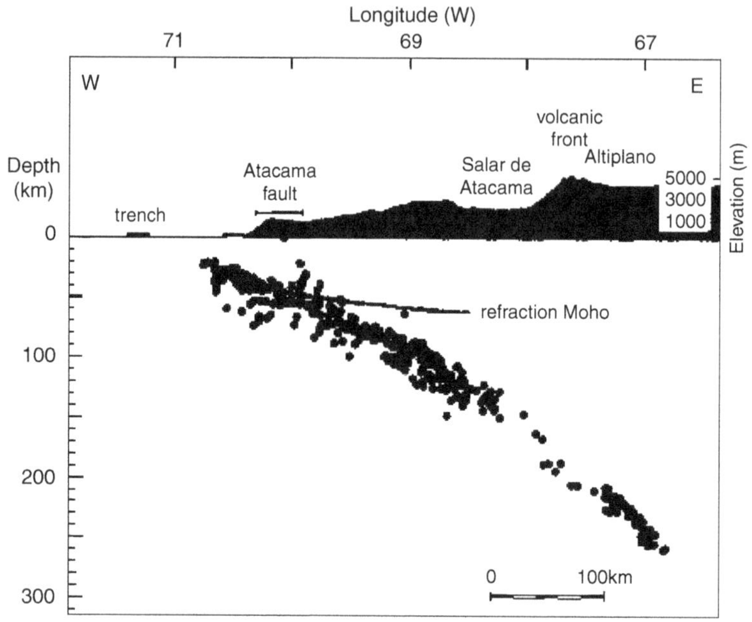

Figure 2.10 Vertical distribution of earthquakes in Chili, along a Wadati-Benioff zone. Modified from Delouis, B., Cisternas, A., Dorbath, L., Rivera, L., Kausel, E., 1996. Tectonophysics, 259, 81–100.

crystallization occurs in magmatic chambers and the residual magma melts the rocks of the overlapping lithosphere. Depending on the nature of the overlapping lithosphere, subduction volcanism generates a variety of silica-rich products ranging from granodiorites to andesites and rhyolites (overlapping continental lithosphere) or basalts (overlapping oceanic lithosphere). Most of the time the overlapping lithosphere is continental in nature and the transition from oceanic to continental crust is abrupt and associated with intense geological activity: converging plate boundaries then represent active continental margins.

The configuration of convergent plate boundaries include (Figure 2.11):

- A bulge of the subducted plate of variable importance related to the friction between both plates.
- A trench where both lithospheric plates are in contact, which represents the trace of the subduction zone at the solid Earth's surface. Trenches are more or less visible, depending on the quantity of sediment carried by the subducted plate and intensity of erosion of the overlapping plate.
- An accretionary wedge where sediments previously deposited in the trench accumulate and are reorganized as subduction progresses. Its importance and morphology depend on the quantity of sediment made available in the trench.
- A forearc basin may develop on the accretionary wedge in relation to tectonic activity and morphology of both accretionary wedge and volcanic arc.

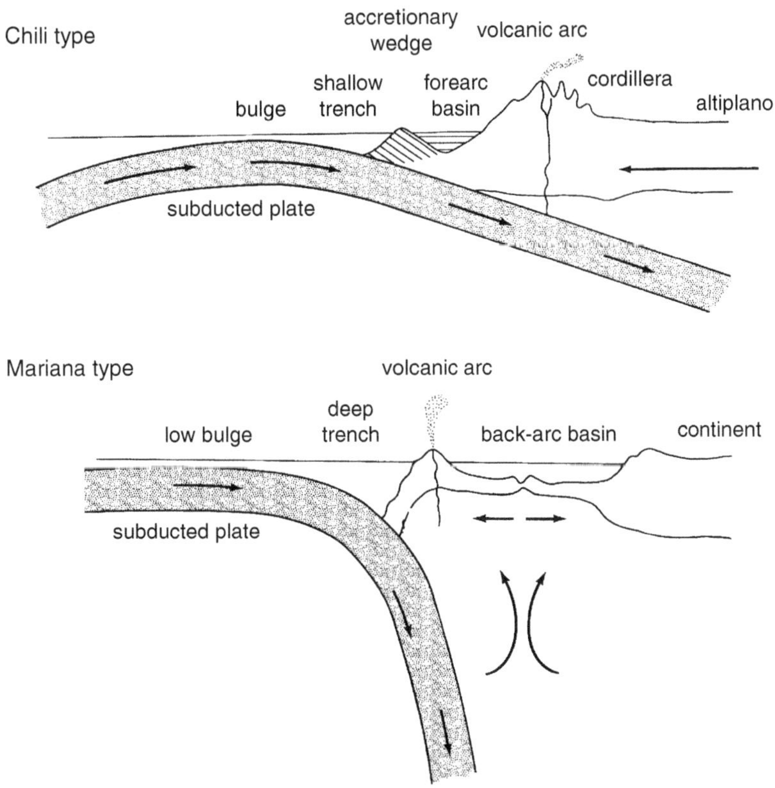

Figure 2.11 Configuration of convergent plate boundaries (active margins). Chili-type convergent boundaries involve young, fast-moving, low-density subducted lithosphere. Mariana-type convergent boundaries involve old, slow-moving, high-density subducted lithosphere. Modified from Kennett, J.P., 1982. Marine geology, Prentice-Hall, Englewood Cliffs, NJ.

- A volcanic arc where a succession of important, mainly explosive volcanoes develop on the tectonically active periphery of the overlapping plate, above the subduction zone.
- A backarc (or marginal) basin on the inner side of the tectonically active border of the overlapping plate. Its development depends on the characteristics of the tectonics and volcanism associated to the subduction.

However, the morphology of convergent plate boundaries shows important regional variability. The relative linear velocity of the converging plates and the age (density) of the subducted plate are probably important factors. Older and dense oceanic lithosphere is easily subducted, especially when the relative velocity of the plates is low. The angle of subduction is high, seismic and volcanic activity are rather low. In this case the volcanic arc is poorly developed, and therefore the production of siliciclastics and volcaniclastics is small. As a consequence the oceanic trench clearly shows in the bathymetry and the accretionary wedge is poorly developed. Backarc distension is important enough to allow formation of oceanic lithosphere, resulting in a succession of marginal basins. Examples include island arcs of the Northwest Pacific from the Kuriles to the Marianas, facing marginal basins from the Okhotsk Sea to the China Sea. Younger and low-density oceanic lithosphere dips at low angle, increasing the friction between converging plates. This results in important seismic, tectonic and volcanic activity, especially when relative velocity of the plates is higher. In this case, a cordillera develops at the periphery of the overlapping plate: high relief and geological activity are associated with intense erosion, filling the trench with siliciclastics and volcaniclastics. This allows formation of a large accretionary wedge. Backarc distension is minor and results in the formation of grabens, sometimes raised to high elevation because of the importance and activity of the cordillera. Examples include the Southeast Pacific where subduction of the young Nazca Plate below South America is associated with the formation of the Andes Cordillera and the grabens of the Altiplano.

The density of lithospheric plates varies with age and lithology and most of them carry both oceanic and continental crusts. When two lithospheres of similar density converge in the oceanic trench, the subduction is obstructed and the plates collide. The most frequent type of collision involves sections of converging lithosphere carrying continental crust. Because of low density, the subducted plate cannot dip into the asthenosphere by gravity anymore. As subduction ceases, the continental crusts shorten, overlap and thicken, due to compression. Subsequent isostatic readjustment raises the colliding lithospheres to altitudes of 5,000–6,000 m on the average. Sometimes the oceanic crust ruptures and may overlap parts of continental crust (obduction). At this stage of convergence, the oceanic lithosphere created during the whole duration of the ocean has been destroyed. Collision episodes involving two given plates have consequences on the motion of other plates, evidenced through changes in rotation poles and velocity.

2.1.6. The Wilson Cycle

Collision belts represent traces of ancient oceans, which opened and closed as a consequence of the motion of lithospheric plates. Wilson in 1966 and Dewey and Bird in 1970 proposed that the terranes exposed in the Appalachians accumulated in a proto-Atlantic Ocean which opened and closed during the Early Paleozoic. Also the Tethys Ocean, which once separated the Laurasia to the north from the Gondwana to the south in the Early Mesozoic, closed partly in the same manner to form the Himalayas, the Caucasus and the Alps. Therefore, oceans open and close within time spans of a few hundred million years, and the Wilson cycle corresponds to the successive stages of ocean evolution: early formation of a continental rift which grows through the creation of oceanic lithosphere between passive margins till rupture and subduction along one of the margins which becomes active, leading to progressive closure and collision. The Wilson cycle leads to periodic formation of a unique supercontinent. The last one was the Permo-Triassic Pangea, surrounded by a unique ocean, the Panthalassa. The current cycle started with the early opening of the Tethys Ocean in the Triassic more than 200 Myr ago and may last for an equivalent time span. Most of the initial breakups occurred during the Cretaceous, an interval of intense magmatic activity which also led to accelerated spreading rates and formation of large igneous provinces: oceanic plateaus (Kerguelen Plateau and Maud Rise in the Southern Ocean, Mozambique Ridge in the Indian Ocean, Sierra Leone Rise in the South Atlantic, etc.), volcanic rifted margins (Argentina and South Africa margins in the South Atlantic) and also continental flood basalt provinces (Karoo lavas of Africa, Deccan traps of India, etc.).

Initially a wide gulf of the Pangea largely open to the Pacific, the Tethys Ocean continuously opened for more than 100 Myr, from the Triassic through the Middle Cretaceous. Spreading progressed westward during the Jurassic, turning the Tethys into a wide ocean of low latitudes and east–west orientation, communicating on both extremities with the Pacific (Figure 2.12). Opening also progressed to the south, separating Arabia and Africa from India and Antarctica. The Indian Ocean developed later in the Middle Cretaceous, with final separation of India from adjacent continents, as convergence started in the eastern Tethys. At the same time, seafloor spreading progressed from south to north in the South Atlantic and started separating Europe from North America in the North Atlantic. In the Late Cretaceous, seafloor spreading in the Southern Ocean started separating Australia from Antarctica. Also, final separation of Africa from South America and Greenland from northern Europe progressively turned the Atlantic into a wide ocean of north–south orientation, including part of the western Tethys. In the Eocene all major oceans were already open, with the Tethys rapidly decreasing in size (Figure 2.13). A series of collisions of greater India and Eurasia in the Eocene and Oligocene progressively closed the eastern Tethys. At the same time seafloor spreading accelerated in the Southern Ocean, leading to final separation of Australia and Antarctica in the earliest Oligocene. By the Miocene, convergence and multiple collisions had reduced part of the western Tethys Ocean to the size of the Mediterranean Sea, while the Himalayas, the Caucasus and the Alps progressed rapidly. To the south, active spreading in the Southern Ocean and rapid northward drift of Australia had shaped a large ocean at high latitudes and constricted the Indonesian Seaway near the equator. Most of the tectonic events that led to the modern configuration of the oceans were concentrated during intervals of plate reorganization. For example, the first subduction of tethyan lithosphere below Eurasia was followed by increased spreading rates in the Indian Ocean. Also, the first collision of India and Eurasia around 50 Ma in the Eocene coincided with changes in rotation poles of several lithospheric plates and increased spreading in the Southern Ocean, which in turn led to final separation of Australia and Antarctica near the Eocene/Oligocene boundary at about 33 Ma.

The modern world includes oceanic basins at different stages of their evolution: (i) the East African Rift may represent an embryonic ocean; (ii) the narrow Red Sea lacking a mid-oceanic ridge and characteristic continental shelves may represent a young ocean stage; (iii) the Atlantic and Indian oceans represent a mature stage of ocean evolution, with fully developed mid-oceanic ridges and continental shelves; (iii) the Pacific is a declining ocean, bounded by active margins and (iv) the Tethys Ocean is in a terminal stage where small oceanic basins (Mediterranean Sea) alternate with collision belts (Alps). From geophysical criteria, the oceans are defined by the presence of oceanic lithosphere. However, seawater fills the most depressed areas of the Earth's surface. Usually they both coincide, because of lower

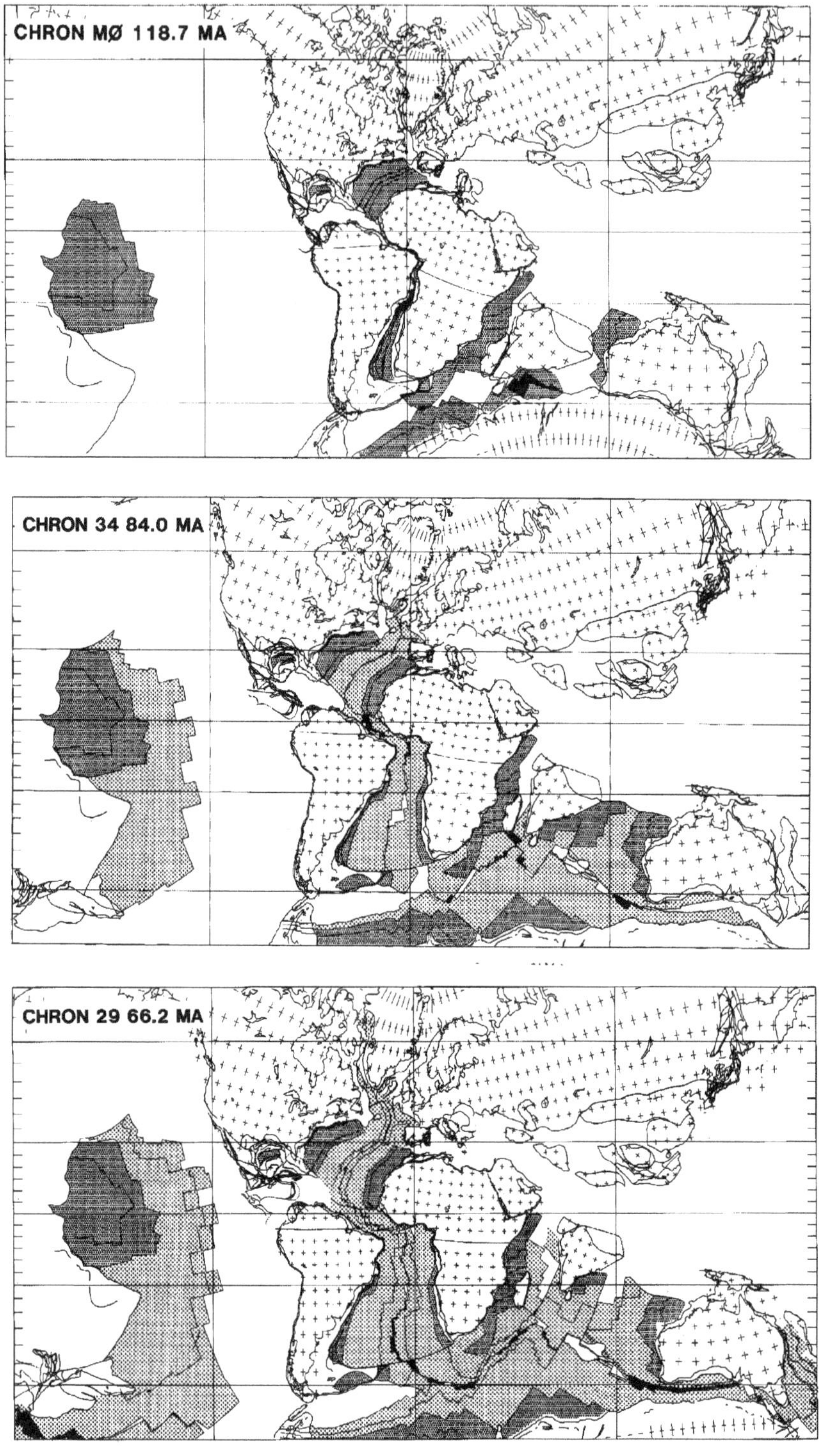

Figure 2.12 Breakup of Pangea and configuration of the continents and oceans during the Cretaceous. Note the early development of the Tethys Ocean between Laurasia (Eurasia and North America) and Gondwana (Africa, India and Australia), later followed by the meridional oceans (Atlantic and Indian). Reprinted from Scotese, C.R., Gahagan, L.M., Larson, R.L., 1988. Tectonophysics, 155, 27–48.

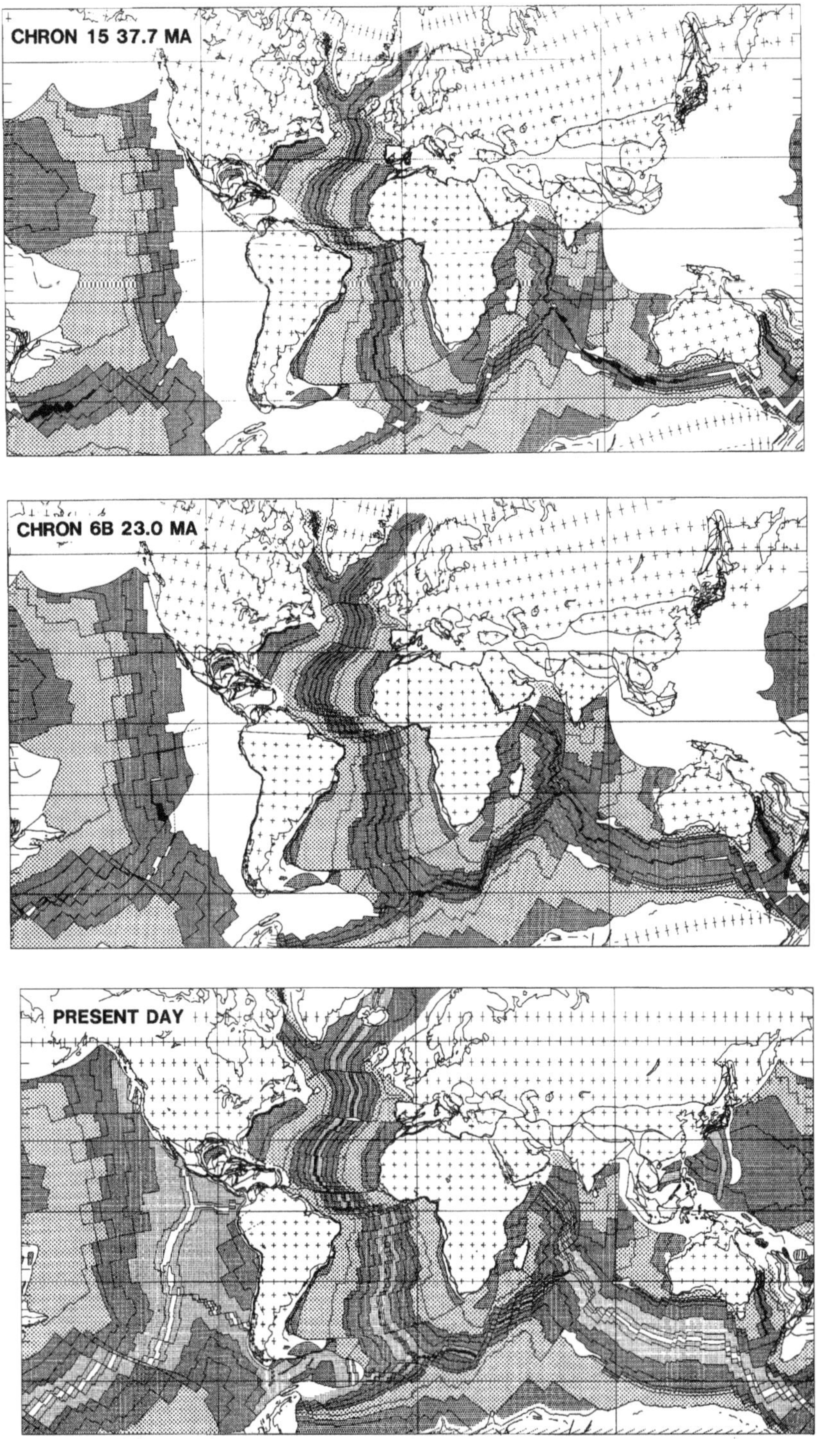

Figure 2.13 Configuration of the continents and oceans during the Cenozoic. Note concomitant development of the Southern Ocean at high latitudes and closure of the Tethys Ocean at low latitudes. Reprinted from Scotese, C.R., Gahagan, L.M., Larson, R.L., 1988. Tectonophysics, 155, 27–48.

buoyancy of oceanic crust relative to continental crust. However, there are some places where oceanic lithosphere outcrops in subaerial conditions because of magmatic activity (Iceland) or presence of morphological barriers (Afar), and many places where continental lithosphere is below sea level because of crustal thinning and breaking (most continental margins, some rift areas such as the Gulf of Suez), and/or plate deformation (English Channel, North Sea).

2.2. Oceanic Waters and Their Interaction with Global Climate

2.2.1. Incoming Energy at the Earth's Surface

Energy fluxes from the Earth's interior are locally important (along mid-oceanic ridges, for example), but average value at the surface under normal conditions is about 0.05 W/m^2, which is too low to play a significant role in climate. The bulk of energy comes from the Sun. Energy in the form of electromagnetic radiations is emitted by the "cool" (6,000°C) photosphere which is the visible surface of the Sun, while energy in the form of a plasma of hot gases (an ensemble of positively charged nuclei and negatively charged electrons), the solar wind, is produced in the "hot" (1,000,000°C) corona. The solar wind travels at a speed of 1.5×10^6 km/h through space following the open curves of the Sun's magnetic field, the Earth being protected by its own magnetic field. The electromagnetic radiations penetrate the external envelopes of the Earth.

The quantity of energy emitted by a body is provided by the equation

$$F = e\sigma AT^4 \text{(Stefan's law)}$$

where F is the flux of energy emitted by the body, e the emissivity (which varies from 0–1), σ a constant, A the area of the body and T its absolute temperature. Also, the wavelength for maximum emission (λ_{max}) varies to the inverse of the absolute temperature T of the body

$$\lambda_{max} = \frac{2.898 \text{ mm} \times K}{T} \text{(Wien's law)}$$

where K is a constant. When the temperature of a body increases, the wavelength of the emitted radiation decreases and the flux of energy increases (and inversely). The Sun behaves approximately as a black body ($e = 1$), that is a body which absorbs all the incoming energy and emits as much energy as possible according to its temperature. The estimated absolute temperature of the photosphere (5,800 K) allows a flux of energy of 2.33×10^{25} kJ/mn. The wavelengths extend from the ultraviolet to the infrared domains ($\lambda < 4$ μm), with a maximum at the transition from the ultraviolet to the visible domains. Energy is being transported in the form of heat and light. Available energy being inversely proportional to the square of the distance from source (here 150 millions km), only a small fraction of solar energy reaches the approaches of the Earth: about 1,370 W/m^2 (the solar constant).

As solar electromagnetic radiations penetrate the outer envelopes of the Earth, they interact with the molecules of the atmosphere: O_3, O_2, CO_2, H_2O, etc.

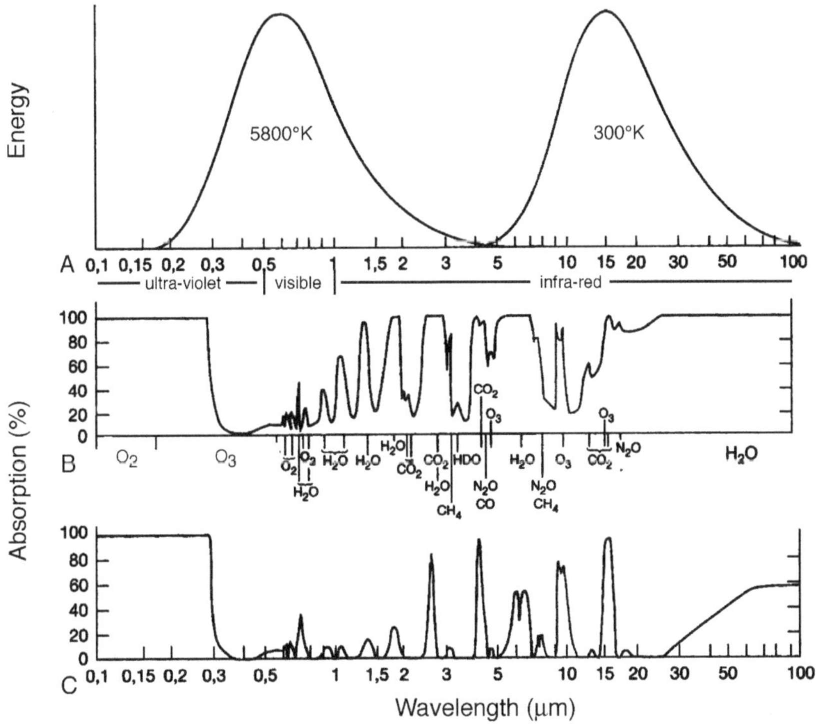

Figure 2.14 Absorption of radiation in the atmosphere. (A) Energy spectra for blackbodies of temperature 6,000 K (Sun) and 256 K (Earth). (B) Absorption of radiation by atmospheric gases for clear skies near ground. (C) Absorption of radiation by atmospheric gases in the stratosphere. Modified from Brahic, A., Hoffert, M., Schaaf, A., Tardy, M., 1999. Sciences de la Terre et de l'Univers, Vuibert, Paris.

(Figure 2.14). This interaction (diffusion) represents a complex assemblage of reflection, refraction and scattering of incoming radiation by interfaces, here the composite surface of gas molecules. This is especially efficient when the size of interacting molecules is roughly equivalent or bigger than incoming wavelengths (conditions of Mie). As a result, part of the radiation is trapped (absorption) or reflected (albedo) and the quantity of transmitted energy decreases. The intensity of diffusion, as well as the relative proportion of absorption and reflection, varies with the density and complexity of the assemblages of interacting molecules. Due to significant presence of O_3 molecules in the outer atmosphere most of the ultraviolet radiation is diffused there, whereas the variable presence of H_2O and other aerosols (including eolian dust) in the lower atmosphere modulates the importance of some visible (and infrared) wavelengths. About 55% of the solar constant reaches the Earth's surface, average quantity of energy on the half sphere exposed to the Sun being 342 W/m^2.

The solid Earth and oceans also behave like a black body, of average absolute temperature 300 K. The flux of emitted energy is roughly similar to that received from the Sun (the quantity of geothermal energy being currently negligible on the average), and the spectrum of radiation is entirely within the infrared domain

($\lambda > 3\,\mu m$). Aerosols made of CO_2, H_2O, but also CH_4 and other molecules and particles about the size of infrared wavelengths are abundant in the lower atmosphere (troposphere), where part of the energy re-emitted by the Earth's surface is diffused. To compensate for the deficit of outcoming energy the temperature of the troposphere increases, according to Stefan's law. As a consequence, average temperature at the Earth's surface is +15°C, whereas it should be −18.2°C in the absence of diffusion of re-emitted infrared radiation by the lower atmosphere. The energy trapped there is at the origin of the greenhouse effect, which varies with the quantity of interacting molecules and aerosols. For example, the primitive atmosphere of the Earth, enriched in CH_4, CO_2, H_2O and NH_3, was probably associated to an enhanced greenhouse which allowed average surface temperatures up to 58°C, despite lower solar luminosity than now.

2.2.2. Variability of Incoming Energy and Distribution at the Earth's Surface

The quantity of available solar energy as well as the greenhouse vary regionally. First of all, the quantity of incoming solar energy prone to be absorbed by the Earth's surface varies strongly from day to night. This energy is highest at low latitudes and decreases with increasing latitude. It is minimum at high latitudes where high obliquity of the solar beam results in important reflection and higher diffusion than at low latitudes, and duration of insolation is irregular.

The gravitational influence of the Sun and planets alters the orbital parameters of the Earth and therefore the quantity and distribution of incoming solar energy (Figure 2.15). One parameter to be modified is the shape of the orbit, which varies from almost circular to slightly elliptical with a main period of 413 kyr and another one around 100 kyr. From minimum to maximum eccentricity of the orbit, the distance from Earth to Sun varies by about 18.3 million km, but the corresponding variation of energy is proportional to the inverse of the square of this value. Moreover, the velocity of planets along their orbits vary with distance from the Sun, according to the second law of Kepler (the radius vector describes equal areas in equal times). For example, the Earth moves faster when nearer the Sun and slower when more distant, modulating the quantity of incoming energy on an yearly basis. In the end, the eccentricity of the orbit induces only small changes of less than 0.2% of the incoming energy (i.e., less than $0.7\ W/m^2$).

The obliquity (tilt) is the angle of the rotational axis of the Earth relative to the perpendicular to the plane of the orbit (ecliptic). Variations of the obliquity are caused by the gravitational influence of large planets such as Jupiter. Obliquity varies from 22° to 24.5° with a period of 41 kyr. In theory its variations influences the seasonal contrast only, the summer hemisphere receiving more radiation and the winter hemisphere less when obliquity is higher. However, summer increases of insolation cannot be compensated by winter decreases at high latitudes, where absolute variations of insolation up to $17\ W/m^2$ are high enough to significantly affect the balance of energy and climate.

The precession has two major components: (i) the axial precession corresponds to the wobbling motion of the rotational axis of the Earth which covers an angle of

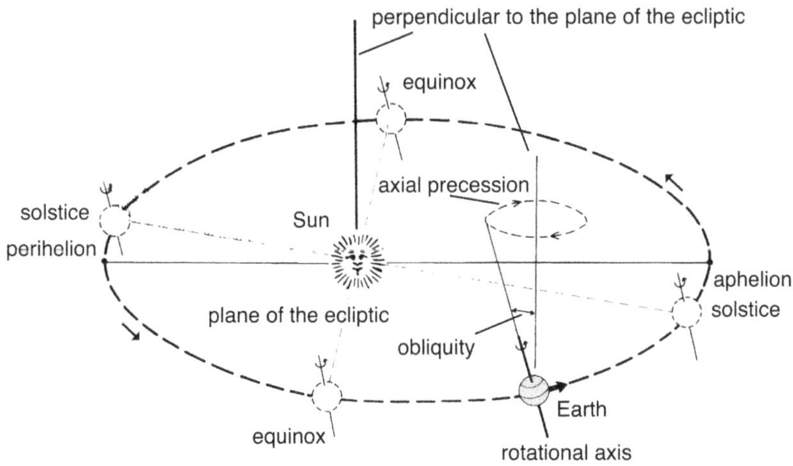

Figure 2.15 Representation of the Earth's orbit, with the main orbital parameters which modulate the quantity and distribution of incoming solar energy. Modified from Wells, N., 1986. The atmosphere and ocean: A physical introduction, Taylor and Francis, London.

23.6° for a period of 26 kyr and (ii) the orbital precession corresponds to the wobbling motion of the major axis of the Earth's orbit which has a period of 22 kyr. The cause of precession is the attraction of Sun, Moon and other planets on the equatorial bulge of the Earth. One consequence of precession is the motion of equinoxes and solstices along the orbit. Precession also decreases summer insolation and increases winter insolation in one hemisphere, and the opposite in the other hemisphere. Currently, increased winter insolation occurs in the northern hemisphere with a perihelion (Earth's annual closest approach to the Sun) early in January. The influence of precession is maximum at mid-latitudes, where incoming energy may vary up to 8% (i.e., 40 W/m^2). However this is not an absolute variation, since summer increases of incoming energy are compensated by winter decreases and northern hemisphere increases by southern hemisphere decreases. Precession mostly affects the seasonal and meridional distribution of incoming energy.

Most of the variability of incoming solar energy results from orbital parameters and is relative. During intervals of higher solar activity marked by increased sunspots, the solar wind increases as well as the energy emitted at some wavelengths of the ultraviolet domain. The Earth's magnetic field and distance from the Sun strongly reduce the impact of such events, the solar constant fluctuating by 0.1% only between minima and maxima of solar activity.

In contrast, infrared radiation is constantly re-emitted by the Earth's surface, with minor circadian variations. However, the efficiency of the greenhouse varies regionally with the density of molecules and aerosols prone to diffuse the infrared radiation. This includes H_2O in relation to the cloud cover, CO_2, CH_4, etc. Some of them are now extensively produced through human activities and are concentrated over densely populated areas, together with other pollutants such as ozone and hydrocarbons.

The quantity of energy actually absorbed at the Earth's surface varies according to its nature and composition. The reflection (albedo) over the oceans represents 2–10% of the incoming solar energy, but increases on the continents from 9% to 20% over more or less densely forested areas and up to 90% over fresh snow. However, the efficiency of the absorption also depends on the heat capacity of the materials, which is the quantity of energy required to raise the temperature of one unity of mass by 1°C. Raising the temperature of seawater (heat capacity of 4,184 J/kg/K) by 1°C requires five times more energy than for dry sand (heat capacity of 840 J/kg/°C). Therefore, continents react rapidly to changes of incoming energy and their temperatures vary strongly with the seasons. The oceans react more slowly, seasonal differences of temperature being relatively minor by comparison with the continents. On the average, oceans are cooler in summer and warmer in winter than adjacent continents. Also, the northern hemisphere, where continents dominate, has warmer summers (22.4°C average) and cooler winters (8.1°C average) than the southern hemisphere where oceans dominate (17.1°C and 9.7°C average, respectively).

2.2.3. Redistribution and Transport of Energy

Variations of incoming solar energy, albedo, absorption and related greenhouse generate regional surpluses (e.g., in and over tropical oceans) and deficits (e.g., at high latitudes) of energy on an yearly basis (Figure 2.16). A redistribution of available energy is therefore necessary to compensate for these gradients. Energy is being transported principally in the form of heat. The atmosphere, oceans and continents form a complex assemblage of heat reservoirs. Thermal imbalance between heat reservoirs drives heat transfer, mostly from low to high latitudes, and between oceans and continents.

Energy can be transported in the form of sensible heat by conduction, that is transmission of molecular activity through a substance. In this case, heat transfer varies with the heat capacity and temperature of the materials. Conduction is mostly efficient on continents made of solid materials where molecules are closely associated and minimum in the atmosphere where molecules are dispersed. Energy is mostly transported by convection in fluids, which are able to move and transfer heated parts of their mass. In this case, energy is transported in two different ways: (i) as sensible heat by warm air and water masses in motion and (ii) as latent heat through evaporation and condensation.

Evaporation occurs when effective vapor pressure of the air is below saturation. Saturation vapor pressure increases with temperature and evaporation rate increases with the difference between effective and saturation vapor pressures. Energy is required from the environment to overcome molecular attraction (600 cal/g for seawater at 0°C), and this loss of energy decreases the temperature of the remaining water mass. Condensation occurs when effective vapor pressure of the air is above saturation. When an air mass moves to cooler areas its temperature decreases and its saturation vapor pressure may drop below its effective vapor pressure, triggering condensation. Also, air masses moving to areas of lower pressure increase in volume, the required energy being drawn from the environment (adiabatic expansion). This decreases the temperature of the air and its saturation vapor pressure which may drop

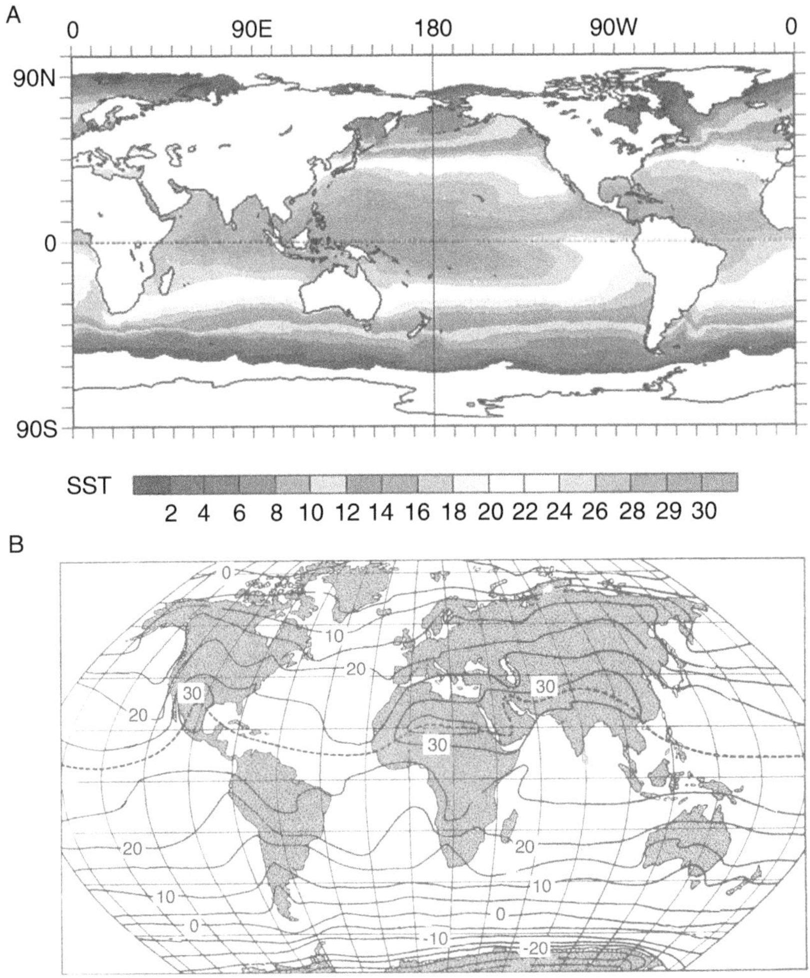

Figure 2.16 Sea surface temperatures (A) and air temperatures (B) during the northern hemisphere summer. Note maximum temperatures at low latitudes west of the oceans and over the tropical continents, and differences in the equator to poles gradients of temperatures for the oceans and atmosphere. Modified from http://www.cdc.noaa.gov/map/images/sst and Barry, R.G., Chorley, R.J., 1992. Atmosphere, weather and climate, Routledge, London.

below its effective vapor pressure, triggering condensation (conversely, reduction of volume in high pressure areas is associated to increases of temperature and saturation vapor pressure). Condensation is favored by the presence of hygroscopic nuclei such as aerosols, ice crystals and eolian dust. The size of the drops then increases by coalescence, collision or sublimation, and in some cases may increase up to 1,000 μm within a few minutes. The smallest drops remain in suspension, but fall by gravity as they increase in size. The energy drawn from the ocean for evaporation is then returned to the environment, increasing its temperature.

Both ocean and atmosphere are involved in heat transfer to areas of thermal deficit, which is maximum at high latitudes. The role of the ocean is most important from low to mid-latitudes, and the role of the atmosphere increases from mid- to high latitudes. The intensity of heat transfer varies mostly with equator to pole gradient of temperature. Both are at their maximum in winter, when marine and atmospheric circulation is more effective.

2.2.4. Role of the General Atmospheric Circulation

The equator to pole gradient of temperature is the major trigger of atmospheric circulation. Available potential energy at low latitudes is transformed into kinetic energy by moving air masses (sensible heat), part of this energy being dissipated by friction and turbulences. Circulation should decrease in intensity as energy is returned to the environment, lowering the meridional gradient of temperature. However, the atmosphere moves with the solid Earth around its rotation axis. Its angular momentum varies with angular velocity and distance from the rotation axis. The rotation being uniform, the angular momentum of the atmosphere is maximum at the equator and null at the poles. As to ensure the conservation of angular momentum air masses increase in velocity as they move poleward, with an eastward deviation due to the Coriolis effect. This compensates for progressive loss of energy of north-bound air masses. The atmosphere transports energy and momentum via vertical and horizontal circulation.

Warm air masses enriched in water vapor through intense evaporation (because of high saturation vapor pressure) raise near the equator and generate low pressures (the intertropical convergence zone or ITCZ), while their adiabatic expansion (reduction of saturation vapor pressure) favors intense precipitation (Figure 2.17). They lose their energy as they move to higher latitudes and to the upper

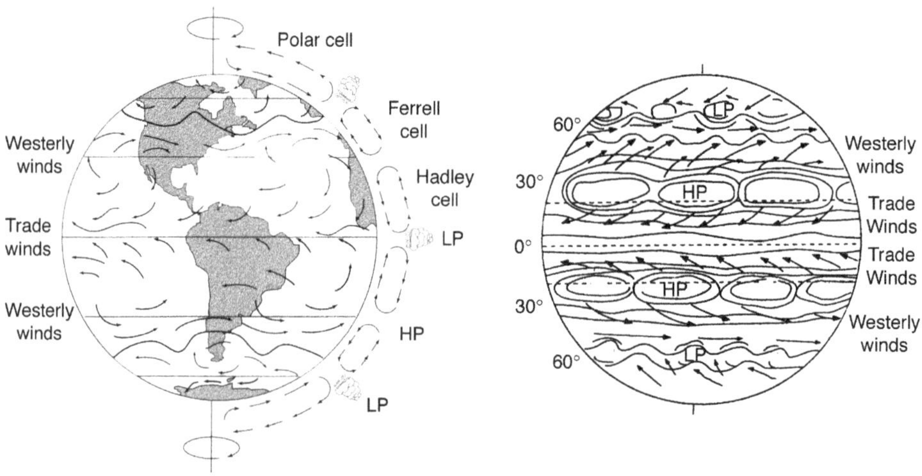

Figure 2.17 (Left) Vertical circulation of the atmosphere and surface winds. (Right) Horizontal circulation of the atmosphere. HP, high pressures and LP, low pressures. Modified from Brahic, A., Hoffert, M., Schaaf, A., Tardy, M., 1999. Sciences de la Terre et de l'Univers, Vuibert, Paris.

troposphere, cool and subside. Subsiding air masses with low effective vapor pressure generate high pressures, and decrease in volume while their saturation vapor pressure increases. Maximum subsidence occurs in the twenties and thirties of latitude, where they favor the development of tropical deserts. The gradient of pressure with lower latitudes generates surface winds to the equator, with a westward deviation due to the Coriolis effect: the trade winds. Air masses warm, absorb water vapor and raise as they approach the equator. Gradient of temperature with higher latitudes and conservation of angular momentum also generate surface winds at higher latitudes. Air masses warm and absorb water vapor as they travel above the oceans. They raise above dense and cold air from the fifties to the sixties of latitude, generating low pressures. Rapid decrease of temperature in these areas of thermal deficit and adiabatic expansion lower the saturation vapor pressure and increased condensation fosters precipitation. Cold air subsides again in areas of polar high pressure. The vertical atmospheric circulation includes three successive meridional cells (Hadley, Ferrell and polar), the most important being the tropical Hadley cell (Figure 2.17). Their relative importance varies with the seasons. However, the Hadley cells are not continuous around the globe, but fragmented. Air masses raise over areas of thermal surplus (oceanic warm pools, continents in summer, etc.) and subside over areas of thermal deficit (cool ocean surface waters, continents in winter, etc.), creating zonal vertical cells (Walker circulation). The Walker circulation plays a significant role in periodic climatic and oceanographic variations, for example the El-Nino Southern Oscillation and regional monsoons.

The horizontal circulation has maximum impact from the thirties to the sixties of latitude. Subsidence of cold air in polar and tropical high-pressure areas is associated to westerly circulation (a consequence of the Earth's rotation) which forms undulations. Under influence of the Coriolis force these undulations may increase in size, as to form isolated anti-cyclonic eddies at low altitude. In winter, some of these anticyclonic cells may travel as far as the thirties of latitude. Tropical and polar high-pressure anticyclonic cells are separated by cyclonic corridors of low pressure which carry warmer air to higher latitudes. Transitions from anticyclonic cells to cyclonic corridors of depression are associated to frontal mechanisms which trigger precipitation (Figure 2.18): at warm fronts incoming warm air masses raise above cold high pressures, decreasing in temperature and saturation vapor pressure, whereas at cold fronts incoming cold air masses decrease the temperature and saturation vapor pressure of warm air masses. Warm fronts are usually more active than cold fronts. The low altitude horizontal circulation strongly interacts with the morphology of the Earth's surface. Warm air is constantly enriched in water vapor above the oceans, where frontal mechanisms are active. They increase in intensity on continents where relief and changes in temperature favor condensation and precipitation. Continental relief also favors the subsidence of high-pressure cells, initiating strong regional winds.

2.2.5. Role of the General Oceanic Circulation

Seawater covers about 71% of the Earth's surface, essentially in low areas of dense oceanic crust, and 60% of these areas are located in the southern hemisphere.

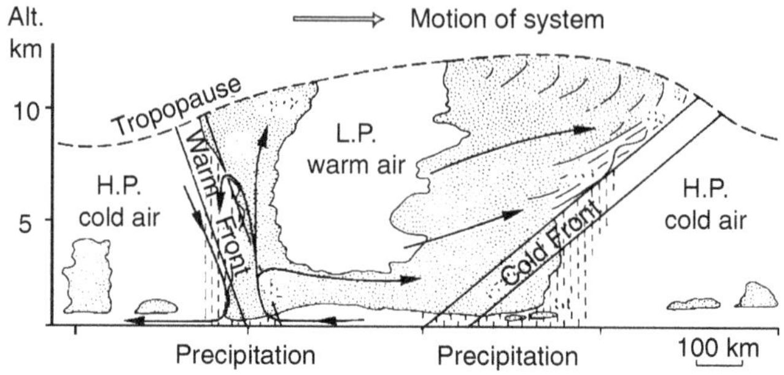

Figure 2.18 Anticyclonic cells separated by a cyclonic corridor of depression and related frontal mechanisms and precipitation. Modified from Barry, R.G., Chorley, R.J., 1992. Atmosphere, weather and climate, Routledge, London.

Oceanic circulation is strongly controlled by the morphology of the Earth's surface, as currents have to find their way around landmasses, mid-oceanic ridges and oceanic plateaus, and to adjust to strait areas. Therefore, tectonic processes play a major role in the control of oceanic circulation and climatic consequences at geological scale.

The quantity of energy made available within the ocean is more important at low latitudes, and this shows clearly when looking at sea surface temperatures. Temperatures above 20°C are common at low latitudes, but drop to 8°C around 500 m water depth. The thermocline marks the limit of the surface water mass, where most biological activity is concentrated because of the availability of light (photosynthesis), oxygen and nutrients. In the modern ocean, surface waters only represent a thin layer of warmer waters, but nevertheless play a major role in poleward heat transfer. Below, oceanic circulation is mostly driven by differences of density, which result from changes in salinity and principally temperature (thermohaline circulation). Locally the role of salinity may increase, as observed in the Mediterranean outflow to the Atlantic. The role of salinity currently remains of minor importance but may have been dominant during past geological intervals of deep saline waters (halothermal circulation).

The energy from the wind is first used to generate waves at the ocean surface, but as their speed approach 30% of that of the wind, the friction starts generating a surface current. The energy from the wind is transmitted downward as a moment and is best illustrated by a succession of frictions at the upper and lower limits of an infinite number of virtual layers of seawater (theory of Ekman). Because they are moving, all the layers are influenced by the force of Coriolis. The velocity (U) of the resulting surface current is

$$U = \tau/\nu\rho 2\Omega \sin\Phi$$

where τ is the friction exerted by the wind, ν the viscosity of seawater, ρ the density of seawater and $2\Omega \sin\Phi$ the Coriolis parameter (with Ω the angular velocity and Φ the latitude). Also, the direction of the resulting surface current deviates by 45° from the wind direction. Due to successive frictions, the velocity of the current decreases exponentially with depth and its deviation from the wind direction increases, and the successive vectors of current describe a spiral (Ekman's spiral). At a depth which varies with latitude and other characteristics, the flow is opposite to its initial direction. The Ekman transport represents the average motion of the surface, wind-driven layer (Figure 2.19). At low latitudes, surface winds that compensate for the gradient of pressure induced by the Hadley circulation deviate by 45° from the gradient (trade winds) and generate surface currents that depart by 45° from the direction of the wind: the northern and southern equatorial currents.

Continental areas block the circulation of surface waters which accumulate near these barriers. Warm waters carried westward by the equatorial currents accumulate west of the oceans to form warm pools, off New Guinea and Indonesia for the Pacific, off Amazonia and in the Caribbean for the Atlantic (Figure 2.20). Warm pools are also areas of higher sea level and hydraulic pressure, where warm surface currents are produced to compensate for the resulting gradients. A small part of this water flows back eastward in the equatorial doldrums (the low pressure areas of the ITCZ) as to compensate for the hydraulic gradient: the equatorial counter-currents. Most of the warm surface waters form narrow and fast current systems that transport sensible heat to areas of higher latitude: the western boundary currents. For example, the waters of the Caribbean warm pool flow to the North Atlantic through the Straits of Florida to form the Gulf-Stream, the North Atlantic current and then the Norwegian current, carrying warm waters to the high latitudes of Europe. As for the atmosphere, the conservation of angular momentum compensates for the loss of energy at higher latitudes, and keeps the oceanic waters moving, with a tendency to form eddy circulation.

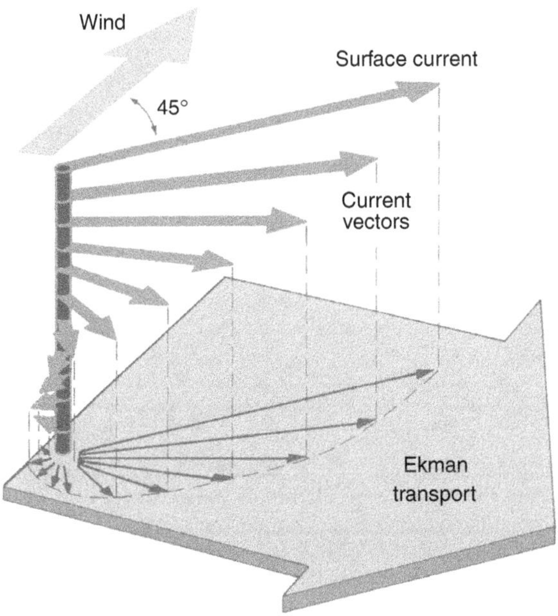

Figure 2.19 The Ekman spiral. Modified from Kennett, J.P., 1982. Marine Geology, Prentice-Hall, Englewood Cliffs, NJ.

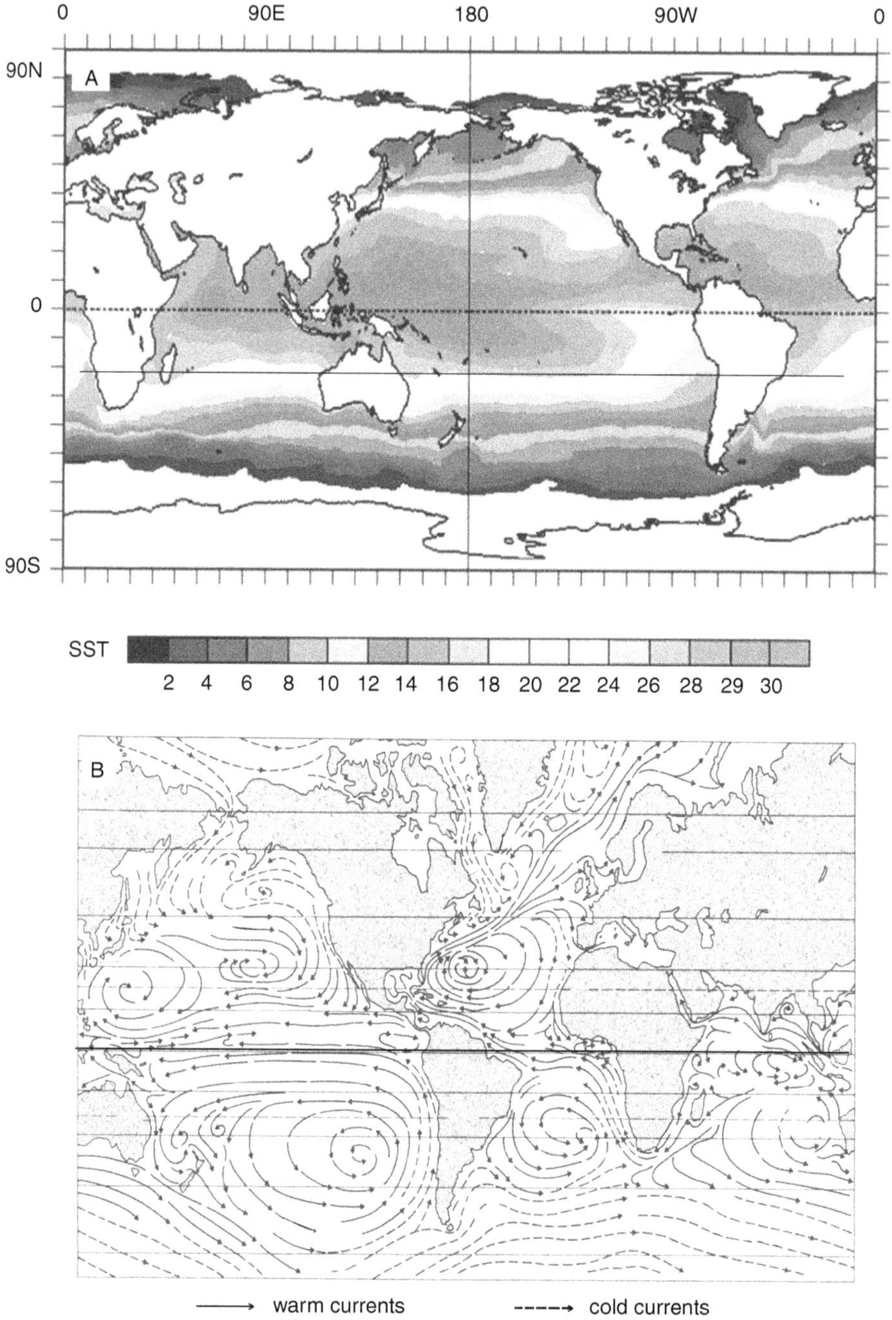

Figure 2.20 Average distribution of sea surface temperatures in summer (A) and main pattern of sea surface circulation (B). Note accumulation of warm waters West of the ocean, poleward transport of warm waters by western boundary currents and equatorwards transport of cool waters by eastern boundary currents. Modified from http://podaac.jpl.nasa.gov/sst/images/clim.gif and Pickering, K.T., Hiscott, R.N., Hein, F.J., 1989. Deep-marine environments, Unwin-Hyman, London.

When the force induced by a gradient balances the Coriolis force, the resultant current is in geostrophic equilibrium and progressively flows perpendicular to the gradient. As a result, the surface circulation forms vast vortices within oceanic basins: for example the North Equatorial current, Gulf-Stream off North America, North Atlantic current to Europe and Canary current off North Africa in the North Atlantic, and the South Equatorial current, Brazil current off South America, Antarctic circumpolar current and Benguela current off South Africa in the South Atlantic. East of the oceans, cold waters are being transported from high to low latitudes, within large and slow current systems: the eastern boundary currents.

East of the oceans, the friction of the trade winds moves surface waters away from coastal areas, where low hydraulic pressures develop. The resulting pressure gradient generates a perpendicular geostrophic flux to the equator. The eastern boundary flow to low latitudes progressively accelerates and deviates westward, while deeper and cooler waters are upwelled. Wind and surface water activities regulate the intensity of the coastal upwellings. Upwelled waters are enriched in nutrients, and upwelling areas support an intense biological activity, leading to high production of organic matter and mineral biogenic particles.

The distribution of sea surface temperatures is modified as a result of oceanic circulation (Figure 2.20): for similar latitudes, higher temperatures are observed in the western parts of the oceans (east of the continents) and lower temperatures in the eastern parts of the oceans (west of the continents). This has deep implications on climate, as well as on the quality of oceanic sediments. At low to mid-latitudes of both hemispheres, the presence of warm waters in the western parts of the oceans is associated with dominant heat transfer from the ocean to the continent, via the atmosphere. Air masses warm and increase their vapor pressure as they travel over the ocean and release their latent heat (condensation and precipitation) over continental areas. This is especially the case in winter when the gradient of temperature is highest (lowest temperatures on continent). Higher precipitation there promotes chemical weathering, as well as erosion by runoff. Siliciclastic particles are then carried by rivers to specific shoreline areas. For example, the relatively short Fly River of New Guinea carries to the westernmost Pacific Ocean the most important terrigenous load of the world. The presence of cold waters in the eastern parts of the oceans is associated with dominant heat transfer from the continent to the ocean. Dry air masses blow from the continent to the cool ocean, sustaining regional aridity and coastal upwellings. This is especially the case in summer, when the gradient of temperature is highest (highest temperatures on continents). Arid conditions on continental areas facilitate physical weathering and eolian erosion. Eolian particles are then largely dispersed by the winds over wide oceanic areas. For example, particles eroded by desert storms from Saharan areas are then widely dispersed over the Mediterranean and the entire Central Atlantic. Poleward heat transfer has deep implications on oceanic surface circulation and regional climates. This in turn has major consequences on natural processes including continental weathering, erosion and oceanic productivity, and therefore on the quality of oceanic sediments.

Heat transfer from the ocean to the atmosphere increases near the subtropical convergence and the polar front where topmost waters decrease in temperature and subside. Surface waters develop vertical convection cells of latitudinal extension similar to that of the vertical atmospheric circulation. However, heat transfer to the intermediate waters below the thermocline is very limited. Oceanic circulation below the thermocline is mostly driven by the relative differences in density of the water masses, a consequence of changes in their temperature (colder waters increase in density) and salinity (more concentrated waters increase in density). Temperature of ocean waters is minimum at high latitudes. Salinity of surface waters is maximum from the twenties to the forties of latitude where evaporation is maximum and freshwater supply low, especially in semi-enclosed basins such as the Mediterranean and the Persian Gulf. Salinity decreases in areas of higher precipitation and/or ice melting, that is near the equator and at mid- to high latitudes.

One of the major density-driven water mass is the Antarctic Intermediate Water (AAIW) which forms near the southern polar front (Figure 2.21). There, cooled surface waters subside especially in winter and are intensively mixed with circumpolar waters. The resulting water mass flows below surface waters and reaches as far as the mid-latitudes of the northern hemisphere where its trace (from temperature and salinity criteria) is lost. The North Atlantic Deep Water (NADW) mostly forms in a similar manner but results from an assemblage of dense waters: the Northeast Atlantic Deep Water (NEADW) which subsides in the southern Norwegian Sea; the Denmark Strait Overflow Water which carries cold and high

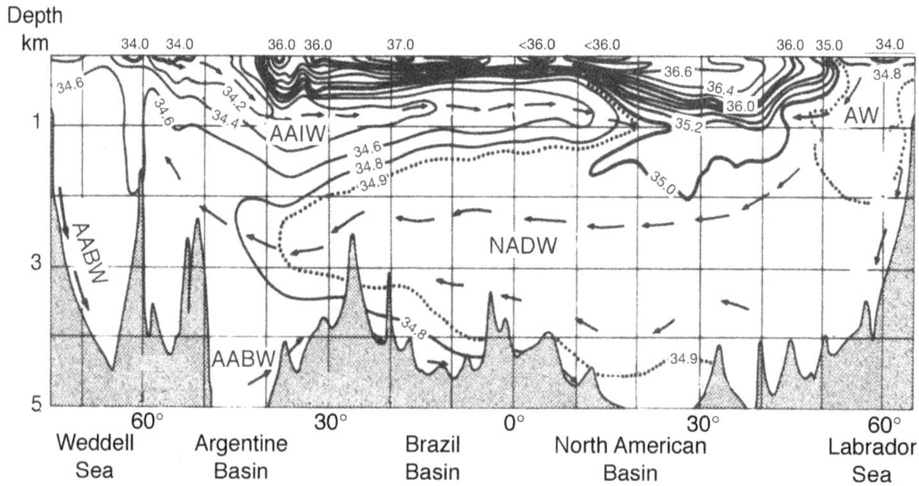

Figure 2.21 Characterization of the main water masses based on salinity and their distribution in the Atlantic Ocean. AABW, Antarctic Bottom Water; AAIW, Antarctic Intermediate Water; AIW, Arctic Intermediate Water; NADW, North Atlantic Deep Water. Note subsidence of AABW along the Antarctic margin and AAIW at about 50°S near the Antarctic convergence, subsidence of warm, more saline surface waters and combination with cold arctic waters at about 50°N in the North Atlantic to form NADW, and interaction of deepwater masses with ocean basins morphology. Modified from Tchernia, P., 1978. Océanographie régionale: description des océans et des mers, ENSTA, Paris.

salinity Arctic waters along the Greenland margin and the Davis Strait Overflow, along the Labrador margin. The NADW fills most of the North Atlantic basins and flows around 2,000–3,000 m water depth across the western Atlantic (because of the Coriolis force) to the southern fifties, and its path is then followed across the other major oceans. The Antarctic Bottom Water (AABW) mostly dives along the Antarctic continental margin (especially in the semi-enclosed Weddell Sea and Ross Sea basins) and fills the southernmost basins of the Southern Ocean. Some AABW also subsides from the deep circumpolar current, and a mixture of both components flows through the deepest parts (below 4,000 m water depth) of the ocean basins of both hemispheres. The path of the AABW is highly controlled by the morphology of the ocean floor: for example, its circulation to oceanic basins is constrained by mid-oceanic ridges and oceanic plateaus, but favored by deep axial valleys of transform fault areas.

Some coastal or small oceanic basins are more or less isolated from the open ocean by silled straits: exchanges of water masses are possible above sill depth only and limited by the depth and width of the silled strait. In such basins, the vertical distribution of hydrologic parameters (temperature, salinity, oxygen content and density) is very different from the one observed in open oceanic areas: below sill depth the water column is rather uniform and exchanges are possible through vertical mixing only, providing that the vertical contrast of densities is not too high. When evaporation over the basin is low and freshwater supply is high because of intense precipitation (Figure 2.22), important river discharges or wide continental drainage basins, low-salinity waters outflow in surface to the open ocean, while denser marine water enters at sill depth (estuarine circulation). Such basins are generally stratified, with deeper waters depleted in oxygen and enriched in organic matter and sulfides. One example is the Black Sea, which collects several of the most important European rivers and is open to the Aegean Sea via the narrow (650 m) and shallow (30 m) Strait of Bosphorus only. When evaporation over the basin increases, surface waters increase in salinity and density, sink and fill the deepest parts of the basin (Figure 2.22). Resulting lowering of sea level is compensated by an incoming flux from the open ocean in surface, while denser waters outflow at sill depth. Another example is the Mediterranean Sea which is in great part located in the northern thirties of latitude, of semiarid climate. Mediterranean waters mostly form in the northwest basin, but also in the Adriatic and Aegean. There, strong and dry northerly winds cool surface waters and increase evaporation rates. Vertical mixing allows still relatively warm and more saline waters to sink and fill the deepest basins. The production of dense Mediterranean water is most important in winter, when wind activity is at its maximum. Dense Mediterranean waters overflow at sill depth (300 m) through the Strait of Gibraltar. The Mediterranean water outflow, which spreads across the northeast Atlantic at intermediate water depth, is relatively warm and of high salinity.

The modern world is characterized by the importance of high-latitude oceans, especially in the southern hemisphere, and the presence of ice caps. This configuration is highly favorable to the formation of cold dense waters and the development of the thermohaline circulation across the oceans. The characteristics and circulation of dense, cold water masses change with alternating glacials and interglacials. For

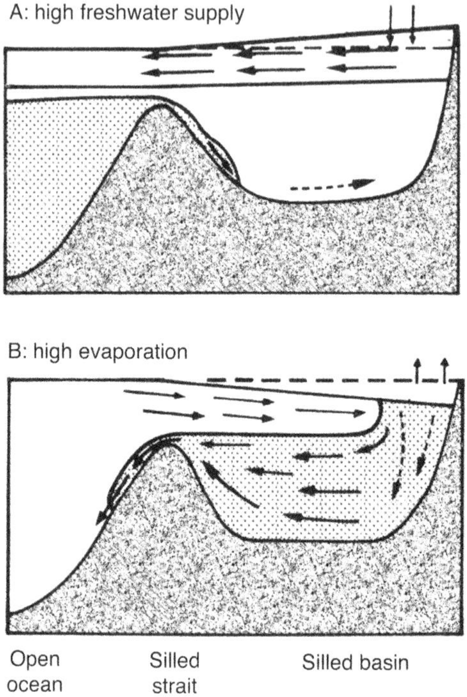

Figure 2.22 Distribution of waters and circulation in semi-enclosed, silled basins. (A) High freshwater supply to the basin, outflow of low-salinity waters in surface, inflow of denser marine waters at sill depth. (B) High evaporation in the basin, formation and subsidence of high-salinity waters which outflow at sill depth, inflow of lighter marine waters in surface. Modified from Tchernia, P., 1978. Océanographie régionale: description des océans et des mers, ENSTA, Paris.

example, the arctic components of the NADW are shut down during glacials as ice cover increases at northern high latitudes, while the production and circulation of its NEADW components increase in proportion and move southward. It is likely that production rates and characteristics of cold dense waters were somewhat different from the modern ones early in the Cenozoic, as the North Atlantic and Southern Ocean were of limited extension with opening processes not completed yet at southern high latitudes between East Antarctica and Tasmania and between West Antarctica and South America. The Mediterranean outflow demonstrates that evaporation may also play a role in the development of dense water masses. Halothermal circulation is currently of extremely minor importance because of the limited extension of oceanic areas and semi-enclosed silled basins in low latitudes of high evaporation, and the large dominance of thermohaline circulation. The production and circulation of warm, dense waters were probably more important before the Southern Ocean reached full development, in the Paleogene (Figure 2.23). This is suggested by low values and reversed gradients of the benthic oxygen isotopes at southern high latitudes of Kerguelen Plateau and Maud Rise. Halothermal circulation was probably a major feature of the Jurassic and Cretaceous as the Tethys

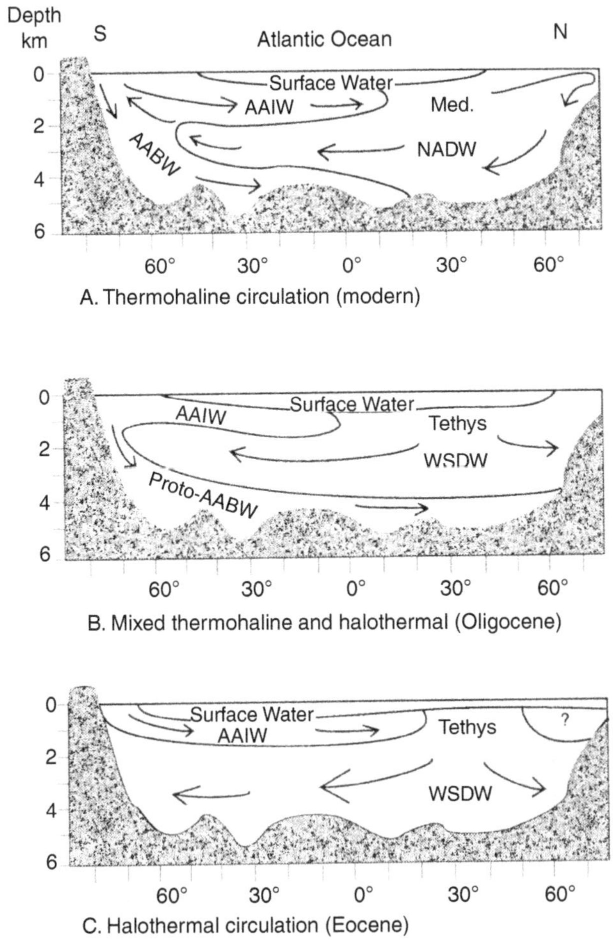

Figure 2.23 Evolution of intermediate and deepwater circulation during the Cenozoic. AABW: Antarctic bottom water; AAIW: Antarctic Intermediate Water; MED: Mediterranean Water; NADW: North Atlantic Deep Water; WSDW: warm saline deep water. Halothermal circulation prevailed during some intervals of the Early Paleogene, while thermohaline circulation prevailed in the Neogene. Concurrent presence of both thermohaline and halothermal circulations probably characterized most of the Paleogene. Modified from Kennett, J.P., Stott, L.D., 1990. Proceedings of the Ocean Drilling Program, Scientific Results, College-Station, TX, volume 113, pp. 865–880.

Ocean expanded at low latitudes: its passive margins sustained carbonate platform environments favorable to dense water formation and the deep Tethys probably acted as a large reservoir of energy for poleward heat transfer.

The configuration of oceans and continents, the volume of the oceans, the morphology of the ocean floor including nearshore areas, the quantity of available energy and atmospheric parameters are important factors that rule the production and circulation of dense oceanic waters. Most of these factors vary through time, as the oceans evolve according to the Wilson cycle: strong links may exist between plate tectonic processes, ocean circulation and climate at geological scale.

2.3. Oceanic Sediments: Sources, Dynamics, Classification and Transformation

It is well known since the Challenger expedition that surface oceanic sediments are a mixture of particles of different origin: some are transported from the continents by rivers, winds or ice, some are organic or mineral residues of biological activity, while others formed directly on the seafloor.

2.3.1. Sources of Terrigenous Sediments

Terrigenous particles are eroded from the continents, where they are mostly produced through weathering of parent rock that outcrop at the surface. Continental weathering corresponds to an array of mechanisms that separate the particles from their substrates before erosion. Physical weathering processes mostly rely on important changes of the pressure exerted on the mineral components of the substrates to separate particles. This includes differential changes in the volume of minerals associated to high variations of temperature (circadian and seasonal) and the crystallization of water or salts in the pore cavities of the parent rock. Chemical weathering processes rely on the differential solubility of the constituents of the parent rock in freshwater. Some chemical elements are removed by runoff, while others are combined to form new minerals. Chemical weathering processes require the availability of water and vary in intensity with temperature, as any chemical reaction. Physical weathering (Figure 2.24) is especially active in tropical deserts which are areas of high atmospheric pressure (sinking air from the Hadley cells) and dominant continent to ocean heat transfer (cold surface waters and upwellings). There, important temperature changes from day to night favor the fragmentation of the parent rocks, and wind systems (trade winds and corridors of depression) disperse the particles over the oceans.

Physical weathering largely dominates at high latitudes where widespread availability of freshwater and low temperatures favor ice formation and, in turn, the fragmentation of the parent rocks. There, rivers, ice-flows and glaciers carry the terrigenous particles to the shoreline. Continental morphology and tectonic activity also play a significant role in the control of physical weathering, as ice formation increases with altitude and high relief increases the mechanical effect of running waters. Chemical weathering being highly dependent on moisture and warm temperatures (Figure 2.24), it is especially important at low latitudes, but also in temperate areas of low relief where low temperatures are occasional or seasonal. There, large rivers carry the particles to the shoreline. It is of special interest to note that rivers may carry huge quantities of terrigenous particles to specific points of the coastline where they accumulate rapidly, whereas most nearshore areas receive only minor quantities and that wind activity disperses

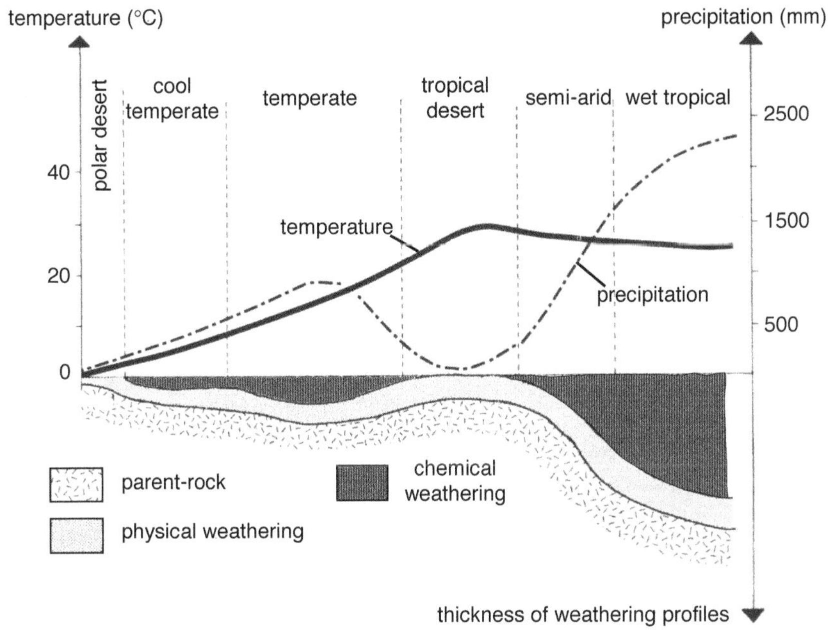

Figure 2.24 Meridional distribution of chemical and physical weathering. Thickness of weathering profiles is proportional to the intensity of chemical weathering. Note relationships between intensity of weathering, precipitation and temperature. Modified from Renard, M., Pomerol, C., 2000. Elements de géologie, Colin, Paris.

terrigenous particles over wide oceanic areas. Moreover, 70–80% of river-borne particles are supplied to the low latitudes, especially on southern and eastern sides of Asia. Besides the composition of the parent rock, it is clear that tectonic activity (through its impact on morphology), climatic conditions (temperature and precipitation) and the vegetation cover which protects from erosion, play a major role in the control of the quality and quantity of terrigenous particles and their transport to the ocean. The role of plate tectonic processes is especially important at geological scale, since active stages of the Wilson cycle involving continental lithosphere (continental rift, active margins) are associated with intense erosion: terrigenous particles fill related depositional foci, that is subsiding narrow rift basins and oceanic trenches, respectively. In the modern ocean, terrigenous particles are also abundant in narrow young oceanic basins (Red Sea) and near continental areas (shelves). They decrease with distance from source areas and on isolated ridges and plateaus, suggesting dominant transport near bottom: although their abundance is sometimes very low (below 5%), terrigenous particles are nevertheless present in almost all oceanic areas.

2.3.2. Sources of Biogenic Sediments

The quantity of living organisms is directly related to the availability of light and nutrients. Therefore, biological activity develops mainly from surface to the limit of

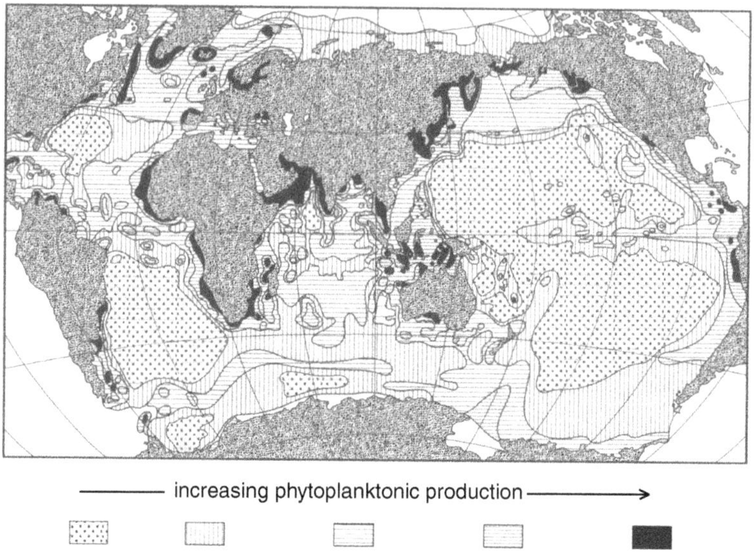

Figure 2.25 Variations of phytoplanktonic production in the ocean. Note maximum production in areas of coastal upwelling and minimum production in the center parts of the oceans. Modified from Biju-Duval, B. 1999. Géologie sédimentaire, Technip, Paris.

solar energy penetration and related photosynthesis at 100–200 m water depth (the photic zone), in areas where nutrients (Si, P, K, SO_4, NO_3, Fe, Mo, etc.) are made available from the deep ocean by upwellings, and from the continents by runoff. While some nutrients are widely available in the ocean, others are present in low concentration (Si, Fe, P, etc.) and have a limiting effect on oceanic productivity. Therefore, areas of high productivity (Figure 2.25) include regions of oceanic divergence (equatorial divergence, for example) and of coastal upwellings east of the oceans (Benguela current system east of the South Atlantic, for example). Biological activity also varies with the diversity and expansion of species, in close connection with the characteristics of the water masses: temperature, salinity and oxygen content principally. For example, marine organisms usually tolerate salinities between 30% and 40% (with only a few exceptions, among them some ostracods) and are absent from hypersaline as well as from hyposaline coastal lagoons. Also, the diversity of species decreases from low to high latitudes, cold species being more tolerant to changes in temperature.

Marine organisms include benthic and pelagic species. Benthic organisms are static (sponges, corals, bryozoans, brachiopods, etc.) or mobile (crabs, urchins, worms, etc.). Some species live on the seafloor or on other organisms (epifaunal species), while others dwell in the sediment (infaunal species). Benthic activity is particularly concentrated in coastal areas, where the seafloor is within the photic zone because of shallow water depth. There, benthic activity is strongly influenced by the nature of the substrates: (i) rocky substrates are mostly characterized by the development of static epifaunal (and sometimes infaunal) species, algae and sometimes coral reefs and local accumulation of debris; (ii) sandy substrates are characterized by the absence of static epifaunal species because of substrate instability and the abundance of mobile, infaunal species protected by mineral shells and (iii) muddy substrates mostly made of fine particles are characterized by burrowing infaunal species (worms) which produce abundant fecal elements. Beyond coastal areas in the deep ocean, marine organisms mostly consist of pelagic forms which develop in the upper 200–300 m of the water column in the photic zone and mostly include phytoplanktonic and zooplanktonic species. In volume, planctonic organisms contribute to the most important part of marine biological activity. The vast majority of marine organisms is made of organic matter only, and the probability that they are preserved is very weak. However, some groups are characterized by mineral elements: bioconstructions, shells, skeletons and tests of microorganisms, etc. Biogenic particles consist of mineral fragments which are secreted by living organisms using chemical elements from seawater and preserved after their death. They are mostly made of calcium carbonate (calcite, Mg-calcite, aragonite, etc.), especially in areas of warmer oceanic waters. For example, calcareous algae, corals and other benthic organisms may construct enormous reef systems in nearshore tropical areas. In the pelagic realm, calcareous microorganisms dominate in areas of warmer surface waters from the tropics to the polar fronts. However, silica (opal) secreting microorganisms dominate in colder areas of upwelled waters, and especially in high latitude areas beyond the polar fronts. In fact, the abundance of opal secreting microorganisms is limited by the availability of silica in seawater.

Photosynthesis is the primary source of organic matter, when carbon dioxide and water are combined to form carbohydrates and oxygen, for example, using energy from the Sun. The reaction is done by autotroph organisms (algae, bacteria, higher plants, etc.). These organisms feed heterotroph organisms which use the energy from carbohydrates and other components for their vital processes (growth, respiration, etc.) and fabricate new components (among them proteins). The organic matter includes all these components, together with related by-products such as secretions and dejections. Primary production is mostly driven by the availability of solar energy and develops more rapidly when insolation increases. In the ocean, seasonal blooms are especially important in areas where nutrients are most abundant: coastal upwellings and divergence areas, and proximity of some river mouths. Organic matter also originates from continents, where soils and higher plants are the main contributors. The quantity of continental organic matter carried to the ocean by runoff varies through time and is most important during flooding events. Components of continental origin represent about 30% of the whole organic matter in the modern ocean.

2.3.3. The Water Column

Seawater is close to neutrality, with a slightly alkaline pH around 8. Seawater contains chemical elements in solution, principally Cl and Na which make about 85% of the total. The remaining 15% mostly include SO_4, Mg, Ca, K, CO_3 and Br. Chemical elements are derived from volcanic activity and continental weathering via runoff principally and accumulated during geological times. It is noteworthy that some chemical elements which are widely distributed on continents such as Si, Fe and Al make less than 0.5% of all chemical elements in seawater.

Most of the particles in the water column consist of organic matter, which represents about 50% of the total on average. Organic matter is principally of aquatic origin and is concentrated in the photic zone, including shelves and nearshore areas of river mouths. Less than 20% of the total organic matter sink below the photic zone and less than 10% reach the seafloor, the rest being oxidized in the water column. However, organic particles are more abundant in and below areas of high productivity, that is oceanic divergences and coastal upwellings where the availability of dissolved oxygen in seawater is not important enough to allow efficient oxidation of high quantities of organic matter.

Terrigenous particles represent about 30% of the particles in suspension in the water column. The majority are derived from ice flows and icebergs (about 20%) because of the high abrasion power of ice, the rest (about 10%) being derived from rivers. Terrigenous particles mostly consist of minerals which are resistant to continental weathering, principally quartz and also some feldspars, together with trace amounts of heavy minerals such as garnet, epidote, pyroxenes, amphiboles, etc. They are associated to minerals derived from continental weathering, that is clay minerals. Terrigenous particles are concentrated near continents, especially at low latitudes and west of the oceans which are areas of high rainfall and runoff. They decrease in abundance to the center parts of the oceans where the relative proportion of the finest terrigenous elements, that is clay minerals, increases. Eolian dust accounts for only less than 0.5% of the particles in the water column, but wind activity brings fine particles (below 30 μm in size) directly to the center parts of the oceans. Eolian particles are concentrated off arid and semiarid areas and also in some deep basin areas where most particles of biologic origin are dissolved.

About 20% of all particles in the water column are mineral particles of biological origin and are mostly concentrated in the photic zone. The majority (15%) of these particles are siliceous, but dissolve rapidly because of undersaturation of silica in the upper water column. On average, less than 1% of them reach the seafloor, but the proportion varies regionally from 0% to 80%. They mostly consist of diatoms (phytoplankton) and radiolarians (zooplankton). The rest (5%) of the mineral biogenic particles are calcareous, but their abundance decreases in cold and/or poorly oxygenated waters because of dissolution, and on average about 20% of them reach the seafloor. They mostly consist of coccoliths (phytoplankton) and foraminifers (zooplankton).

The water column is more turbid near surface and seafloor (Figure 2.26). From the photic zone, the concentration of particles decreases with depth to a minimum around 3,000 m below sea level. The turbidity increases again below 3,000 m, in relation with the characteristics of the water masses: concentrations are lower below oceanic gyres and higher on the path of the main bottom currents (nepheloid layer). Abyssal currents may transport particles over distances of more than a thousand kilometers by steps, through a succession of depositions and resuspensions according to the velocity of the bottom currents. For example, Antarctic and sub-Antarctic diatoms and coccoliths have been recognized up to low latitudes.

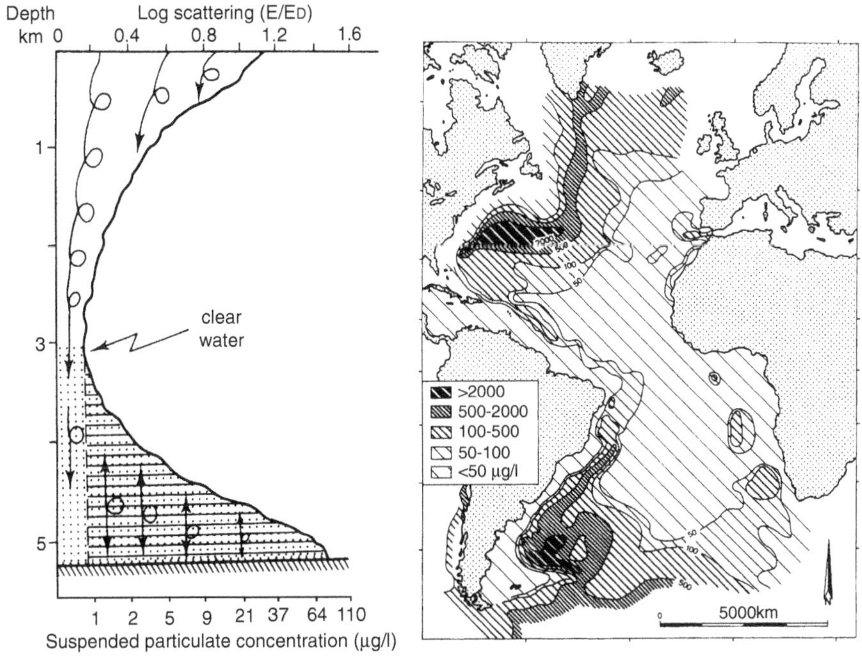

Figure 2.26 Schematic depth profile of the concentration of suspended matter in the ocean (left) and concentration of suspended matter in the bottom nepheloid layer in the Atlantic Ocean (right). Note the variability of concentration in the bottom nepheloid layer. From Biscaye, P.E., Eittreim, S.L., 1977. Marine Geology, 23, 155–172.

On average, the concentration of suspended matter is higher west of the oceans near areas of higher precipitation than east of the oceans near arid to semiarid regions, despite higher productivity there. Concentration of particles is also higher at high latitudes, because of intense continental erosion by ice and high productivity in the ocean. In addition, suspended matter concentration is three times higher in the larger Pacific Ocean bounded by active margins than in the Atlantic Ocean bounded by passive margins.

2.3.4. Other Sources of Sediments

Mid-oceanic ridge volcanism and hot-spot volcanism (anhydrous fusion of the mantle) principally contribute fluid basaltic lavas which may partly alter in surface while cooling in seawater. When emitted in subaerial areas, these lavas are weathered in accordance with local climatic conditions, like other substrates. However, the nature of chemical weathering products is largely influenced by the composition of the lavas, generally enriched in Fe, Mg, etc., by comparison with the average chemical composition of the continental lithosphere. The particles are then carried to the ocean by runoff or wind activity, like those produced from other subaerial substrates. Subduction volcanism (hydrated fusion) is explosive and contributes huge quantities of glass and ashes. Volcanic glass and ashes accumulate rapidly during eruptions, to be later reworked by runoff. However, volcanic glass and ashes are also transported by winds and distributed over large oceanic areas where they settle to form distinct sedimentary layers.

Although authigenic sediments are not widespread in the modern ocean, authigenic particles are observed in many types of sediments where they form near the water/sediment interface through chemical and biochemical processes. Chemical elements are provided by seawater, biogenic and terrigenous particles and hydrothermal activity. Biological activity plays a major role in the formation of authigenic elements such as glauconite and phosphates, whereas hydrothermal activity looks closely associated to the development of metalliferous oxides such as crusts and nodules. Authigenic components are currently more abundant in areas of low deposition such as shallow and high-energy parts of continental shelves for glauconite and phosphates, and deep isolated parts of the largest oceanic basins for metalliferous crusts and polymetallic nodules.

2.3.5. Distribution of Sediments in the Modern Ocean

Because of the good ventilation of the oceans, organic particles which account for most of the oceanic suspended matter are generally destroyed by oxidation in the water column and therefore are not incorporated in most oceanic sediments. However, there are some exceptions in areas of low oxygen content: below coastal upwelling and divergence areas where most of the oxygen is being consumed through biological processes (respiration, etc.), and in poorly ventilated areas where the morphology of the basins (California borderland) and/or water stratification (Black Sea) impede the oxygenation of the deeper waters. Therefore, terrigenous and biogenic particles are major components of modern oceanic sediments, with organic particles being locally important, and authigenic particles as minor components.

The composition of oceanic sediments (Figure 2.27) is modulated by the relative importance of the main sources of particles (runoff, productivity, etc.) and by oceanic processes (transport by currents, dissolution, etc.). The importance of the sources also modulates the quantity of sediment to be produced (mass accumulation rates). In areas or intervals of high sedimentation rate, particles from a dominant source dilute other components. For example, off river mouths, there is a dilution of all types of particles by a dominant terrigenous source and deposition of terrigenous sediments with high sedimentation rates. Also, high productivity in upwelling areas results in deposition of biogenic sediments with high sedimentation rates. In areas or intervals of low sedimentation rate, the low contribution (or absence) of some sources favors the concentration of other components. For example in the deep central Pacific Ocean, low productivity, dissolution of biogenic particles and distance from major river mouths favor the concentration of eolian and authigenic particles in the sediment.

Terrigenous sediments dominate near continental areas because of the proximity of sources. Their accumulation favors the formation of continental shelves on the passive margins of divergent oceans such as the Atlantic and fills oceanic trenches in

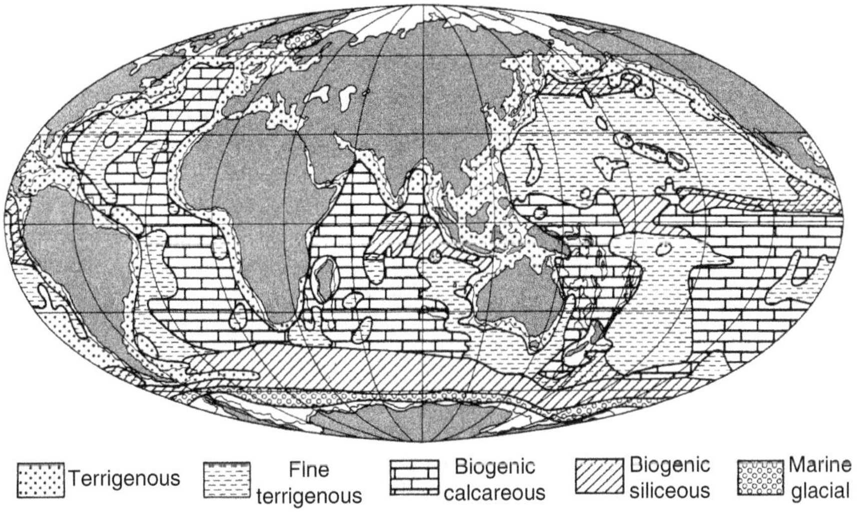

Figure 2.27 Distribution of the major types of sediments in the modern ocean. Note the dominance of terrigenous sediments near continental areas, fine terrigenous sediments in the deepest parts of the ocean basins, biogenic calcareous sediments in the center parts of the low to mid-latitudes oceans and biogenic siliceous sediments at high latitudes. Modified from Kennett, J.P., 1982. Marine geology, Prentice-Hall, Englewood Cliffs, NJ.

some active margin areas such as the Chile margin. However there are some exceptions, for example off areas of low erosion such as northeastern Australia where coral reefs develop in warm coastal waters. Terrigenous components decrease in abundance with distance from the continent, as does the dilution of biogenic components: roughly comparable abundances of terrigenous and biogenic components form hemipelagic sediments, which are commonly found below areas of coastal upwelling, on the continental rise and in shallow oceanic basins. The proportion of biogenic components increases to the center parts of the oceans, leading to pelagic biogenic sediments. Calcareous biogenic sediments dominate in shallower areas (mid-oceanic ridges, oceanic plateaus, shallow basins) but are absent from areas bathed by CO_2-rich waters such as high latitude areas and the deepest parts of oceanic basins where cold waters formed at high latitudes circulate. Calcareous biogenic sediments are replaced by siliceous pelagic sediments in areas of cold waters favorable to the preservation of silica, at high latitudes and below oceanic divergences. However, fine terrigenous (together with authigenic components) sediments are present in deep oceanic basins below areas of low productivity and intensive dissolution of both silica and carbonates. The distribution of modern sediments mimics the morphology of the oceans and the pattern of thermohaline circulation.

2.3.6. Coastal Areas

Coastal areas are the interface between continental and marine environments, where terrigenous input from the continent and benthic biological activity interact with surface conditions (waves, tides, currents, etc.). They include deltas and estuaries, beaches and tidal flats, coral reefs and lagoons. Coastal areas are in constant evolution, as they move back and forth with absolute and relative variations of sea level. Absolute variations of sea level are closely associated to the changing volume of polar ice and coastal areas shift with the alternating glacial and interglacial intervals. For example, sea level was about 130 m below zero during the last glacial maximum around 20 ka: the English Channel was then subaerial, and most modern rivers of northwest Europe were tributaries of a major stream which flowed directly into the Atlantic Ocean. Coastal areas moved inland with the subsequent latest Pleistocene and Holocene sea level rise at a maximum rate of 3.7 cm/yr and stabilized around 6 ka (Figure 2.28). Relative variations of sea level are associated to regional and global tectonics. Coastal areas shift with uplift and subsidence of the lithosphere and with plate tectonic events. This is especially important at geological scale, but may occur during abrupt events. For example, the shoreline locally moved seaward by a few tens of meters during an earthquake which raised a submerged block at 1.5 m above sea level in New Zealand during the late seventies.

Terrigenous sediments are dominant in coastal areas near river mouths. When the terrigenous load of the river is important a delta develops, progressively creating a coastal plain and expanding the shelf. For example, the delta of the Rhône river which drains the western Alps expanded by several tens of kilometers during the Holocene (Figure 2.29). In other areas the terrigenous input is low, restricted to local erosion and reworking of nearshore sediments and substrates. Biological activity in coastal waters is principally benthic in nature, and especially important in low-latitude areas of low-terrigenous input where constantly warm waters favor the development of carbonate-secreting species. They include algae, corals, molluscs, foraminifers, and the accumulation of their remains produces enormous quantities of biogenic sediment. This is the case, for example, off eastern Australia along the

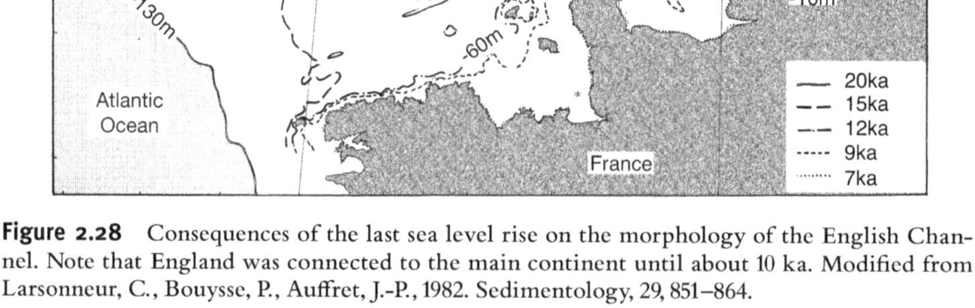

Figure 2.28 Consequences of the last sea level rise on the morphology of the English Channel. Note that England was connected to the main continent until about 10 ka. Modified from Larsonneur, C., Bouysse, P., Auffret, J.-P., 1982. Sedimentology, 29, 851–864.

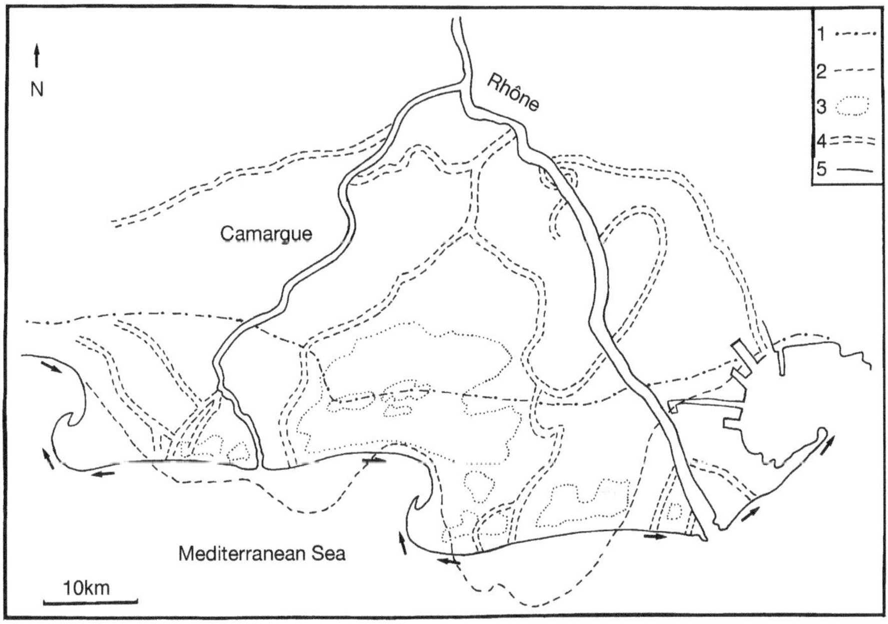

Figure 2.29 The Rhône delta in the Gulf of Lions (Western Mediterranean Sea). 1: roman shoreline; 2: 18th century shoreline; 3: coastal lagoons; 4: abandoned channels; 5: modern shoreline and channels. Arrows indicate major directions of sediment transfer. Note rapid extension of the Rhône delta during historical times and modern erosion in some areas. Modified from Blanc, J.J., 1982. Sédimentation des marges continentales actuelles et anciennes, Masson, Paris.

Great Barrier, Madagascar, Belize, etc. In most coastal areas, the temperature of seawater varies strongly with the seasons, with consequences on both the abundance and diversity of species (destruction of eggs and larvae in winter, for example). Globally, both the abundance and diversity of coastal marine species decrease with seawater temperature, and coastal benthic activity is minimum at high latitudes (Figure 2.30). As a result, terrigenous coastal sediments increase in importance with latitude because of dilution.

Beaches are accumulations of non-cohesive sediments made of sand and gravel principally. The morphology of beaches may vary rapidly with wave conditions, the energy of the waves being partly used to move the sediment. A back and forth motion of sediment particles is produced at water depths below half the wavelength (Figure 2.31). This motion is proportional to the orbital velocity of the waves, which varies with the water depth: higher and directed onshore for the crests and lower and directed offshore for the deeps. The capacity of wave-induced currents to move sediments is higher for the crests (coarse and fine particles are transported onshore) than for the deeps (only fine particles are transported offshore). This dissipates only a small part of the waves energy, but significantly decreases their velocity. The rate of propagation of the waves energy being proportional to their height and velocity and constant, decreases in velocity are compensated by increases in height. The water percolates into non-cohesive

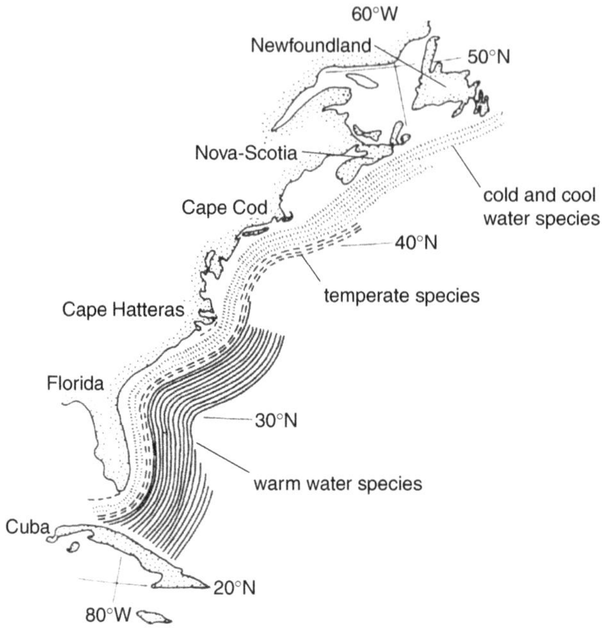

Figure 2.30 Diversity of benthic lamellibranch species in coastal areas of the NW Atlantic. Surface areas of symbols are proportional to the extension and number of species. Note ubiquity of cold and cool water species and variety of warm water species. Modified from Seibold, E., Berger, W.H., 1996. The sea floor, Springer, Berlin.

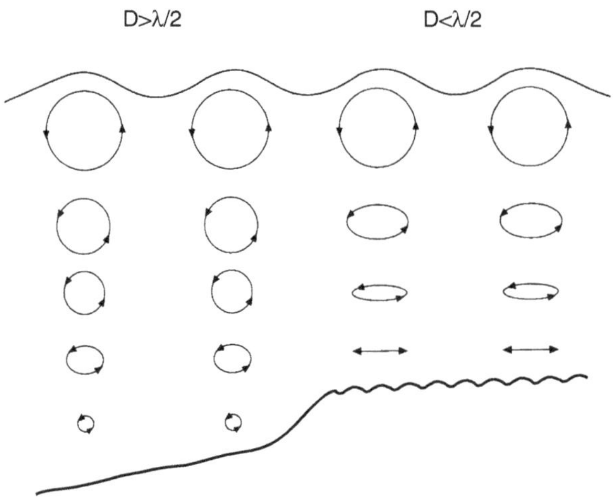

Figure 2.31 Interaction of waves and seafloor. For water depths (D) below half the wavelength (μ/2), wave-induced currents move sediments and start forming ripple-marks. Modified from Nichols, G., 1999. Sedimentology and stratigraphy, Blackwell, London.

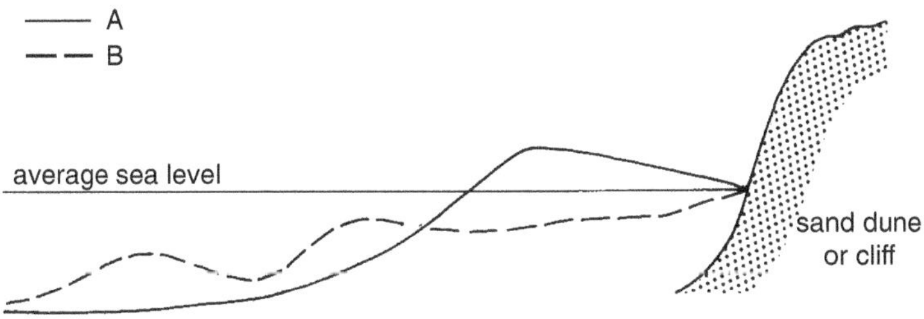

Figure 2.32 Relationships between the morphology of beaches and the characteristics of incoming waves. (A) Beach profile for long wavelength, low-frequency waves and (B) beach profile for short wavelength, high-frequency waves. Modified from Chamley, H., 2000. Bases de sédimentologie, Dunod, Paris.

sediments in the swash zone, further decreasing the capacity of receding water to move coarse particles offshore. The impact of the waves results in a net accumulation of the coarsest particles on the beaches, the finest particles being abundant only at water depths greater than the deepest possible influence of the waves, that is usually 10–20 m.

The influence of the waves on nearshore sediments also varies with their characteristics and direction. Waves of low frequency and long wavelength break over long distances. Their energy slowly decreases and percolation along the beach is important: coarse particles are pushed to the upper beach and accumulate, accreting the subaerial part of the beach. Waves of high frequency and short wavelength break over small distances. Their energy is rapidly and incompletely released, and percolation along the beach is of minor importance: receding waters still have enough energy to sweep coarse particles offshore and erode the upper beach (Figure 2.32). When incoming waves are oblique to the coastline, they induce longshore currents which carry sediments along the beach and modify its morphology. In areas where oblique wave direction is dominant through time, lateral transport of sediment often results in the formation of sand drifts which may isolate coastal lagoons (Figure 2.29).

Tidal flats are areas where the waves do not break long enough in the same place to shape the coastline because of the dominant influence of the tides. This is the case in some estuaries and bays protected from the dominant waves (Figure 2.33). As the flood tide progresses, seawater fills first the tidal channels and then spills over the flats. The energy of the flood is high enough to transport coarse particles in the main tidal channels but decreases rapidly as the flats are submerged, and only fine particles (silts and clays) are being transported to the shore at high tide. They form cohesive sediments, very resistant to the erosion. During the ebb tide, seawater is drained back to the tidal channels and then to the ocean. In the uppermost flat which is submerged only a few days per year during very high tides, salt marshes may develop in temperate mid-latitude areas, while mangrove swamps expand in

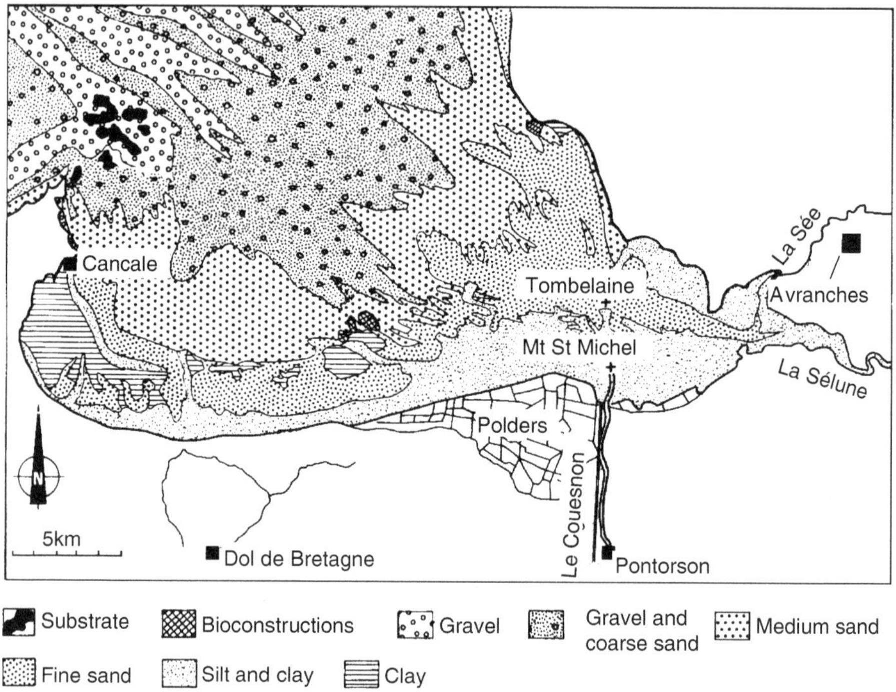

Figure 2.33 Tidal flats in the bay of Mont Saint Michel (NW France), English Channel. Note that grain size decreases nearshore but is coarser in river channels and that accretion favored the construction of polders in the southern part of the bay. Modified from Chamley, H., 1988. Milieux de sédimentation, BRGM, Orléans.

humid areas of low latitudes and sabkhas in highly evaporative environments. Tidal flats are areas of sediment accretion.

2.3.7. Sediment Particles in Seawater

Erosion, transport and deposition of a particle by a fluid are controlled by the physical properties of both. For sediment grains, individual physical properties (mass, density, size, morphology, etc.) have to be considered together with those of the whole sediment (sorting, porosity, cohesion, fabric, etc.).

To simplify, sediment particles are sometimes compared to spheres. In this case, the mass of particles of similar density varies with the cube of the radius. A sphere of radius 5 mm is 5 times greater than a particle of radius 1 mm, but 125 times heavier and requires 125 times more energy for being removed. The mass also varies with the density of the particles. Most of the minerals commonly found in oceanic sediments have densities between 2,000 and 3,500 kg/m^3, but some heavy minerals have densities around 5,000 kg/m^3, and the energy required to remove them varies accordingly. The quantity of energy required to remove particles of any given size is smaller when the sediment is relatively homogeneous in grain size, of high porosity

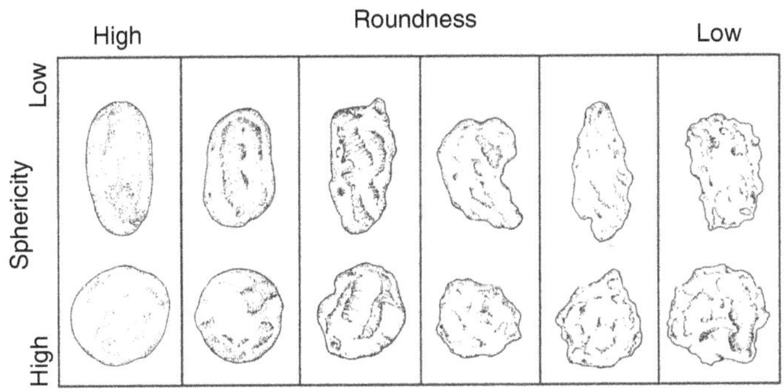

Figure 2.34 Indices expressing the morphology of individual grains. Note that the hydraulic behavior of particles that deviate the most from high roundness and sphericity also deviate the most from theory. Modified from Chamley, H., 2000. Bases de sédimentologie, Dunod, Paris.

and non-cohesive. More energy is required when homogeneity of grain size and porosity decrease, and cohesion increases. The morphology of the particles influences their hydraulic performances, but this decreases with the size of the grains. This is expressed through specific indices such as sphericity and roundness of individual grains (Figure 2.34). Generally the floatability of the particles increases as they deviate from a sphere, and is maximum for platy particles such as sheet silicates (micas, clay minerals, etc.).

The capacity of a fluid to transport particles is determined by its density, viscosity and velocity among other parameters. Differences in the quality and quantity of sediment particles transported by seawater and air are understandable from their differences in density (1,030 and 1 kg/m^3, respectively) and viscosity (1×10^{-3} and 1.78×10^{-5} Ns/m^2, respectively). Fluid motion can be laminar (molecules move uniformly downstream, parallel to each other) or turbulent (molecules move in all directions, with a net displacement downstream). Laminar flows mostly characterize very slow or highly viscous fluids. For any given fluid, the transition to turbulent flow occurs when the velocity of the fluid or the roughness of the substrate increase, or the thickness of the fluid layer decreases. This is expressed by the Reynolds number

$$Re = \frac{du\rho}{\mu}$$

where d is the depth of flow, u its velocity, ρ its density and μ its viscosity. Fluid flows are laminar for low Reynolds numbers (below 500) and turbulent for high Reynolds numbers (above 2,000). For very thick fluid layers, turbulent movement may occur only in the part of the fluid adjacent to the substrate, the boundary layer. Turbulent flows increase the capacity of a fluid to erode and transport particles. For low-viscosity and low-density fluids such as water and especially air, turbulent flow conditions are reached rapidly through increased velocity.

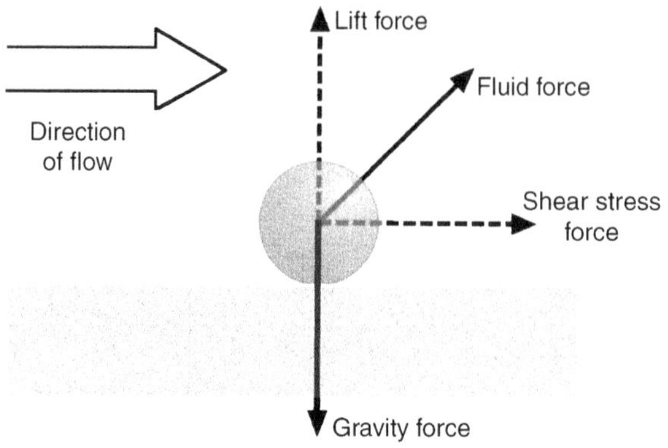

Figure 2.35 Forces exerted by a flow on a single grain. The grain is eroded and entrained when the lift force component compensates for the gravity force. Modified from Nichols, G., 1999. Sedimentology and stratigraphy, Blackwell, London.

Particles in a fluid are subjected to both gravity and fluid forces (Figure 2.35). The fluid force is composed of a shear stress force that is parallel to the flow and a lift force that is perpendicular to the flow. The Bernoulli equation states that the quantity of energy of a portion of moving fluid is made of the sum of potential (ρgh), kinetic ($\rho u^2/2$) and pressure (P) energies. Heat loss due to friction being negligible, the conservation of energy principle implies that this quantity is constant

$$\rho gh + \rho u^2/2 + P = cst$$

where ρ is the density of the fluid, g the gravity, h the fluid depth, u the fluid velocity and P the pressure. For any given fluid, the potential energy does not change as long as the depth remains the same. This condition being fulfilled, any change in fluid velocity results in a change of pressure, which is the source of the lift force. When an object is placed in a moving fluid, the principle of mass conservation states that the quantity of fluid must be the same upstream and downstream of the object. Therefore, the fluid must move at greater velocity around the object, decreasing the pressure exerted by the fluid and increasing the lift force. A particle placed on a substrate is moved (according to the shear stress force exerted by the fluid) when the lift force compensates for the gravity, and as long as these conditions persist. However, sediments are complex mixtures of particles of different size, density, morphology and fabric, and this influences their ability to be moved. For example, particles are more easily moved when resting on an assemblage of smaller grains than when embedded within other grains of similar or greater size.

In the ocean, the friction exerted by a flow on the seafloor results in decreased velocity in the boundary layer (Figure 2.36). There, the velocity is near zero in the uppermost sediment (adsorbed layer) and increases to the main flow velocity over distances of 1–10 m. Conversely, the friction (shear-stress force) is maximum at the seafloor and decreases with distance. The actual velocity and shear stress force in the lower part of the boundary layer are important parameters to determine the capacity

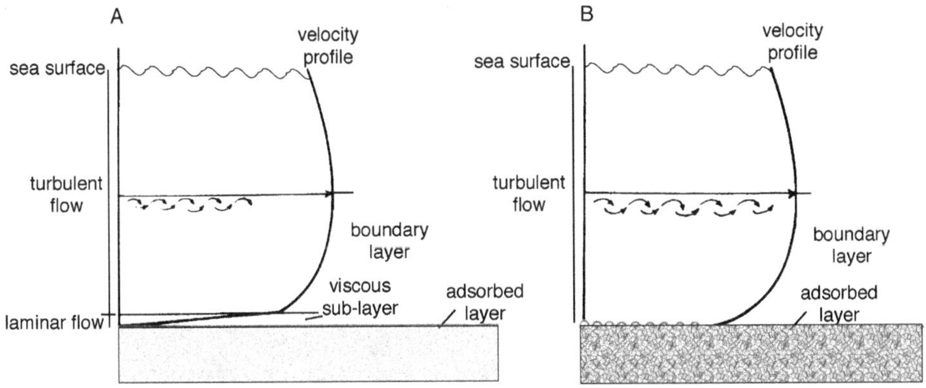

Figure 2.36 Interaction between flow and seafloor. The velocity of the turbulent flow decreases in the boundary layer. (A) Smooth boundary (fine sediment and/or low velocity) with development of laminar flow in the viscous sublayer and (B) rough boundary (coarse sediment and/or high velocity) with extension of turbulent flow to the seafloor. Modified from Shepard, F.P., 1963. Submarine geology, Harper's, New York, 1963.

of a current to erode and transport sediment particles. When the seafloor is made of very fine particles and the velocity is very low, an almost laminar flow, a few millimeters thick, may develop at the sediment/ocean interface: the viscous sub-layer where the gradient of velocity is linear, which tends to protect the seafloor from erosion. When the diameter of particles increases above about one-third the thickness of the viscous sub-layer, the flow progressively becomes turbulent, increasing its capacity to move particles.

The fluid velocity necessary to move particles is the critical velocity. However, the lift force required to bring a particle into the flow, as well as the shear stress force required to move this particle along the flow, increase with the mass of the particle which is related to its size and density. Sediment particles are moved in a fluid either as bedload or suspended load. Transport as bedload includes: (i) materials which are moved by traction, either rolled or dragged without losing contact with the seafloor and (ii) materials which are moved by saltation, being periodically lifted into the fluid and carried short distances before returning to the seafloor. Transport as suspended load includes those particles which are lifted higher in the flow by turbulent motion and are advected downstream over great distances. Particles may alternate between bedload and suspended load according to the velocity of the flow. For a given fluid velocity, platy particles of large surface have a better buoyancy and can be kept in suspension more easily. Sediment particles show a tendency to settle when the velocity of the flow decreases as the gravity force exerted on the particle surmounts the lift force of the fluid (critical settling velocity): particles transported as bedload stop rapidly, whereas those transported as suspended load start sinking to the floor. The Stoke's law is commonly used to find the settling velocity (V) of suspended particles according to their size and density

$$V = CD^2$$

where D is the diameter of the particle and C a constant for a given fluid and category of particle

$$C = \frac{(\rho_s - \rho_f)g}{18\mu}$$

where ρ_s is the density of the particles, ρ_f the density of the fluid, g the gravity and μ the viscosity of the fluid. Settling velocities vary with the diameter of the particles. As is, Stoke's law applies to particles below 0.2 mm in size, which are the most common in the ocean. Above 0.2 mm the settling velocity is proportional to the progressive variation of the power of the diameter, and to the square root of the diameter above 2 mm. Stoke's law essentially describes the settling of spheres of same density in a still environment, and does not account for the variety of sizes, morphologies and densities of sediment particles, nor for the turbulent motion of oceanic water masses. It provides a rough estimate of the relative settling of fine sediment particles.

The relationships between grain size, velocity of the flow and transport of particles is illustrated by a diagram initially developed by Shields in 1936, later improved by Hjulström, and subsequently modified by many workers, but known as the Hjulström diagram (Figure 2.37). The upper curve shows the velocity required to move a particle. This velocity decreases with the size of the particles for coarse particles above 0.1 mm, but increases with decreasing size for fine particles. Identical flow velocities are required to move particles 20 and 0.002 mm in size. This is partly due to the low rugosity of fine-grained surfaces, which may be protected from erosion by the formation of viscous sub-layer for low flow velocities. This is also due to the cohesion of the finest sediment particles once they are deposited. The lower curve shows the settling velocity of particles, for which bedload comes to a rest and suspended load starts sinking to the floor. Changes in the velocity of the flow are reflected by changes in the size of particles to be deposited: a decrease in flow velocity through time results in an upward decrease of the sediment grain size (normal grading), while an increase in flow velocity through time results in an upward increase of the sediment grain size (reverse bedding). However, it is clear that particles below 0.01 mm can settle only in perfectly still environments. The finest particles below 0.005 mm in size may need months to years to reach the seafloor according to Stoke's law. The Hjulström diagram clearly shows the different behaviors of coarse and fine particles relative to fluid forces.

Coarse sediments are non-cohesive. Individual particles are removed, transported and deposited according to above described principles. The behavior of particles below 0.2 mm in size progressively deviates from these principles, to a maximum deviation for particles below 0.01 mm. Most of the suspended matter in the deep ocean is flocculated. Flocculation results from molecular attraction by the forces of van der Waals which are weak and vary inversely as the square of the distance between particles. As for the forces of van der Waals to be efficient, particles must be maintained in close proximity. Oceanic waters contain a proportion of organic matter, which is charged negatively. Terrigenous particles include a proportion of clay minerals which dominate the finest part of the

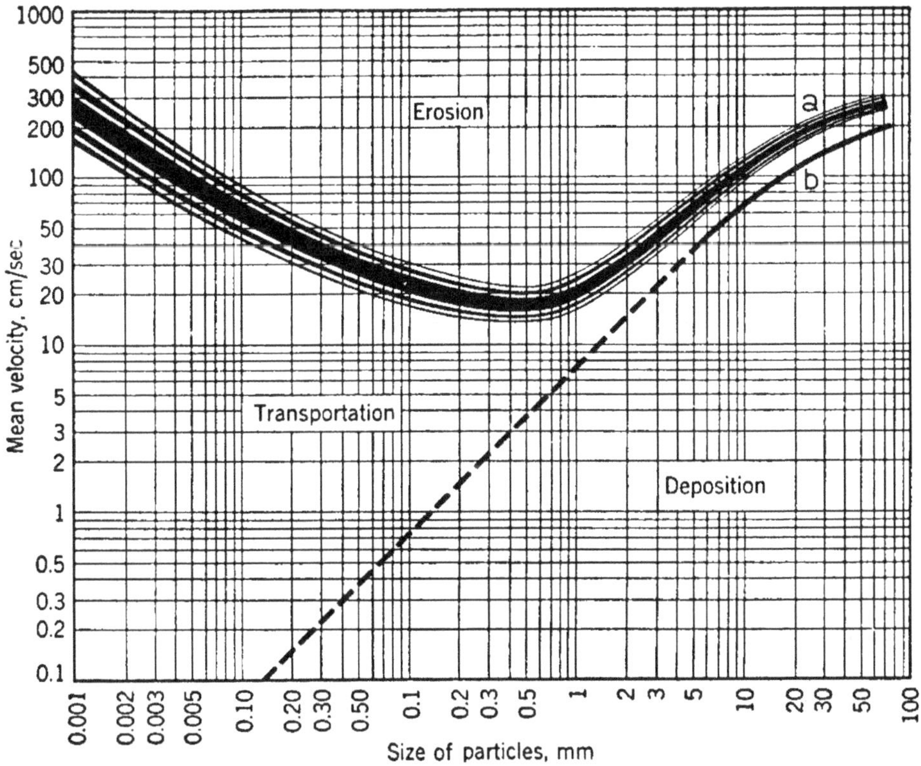

Figure 2.37 Hjulström diagram. Note that particles below 0.5 mm require higher flow velocities to be removed as they decrease in size (because of cohesion), and that particles below 0.01 mm in size are being transported for very low flow velocities under 0.1 cm/s. Reprinted from Bouma, A.H., Brouwer, A., 1964. Turbidites, Elsevier, Amsterdam.

suspended terrigenous load and are also charged negatively. For electrostatic reasons, these particles should be kept away from one another. However, seawater also contains free cations charged positively, principally Na but also Ca, Mg, K, etc. Interaction of clay minerals and organic matter with free cations reduces the negative charge of the particles as to allow molecular attraction providing that particles are brought in close proximity. Flocculation is triggered in those areas of the ocean where organic and/or clay particles are in close proximity because of high concentration, turbulent motion or ingestion by living organisms. These conditions are, for example, fulfilled in the bottom nepheloid layer of the ocean basins. Flocculation may lead to the formation of agglomerates (flocs) up to 10 mm in size, which have the hydraulic behavior of single particles of similar size and density. Flocculation lasts as long as the forces of van der Waals are active. Relative proportions of mineral and organic material control the size and physical properties of flocs, which generally increase in size and decrease in density for higher contents in organic matter. The strength of flocs being related to their density, highly condensed flocs produce stable sediment layers, whereas loosely formed flocs rather produce easily erodible sediment layers. In the sediment, molecular attraction extends to adjacent flocs and may be reinforced by the activity of some bacteria producing organic films. This results in the formation of cohesive sediments, which are far more difficult to erode than non-cohesive deposits. However, the cohesion of sediments is improved when they contain 5–10% of flocculated components or more.

2.3.8. Gravity Flows

Once they have been deposited, young sediments of high water content are unstable and can be easily reworked. This is especially the case in areas of submarine relief where some instability may result from the combination of morphological gradients with accumulation of sediment and/or seismic activity: young and narrow oceanic basins, continental shelves and slopes of passive margins and oceanic plateaus, oceanic trenches and forearc and backarc basins of active margins principally. In such cases the particles are transported either as a mass of sediment, or a mixture of sediment and seawater, under the influence of gravity. Gravity flows are considered as fluids of high density (usually between 1,500 and 2,600 kg/m^3) and very high viscosity (usually between 1×10^2 and 1×10^3 Ns/m^2) of either laminar or turbulent behavior. They include gravitational instabilities (slides, slumps), laminar viscoplastic flows (debris flows, mud flows) and turbulent hyperconcentrated flows (turbidity currents).

A mass of sediment that rests on a slope of angle α is exposed to a shear force tending to cause slip along contact surfaces, which opposes a friction force resisting the movement. The force due to the mass of sediment

$$W = \rho g h$$

where ρ is the density of the sediment, g the gravity and h the thickness of the sediment mass has two components: one component ($s = \rho g h \sin \alpha$) is parallel to the slope (shear force) and the other ($n = \rho g h \cos \alpha$) is perpendicular to the slope (normal force). The shear component increases with depth due to the weight of overlying sediment, but also with the angle of the slope and the density of the sediment. At some depth below surface the shear equals the friction force, and this forms a potential slip plane of failure. At this depth, the friction coefficient which is the ratio of the friction force and the normal force due to the weight of sediment (Coulomb's law) is

$$f = \rho g h \sin \alpha / \rho g h \cos \alpha = \tan \alpha$$

These conditions can be reached either by increasing the shear component or decreasing the normal component. However, the formation of a potential slip plane of failure in a mass of sediment is also conditioned by the global physical properties of the sediment: fabric, porosity and permeability, degree of cohesion, water content, etc. For example, critical conditions are reached more easily for non-cohesive sediments (through simple accumulation) than for cohesive deposits. Most of the time, critical conditions have to be triggered by other mechanisms. For example, ground motion related to earthquake activity may increase the shear component as to reach critical conditions. Also, upward pressure exerted by interstitial fluids (water or other fluids such as methane) expelled from the sediment as accumulation progresses may compensate for the normal force (and decrease the friction force of the substrate) as to reach critical conditions.

Gravitational instabilities may result from the development of a slip plane deep below the sediment surface. They are often triggered by earthquakes and may occur on gentle slopes. Their distance of transport may reach up to a few kilometers, being slightly shorter for the slides which maintain the internal structure of the sediment with little disturbance and slightly longer for the slumps which show important internal disturbance including sediment folding and a succession of slip surfaces.

Viscoplastic flows are dense, viscous mixtures of sediment with water. Such mixtures have low Reynolds numbers and their flow is predominantly laminar. Sediments are transported over distances in the range of tens and exceptionally hundreds of kilometers. The internal structure of the mass of sediment is destroyed but its coherency is maintained and resulting deposits are very poorly sorted. However the top part of the flow may become dilute, resulting in some gradation of deposits and transport of fine material as suspension. Debris flows may occur in sediments of varied composition but high water content, when excessive water pressure reduces the normal component of gravity and the friction force of the substrate is in excess. Debris flows may carry non-cohesive material of any size. They mostly occur nearshore in alluvial fan areas and in volcanic areas following pyroclastic eruptions. Mud flows mostly consist of fine sediment of terrigenous or hemipelagic composition and may concern low-cohesive material.

Hyperconcentrated flows are mixtures of sediment with water, containing more than 50% water. They are less dense than viscoplastic flows and have relatively high Reynolds numbers, and therefore are turbulent. They include a variety of flows depending on their density, which especially varies with the relative proportion of sediment and water. Some flows have a low density close to those of seawater: this is the case of mixtures of freshwater and sediment particles which circulate at the surface of the ocean (hypopycnal flows) or at interfaces of water density (hyperpycnal flows) in coastal areas near river mouths. Hyperconcentrated flows which move under gravity due to their contrast of density with seawater are turbidity currents (Figure 2.38). They are especially frequent off major river mouths and in areas of explosive volcanism where they rework pyroclastic deposits. Turbidity currents move down the slopes of continental shelves or oceanic trenches which provide potential energy, but then can circulate over flat areas as long as the contrast of density with seawater persists. Sediments may be transported up to a thousand kilometers. Turbidity currents lose density by deposition of the coarsest particles and decelerate to a stop when the density contrast is insufficient to maintain momentum. Therefore, they separate coarse material, which is deposited first, from fine material, which stays longer in suspension. Resulting accumulations of sediments, known as turbidites, are normally graded and rest on a scoured base, as turbidity currents erode their

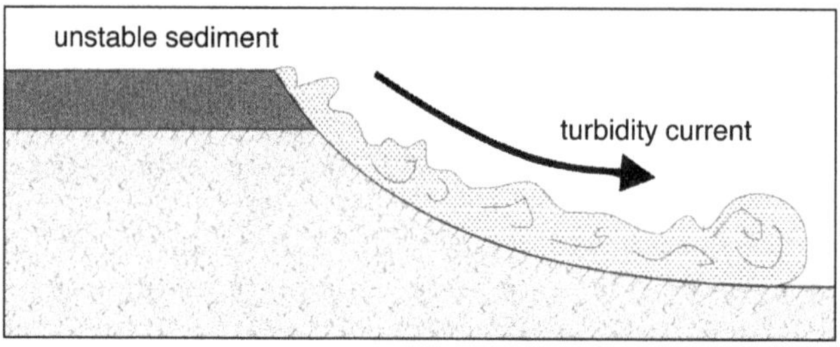

Figure 2.38 Turbidity current. A mixture of destabilized sediment and water moves down slopes and even relatively flat areas, driven by gravity and contrast of density with seawater. Modified from Nichols, G., 1999. Sedimentology and stratigraphy, Blackwell, London.

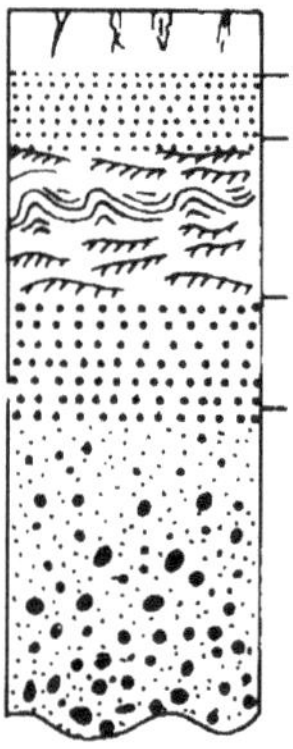

Figure 2.39 Turbidite (Bouma units). Note that sediment classification (sand, silt, clay) refers to grain size. Modified from Chamley, H., 2000. Bases de sédimentologie, Dunod, Paris.

substrate. Detailed composition, grain size, texture and structure of turbidites have been described by Bouma in 1962 and their succession is known as the Bouma sequence, which includes (Figure 2.39):

- Poorly sorted coarse sands transported at the base of the current where turbulence is reduced.
- Finer and better sorted laminated sands generated by separation of particles in the turbulent upper flow.
- Cross-laminated medium to fine sands with ripples indicative of lower flow velocity;
- Fine sand and silt with some lamination, indicative of weakening flow.
- Fine-grained silt and clay, deposited from suspension as the current ceases.

In reality the Bouma sequence is rarely complete, and helps to make distinction between proximal (near source) turbidites that are mostly sandy and contain the first (lower) units of the sequence, and distal (far from source) turbidites that are mostly muddy and contain the last (upper) units of the sequence. Reworking of fine, terrigenous, hemipelagic or volcaniclastic sediments may produce mud turbidites characterized by a sharp basal contact, more or less distinct normal grading and some bioturbation at their upper part.

2.3.9. Sediment Classification

Classification of sedimentary deposits by physical, chemical or biological analysis has been a constant preoccupation since the early days of marine geology. The proposed classifications have been progressively combined, as to address two main fields of investigation: the dynamics of sediment formation and the origin of sediments. They are based on the size and the nature of sediment particles, respectively.

Classifications based on the size of the particles have been conceived as to reflect observed changes in the physical properties and dynamics of sediments: settling velocities, permeability, fabric, etc. Most of these parameters were found to vary as some power of particle size rather than directly with size, and a logarithmic scale was adopted for describing sediment distribution. The first scale was proposed by Udden (1898), later improved by Wentworth (1922) and Krumbein (1936) principally. Most sediments are classified as clay, silt, sand and gravel:

Sediment		Scale (mm)
Boulder		Above 256
Gravel	Coarse (cobble)	256–64
	Medium (pebble)	64–4
	Fine (granule)	4–2
Sand	Very coarse	2–1
	Coarse	1–0.5
	Medium	0.5–0.25
	Fine	0.25–0.125
	Very fine	0.125–0.063
Silt	Coarse	0.063–0.031
	Medium	0.031–0.016
	Fine	0.016–0.004
Clay		0.004–0.0002
Colloid		Below 0.0002

Clay- and silt-sized particles of both terrigenous and biologic origin dominate largely in the deep ocean, sand and gravel being especially important in coastal areas.

Classifications based on the nature of the particles have been designed to reflect the processes leading to sediment formation. The first classification based on the nature of sediment particles has been proposed by Murray and Renard in 1891, from the results of the Challenger expedition. They observed the compositional complexity of most sediments and based their classification on major components. The word "ooze" was used for sediments composed principally of shells and mineral remains of pelagic organisms, and the word "mud" for sediments principally made of mineral products derived from the disintegration of land. The dominant component of the sediment was then used as a discriminant to name the sediment. This classification reflects the dominant biogenic or terrigenous nature of the sediments and their distribution in the ocean, and include principally:

- Globigerina ooze, which is the most extensive of all deep-sea deposits (especially in tropical areas) and contains remains from other organisms as well as a few terrigenous components.
- Diatom oozes, found in the Southern Ocean and some parts of the Indian and Pacific oceans outside the tropical domain, which contains some other biologic and terrigenous elements.
- Radiolarian ooze, confined to the greater depths of the ocean, which may also contain some pumice, Fe–Mg minerals, micronodules and cosmic spherules.
- Red clay, interpreted either as a product of the disintegration of land further distributed by oceanic currents, an insoluble residue of the Globigerina ooze or an alteration of volcanic products, and contains other elements very similar to those found in the Radiolarian ooze.
- Blue mud, found near continents as well as in the Mediterranean sea and the Arctic ocean, which is dominated by quartz fragments and contains a proportion of organic matter and iron sulfide.
- Red mud, found near continents of low latitude, which contains some Fe products from continent.
- Green mud and sand, which are also found near continents of low latitude, but in exposed areas.
- Volcanic mud and sand, which come from the disintegration of volcanic products and are found around oceanic islands of volcanic origin.
- Coral mud and sand, found off coral reefs and islands where their importance decreases with distance.

Some categories of this classification are still commonly used.

Further classifications were designed to introduce more precision in the description of the sediment, using dominant but also other important components. Early DSDP cruises used visual core description and smear-slide estimates to define lithologic units and formations based on arbitrarily chosen criteria. The necessity to compare results from different cruises and areas led to the definition of class limits using smear-slide quantification with 25% increments. For simple mixtures of terrigenous clay and calcareous coccoliths (nannofossils):

Clay (%)	Nannofossils (%)	Name of sediment
0–25	75–100	Nannofossil ooze
25–50	50–75	Clayey nannofossil ooze
50 75	25 50	Nannofossil clay
75–100	0–25	Clay

Nannofossil ooze is a pelagic sediment, clayey nannofossil ooze and nannofossil clay are hemipelagic sediments and clay is a terrigenous sediment. By the same time, additional precision was introduced on the degree of diagenesis: soft sediments are named ooze or mud, firm sediment is chalk and hard sediments are identified by the addition of the suffix "stone" (limestone) or "ite" (radiolarite).

This classification has been further modified to bring additional precision (by taking into consideration abundances of 10–25%) and fit all types of sediments. As a rule, sediment names include:

- the sediment type (ooze, mud, chalk, etc.);
- major modifier(s) in the abundances 25–100%;
- minor modifier(s) in the abundances 25–50%, attached to the suffix "-bearing".

The modifiers are listed in order of increasing abundances to name the sediment. For example, a soft sediment made of 20% foraminifers and 80% coccoliths (nannofossils) is a foraminifer-bearing nannofossil ooze. Current ODP classifications (Figure 2.40) are based on three end members (biogenic siliceous, biogenic calcareous and terrigenous siliciclastic), the texture (size) of siliciclastic components (clay, silt and sand-sized grains) and the degree of induration. Relevant information is deduced from visual core description, smear-slide observations and other analyses (carbonate content, etc.). Components below 10% in abundance are neglected. Figure 2.40 shows the variability of classes for mixtures of siliceous and calcareous biogenic particles modified (diluted) by variable amounts of terrigenous siliciclastics. Note that the words "siliceous" and "calcareous" are used only when the biogenic debris are not identifiable. When identifiable, they are replaced by the appropriate describer (foraminifer, diatom, etc.). Some examples are provided below:

- A firm sediment composed of 5% foraminifers, 15% siliciclastic silt, 20% diatoms and 60% nannofossils would be termed a silt- and diatom-bearing nannofossil chalk.

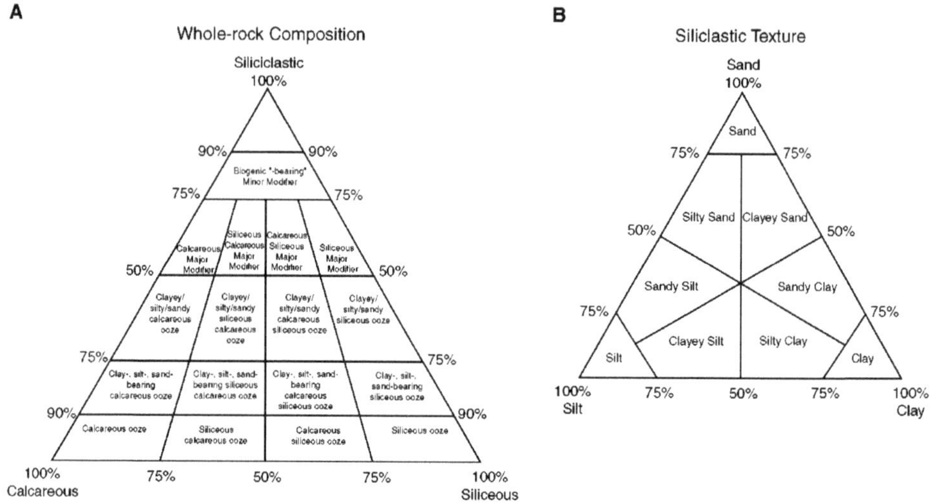

Figure 2.40 Sediment classification, Ocean Drilling Program. (A) Classification for mixtures of biogenic and terrigenous particles and (B) classification for mixtures of pure terrigenous particles. From Exon, N.F., Kennett, J.P., Malone, M.J. et al., 2001. Proceedings of the Ocean Drilling Program, Initial Reports, volume 189. Ocean Drilling Program, College Station, TX.

- A soft sediment composed of 5% siliciclastic clay, 20% nannofossils, 35% radiolarians and 40% diatoms would be termed a nannofossil-bearing radiolarian diatom ooze.
- A sediment composed of 15% foraminifers, 25% glauconite and 60% sand would be termed a foraminifer-bearing glauconitic sand.
- A sediment composed of 35% silt and 65% gravel would be termed a silty gravel.
- A sediment composed of 20% volcanic ash, 30% silt and 50% sand would be termed an ash-bearing silty sand.

However, this classification needs occasional adaptation to better fit the objectives of specific projects.

2.3.10. The Accumulation and Diagenetic Alteration of Oceanic Sediments

Biological, chemical and physical processes within a sedimentary column cause the diagenetic alteration of the components, which progressively transforms soft sediments into sedimentary rocks (Figure 2.41). Biological processes dominate in the early stages near the transition from water to soft sediment, but chemical and physical processes increase in importance with depth. Biological, chemical and physical processes intervene in the composition and circulation of pore fluids, which facilitate the dissolution of adjacent minerals, the migration of the chemical elements and the precipitation of new mineral phases. Therefore, pore fluids play a major role in sediment diagenesis.

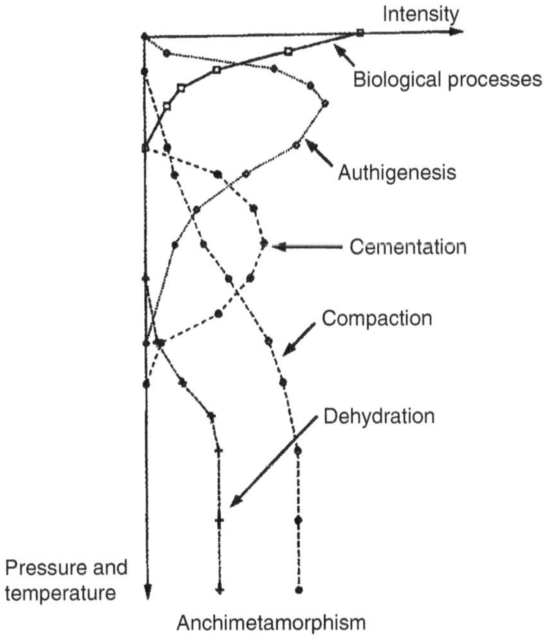

Figure 2.41 Succession and relative importance of biological, physical and chemical processes, which control the diagenetic alteration of oceanic sediments. Modified from Chamley, H., 2000. Bases de sédimentologie, Dunod, Paris.

Among early diagenetical processes, bioturbation is especially important in areas where nutrients and oxygen sustain actively burrowing organisms, but its impact on the sedimentary column decreases in importance with increasing sedimentation rates. Bioturbation generally occurs within a few centimeters of the uppermost sediment (where oxygen is available), but some burrows may exceptionally reach sub-bottom depths of about 1 m. Bioturbation locally changes the pattern of sediment accumulation and creates strong gradients in the content of organic matter and the composition of pore waters. The related alteration focuses on microenvironments and adjacent sediment principally. For example, the anoxic decomposition of organic molecules which accumulate in burrows may produce ammonia. Ammonia accumulates in anoxic microenvironments where it is adsorbed at the surface of sediment particles, but may also migrate into oxygenated interstitial waters of the uppermost sediment where it is transformed to nitrite and nitrate through bacterial metabolism. In the absence of oxygen, that is beneath the bioturbation zone and in dysoxic areas, the early stages of sediment diagenesis may include for example bacterial sulfate reduction, which involves a community of interacting microorganisms: some microorganisms break down complex organic matter into simple molecules, whereas other bacteria use sulfate from pore waters as an energy source for disintegrating simple organic molecules into hydrogen sulfide and carbonic acid. Hydrogen sulfide may react with Fe-rich sediment particles to produce pyrite and other sulfides. As sulfate comes from seawater principally the reactions cease as burial increases, after all sulfate has been removed from pore waters (generally 1–1.5 m below seafloor).

Being largely related to biological activity, early diagenetical processes principally focus on the alteration of organic elements of the sediment. Below the sulfate reduction zone, a sequence of bacterial fermentation reactions produce methane from the remaining organic matter, upward migrating methane providing simple organic molecules to sulfate-reducing bacteria. In areas of high sedimentation rates and/or high contents of organic matter, early diagenetic processes are however insufficient to remove and transform significant proportions of organic compounds from the sediment.

The pressure exerted by freshly deposited sediments onto older, underlying deposits, increases as sediment series accumulate. The lithostatic pressure increases by 20–30 bars/10 m on the average in basin and passive margin areas, but much higher values are recorded in active margin areas where pressure results from the combination of sediment burial and plate convergence. Pore fluids are expelled upwards as pressure grows, facilitating the compaction of the sediment which loses porosity while increasing in density. Highest porosities of 80–90% are recorded in freshly deposited biogenic siliceous oozes, porosity losses being still limited at burial depths of a few hundred meters. Lower porosities are recorded in biogenic carbonates (70–75%) and siliciclastic sediments (60–80%), rapidly decreasing to 40–60% at burial depths of a few hundred meters. Compaction causes a reduction in thickness of sediment sections (Δh) and the degree of compaction is estimated from initial and final porosities of sediment sections

$$\Delta h = \frac{1 - \mathrm{Po_d}}{1 - \mathrm{Po_i}}$$

where $\mathrm{Po_d}$ is the porosity at depth and $\mathrm{Po_i}$ the initial porosity of the sediment. Initial porosities being unknown, porosities of modern equivalents are used instead.

The consequences of compaction are hardly visible in homogeneous sediments. For instance, electron microscope observations only can evidence the preferred orientation of sheet silicates in mudstones (see Section 10.4). In contrast, the consequences of compaction clearly show in series associating sediment of different nature and initial porosities (differential compaction). For example, dominant clay and silt-sized terrigenous sediments of high porosity coexist with sandy channel deposits of lower porosity in deltaic and deep-sea fan environments. The potential for porosity loss is therefore higher for clays and silts, and sandy channel deposits gradually form sandy lenses within claystone to siltstone series as compaction proceeds.

As burial and/or convergence increase, the related stress may result in sediment deformation. The stress field is described by a matrix or, to simplify, by an ellipsoid where the three axis define major stress directions. The mean value of the stress field defines the hydrostatic component of the stress field, described by a sphere. The

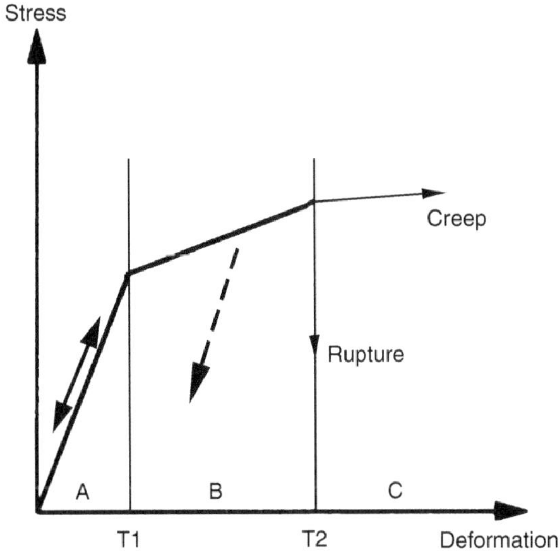

Figure 2.42 Sequence of deformation of sediment series with increasing stress. T1 and T2, thresholds. (A) Domain of reversible deformation; (B) domain of non-reversible deformation, even if stress decreases; and (C) domain of fracturation and creep. Modified from Biju-Duval, B., 1999. Géologie sédimentaire, Technip, Paris.

intensity of the stress being identical in all directions, the hydrostatic component of the stress field only reduces the volume of the sediment, with no deformation. The difference between intensities along major stress directions and the mean stress value is termed a deviator, which is responsible for sediment deformation. For example, deviator components of the stress field may cause the distortion of macrofossils within compacted sediment series. For low intensities of the stress field, deformation is of minor importance and reversible (Figure 2.42). For higher intensities of the stress field and beyond a threshold, minor variations of the stress produce significant and non-reversible deformation until a further threshold is reached. Beyond, brittle sedimentary rocks are fractured whereas ductile sedimentary rocks creep. The sequence of deformation and thresholds being controlled by the elasticity and Poisson modulus of the sediment principally, they may vary with the nature and composition of sediment series. Also, the threshold for rupture is reached more rapidly under tensional stress than under compressional stress.

The compaction of sediment series is associated to the release of pore waters, which initially consist of seawater trapped within the pore space of the sediment. In some cases (e.g., in clays and claystones) connection within the pore space is limited, the fluids hardly circulate and the sediment resists compaction as indicated by minor changes in porosity as burial progresses. In most cases however, connection between porous intervals is sufficient to allow the fluids to circulate and the compaction to proceed. The capacity of sediment series to release pore fluids is termed permeability, the intensity of fluid discharge being expressed by Darcy's law

$$Q = \frac{S}{v} \times \frac{dP}{dx} \times K$$

where Q is the quantity of fluid released, K the permeability of the sediment series, v the viscosity of the fluid, S the area of the section, dx the length of the section and dP the pressure gradient. At regional scale, the release of pore fluids is also facilitated in areas of brittle fracturing (fracture permeability). In fine, the permeability of sediment series is controlled by their composition as well as their degree of compaction and deformation. This is the reason why fluid release is especially important within accretionary wedges of active margin areas. For example, about 2,000 m^3 of pore fluids/km are expelled from the Barbados accretionary wedge to the seafloor in a single year.

The compaction of oceanic sediments and related circulation of pore waters are associated to chemical reactions which play a central role in the diagenetic alteration of sediment series. The intensity of chemical reactions increases with pressure and especially temperature. Temperatures increase by about 1°C/30 m on the average in basin and passive margin areas, but much higher gradients are recorded in areas of intense heat flow. Regions of maximum heat flow are generally associated to volcanic and hydrothermal activity and include rift basins, active mid-oceanic ridges and active continental margins. Sediments in those areas are therefore likely to experience rapid and significant chemical diagenesis, especially active margins where high heat flow and compressional stress coexist. Most oceanic sediments being made of fine-grained elements a few micrometers in size, they provide large surface areas for exchanges of chemical elements between particles and pore waters.

The solubility of minerals being higher under high pressure, dissolution starts at contact points between sediment grains and the concentration of pore waters in chemical elements taken from adjacent particles increases (pressure solution). Also, circulating pore waters may dissolve adjacent minerals until conditions of chemical equilibrium are reached, in accordance with local conditions of pressure and temperature. As the quantity of ions in pore fluids increases, they are more likely to be attracted to the surface of a solid. The forces of van der Waals and the formation of chemical bondings participate in the adsorption of chemical elements onto solid surfaces. The adsorption of ions being related to their electrical potential, some chemical elements are more easily adsorbed than others.

Moreover, chemical elements at the surface of organized crystals (where bonding is limited) have more free energy than those within the crystals. Also, one mole of small-sized crystals has more free energy than one mole of coarse crystals, because of differences in their respective surface areas. Therefore, a relatively high degree of free energy is required for initial formation of crystals from pore waters (nucleation) because of the relative importance of surface conditions in nuclei. In sediment series, nuclei most frequently develop at the surface of existing particles which provide part of the required free energy. During the subsequent phase of crystal growth, significantly lower levels of free energy are required. Nucleation rates are higher in supersaturated pore waters, leading to the formation of abundant small-sized crystals or amorphous material. For example, increased dissolution of siliceous biogenic elements as burial progresses and related high contents of silica in pore waters frequently initiate the formation of microcrystalline opal C–T (cristobalite–tridymite). Smaller quantities of nuclei are produced at lower concentrations, leading to the formation of fewer and larger crystals providing that crystal growth is not limited by changing environmental conditions. Crystal formation is facilitated when precipitate and pre-existing particles are of similar mineralogies, the free energy of nucleation being relatively low and close to those for crystal growth conditions in this case. This is, for example, the case when crystals of calcite grow at the periphery of calcareous microfossils under shallow burial depths of a few hundred meters. Crystal formation is also facilitated when pore waters circulate to areas of changing lithologies and/or pressure and temperature, where new conditions of chemical equilibrium may result in supersaturation and precipitation of specific elements. For example, cold and CO_2-rich pore waters may dissolve carbonate particles and circulate to areas of lower pressure where carbonate cements or concretions precipitate.

Crystal formation (authigenesis) generally occurs within the original pore space of the sediment (cementation) or within the pore space created by the dissolution of pre-existing particles (replacement), progressively lowering the porosity and permeability of the sediment series. In some cases however, chemical diagenetic reactions may strongly alter the physical properties of the sediment series in different ways. For example, the replacement of aragonite by calcite is associated to an increase in volume of 8.25%, which modifies local pressure conditions. On the contrary, the replacement of calcite by dolomite is associated to a decrease in volume which locally increases the porosity of the sediment series. In rare circumstances, chemical diagenetic reactions do not alter pre-existing structures. For example, the morphology and structure of ammonites is preserved during the replacement of calcite by pyrite which is not associated to perceptible changes in volume.

As long as sediment deformation and chemical diagenetic reactions are limited, and providing that an accurate age model can be deduced from microfossil assemblages as well as from radiometric and paleomagnetic measurements, it is possible to describe the variability of sediment deposition in the ocean through time. Linear sedimentation rates (LSR) correspond to the thickness of sediment section per time interval (usually in cm/kyr) but do not take into account the compaction of the sediment section. This is resolved by integrating the density of the sediment in the description of mass accumulation rates (usually in $g/cm^2/kyr$) of bulk sediment

$$\mathrm{MAR_{bulk}} = \mathrm{LSR}^{*}\rho_{\mathrm{dry}}$$

where ρ_{dry} is the density of the dry sediment, or

$$\mathrm{MAR_{bulk}} = \mathrm{LSR}(\rho_{\mathrm{wet}} - 1.026^{*}\mathrm{Po}/100)$$

where ρ_{wet} is the density of the wet sediment, 1.026 the average density of pore waters in upper oceanic sediment series and Po the porosity of the sediment. The density and porosity of the wet sediment can be both measured on freshly recovered sediment cores. Mass accumulation rates of individual components are then deduced from mass accumulation rates of the bulk sediment and percentages of individual components in the sediment

$$MAR_{component} = MAR_{bulk}(\%_{component}/100)$$

FURTHER READING

Allen, P.A., 1997. Earth surface processes. Blackwell, London.

Barry, R.G., Chorley, R.J., 1992. Atmosphere, weather and climate. Routledge, London.

Berner, R.A., 1980. Early diagenesis, a theoretical approach. Princeton University Press, Princeton, NJ.

Biju-Duval, B., 1999. Géologie sédimentaire. Technip, Paris.

Boillot, G., Huchon, P., Lagabrielle, Y., 2003. Introduction à la géologie: La dynamique de la lithosphere. Dunod, Paris.

Chamley, H., 2000. Bases de sédimentologie. Dunod, Paris.

Chamley, H., 1988. Les milieux de sédimentation. BRGM, Orléans.

Einsele, G., 1992. Sedimentary basins, evolution, facies, and sediment budget. Springer, Berlin.

Fowler, C., 2000. The solid Earth: An introduction to global geophysics. American Geophysical Union, Washington, DC.

Kennett, J.P., 1982. Marine geology. Prentice-Hall, Englewood Cliffs, NJ.

Nichols, G., 1999. Sedimentology and stratigraphy. Blackwell, London.

Pickering, K.T., Hiscott, R.N., Hein, F.J., 1989. Deep-marine environments. Unwin-Hyman, London.

Seibold, E., Berger, W.H., 1996. The sea floor, an introduction to marine geology. Springer, Berlin.

Shepard, F.P., 1963. Submarine geology. Harper's, New York.

Tchernia, P., 1980. Descriptive regional oceanography. Pergamon, Oxford.

Wells, N., 1986. The atmosphere and ocean, a physical introduction. Taylor and Francis, London.

Other references used in this chapter:

Biscaye, P.R., Eittreim, S.L., 1977. Suspended particulate loads and transports in the nepheloid layer of the abyssal Atlantic Ocean. Marine Geology, 23: 155–172.

Chen, M.S., Wartel, S., Temmerman, S., 2005. Seasonal variation of floc characteristics on tidal flats, the Scheldt estuary. Hydrobiologia, 540: 181–195.

Exon, N.F., Kennett, J.P., Malone, M. et al., 2001. Proceedings of the Ocean Drilling Program, Initial Reports, volume 189. Ocean Drilling Program, College Station, TX.

Maxwell, A.E., von Herzen, R. et al., 1970. Initial Reports of the Deep Sea Drilling Project, volume 3. U.S. Gov. Print. Office, Washington, DC.

Plafker, G., Galloway, J.P. (Editors), 1990. Lessons learned from the Loma Prieta, California, earthquake of October 17, 1989. U.S. Geological Survey Circular 1045. U.S. Gov. Print. Office, Washington, DC.

Sawyer, D.S., Whitmarsh, R.B., Klaus, A. (Editors), 1994. Proceedings of the Ocean Drilling Program, Scientific Results, volume 149. Ocean Drilling Program, College Station, TX.

Scotese, C.R., Gahagan, L.M., Larson, R.L., 1988. Plate tectonic reconstructions of the Cretaceous and Cenozoic ocean basins. Tectonophysics, 155: 27–48.

Van Andel, T.H., Heath, G.R., Moore, T.C., 1975. Cenozoic history and paleoceanography of the Central Equatorial Pacific. Geological Society of America, Memoir 143, Boulder, Co.

Whitmarsh, R.B., Beslier, M.-O., Wallace, P.J. (Editors), 1998. Proceedings of the Ocean Drilling Program, Scientific Results, volume 173. Ocean Drilling Program, College Station, TX.

Winterwerp, J.C., 2002. On the flocculation and settling velocity of estuarine mud. Continental Shelf Research, 22: 1339–1360.

Part 2: Major Types of Sedimentary Basins in Oceans History

Oceanic sedimentary basins evolve with the oceans. Sedimentary basins are generally continental during early stages of rift systems, but later fall below sea level as thermal subsidence progresses. Oceanic sedimentary basins vary in shape, extension and depth as the oceans open and close during the Wilson cycle, as do sedimentary processes and general characteristics of oceanic sediments. This is especially the case for the distribution of sedimentary units, which is influenced for example by tectonic activity, relative changes of sea level and currents. This is also the case for the dominant nature of the sediment, which is influenced for example by the extension of the drainage basins and the climate (which regulate the fluxes of terrigenous particles), as well as marine productivity and dissolution (which regulate the fluxes of biogenic and organic particles). Oceanic sedimentary basins are parts of lithospheric plates, and are therefore exposed to lithospheric processes. Chapters 3–9 deal with the general characteristics of the sedimentation during the main stages of the Wilson cycle and their evolution.

CHAPTER THREE

Rift Systems

3.1. Structure and Tectonics of Rift Systems

Because of tensional shear within the lithosphere, divergence and sliding movement may affect portions of lithospheric plates, average velocity being less than a few millimeters per year (Figure 3.1). Lithospheric plates diverge in rift areas, where geological activity is concentrated: this includes volcanism, high heat flow, seismicity and anomalous lithosphere. There, a thinned crust is associated to uplifted and decompressed upper mantle and asthenosphere (Figure 3.2). The crust decreases in density because of an abnormally high heat flow, the whole rift area being raised to reach isostatic equilibrium. The elevation of the thermal bulge varies with the activity of the rift system, but is usually about 1.5–2 km above regional average. Adiabatic decompression and related anhydrous fusion of upper mantle rocks generate a magma which raises and combines with melted elements from the crust to produce an alkali-rich, silica-poor syn-rift volcanism. Syn-rift volcanism varies in importance with the origin and activity of the rift system, and generates a variety of products which range from gabbro intrusions and basalt lava flows to rhyolites and carbonatites. Rift systems are linked to dynamic processes in the lithosphere and asthenosphere, and represent the initial stage of continental breakup.

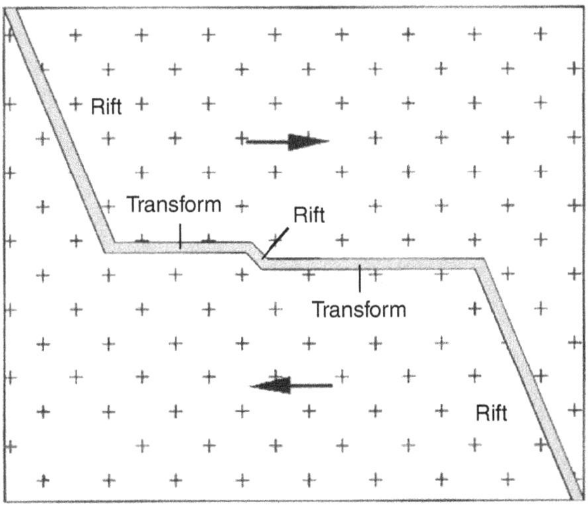

Figure 3.1 Creation of rift (R) and transform (T) areas, as a result of tensional shear within a portion of continental lithosphere. Modified from Boillot, G., Coulon, C., 1998, La déchirure continentale et l'ouverture océanique, Overseas Publishers Association, Amsterdam.

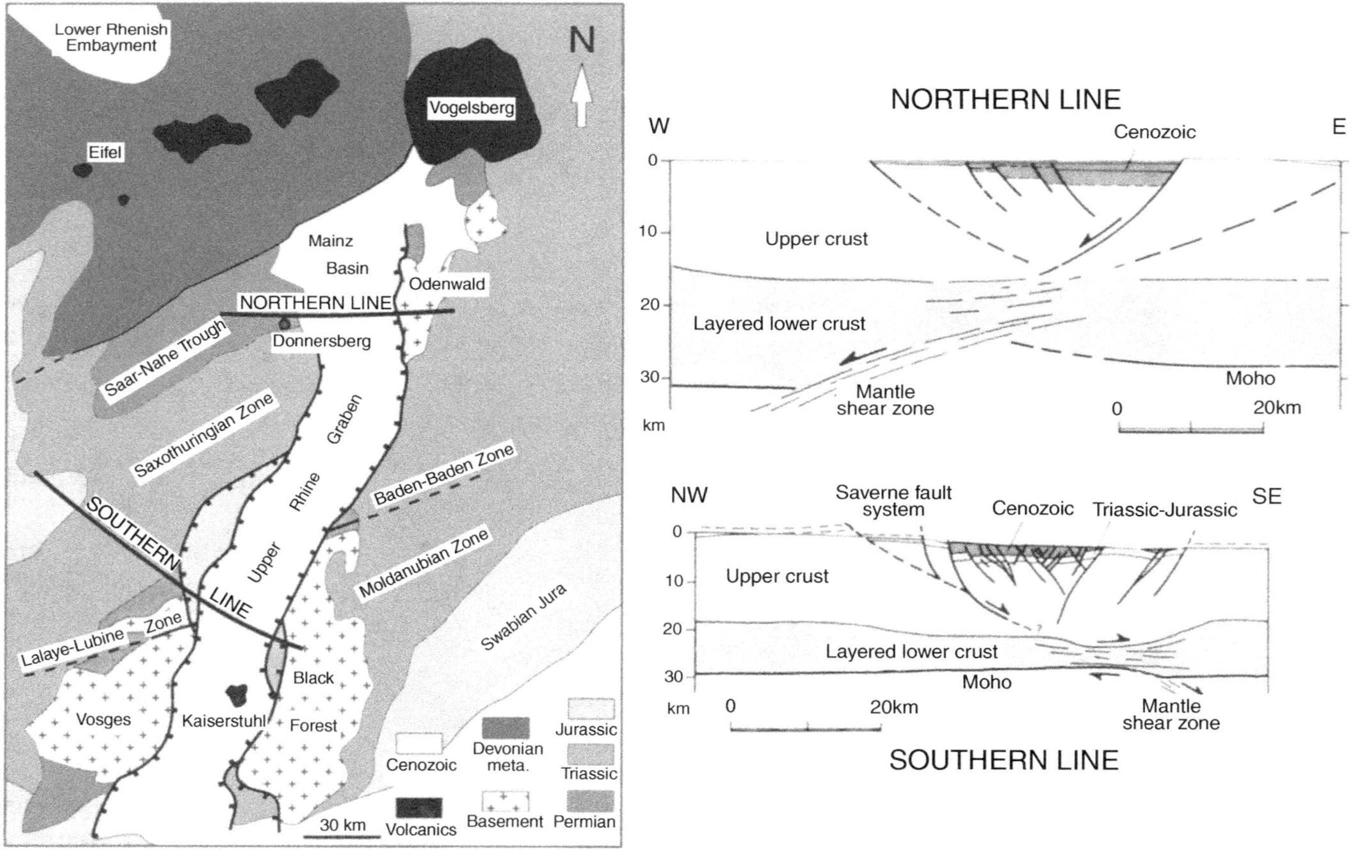

Figure 3.2 Deep structure of the Rhine Graben (ECORS Project). Note the thinning of the continental crust, asymmetry of structure and variability of structure from North to South. Modified from Brun, J.P., Gutscher, M.A., DEKORP-ECORS Team, 1992. Deep crustal structure of the Rhine Graben from DEKORP-ECORS seismic reflection data. Tectonophysics, 208, 139–147.

The morphology of rift systems shows an elongate central depression bounded by normal faults and elevated crest lines. The active parts of rift systems consist of a succession of grabens which are frequently 30–100 km wide, bounded by tilted crustal blocks. The crustal blocks are separated by listric or planar and detachment faults, and fragmented into smaller blocks. Fault activity and graben formation in the upper brittle crust (and upper brittle mantle) and stretching of the ductile parts of the crust and mantle result in a thinned lithosphere (Figure 3.3). Motion

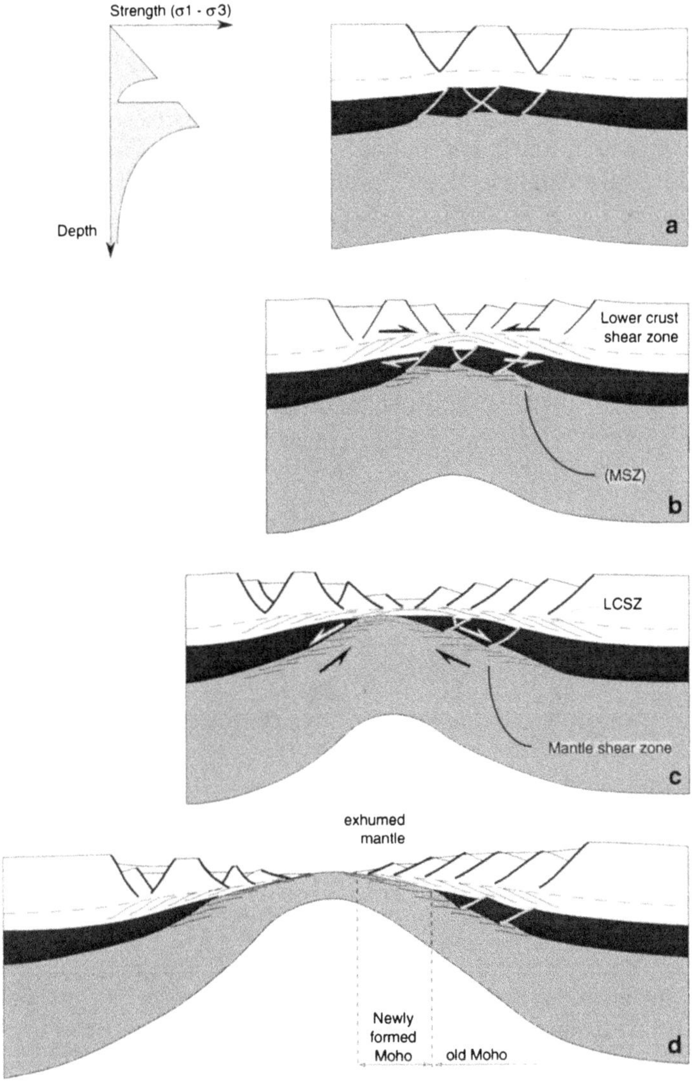

Figure 3.3 Mechanisms of lithosphere thinning and extension during different stages (a–d) of rift development. Note alternances of faulting in brittle layers and stretching in ductile layers of the lithosphere, and the uplift of the astenosphere. Reprinted from Brun, J.P., Beslier, M.-O., 1996. Mantle exhumation at passive margins. Earth and Planetary Science Letters, 142, 161–173.

occurs during earthquakes, which are mostly concentrated in the brittle upper crust (<15 km depth) but also take place in the brittle part of the upper mantle. One consequence of the shortening of the lithosphere is the subsidence of the grabens in the central part of the rift (initial subsidence). Another consequence is the elevation of the rift periphery where the lithosphere is thicker, a combination of mechanical (uplift induced by the formation of the bulge), thermal (decreased density and increased volume of the lithosphere due to elevated heat flow) and isostatic (increased buoyancy of the heated lithosphere) reactions to rift activity.

The structure of the rift systems changes regionally, and several models have been proposed to account for the variety of observations and data. In active rifting, the shortening of the lithosphere is a response to a thermal upwelling of the asthenosphere (hot spot). Abnormal areas of the asthenosphere are characterized by temperatures above 1,400°C and densities below 3.25, generating mantle plumes and elevated heat flows (Figure 3.4). Consequences include thermal thinning of the lower lithosphere by absorption into the asthenosphere, adiabatic decompression, anhydrous fusion of mantle peridotites, formation of a thermal bulge and correlative elevation of the Earth's surface. This in turn weakens strongly and rapidly the upper lithosphere, facilitating the formation of tilted blocks and grabens, and volcanic activity. Syn-rift deposits principally include volcaniclastics and abundant lava flows (trapps). Examples of active rifting include the Yemen–Eritrea–Ethiopia volcanic province of the southernmost Red Sea and East African Rift.

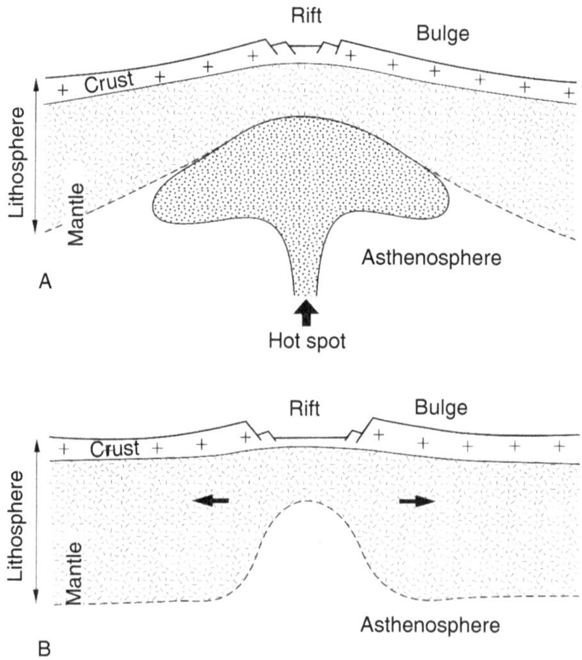

Figure 3.4 Schematic models of active (A) and passive (B) rifting. Modified from Boillot, G., Coulon, C., 1998. La déchirure continentale et l'ouverture océanique. Overseas Publishers Association, Amsterdam.

In passive rifting the shortening of the lithosphere is a response to a regional stress field originating from plate boundary forces (Figure 3.4). The trigger mechanism of the rifting is therefore distant and may originate in changes in plate kinematics and velocity. In passive rifting the shortening of the lithosphere is a response to stretching, progresses rather slowly and is the cause for the rise of the asthenosphere. This in turn generates decompression melting, high heat flow, volcanic activity and thermal bulge. Examples of passive rifting include the Gulf of Suez. Active and passive rifting models do not account for all the observations, and compromises between both models seem possible.

In some cases, tensional stress in a homogene brittle lithosphere results in symmetric faulting (pure shear model). The resulting rift system includes symmetrically tilted crustal blocks dipping toward each other (Figure 3.5). Syn-rift deposits are predominantly terrigenous because of limited and intermittent volcanic activity. They are generally well developed because the morphology of the rift and continuous isostatic adjustment facilitate the erosion of active crustal blocks

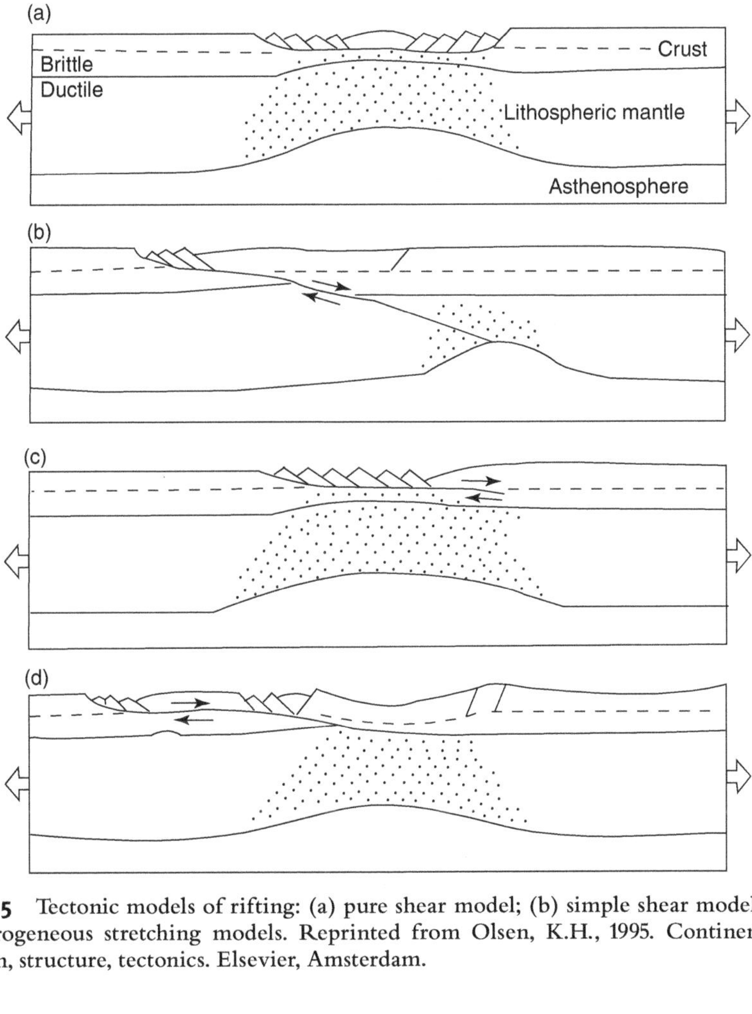

Figure 3.5 Tectonic models of rifting: (a) pure shear model; (b) simple shear model; (c) and (d) heterogeneous stretching models. Reprinted from Olsen, K.H., 1995. Continental rifts: Evolution, structure, tectonics. Elsevier, Amsterdam.

and the accumulation of terrigenous sediment in subsident grabens. Examples of such symmetric rifting are visible in the Rhine Graben and parts of the Gulf of Suez. In most cases, tensional stress is associated to shear stress in weakened areas of the lithosphere like pre-existing fault systems and discontinuities (simple shear and heterogeneous stretching models). The resulting rift system is asymmetric, and associated to detachment faults. The rift may propagate away from the initial active area through time as do the main areas of sedimentation. Syn-rift deposits vary in thickness and are generally complex, with many unconformities. Examples of asymmetric rifting include most parts of the East African Rift.

Shearing stress and related motion between sliding lithospheric plates is generally associated to tensional or compressive stress, according to the location. Also, the motion does not occur along single fault lines but along series of more or less parallel, offset faults, separated by distances of a few to a few tens of kilometers. Pull-apart basins consist of relatively short grabens, which develop in those areas where offset fault systems accommodate both shearing and tensional stress (Figure 3.6). There, the shortening of the lithosphere occurs progressively in the direction parallel to the main fault system, accommodating the tensional stress via the formation of tilted blocks. The transition is rather abrupt along the main faults, which accommodate the shearing stress and sliding motion. Pull-apart basins are similar to rift systems in their structure, processes and evolution. Examples of pull-apart basins include submarine deeps of the Gulf of Aqaba, the Dead Sea graben and the Sea of Galilea graben along the Jordan fault which separates the Arabian plate from the Levantine plate, and the Salton Trough of Southern California and submarine deeps of the northernmost gulf of Baja California along the San Andreas fault which separates the Pacific plate from the North American plate.

The life span of rift systems generally varies from about 5 to about 30 Myr. According to the Wilson cycle, rift systems are the earliest stage of ocean evolution. The rift stage ceases when typically oceanic basalts are being produced in the central graben.

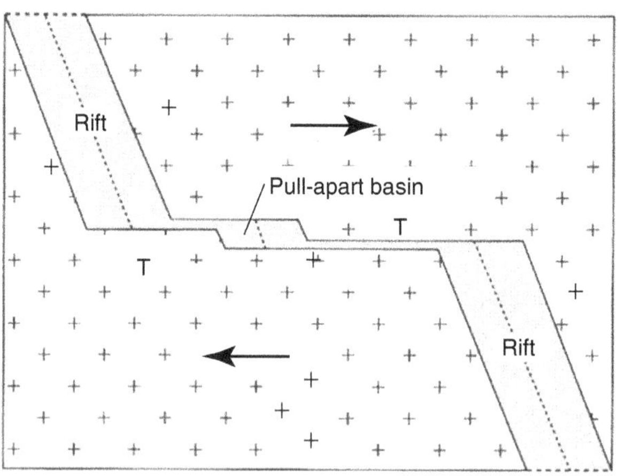

Figure 3.6 Creation of a pull-apart basin, as a result of shearing and tensional stress along a transform area (T). Modified from Boillot, G., Coulon, C., 1998. La déchirure continentale et l'ouverture océanique, Overseas Publishers Association, Amsterdam.

This marks the transition to the young ocean (or crustal fissure) stage characterized by a portion of oceanic lithosphere separating divergent (or rifted) continental margins. In some cases the activity of the rift system declines and may cease before the crustal fissure stage is reached, either because of a decreased importance of the abnormal asthenosphere and related mantle plumes, or because of weaker tensional stress within the lithosphere. The density of the lithosphere increases and its volume decreases as the heat flow weakens, and the morphology of the rift system progressively recedes as a result of thermal subsidence. The rift system therefore remains part of the initial lithospheric plate. In the case of active rifting, this leads to the formation of large igneous provinces such as the Deccan Traps of India. In the case of passive rifting, this leads to the formation of intraplate basins of thinned lithosphere such as the Paris and London basins and the North Sea basins of the European plate.

3.2. Sedimentation of Rift Systems

The pre-rift regional morphology and structural geology control important features such as the regional average elevation, the accommodation of tensional stress and the path of major rivers which have a strong impact on graben morphology and filling. Other major influences include:

- The magmatism, which controls thermal effects and volcanic activity. This includes principally the importance of the thermal bulge, the formation of volcanoes, the nature and importance of volcaniclastics, lava flows and hydrothermalism, as well as divergence rates and the life span of the rift system.
- The tectonism, which regulates the accommodation of tensional stress and isostatic adjustment. This in turn influences the morphology of the grabens, the foci of deposition and erosion rates principally.
- The climate, which regulates the degree of weathering and erosion of continental surfaces, the transport of particles, and sometimes the nature of sedimentary deposits.

All these major controls closely interact, making rift sediments especially complex in their nature and organization. Because of combined influences of pre-rift elevation and thermal bulge, many rift areas are well above sea level and rift sediments are continental (East African Rift). Where pre-rift morphology was low and thermal activity relatively weak, rift areas are partly below sea level as a result of lithosphere thinning and initial subsidence, and rift sediments are lagoonal to shallow marine (Gulf of Suez). The main features of the sedimentation in the East African Rift and the Gulf of Suez are detailed in Sections 3.3. and 3.4.

The active rift being narrow, most siliciclastics eroded from the thermal bulge are transported by rivers outside the rift system, at least during early rift development. Because of active magmatism and tectonism, the rift shoulders and tilted crustal blocks are exposed to continuous erosion and an increasing quantity of siliciclastics accumulates in the subsident central grabens as the rift system progresses. The removal of significant quantities of low density surface terranes induces isostatic adjustment: each time 100 m of continental lithosphere where low-density crustal rocks ($d = 2.8$) overlay abnormal mantle rocks ($d = 3.25$) are being eroded, isostatic adjustment generates an elevation (h) of the continental relief of

$$h = 100 \times 2.8/3.25 = 86.1 \text{ m}$$

In theory, the thickness of terranes to be eroded before the elevation of the rift shoulders decreases of 1.5 km (which is their average elevation relative to regional relief) is 10.8 km, thus providing an impressive quantity of terrigenous materials for rift sediments. The subsidence of the central grabens is a combination of initial subsidence (which is closely related to the degree of crustal thinning) and load of rift deposits. The degree of thinning can be approximated from the ratio between initial and final thicknesses. For maximal crustal thinning at the end of the rift stage (transition to oceanic crust derived from the mantle), the maximum theoretical thickness of rift deposits is 13.6 km.

The theoretical values provided here are derived from very simple models which do not account for many aspects of the complex reality, but they clearly indicate that: (i) erosion is continuous and very active in rift systems, but does not affect significantly their morphology; (ii) terrigenous elements are an important characteristic and form a significant part of rift deposits, together with volcanic products and (iii) rift grabens may contain huge quantities of deposits. In fact, known thicknesses of rift deposits commonly reach up to 8 km in some parts of the East African Rift as in parts of fossil rift systems buried below passive continental margins and sedimentation rates are highly variable, from about 100 m/Myr up to 1 km/Myr and more.

Whereas pre-rift sediments are incorporated to the tilted crustal blocks, syn-rift deposits which accumulate in central grabens are subjected to ongoing active tectonism: syn-rift deposits are commonly faulted at all scales, from regional fault systems to microfaulting within sediment layers (Figure 3.7). Detachment faults may also occur in ductile, undercompacted series (e.g., clays or evaporites) and/or fluid-rich sediments, leading to the formation of diapirs and/or "roll-over" structures (Figure 3.8) as tectonism and deformation progress through time.

Syn-rift deposits are dominated by siliciclastics, volcaniclastics and lava flows as a result of the major control of magmatism and tectonism over rift processes, but their relative proportion may vary locally as well as through time according to the respective importance of the controls. Siliciclastics are transported by gravity along slope and by run-off principally. They accumulate in sub-aerial conditions most of the time (East African Rift, Dead Sea, Salton Trough) where they form colluvium, alluvial fans and fluvial deposits. They also accumulate in marine environments where central grabens subside below sea level and communicate with the ocean (Gulf of Suez, northernmost gulf of Baja California), where they often build delta plains and submarine fans. Because of the mobility of the graben morphology which evolves with magmatism and tectonism through time, the drainage systems may vary strongly and allow formation of intermittent or more permanent lacustrine environments. The importance and characteristics of lacustrine environments and deposits are strongly controlled by climatic conditions which, together with the nature of the substrates and hydrothermal activity, may lead to a

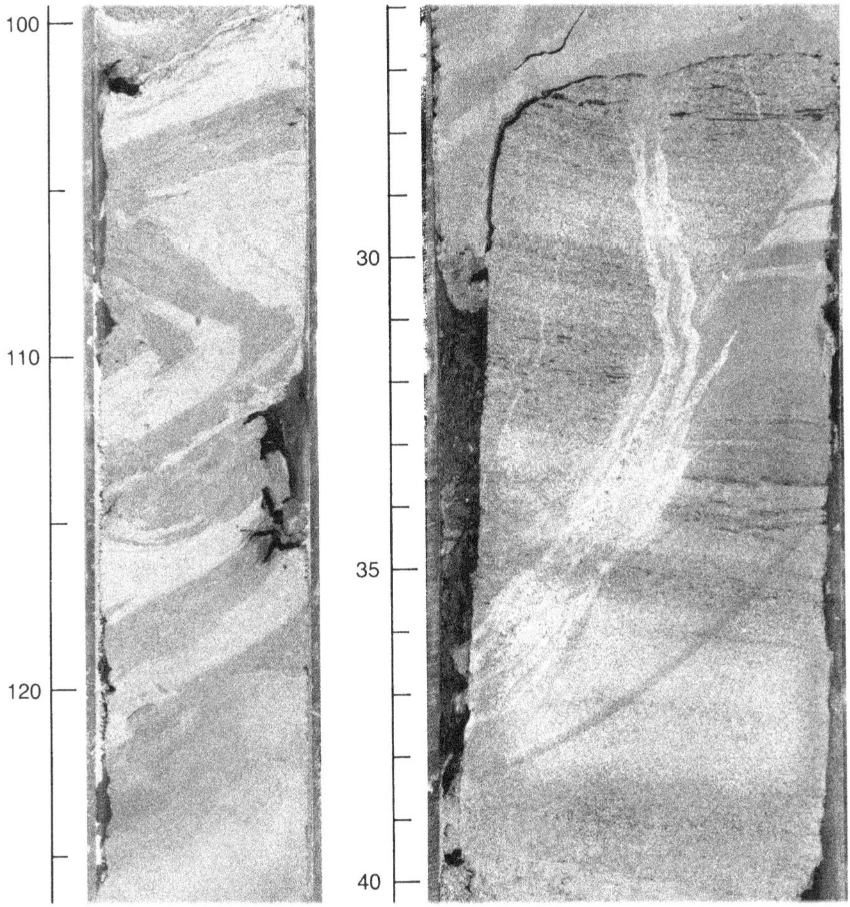

Figure 3.7 Deformation of syn-rift deposits, ODP Site 638C, Galicia margin. Left: slump-fold in a claystone and marlstone unit. Right: Microfault in sandstone, sutured by calcite cement and cut by calcite veins. Scale in centimeter. From Boillot, G., Wnterer, E.L., Meyer, A.W. et al., 1987. Proceedings of the Ocean Drilling Program, Initial Reports 103, College Station, TX.

variety of terrigenous, biogenic and authigenic sediments. For example, fine siliciclastics enriched in organic matter favorable to hydrocarbon formation, diatomites and playa-lake evaporites are commonly found in rift lacustrine series. Because of significant differences in sedimentary processes, the examples below detail rift sediments of continental and marine environments.

3.3. Example of Rift Environments and Sediments in a Continental Context: The East African Rift

The East African Rift is about 40–60 km wide and extends from the Southernmost Red Sea to the Zambeze (Figure 3.9) over more than 6,000 km.

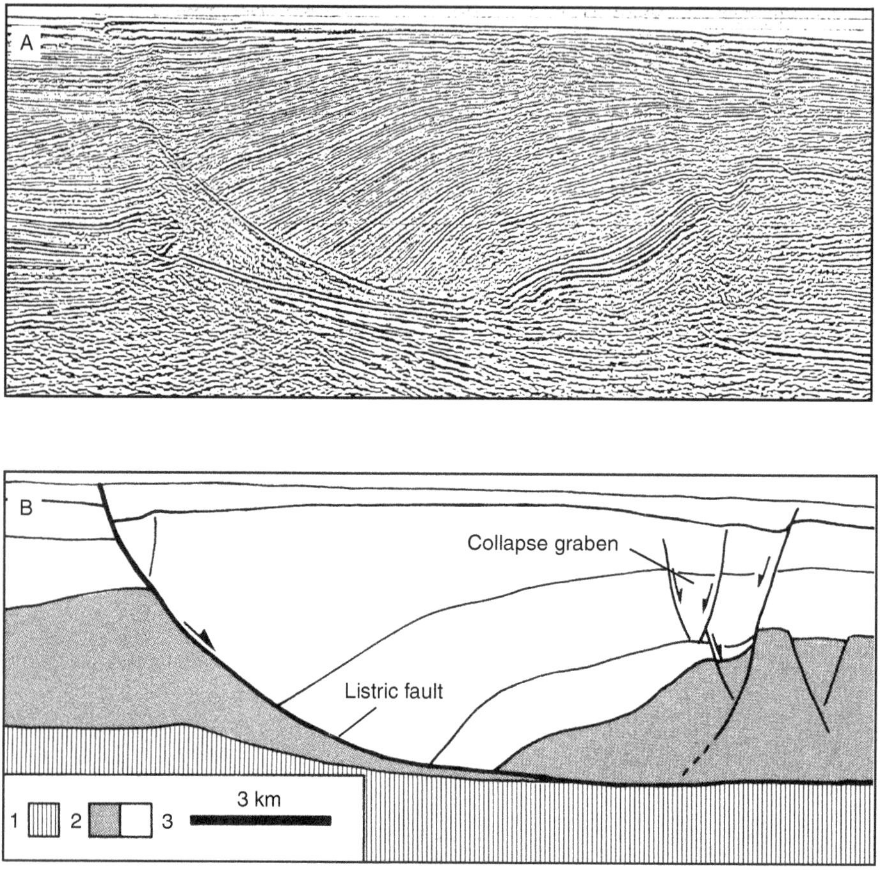

Figure 3.8 Development of a listric fault and roll-over structures on an evaporitic unit. Initial seismic profile (A) and its interpretation (B). 1: early syn-rift unit; 2: evaporites unit; 3: upper syn-rift unit. Modified from Faure, J.-L., Chermette, J.-C., 1989. Deformation of tilted blocks, consequences on block geometry and extension measurements. Bulletin de la Société Géologique de France, 5, 461–476.

It consists of two main branches separated by the Assoua shear zone which acts as a transform fault. The elevation of the rift is usually of 2,000–2,500 m, but may locally reach 3,000 m with a few summits up to 5,000 m. The regional seismicity is important, and earthquakes occur more frequently in the active part of the rift where the hypocenters are concentrated at shallower depth of the crust in the East branch than in the West branch, because of differences in the degree of lithosphere thinning. The tensional stress being accommodated along detachment faults for a part, the East African Rift is also asymmetric (Figure 3.10). The East branch of the rift is characterized by a very important magmatism, especially to the North where an alkaline volcanism enriched in sodium developed. By comparison, volcanism is relatively minor along the West branch where an active tectonism led to the formation of a complex succession of deep grabens, especially to the South where the tensional stress is partly accommodated along the Tanganyika–Rukwa–Malawi

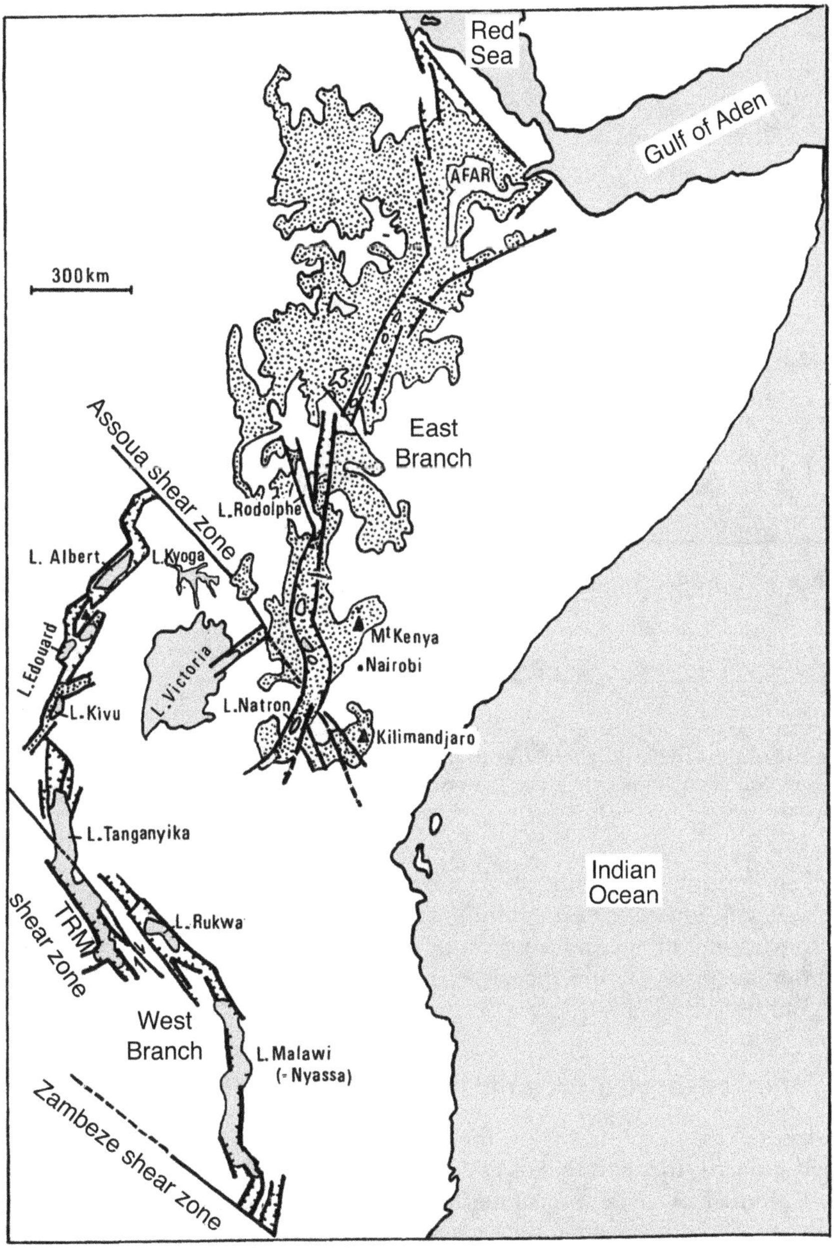

Figure 3.9 The East African Rift System: location map. TRM shear zone: Tanganyika–Rukwa–Malawi shear zone. Modified from Debelmas, J., Mascle, G., 2000. Les grandes structures géologiques. Dunod, Paris.

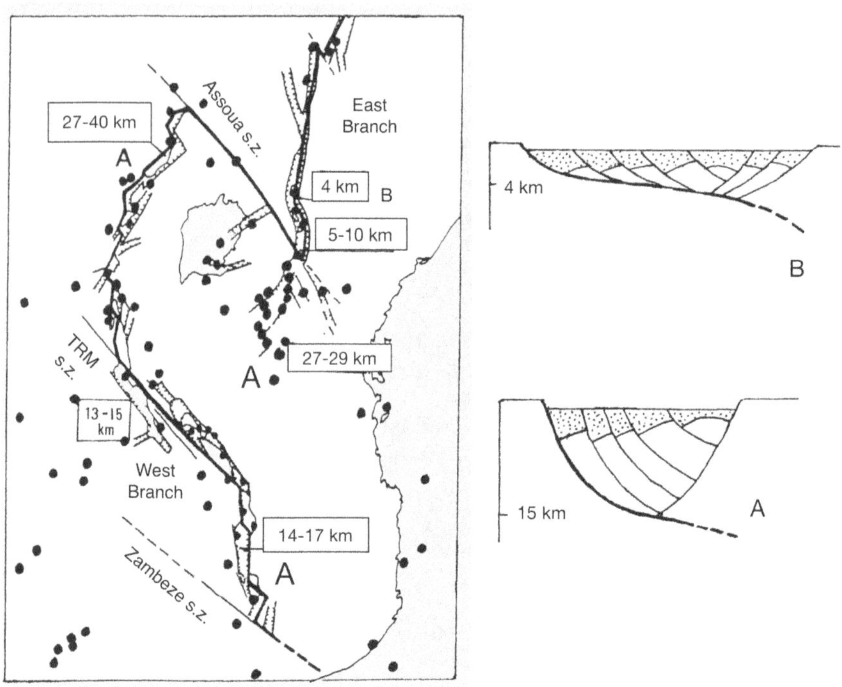

Figure 3.10 Seismic activity of the East African Rift System: distribution and depth of hypocenters, and relation to the deep structure of the rift. TRM, Tanganyika–Rukwa–Malawi and s.z., shear zone. Modified from Debelmas, J., Mascle, G., 2000. Les grandes structures géologiques. Dunod, Paris.

(TRM) shear zone (transtension). The East African Rift has been active since the Late Oligocene and progressed from North to South, where initial syn-rift deposits are Late Miocene in age. Because of the complexity of the East African Rift, the description focusses on specific areas, namely the Tanganyika area of the West branch, and the Baringo–Bogoria area of the East branch.

3.3.1. The Tanganyika Area of the East African Rift

Early stages of rifting (9–12 Ma) in the lake Tanganyika area led to the formation of two half-grabens; the actively subsiding South Rusizi half-graben and the nascent North Kigoma half-graben, separated by a structural high subjected to continuous erosion. Rift activity extended to the North later in the Late Miocene. Seismic and tectonic investigations of the northern lake Tanganyika (Figure 3.11) have revealed a succession of six major sedimentary sequences since the early stages of rifting, which started there at about 7.4 Ma. A major phase of tectonism with a significant vertical component around 1.1 Ma was associated to local erosion before the sedimentation resumed in both actively subsiding half-grabens. A typical rift morphology was acquired by about 380 ka when subsidence ceased to the South but continued to the North (formation of the Bujumbura and Rumonge sub-basins) and the central horst

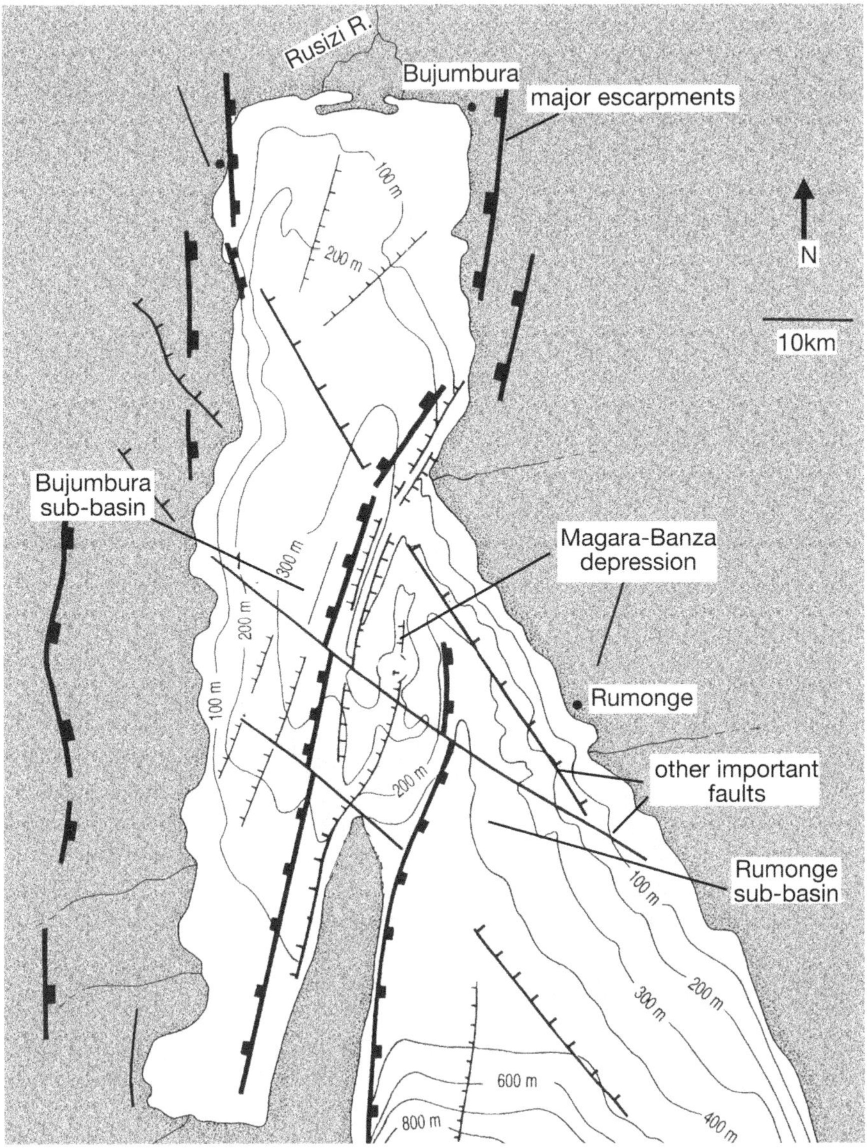

Figure 3.11 Location map of the northern lake Tanganyika area. Note major escarpments which bound two half-grabens separated by a partly collapsed central horst. Modified from Lezzar, K.E., Tiercelin, J.J., De Batist, M., Cohen, A.S., Bandora, T., Van Rensbergen, P., Le Turdu, C., Mifundu, W., Klerkx, J., 1996. New seismic stratigraphy and Late Tertiary history of the North Tanganyika Basin, East African Rift System, deduced from multichannel and high resolution reflection seismic data, and piston core evidence. Basin Research 8, 1–28.

partly collapsed (formation of the Magara depression). After 175 ka the subsidence decreased and roll-over structures developed in the Rumonge sub-basin. Beginning around 380 ka, thick and homogeneous sedimentary units were replaced by recurrent associations of three types of sediments (Figure 3.12): (i) transgressive, chaotic to poorly stratified "basin-fill" units at the base of each sequence; (ii) chaotic and lens-shaped units probably formed from gravity flows, which fill local depressions and channels and may correspond to deep lacustrine fans; and (iii) "sheet-drape" units commonly characterized by parallel reflectors and transparent layers which represent about two thirds of each sequence and correspond to lake highstand conditions. The seismic profiles also illustrate the importance of syn-sedimentary deformation, mainly normal and reversed faults, and undulated sediment layers. Before 380 ka, the sedimentation in the northern lake Tanganyika was mainly controlled by tectonism. As tectonism decreased after 380 ka, lake Tanganyika principally responded to climate change, as expressed from lake level variations and transgressive–regressive sequences.

3.3.2. The Baringo–Bogoria area of the East African Rift

The Baringo–Bogoria area consists of two half-grabens separated by a sill and is located near the intersection of the East Branch (rift Gregory) with the Assoua shear zone (Figure 3.9). Rift activity started here in the Early Miocene with the formation of a tectonic graben (Figure 3.13). Volcanic activity started around 23 Ma (basalt) and has been continuously present since then with peaks of activity in the Middle to Late Miocene; from 14 to 10 Ma when massive production of an estimated volume of 40,000 km^3 of phonolite filled the proto-rift, and from 9 to 7 Ma when about 400 km^3 of phonolite poured into the half-grabens. Other significant volcanic episodes occurred around 6 Ma (trachytes and phonolites) and around 5 and 2 Ma (basaltic dykes). Intervals of maximum tectonic activity have been recorded in the Middle Miocene around 15 Ma when vertical movements increased the morphology of the proto-rift, from 10 to 9 Ma when the half-grabens developed, and from 6 to 4 Ma when a resumption of fault activity on the eastern side of the rift increased its asymmetry. The modern morphology of the Baringo and Bogoria sub-basins has a pull-apart aspect which results from a resumption of activity along the Assoua shear zone for the past 0.2 Myr principally.

Modern sediments from the Baringo–Bogoria area show a variety of facies and structures:

- Piedmont deposits are especially developed in the eastern Bogoria sub-graben where they form small lenses up to 1.5 km^2 in area along the escarpments. There, mega-breccias made of very heterogeneous elements including blocks up to 10 m in size, alternate with conglomerates made of subangular elements about 1–10 cm in size, in a matrix of sand and clay. Breccias and conglomerates are derived from high-viscosity gravity flows (debris flows principally), which have been active during intervals of precipitation. They result from interaction between tectonism (steep relief) and climate.

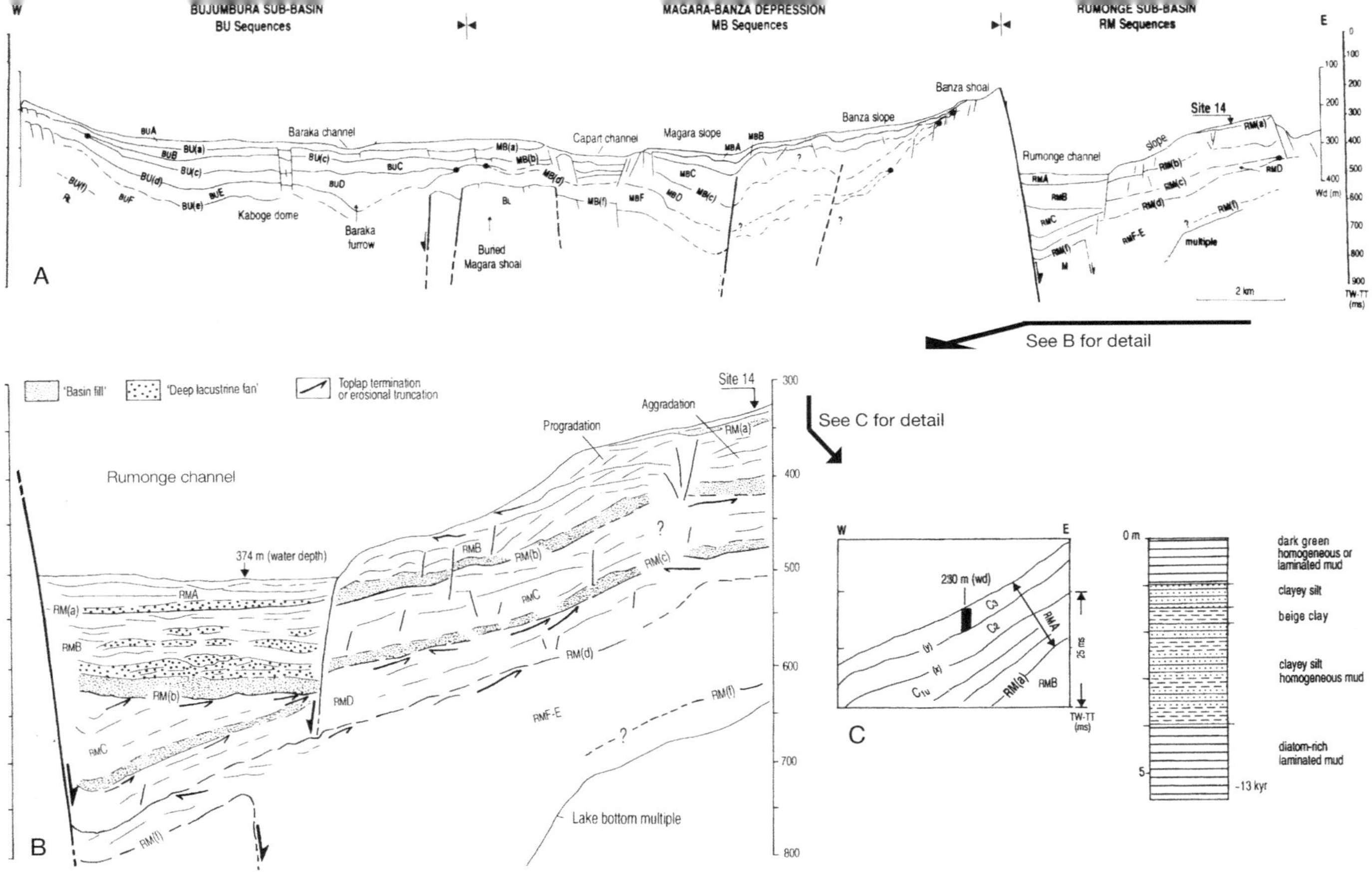

Figure 3.12 Syn-rift sedimentary sequences of the northern lake Tanganyika. (A) Interpreted W–E seismic profile illustrating horst and half-graben structure, syn-rift deformation, and major sequences. (B) Enlargement of the Rumonge sub-basin area, highlighting the importance and distribution of the facies (transparent "sheet-drape" units are in white), the sediment structure and syn-rift deformation. (C) Detail of the upper Rumonge transparent unit, from a piston core. Modified from Lezzar, K.E., Tiercelin, J.J., De Batist, M., Cohen, A.S., Bandora, T., Van Rensbergen, P., Le Turdu, C., Mifundu, W., Klerkx, J., 1996. New seismic stratigraphy and Late Tertiary history of the North Tanganyika Basin, East African Rift System, deduced from multichannel and high resolution reflection seismic data, and piston core evidence. Basin Research 8, 1–28.

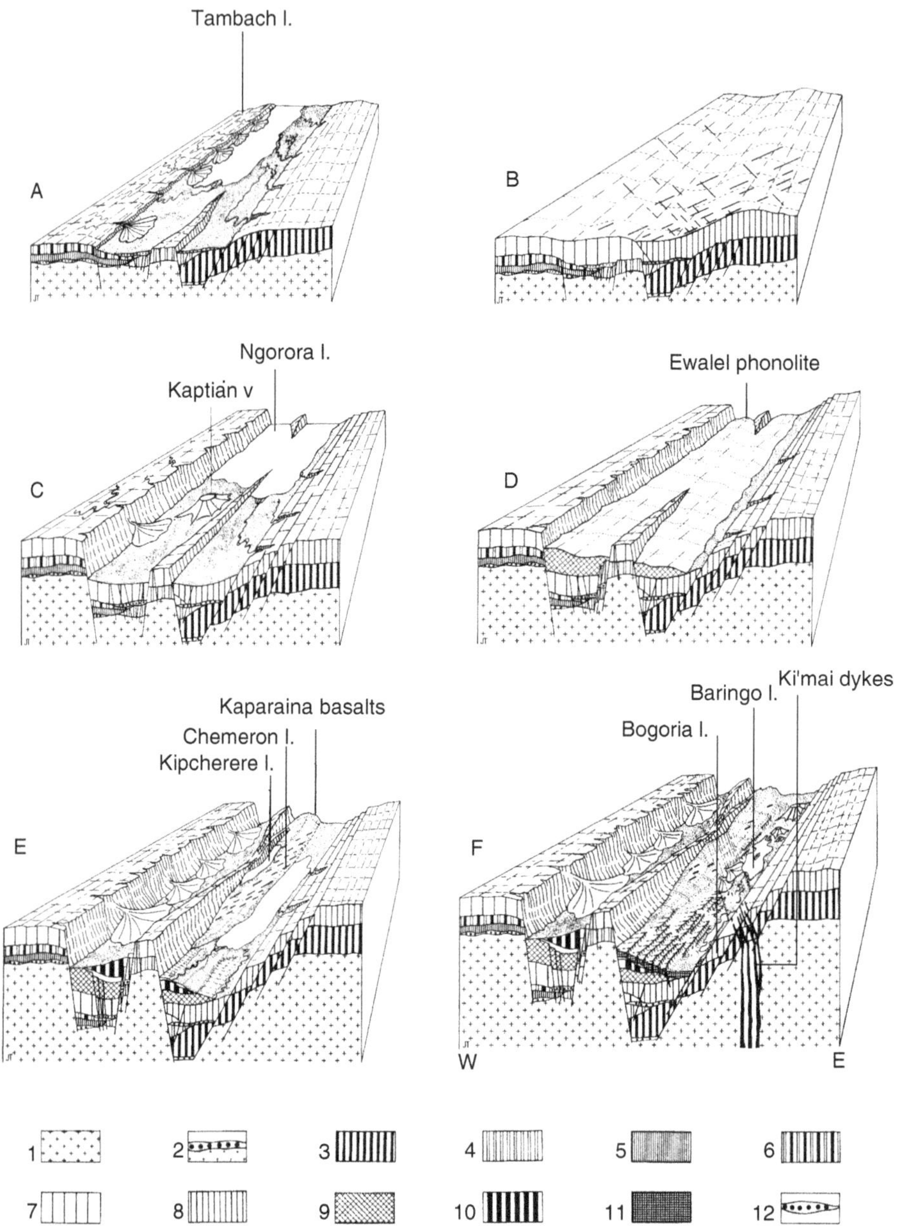

Figure 3.13 Main stages of the evolution of the Baringo–Bogoria rift area since the Early Miocene. (A) Early Miocene; (B) Middle Miocene; (C) Late Miocene; (D) Late Miocene (9–7 Ma); E, Early Pliocene and F, Late Pleistocene. 1, Precambrian substrate; 2, sedimentary units; 3, Samburu basalts; 4, Sidekh phonolites; 5, Elgeyo basalts; 6, Chof phonolites; 7, Uasin phonolites; 8, Rumuruti phonolites; 9, Ewalel phonolites; 10, Kabarnet trachytes and 11, Hannington trachyphonolites. Modified from Tiercelin, J.J. et al., 1987. Le demi-graben de Baringo-Bogoria, Rift Gregory, Kenya: 30000 ans d'histoire hydrologique et sédimentaire. Bull. Centres Rech. Explor. Prod. Elf-Aquitaine, 11, 249–540.

- Fluvial deposits consist of lenses of sand, silt and clay which are disseminated along the path of perennial and intermittent rivers. Sand dominates in upstream areas, whereas silt and clay increase downstream.
- Alluvial and lacustrine fans are especially developed on both sides of the southern Baringo sub-graben (Figure 3.14). Although some conglomerates with a fine matrix are present in the upper part of the fans, they mostly consist of massive and poorly sorted sand. They grade downwards to alternating sand, silt and clay. The more distal units often show typical Bouma units, like parallel lamination and graded bedding. The siliciclastics are mainly derived from physical weathering of adjacent substrates which mainly consist of volcanic rocks, with locally significant chemical weathering as deduced from the mineralogy of the clay fraction (abundant smectites). Most of the fans are now sub-aerial in totality because of unstable morphology (consequence of active tectonism) and variability of lake level (as a result of variable precipitation and climate). Active fans are present on the path of the modern tributaries nearshore.
- The modern lake Baringo is characterized by predominantly terrigenous sediments. Coarse sediments (gravel and sand) are only found nearshore especially off cliff areas, whereas sand and silt dominate near alluvial fans. Silt and clay rapidly increase as to dominate below 2-m water depth. Biological activity is principally controlled by cyanobacteria, but sponges, ostracods and diatoms are also present. However, biogenic elements of the sediment mostly include sponge spicules and some diatoms, with the ostracods being absent below 2-m water depth because of carbonate dissolution as the CO_2 content of the water increases with the preservation of organic matter. Organic elements are mostly of continental origin and principally include lignin and cellulose, together with pollen and spores.
- The modern lake Bogoria is also characterized by an association of terrigenous, organic and authigenic sediments. Terrigenous sediments are concentrated nearshore, down to about 6-m water depth. Sand and gravel dominate from the shore to about 3-m water depth, and grade to dominant silt and clay below. The siliciclastics are transported by the rivers, and then dispersed by wave action and long-shore currents. The sediment progressively grades to organic and authigenic muds below 6 m. Authigenic elements include calcite, sodium carbonates (gaylussite, trona and nahcolite) and sodium silicates (magadiite) which form in the stratified water column. Because of poor ventilation, oxygen content decreases rapidly in the water column and organic elements are preserved. They are mainly derived from local vegetation (cellulose, lignin and pollen) and soils (humus), but also from algae and bacteria. Nearshore on the western side of the lake, blue-green algae and bacteria colonies concentrate near hydrothermal vents, characterized by pH values of 7.8–9.9 and temperatures between 40°C and 92°C, where they participate in the construction of chimneys and pinnacles of calcite and aragonite, associated to zeolites and fluorine, and crusts of amorphous silica.

Although both lakes are located in areas of similar substrates and climate, their hydrology and sediment are quite different (Figure 3.15). Lake Baringo is fed by a large drainage basin, and has a positive water budget. This, together with low hydrothermalism and good ventilation, results in a very low salinity and favors

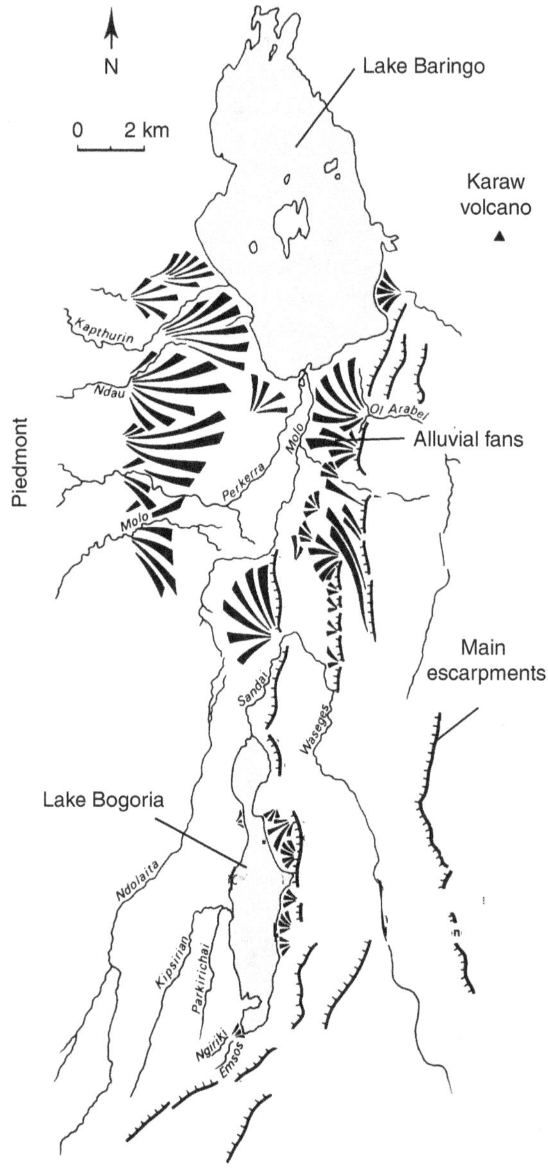

Figure 3.14 Distribution of alluvial fans in the Baringo–Bogoria rift area. Modified from Tiercelin, J.J. et al., 1987. Le demi-graben de Baringo-Bogoria, Rift Gregory, Kenya: 30000 ans d'histoire hydrologique et sédimentaire. Bulletin des Centres de Recherche Exploration-Production Elf-Aquitaine, 11, 249–540.

significant biologic activity and terrigenous sedimentation. Conversely, lake Bogoria is fed by a small drainage basin and has a negative water budget. This, together with significant hydrothermalism and poor ventilation, results in high salinity and favors authigenic sediments as well as the preservation of organic matter. The variability

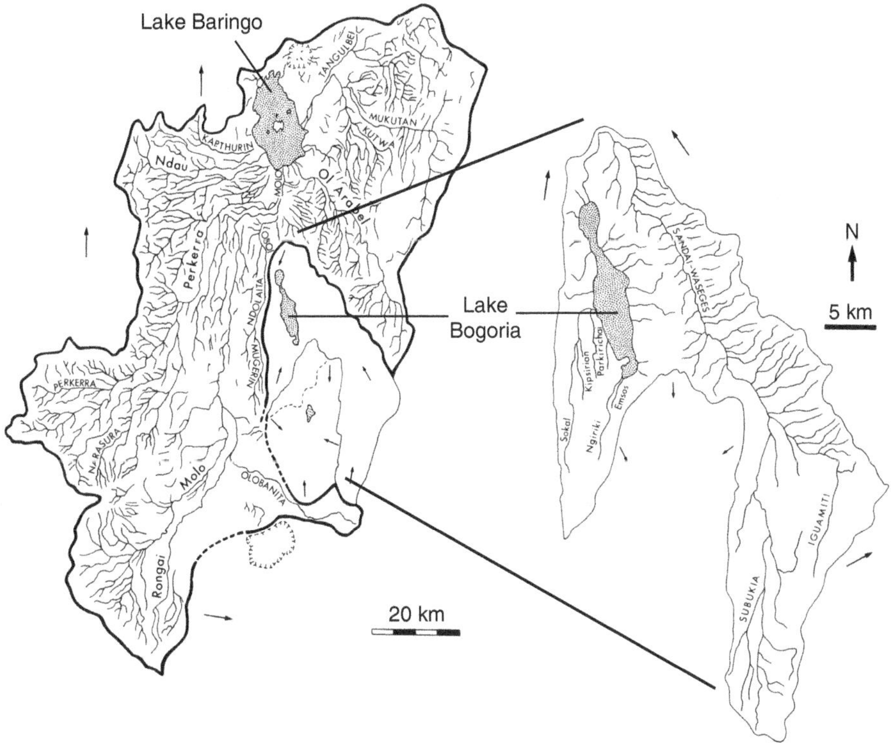

Figure 3.15 Comparison of lake Baringo and lake Bogoria drainage basins. Significantly smaller drainage basin may promote negative water budget, high salinity and mineral authigenesis. Modified from Tiercelin, J.J. et al., 1987. Le demi-graben de Baringo-Bogoria, Rift Gregory, Kenya: 30000 ans d'histoire hydrologique et sédimentaire. Bulletin des Centres de Recherche Exploration-Production Elf-Aquitaine, 11, 249–540.

of Holocene and Late Pleistocene sediments provides evidence for intervals of lake transgression and regression, highest lake levels being recorded during the deglaciation and the Early Holocene.

In the Baringo–Bogoria area of the East African Rift, the sedimentation is mostly controlled by:

- the volcanism, which influences the petrology, the mineralogy and the chemistry of the drainage basins, and
- the tectonism, which controls the extension of the drainage basins, the morphology of the basins and their impact on hydrology.

By comparison, the role of climate looks minor. However, the regime of precipitation plays a significant role in the control of lake level and the frequency of gravity flows, and the role of climate was probably more important during past intervals of high precipitation.

3.4. Example of Rift Environments and Sediments in a Predominantly Marine Context: The Gulf of Suez

The rift of Suez is delimited by the Aqaba fault to the SE but gradually disappears into the Egyptian hinterland to the NW, and is currently 80 km wide on average (Figure 3.16). The rift shoulders reach maximum elevation of 1,750 m to the West and 2,500 m to the East (Sinai) but the deepest parts of the central grabens are below sea level, the maximum water depth being of about 80 m. Tensional stress within the Arabian-African plate during the Early Miocene (Aquitanian, 25–20 Ma) led to shear motion along the Aqaba fault, and distension at its southwestern extremity. The rift of Suez is a typical passive rift, with intraplate tensional stress as the driving force of rifting. It is also asymmetric, with normal faults separating series of horsts and grabens which contain as much as 3,300 m of syn-rift deposits. The rift of Suez contains four major sedimentary units, separated by regional unconformities (Figure 3.17).

Early rift deposits (named Group A) rest on Eocene sediments and consist of siliciclastic red beds and basalt dykes and lava flows which yielded Early Miocene ages around 24 and 21 Ma. In the eastern part of the rift, reddish continental clastics

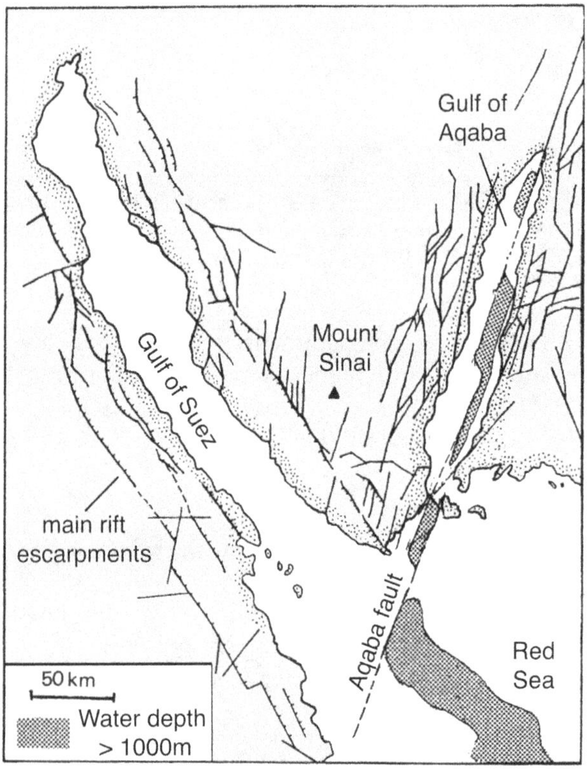

Figure 3.16 The rift of Suez: location map. Modified from Debelmas, J., Mascle, G., 2000. Les grandes structures géologiques, Dunod, Paris.

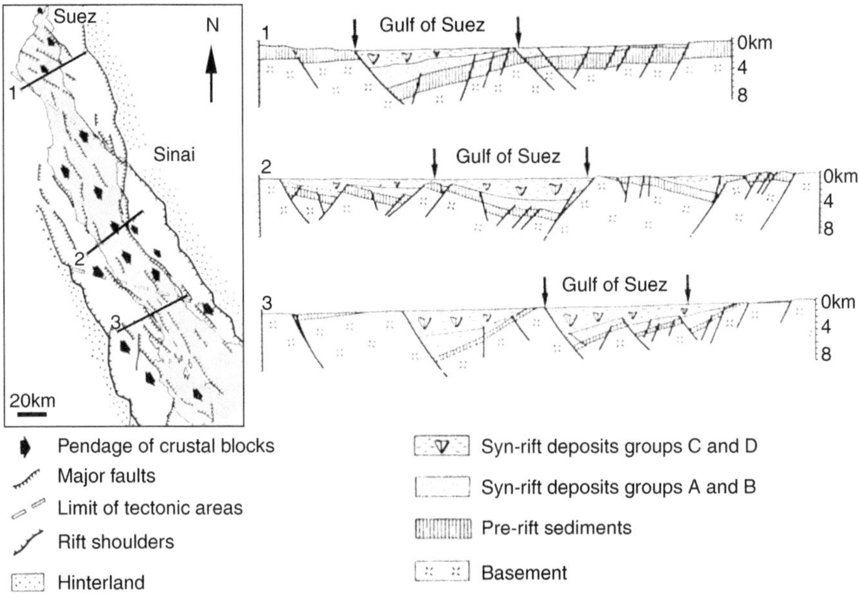

Figure 3.17 Summary of main structures and sedimentary units of the rift of Suez. Gray area represents regions below sea level. Modified from Purser, B.H., Bosence, D.W.J., 1998. Sedimentation and tectonics in Rift Basins: Red Sea, Gulf of Aden. Chapman and Hall, London.

and basalts are overlain by shallow marine deposits which locally show traces of littoral sandbars and grade upward to calcarenites and reef carbonates. In the western part of the rift, reddish continental clastics alternate with sandy and gypsiferous muds and laminated lacustrine carbonates, indicative of alluvial plains with intermittent playa lake environments (Figure 3.18). The sediments of the central part of the rift mostly correspond to channelized fluviatile sandstones grading to alluvial and submarine fans. Thicknesses of Group A deposits are variable, but never exceed 700 m. These characteristics of early rift deposits, together with the presence of fine grained siliciclastics where the clasts are derived from the Eocene surface only, suggest that the rifting started in a poorly drained and low relief (<200 m) area. Rift activity was weak, decreasing in importance and in age to the North.

Rift activity increased during the Burdigalian (19–16 Ma). The uplift of the rift shoulders and axial blocks was associated with an accelerated subsidence, the association of horizontal and vertical movements leading to the formation of a succession of horsts and grabens. Erosion and sedimentation rates of siliciclastics increased significantly as a consequence. Most of the late Early and Middle Miocene sediments of Group B accumulated in marine environments, and rest unconformably on basement, pre-rift and earlier syn-rift deposits (Figure 3.18). In the eastern part of the rift, conglomerates and large, gravelly delta fans developed at the outlet of pre-existing drainage systems. Active tectonism locally exposed Cretaceous and older sandstones which provided elements for sand-sized, shallow marine siliciclastics. In the western part of the rift, delta fans made of relatively coarse siliciclastics alternate with reefs and bioclastic sands. The grain-size of the

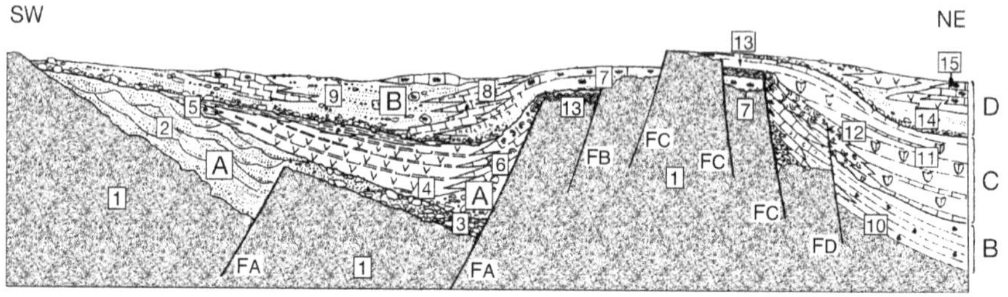

Figure 3.18 Detailed organization of syn-rift deposits, north-western Red Sea. Basement: 1, crystalline and metamorphic rocks. Group A: 2, silts and sands; 3, conglomerates; 4, evaporites and marls; 5, green silty marls and 6, stromatolites. Group B: 7, reefs; 8, reefal talus; 9, clastic fans and 10, Globigerina marls. Group C: 11, evaporites; 12, laminated carbonates and 13, stromatolites. Group D: 14, continental clastics and evaporite lenses and 15, biogenic carbonates. Tectonism: FA, faults active during Group A and sealed by Group B; FB, faults active after Group A and before Group B; FC, faults active during Group B and sealed by Group C and FD, faults and flexures active during and after Group D. Modified from Purser, B.H., Bosence, D.W.J., 1998. Sedimentation and tectonics in Rift Basins: Red Sea, Gulf of Aden. Chapman and Hall, London.

siliciclastics decreases to the central part of the rift where marls deposited in highly subsident grabens, whereas reefs developed on the horsts, providing elements for bioclastic carbonate deposition on the escarpments. The reefs (fringing reefs, biostromes and scattered coral colonies) are of Mediterranean affinity and mostly developed during a relatively brief time interval of high sea level and warm global climate (Langhian to Early Serravallian). Marine conditions became rapidly more restricted in some areas where the reefs receded and the siliciclastics increased in importance, probably because of more important uplift to the North, which limited the exchanges with the Mediterranean Sea.

Because of intensified activity in the Middle Miocene (Serravallian, 15–11 Ma), the eastern and western parts of the rift were uplifted above sea level while the subsidence of the grabens accelerated. The connection with the Mediterranean Sea being restricted, an evaporitic sedimentation (Group C) developed in the central part of the rift, overlapping older deposits. Gypsum, together with other sulfates and stromatolitic carbonates formed at the periphery of the area, grading to halite in the central part of the basin. The evaporites are locally capped by a deformed surface or breccias with evaporite clasts especially to the North, suggesting significant uplift, closure of the Suez isthmus and isolation from the Mediterranean Sea before the beginning of the Messinian salinity crisis. Beginning in the Tortonian (7 Ma) massive evaporites graded to alternating hemipelagic sediments and evaporites to the South, because of increased influx of marine waters from the Red Sea.

Coeval with increased shear activity along the Aqaba fault in the latest Miocene and earliest Pliocene, rift activity strongly accelerated in the Red Sea but remained moderate in the rift of Suez. However, the rift shoulders uplifted to modern heights, and about 1,500 m of Group D sediments accumulated in the marine environments of the subsiding central grabens (Figure 3.18). The sediments almost exclusively consist of siliciclastics, dispersed by marine processes from large delta fans at the mouth of major wadis. They form a succession of sequences separated by hiatuses, because of the combined effect of tectonism and eustatism. The load of terrigenous sediments locally produced salt tectonics in the underlying evaporites, which in turn generated syn-sedimentary deformation of Plio-Pleistocene syn-rift deposits.

By comparison with rift sediments of continental environments, marine rift deposits look more homogeneous in their composition and characteristics. For example, it is probable that the presence of seawater favored widespread formation of evaporites in the Gulf of Suez and beyond in the Red Sea. It is also probable that marine circulation played a major role in the dispersion of siliciclastics. As a consequence, local specifics of climatic and tectonic controls on rift sedimentation are somewhat mitigated. However, marine rift deposits record the major steps of the tectonic evolution of the rift and related changes of rift environments and climate, at a wider scale.

3.5. Ancient Rifts: The South Atlantic Rift Sediments in Brazil and Gabon

Rift systems represent the initial stage of the Wilson cycle, i.e. the phase of continental breakup. As the oceans develop, rift structures and sediments are split and separated by extending areas of oceanic crust, and buried below prograding continental shelves: they are progressively incorporated into passive continental margins.

The opening of the South Atlantic started in the southernmost parts of the African-South American block during the Late Jurassic, and propagated northwards. In the North Brazil–Gabon area (Figure 3.19), the rifting started during the latest Jurassic (Tithonian) and Neocomian time interval from about 146–132 Ma as a result of lithospheric extension and related minor uplift of the astenosphere. The rift system developed along weakened areas of the lithosphere, and the heterogeneous stretching model better accounts for its structure: the intracrustal detachment surface results from the reactivation of a basal thrust plane of the Pan-African fold belt. Lithospheric stretching increased in the Late Neocomian and Barremian (132–125 Ma) when intense faulting occurred in the upper lithosphere and vast quantities of flood basalt were released to the South in the Ponta–Grossa and Walvis area. The rift stage lasted for about 25 Myr.

Regional lithospheric extension first produced a series of rifts: the Tucano/Reconcavo rift system to the West, and the Jacuipe/Sergipe–Alagoas/Gabon/Rio Muni rift system to the East (Figure 3.19). Extension subsequently focused on the East branch leading to the formation of oceanic crust in the proto-South Atlantic, late in the Aptian. Extension subsequently aborted in the West branch, leading to the formation of the onshore Tucano basin. Rift deposits include three units, characterized by different facies associations and structural styles, which are identical in the modern deepest areas of the passive margins and basins of North Brazil and Gabon (Figure 3.20).

The syn-rift unit 1 of latest Jurassic Age fills large depressions, is little affected by faulting, and is relatively thin (<600 m) but rather complex (Figure 3.21): coarse

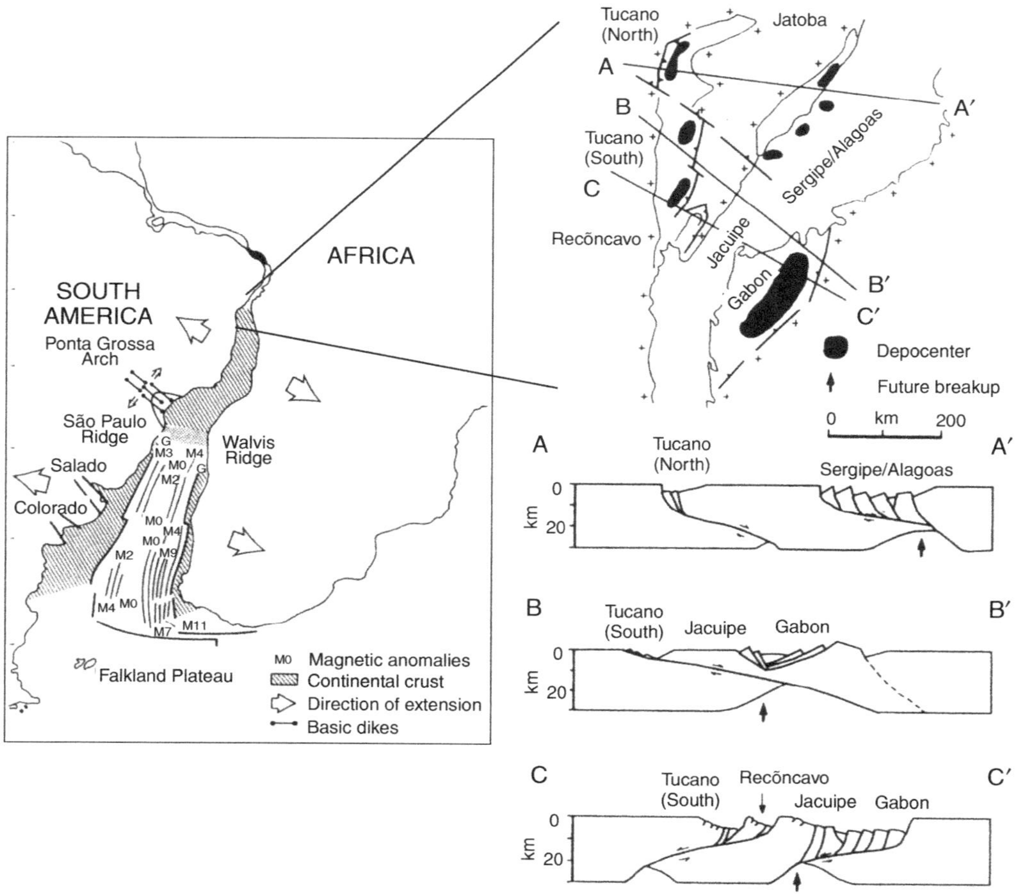

Figure 3.19 The Brazil–Gabon rift system. Left: paleoreconstruction of the Aptian South Atlantic. Upper right: enlargement of the Brazil–Gabon area highlighting the aborted west branch (Tucano basin) and the active east branch (Gabon–Sergipe/Alagoas). Lower right: schematic cross-sections through the Brazil–Gabon rift system. Persistent activity of the east branch led to the formation of oceanic crust in the future South Atlantic. Modified from Chang, H.K., Kowsmann, R.O., Ferreira Figueiredo, A.M., Bender, A.A., 1992. Tectonics and stratigraphy of the East Brazil Rift system: an overview. Tectonophysics, 213, 97–138.

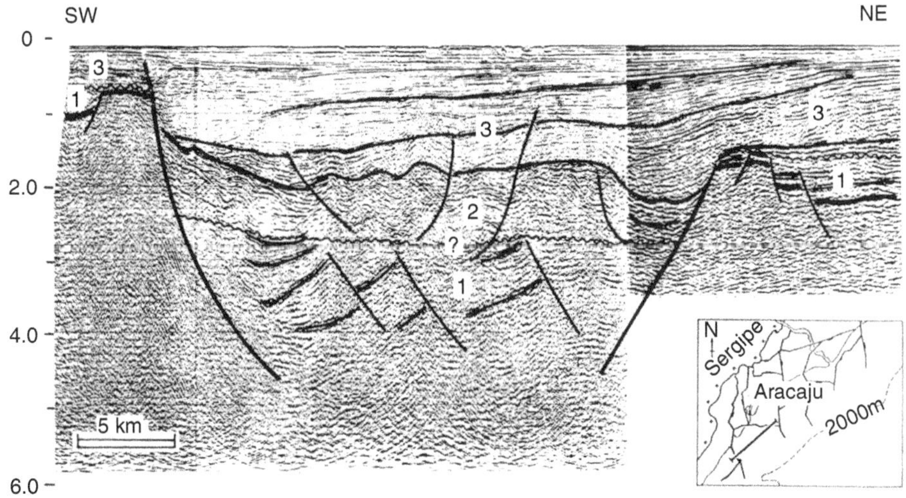

Figure 3.20 Seismic section through the Sergipe–Alagoas Basin (South Atlantic margin) and major structural highs. 1, continental syn-rift sediments; 2, transitional evaporitic unit and 3, post-rift oceanic sediments. Note that rift structures and syn-rift sediments are deeply buried in the continental margin and below oceanic sediments of the Brazilian shelf. Reprinted from Chang, H.K., Kowsmann, R.O., Ferreira Figueiredo, A.M., Bender, A.A., 1992. Tectonics and stratigraphy of the East Brazil Rift system: an overview. Tectonophysics, 213, 97–138.

to fine siliciclastic fluvial and alluvial fan deposits alternate with relatively fine lacustrine siliciclastics and carbonates, and local playa-lake evaporites. Eolian sands are also commonly found in this sequence. The base of the syn-rift unit 1 mainly reflects relatively arid alluvial and fluvial conditions with local salt pans, grading to more significant lacustrine environments where siliciclastics dominate. The upper part of the syn-rift unit 1 mainly reflects alternating lacustrine environments, prograding clastic wedges indicative of active clastic sources, and broad systems of coalescent alluvial and fluvial fans (Figure 3.22). The climatic control of rift deposits is there dominant, although it is not clear whether intense rainfall or tectonics is mainly responsible for intervals of prograding clastic wedges.

The syn-rift unit 2 of Neocomian Age attains thicknesses of 2,700–4,000 m and accumulated in rapidly subsiding grabens where deep and permanently stratified freshwater lakes developed. The lakes were rapidly filled with fine-grained siliciclastics, alternating with turbidites and mass-flow deposits principally. Because of high organic content and deposition in euxinic environments, the fine-grained siliciclastics (shales) represent an important source of hydrocarbons. The lacustrine sediments grade laterally to deltaic fans which prograde through time (Figure 3.21). Deltaic fans dominate the upper part of the unit, where they alternate with periodic intervals of fine-grained siliciclastics and lacustrine carbonates. The entire sequence is deformed by mud diapirs and crustal block tilting. This unit reflects very strong climatic and tectonic controls of rift sedimentation.

The syn-rift unit 3 of Barremian to Early Aptian Age reaches thicknesses around 3,000 m and is characterized by very high sedimentation rates, around 500 to

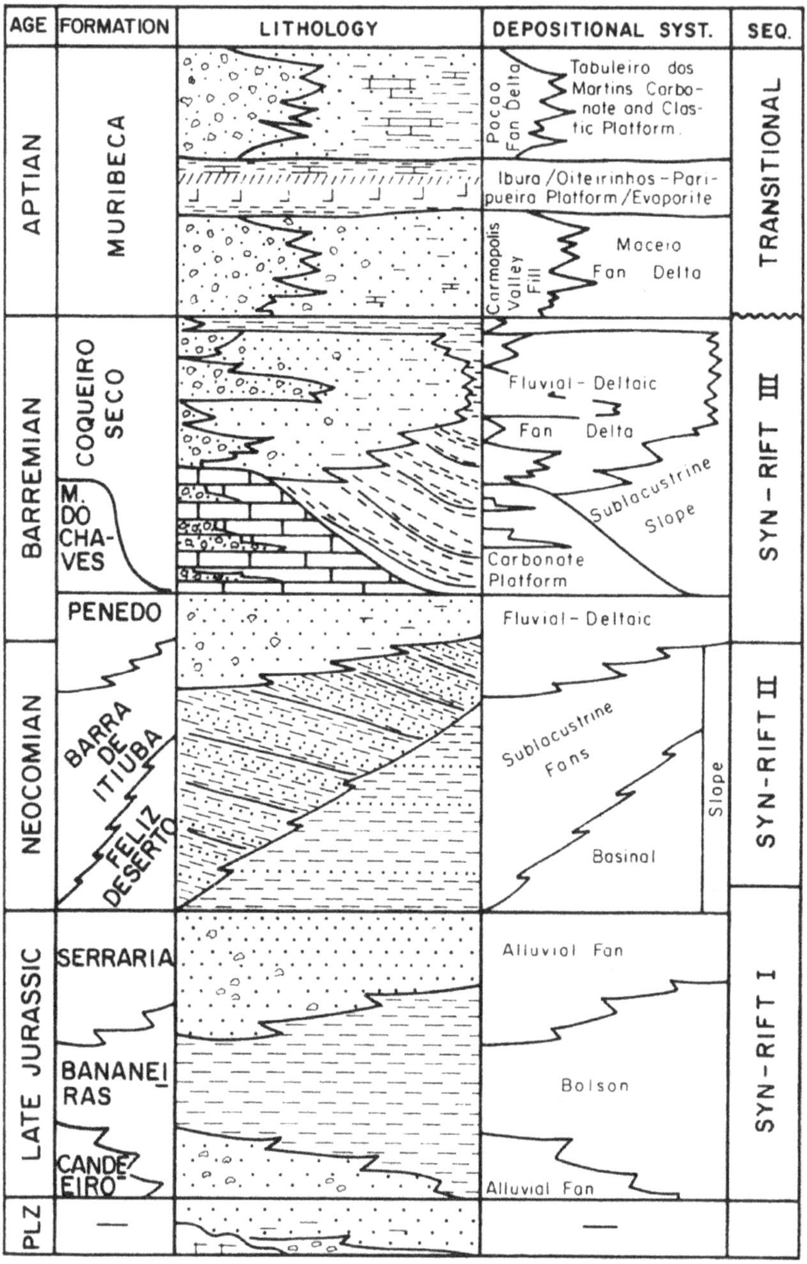

Figure 3.21 Stratigraphy, lithology and depositional models of syn-rift and transitional units in the Sergipe–Alagoas Basin. Reprinted from Chang, H.K., Kowsmann, R.O., Ferreira Figueiredo, A.M., Bender, A.A., 1992. Tectonics and stratigraphy of the East Brazil Rift system: an overview. Tectonophysics, 213, 97–138.

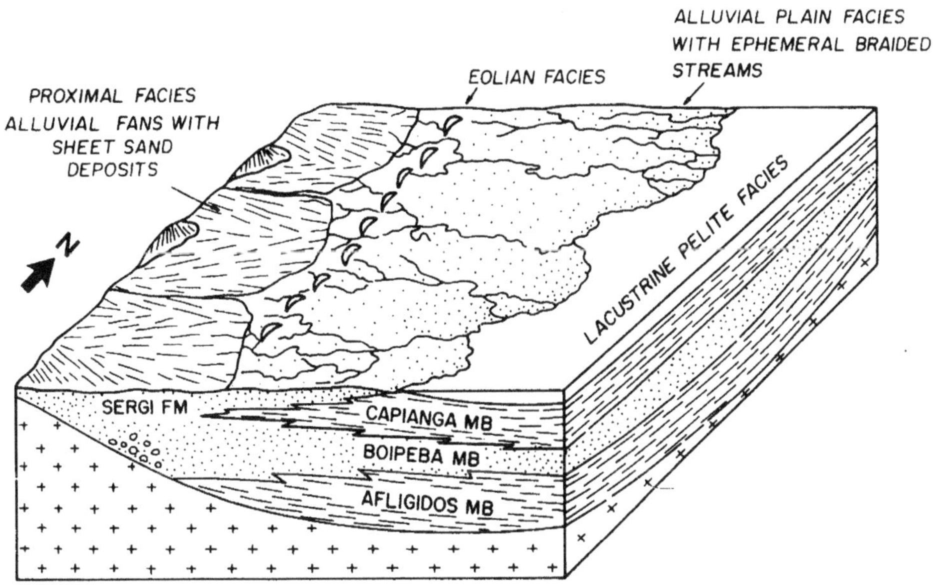

Figure 3.22 Schematic depositional model for Late Jurassic syn-rift unit 1, Recôncavo Basin. Reprinted from Chang, H.K., Kowsmann, R.O., Ferreira Figueiredo, A.M., Bender, A.A., 1992. Tectonics and stratigraphy of the East Brazil Rift system: an overview. Tectonophysics, 213, 97–138.

800 m/Myr (more than a twofold increase by comparison with syn-rift unit 2). This unit consists essentially of alluvial–fluvial sediments and delta fans which fill graben areas with siliciclastics of various sizes, from pebble to clay. They contain sections of highly organic silt to clay-sized siliciclastics (black shales) which represent periodic extension of lacustrine environments. The lacustrine black-shales sometimes grade to lacustrine carbonates which locally developed on structural highs, and to coarse siliciclastics nearshore (Figure 3.21). In the upper part of the unit, dominant alluvial and deltaic sediments locally alternate with gravity deposits. The sediment nature and facies, as well as the sedimentation rates, are principally indicative of significant and continued tectonism during a last pulse of rift activity.

Initial subsidence and sedimentation rates decreased rapidly during the Aptian, and sedimentation terminated abruptly in the West branch of the rift. This drastic change could be related to a shift of the pole of rotation of Africa relative to South America. Sedimentation resumed in the East branch of the rift where an important discontinuity separates syn-rift sediments from a thick section of siliciclastics and evaporites of Late Aptian Age (Figure 3.21), which mark the transition to the young ocean stage and is described in Section 5.4.

The characteristics of the sedimentation and succession of events in the ancient South Atlantic rift look very similar to those observed in modern and recent rift areas. The information is less precise because the South Atlantic rift is known from dispersed outcrops and exploration boreholes only, and many details remain unknown. The South Atlantic in the Brazil–Gabon area is derived from passive rifting and heterogeneous stretching, driven by changes in the rotation of lithospheric plates. However, there is no available information on the elevation of the rift system. Although syn-rift sediments continuously deposited in continental environments, active (hot spot) rifting in the Parana/Walvis area to the South (Figure 3.19) likely produced a morphological barrier, which might have prevented northward penetration of marine waters from the Argentine and Cape basins (magmatic control on rift sedimentation).

FURTHER READING

Boillot, G., Coulon, C., 1998. La déchirure continentale et l'ouverture océanique. Overseas Publishers Association, Amsterdam.

Boillot, G., Huchon, P., Lagabrielle, Y., 2003. Introduction à la géologie: La dynamique de la lithosphère. Dunod, Paris.

Debelmas, J., Mascle, G., 2000. Les grandes structures géologiques. Dunod, Paris.

Einsele, G., 1992. Sedimentary basins, evolution, facies, and sediment budget. Springer, Berlin.

Frostick, L.E., Renaut, R.W., Reid, I., Tiercelin, J.J. (Editors), 1986. Sedimentation in the African Rifts. Soc. Spec. Publ. 25 Blackwell, London.

Mohriak, W., Talwani, M., 2000. Atlantic rifts and Continental margins. Geophysical Monograph, 115. American Geophysical Union, Washington, DC.

Olsen, K.H., 1995. Continental rifts: Evolution, structure, tectonics. Elsevier, Amsterdam.

Ziegler, P.A. (Editor), 1992. Geodynamics of rifting, volume 2: case history studies on rifts, North and South America and Africa. Tectonophysics, 213: 1–284.

Other references used in this chapter

Brun, J.P., Gutscher, M.A., DEKORP-ECORS Team, 1992. Deep crustal structure of the Rhine Graben from DEKORP-ECORS seismic reflection data. Tectonophysics, 208: 139–147.

Chang, H.K., Kowsmann, R.O., Ferreira Figueiredo, A.M., Bender, A.A., 1992. Tectonics and stratigraphy of the East Brazil Rift system: an overview. Tectonophysics, 213: 97–138.

Lezzar, K.E., Tiercelin, J.J., De Batist, M., Cohen, A.S., Bandora, T., Van Rensbergen, P., Le Turdu, C., Mifundu, W., Klerkx, J., 1996. New seismic stratigraphy and late tertiary history of the North Tanganyika Basin, East African Rift System, deduced from multichannel and high resolution reflection seismic data, and piston core evidence. Basin Research, 8: 1–28.

Purser, B.H., Bosence, D.W.J., 1998. Sedimentation and tectonics in Rift Basins: Red Sea, Gulf of Aden. Chapman and Hall, London.

Tiercelin, J.J., Soreghan, M., Cohen, A.S., Lezzar, K.E., Bouroullec, J.L., 1992. Sedimentation in large African lakes: example from the middle Pleistocene – modern deposits of the Tanganyika Trough, East African Rift System. Bulletin des Centres de Recherche Exploration-Production Elf-Aquitaine, 16: 83–111.

Tiercelin J.J., Vincens, A. (Editors), 1987. Le demi-graben de Baringo-Bogoria, Rift Gregory, Kenya: 30000 ans d'histoire hydrologique et sédimentaire Bulletin des Centres de Recherche Exploration-Production Elf-Aquitaine, 11: 249–540

CHAPTER FOUR

Intraplate Basins

4.1. Structure, Tectonics and Sedimentation of Intraplate Basins

Because of lower mantle plume intensity or tensional stress within lithospheric plates, rift activity may decrease and cease before maximum thinning and breakup of the lithosphere allows formation of oceanic crust during the transition to the young ocean or crustal fissure stage. As a result, regional heat flow decreases, volcanic activity stops and upper mantle rocks of lower temperature and higher density are incorporated into the basal continental crust. Because of lower heat flow the density of the lithosphere increases and its volume decreases, and the whole rift area subsides (thermal subsidence). Areas of ancient active rifting may form large igneous provinces (Deccan Traps of India, Parana Province of Brazil), whereas areas of ancient passive rifting may evolve into intraplate basins.

Intraplate basins do not correspond to plate boundaries but are incorporated into lithospheric plates where they show as depressed areas of continental crust. Since erosion and tectonism affected the whole rift system (horsts, grabens and rift shoulders) during its life span, these areas of thinned lithosphere form the basement of the depressed intraplate basin where sediments accumulate. The sediment load increases the subsidence which involves a larger portion of continental lithosphere through time, increasing the importance of the basin (Figure 4.1). To compensate, adjacent portions of lithosphere slightly uplift by a few hundreds of meters as a maximum.

The extension and depth of intraplate basins are therefore clearly controlled by the degree of lithosphere thinning reached during initial rifting and the availability of sediment sources. The subsidence of an ancient rift and depth of the subsequent intraplate basin can be approximated using the degree of crustal shortening and the laws of isostatic equilibrium (Figure 4.2). The degree of crustal shortening β is deduced from the ratio between initial (pre-rift) and final (post-rift) crust thicknesses. To simplify, we consider a continental crust (density $d = 2.8$) about 30-km thick which reaches sea level (elevation 0 m) at isostatic equilibrium, and the lithospheric mantle ($d = 3.3$). The relationship for isostatic equilibrium before and after crustal shortening is

$$2.8 \times 30 = 2.8(30/\beta) + 3.3(30 - z - 30/\beta)$$

where z is the subsidence (or basement depth) in subaerial conditions and

$$z = 4.5(1 - 1/\beta)$$

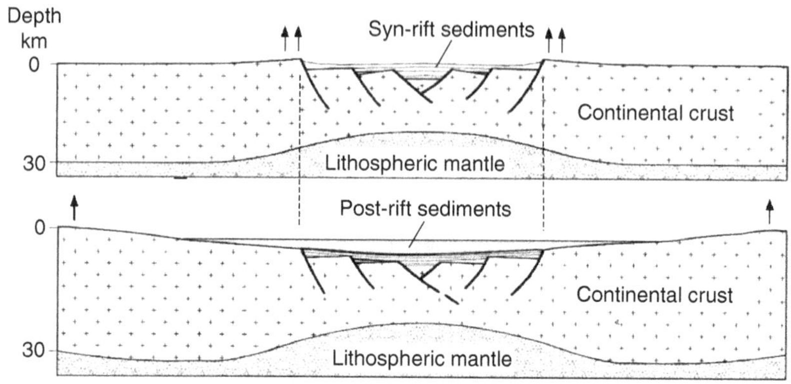

Figure 4.1 Schematic evolution of an aborted rift (top) into an intraplate basin (bottom). Arrows indicate uplifted areas. Note increased subsidence in areas of maximum crustal shortening. Modified from Boillot, G., Coulon, C., 1998. La déchirure continentale et l'ouverture océanique. Overseas Publishers Association, Amsterdam.

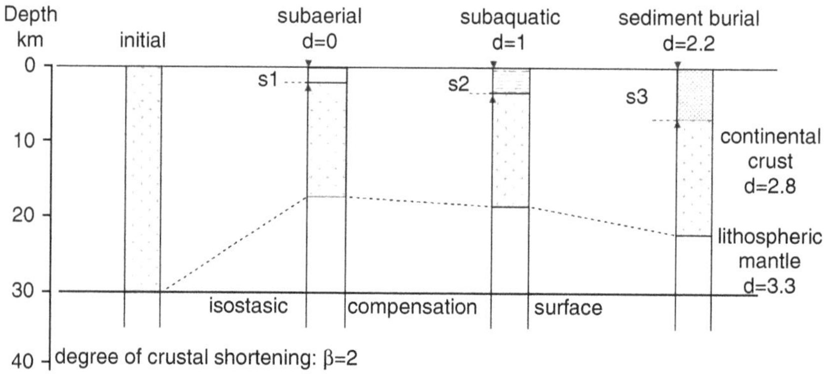

Figure 4.2 Comparative subsidence of a portion of thinned lithosphere (degree of crustal shortening of 2) for subaerial conditions, subaquatic conditions and sediment burial conditions. Note increased subsidence from s1 to s3. Modified from Boillot, G., Coulon, C., 1998. La déchirure continentale et l'ouverture océanique. Overseas Publishers Association, Amsterdam.

As sediment of $d = 2.2$ progressively fills the intraplate basin through time, the relationship evolves to

$$z = 13.6(1 - 1/\beta)$$

Although very simplistic, these relationships give an idea of the importance of the degree of lithosphere thinning and sediment availability in the evolution of intraplate basins. In fact the subsidence, depth of basin and sediment thickness vary locally with the degree of lithosphere thinning: for example, the sedimentary column is thicker above the main grabens of the ancient rift where crustal shortening is maximum, and decreases progressively toward the periphery of the basin (Figure 4.3).

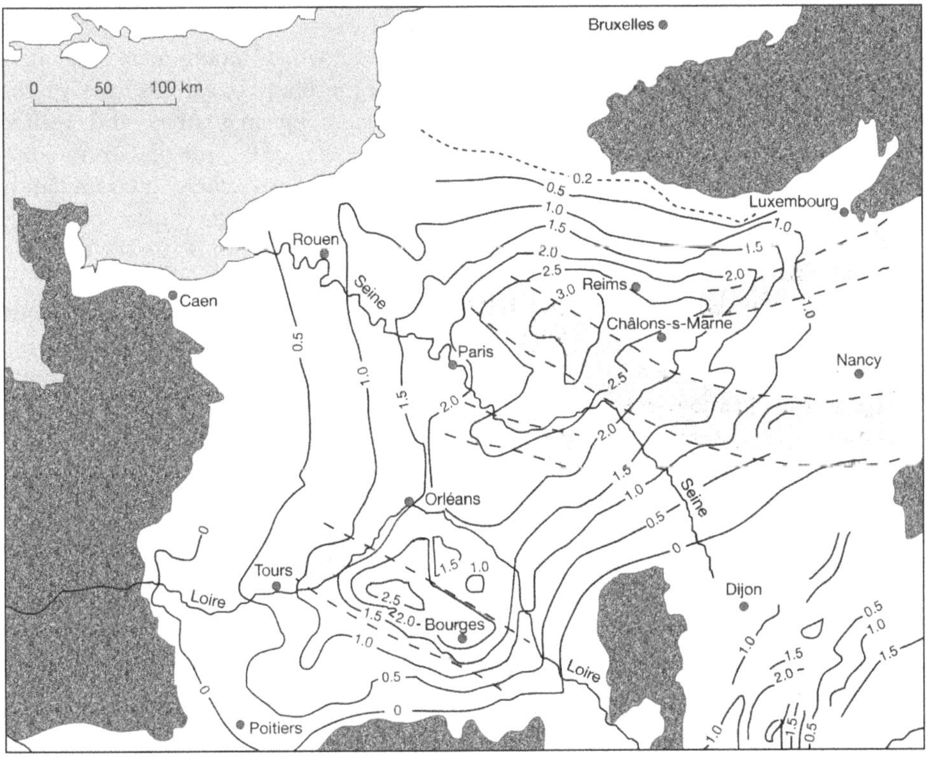

Figure 4.3 Thicknesses of Mesozoic and Cenozoic sediments in the Paris Basin. Note maximum thicknesses above the rift grabens (dotted lines) which are areas of maximum crustal shortening. Modified from Brahic, A., Hoffert, M., Schaaf, A., Tardy, M., 1999. Sciences de la Terre et de l'Univers. Vuibert, Paris.

Intraplate basins being incorporated into lithospheric plates, they move with their lithospheric substrate and react to internal stress transmitted from plate boundaries. Potential sources of internal stress include for example changes in the rotation and velocity of lithospheric plates, and collision events. Tensional or compressional stress within the lithosphere result in intraplate deformation, mainly faulting, folding and uplift, which interfere with subsidence and sediment formation. Intraplate deformation affects both the substrates and sediments of intraplate basins.

Sedimentation in intraplate basins may take place in continental or marine environments, as determined by the degree of lithosphere thinning reached during the initial rift interval as well as by the nature and intensity of intraplate deformation. However, marine environments remain relatively shallow and the water depth never exceeds a few hundreds of meters because of the isostatic response of underlying, low-density continental lithosphere. Marine intraplate basins like the North Sea contain relatively thick sediments (up to a few thousands of meters) because they collect the residue of marine biological activity together with terrigenous particles eroded from adjacent continental areas, and are subjected to an increased subsidence. Sediments accumulate more slowly in continental intraplate basins like the Saharan and Chad basins because of low subsidence and frequent hiatuses, but nevertheless result in thick and little deformed sedimentary sequences over long periods of time of several hundreds of million years. Because of the variability of geologic events at plate boundaries, internal stress and related deformation, many intraplate basins contain a succession of marine and continental sediments: this is the case in the London and Paris basins where predominantly marine Mesozoic sediments alternate with continental deposits. There, predominantly continental to nearshore Cenozoic sediments contain minor marine deposits, the modern English Channel being the result of the post-glacial marine transgression in the currently most depressed area of the basin. Therefore, sediments of intraplate basins record a succession of environmental changes and deformations which are closely related to the tectonic activity of the adjacent lithosphere and lithospheric plates.

4.2. Intraplate Basins of Western Europe: A Brief Summary

Intraplate basins are a common geological feature of Western Europe (Figure 4.4). Because of crustal extension within the Pangea supercontinent during the Permian, pre-existing fracture systems were reactivated and Europe started

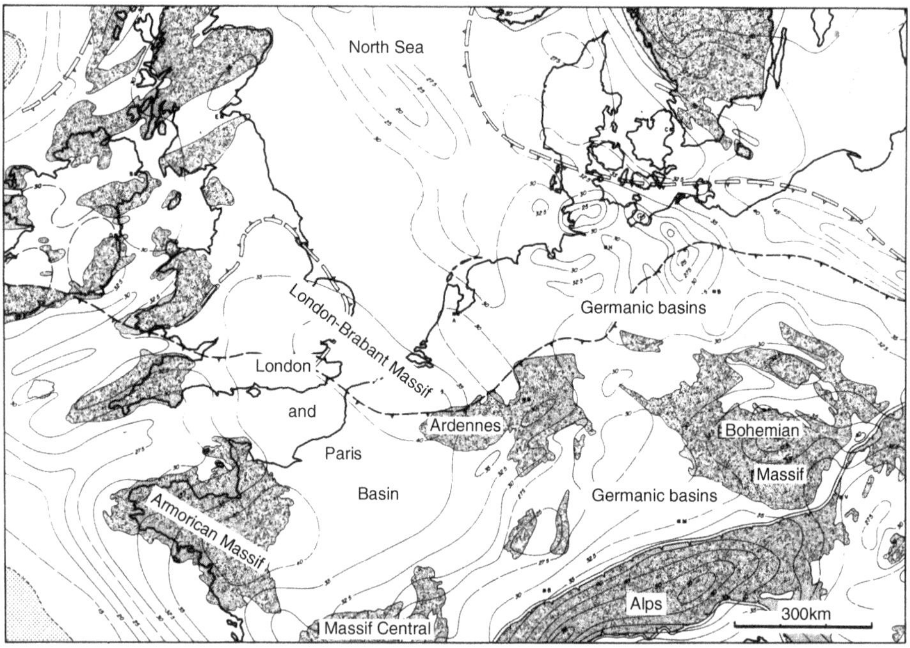

Figure 4.4 Depth of the Moho below Western Europe, highlighting areas of lithosphere thinning and intraplate basins. Areas where basement outcrops are in gray. Black and open barbed lines mark deformation fronts. Modified from Ziegler, P.A. 1992. North Sea rift system. Tectonophysics, 208, 55–75.

moving northward from its tropical position. Passive rifting led to the formation of a succession of small grabens filled with continental siliciclastics (conglomerates, sandstones and claystones principally) to the southwest, and to large basins in Eastern Europe (Germanic basins) and the North Sea where the Rotliegend siliciclastics are overlain by the Zechstein marine carbonates and evaporites, the marine transgression coming from the Arctic shelves. The rifting was especially intense in the Norwegian–Greenland and in the Tethys areas which eventually evolved into oceanic regions.

The Norwegian–Greenland Rift propagated during the earliest Triassic into the North Sea, where 2,000–6,000 m of sediments deposited into the subsiding Viking and Central grabens. Rift activity intensified during the Middle and Late Jurassic, when a major dome and related volcanism formed in the North Sea. The rifting decreased in intensity during the Cretaceous when extension focused on the North Atlantic, but ceased only in the Eocene at about 55 Ma. After a rifting stage which lasted for about 175 Myr, the North Sea has been a marine intraplate basin for the remaining of the Cenozoic (Figure 4.5). However, it is unlikely that thermal subsidence is completed and that the lithosphere of the North Sea is fully stabilized yet.

Although of minor importance by comparison with the North Sea, passive rifting persisted through the Triassic in the Germanic basins and the London–Paris Basin (and further south the Aquitaine Basin) where three similar sedimentary units accumulated in slowly subsiding basins separated by uplifted areas. The lower continental Buntsandstein unit mostly consists of red sandstones and claystones which accumulated in alluvial fans and flooding plains, with local occurrences of playa-lake evaporites and paleosoils. The intermediate shallow marine Muschelkalk unit consists of a variety of biogenic and chemically precipited carbonates slightly diluted by siliciclastics. This unit fills two depressed areas separated by uplifted areas between the London–Brabant Massif and the Bohemian Massif. Facies variations

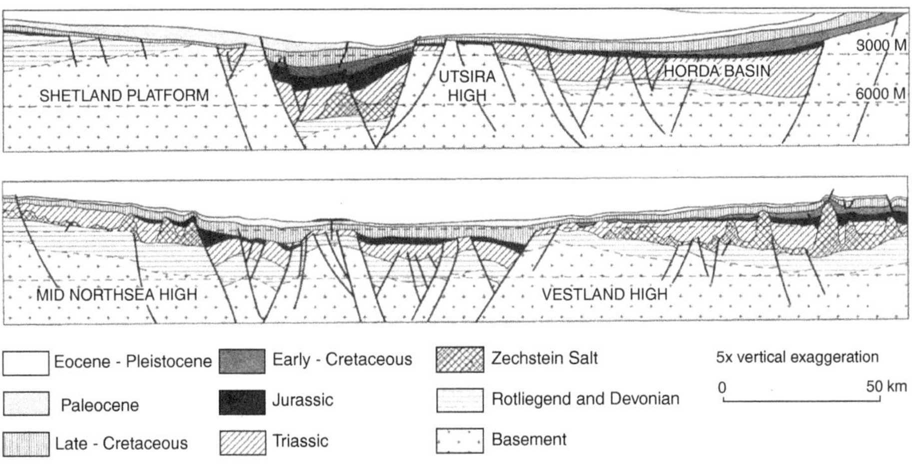

Figure 4.5 Schematic cross-sections of the North Sea in the Central Graben area. Note the distribution of syn-rift (Permian to Early Cretaceous) and post-rift (Late Cretaceous and Cenozoic) sediments. Modified from Ziegler, P.A., 1992. North Sea rift system. Tectonophysics, 208, 91–111.

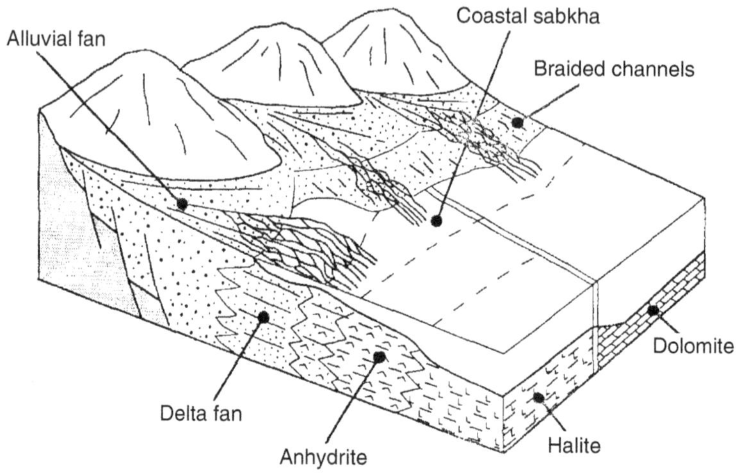

Figure 4.6 Depositional environments of the Keuper series of the Paris Basin. Modified from Bourquin, S., Guillocheau, F., 1996. Keuper stratigraphic cycles in the Paris basin and comparison with cycles in other peritethyan basins (German basin and Bresse-Jura basin). Sedimentary Geology, 105, 159–182.

indicate that the environments of deposition ranged from open marine conditions to carbonate platforms, tidal flats and hypersaline nearshore areas, and were very sensitive to relative sea-level fluctuations. The upper transitional Keuper unit mostly consists of alternating carbonate beds and evaporites with intervals of silty clays. This unit principally accumulated in playa-lake environments and sabkhas, sporadically interrupted by shallow marine transgressions. However, erosion from adjacent highlands was still significant (Figure 4.6). The total thickness of the three Triassic units may reach 800–2,000 m in the Germanic basins. The Triassic units reflect typical passive rift environments followed by a progressive decrease of erosion and subsidence of the rift systems near sea level. However, fluctuations of rift activity and relative variations of sea level led to a variety of sedimentary environments before the rifting definitively ceased at the end of the Triassic.

Thermal relaxation allowed formation of relatively shallow, epicontinental marine environments which connected the Tethys to the Arctic seas and persisted through most of the Jurassic and Cretaceous. The eastern basins were separated from the western basins by structural highs from the London–Brabant Massif to the Bohemian Massif, of variable importance through time (Figure 4.7). The main lines of the sedimentation remained about the same in the western basins till the Early Paleogene when the European Cenozoic Rift System separated the west Germanic Basin from the London–Paris Basin. Related thermal doming led, for example, to significant uplift of the rift shoulders in the Vosges and Black Forest areas adjacent to the Rhine Graben (Figure 4.8). Subsequent erosion removed most Mesozoic sediments from the rift shoulders, where the Triassic Buntsandstein unit and substrate outcrop in some areas. The London–Paris Basin has been divided into several sub-basins during the Cretaceous and especially the Cenozoic, as a result of intraplate deformation.

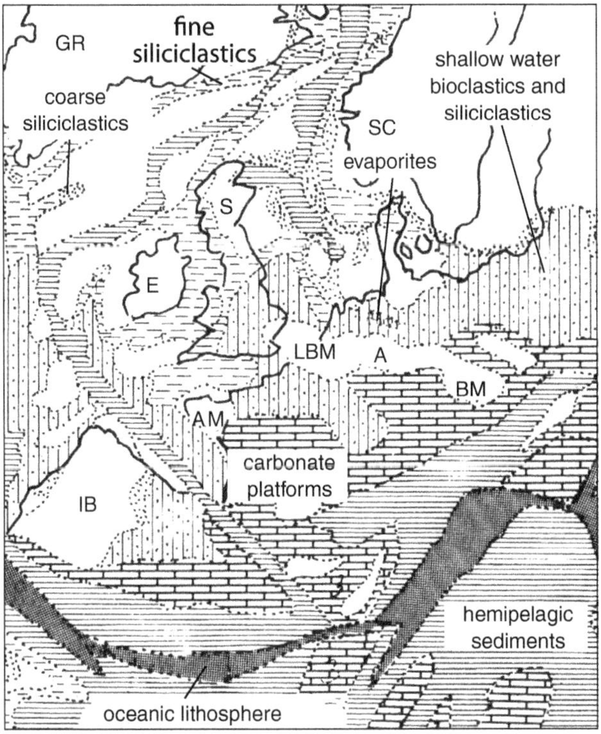

Figure 4.7 Distribution of major sediment facies in the intraplate basins of Western Europe and the Ligurian Tethys in the Late Jurassic. Emerged areas are in white. A, Ardennes; AM, Armorican Massif; BM, Bohemian Massif; E, Ireland; GR, Greenland; IB, Iberia; LBM, London–Brabant Massif; S, Scotland; SC, Scandinavia. Modified from Einsele, G., 1992. Sedimentary basins, evolution, facies, and sediment budget. Springer, Berlin.

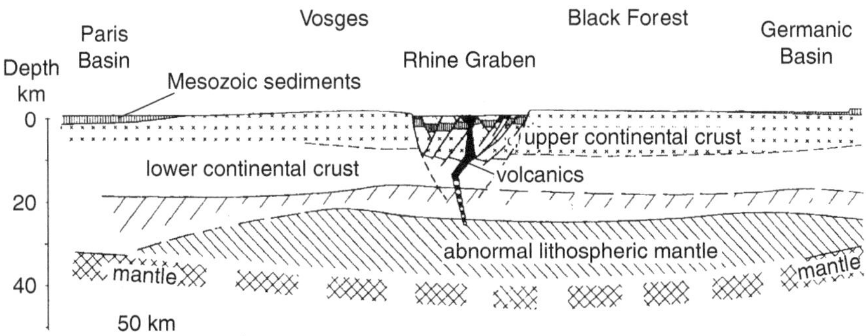

Figure 4.8 Cross-section of the Rhine Graben and adjacent areas from the Paris Basin to the west, to the Germanic Basin to the east. Note the presence of Mesozoic sediments as outcrops in both Paris and Germanic intraplate basins, and as pre-rift sediments in the Rhine Graben. Modified from Debelmas, J. Mascle, G., 2000. Les grandes structures géologiques. Dunod, Paris.

4.3. Example of Intraplate Basin Environments and Sediments: The Paris Basin

The modern Paris Basin is centered on four grabens which developed during a Permian phase of rifting. The basin is bounded by three structural highs of Paleozoic (Variscan) substrate: the Armorican Massif to the west, the Massif Central to the south, the Ardenne Massif to the northeast. The morphology mostly includes flat areas carved by fluvial erosion in the central parts, and well-developed cuestas at the periphery, especially to the east where the sedimentary units have been uplifted (Vosges Massif) and eroded as the European Cenozoic Rift System developed. Up to 3,000 m of sediments fill the Paris Basin, with maximum thicknesses being observed in areas of maximum subsidence and crustal shortening above the rift grabens, and minimum thicknesses in areas of minimum subsidence at the periphery of the basin (Figures 4.3 and 4.4).

4.3.1. Major Sedimentary Units of the Paris Basin

The sedimentary column is highly heterogeneous and includes a diversity of facies, from terrigenous to biogenic and evaporitic, the sedimentary units being separated by discontinuities. The sedimentation reflects several aspects of the geological evolution of the Paris Basin:

- the climatic record integrates changes of global climate and northward migration of the basin from about 25°N during the Triassic to its current position;
- the eustatic record combines sea-level variation of global climatic and geodynamic origins with local changes in the elevation of the European plate; and
- the geodynamic record principally reflects stages of Mesozoic extension as the Tethys and North Atlantic oceans developed, and stages of latest Mesozoic and Cenozoic compression related to the convergence of the African and European plates, closure of the Tethys, and formation of the Alps.

Because of the complexity of the geological history of the Paris Basin, only the main lines of the geodynamic record and typical examples of intraplate basin sediments and sedimentary environments are detailed below.

The Mesozoic sedimentary sequence of the Paris Basin contains a number of unconformities, including five major unconformities which are related to major stages of regional geodynamics and plate tectonics (Figure 4.9). The oldest (Early Cimmerian) major unconformity is of Late Triassic Age (Norian) and separates Keuper evaporites from coastal sabkha sediments. The Early Cimmerian unconformity is related to a westward and northward tilt of the whole Paris Basin and beyond as a result of changes in intraplate stress regime, and correlative migration of the depocenters before rift activity ceased at the end of the Triassic. A second (Middle Cimmerian) major unconformity marks the transition from the Early to the Middle Jurassic. This unconformity is coeval with a stage of ocean opening in the Ligurian Tethys and the beginning of a major interval of thermal doming of the North Sea Rift. Deformation and erosion reflect ENE–WSW

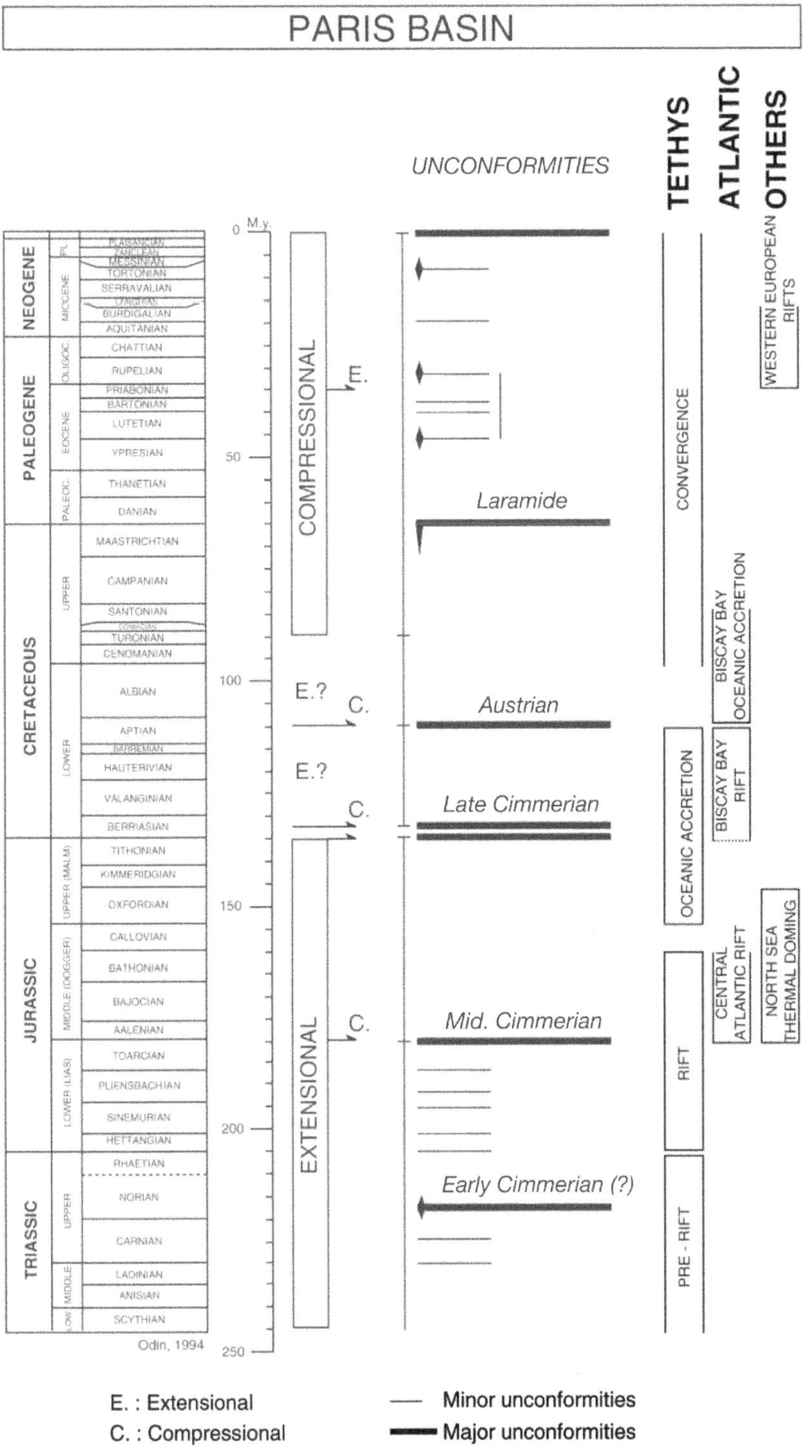

Figure 4.9 Summary of the Mesozoic and Cenozoic geodynamic evolution of the Paris Basin, and correlation with major events at plate boundaries. Modified from Guillocheau, F., et al., 2000. Meso-Cenozoic geodynamic evolution of the Paris Basin: 3D stratigraphic constraints. Geodinamica Acta, 13, 189–246.

compression in the Paris Basin which is located between both areas of active extension. Early Cretaceous times are characterized by two major unconformities of Berriasian Age (Late Cimmerian) and Aptian Age (Austrian) which are consequences of opening processes in the North Atlantic and in the Valais Ocean north of the Ligurian Tethys. Both unconformities result from NE–SW compression with uplift of the western and northern margins of the basin. They are coeval with the extension of the Atlantic Rift north of the Portugal margin beginning in the latest Jurassic, a rifting stage in the Bay of Biscay, and related thermal doming. Syn-sedimentary deformation induced by N–S compression progressively increased during the Late Cretaceous to a paroxysm near the Cretaceous–Tertiary boundary and led to the major Laramide unconformity. Deformation and unconformity coincide with convergence in the Ligurian Tethys, the Valais Ocean and the Bay of Biscay, and the reactivation of Hercynian fault systems and regional uplift in future Alpine areas (eo-Alpine stage).

The Cenozoic history of the Paris Basin is characterized by a series of unconformities and general uplift induced by increased compression in relation to collision events and tectonic activity in the Alps and the Pyrenees. Little is known on the Cenozoic relationships between stratigraphy and tectonics because of the frequency of tectonic events, the quantity of erosional events and the complexity of the sedimentary facies. However, unconformities and deformation were especially important during the Eocene when collision along the northern Tethys margins initiated the Pyrenean tectonic phase (meso-Alpine stage), during the Late Eocene and the Oligocene when the European Cenozoic Rift System developed in western Europe, and during the Miocene (Burdigalian, Tortonian) when major tectonic events occurred in the Alps.

Major and minor unconformities mark the limits of sedimentary cycles which are controlled by intraplate tectonic stress, sea-level variations and sediment fluxes principally. Major cycles are 10–40 Myr long, are bounded by major unconformities, and are driven by the tectonic history of the European plate. They are associated to minor sedimentary cycles which are 1–15 Myr long, and to elementary units which are 0.02–0.4 Myr long. Major sedimentary cycles include a transgressive trend during intervals of tectonic relaxation, subsidence and/or sea-level rise and a regressive trend during intervals of increased tectonism at plate boundaries, intraplate deformation and/or sea-level fall. A total of 12 major cycles have been recognized since the Triassic. Cenozoic cycles are rather complex because of frequent variations in intraplate stress, regional uplift and repetitive erosional events, while Mesozoic cycles are more complete. For example, seven major cycles have been recognized between the Late Triassic and the Turonian (Figure 4.10): the sediments range from siliciclastics to marls and biogenic carbonates, and the environments of sedimentation range from alluvial plain to nearshore and open marine conditions.

The major sedimentary cycles consist either of predominantly siliciclastic or calcareous biogenic sediments. For example, the Bathonian–Oxfordian cycle starts with shallow-water oolites and calcareous bioclastics interfingered with carbonate muds during the transgressive phase. They are succeeded by open marine terrigenous shales at the maximum of the transgression. The regressive trend is characterized by the development of a carbonate platform interfingered with

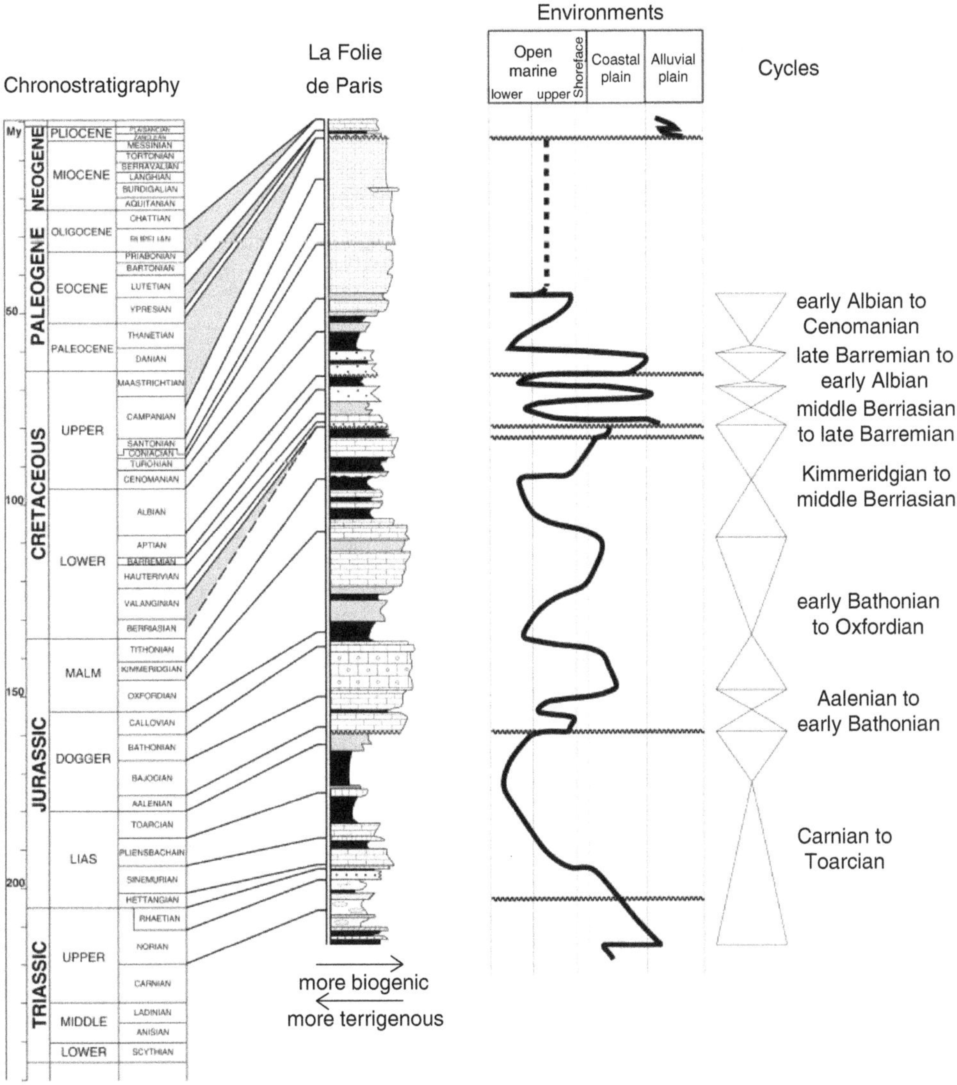

Figure 4.10 Triassic to Late Cretaceous sedimentary cycles in the Paris Basin, as illustrated by the sediment facies recorded in a borehole at site La Folie de Paris, about 100 km east of Paris. Modified from Guillocheau, F., et al., 2000. Meso-Cenozoic geodynamic evolution of the Paris Basin: 3D stratigraphic constraints. Geodinamica Acta, 13, 189–246.

nearshore siliciclastics. The transgressive facies of the subsequent Kimmeridgian–Berriasian cycle include organic-rich fine siliciclastics and muddy carbonates which look similar to tidal-flat deposits and grade to open marine marls at the maximum of the transgression. The regressive facies include shallow water muddy and bioclastic carbonates (Portlandian facies) grading locally to coastal evaporitic deposits (Purbeckian facies).

The sedimentary cycles mostly express the general dynamics of the intraplate basin through time. The diversity of the tectonic, eustatic and climatic influences, and their relationships, are depicted with great detail in a variety of facies and their succession through time. Some examples are detailed below.

4.3.2. Early Jurassic Siliciclastics of the Paris Basin

After the rifting ceased in the latest Triassic, thermal subsidence progressed during the Early Jurassic (Liassic). Because of the rigidity of the upper crust, the subsidence was associated with the reactivation of ancient faults which led to the formation of small sub-basins where up to 500 m of sediments accumulated. The progression of the thermal subsidence at a time of increased sea level produced a marine transgression which came from the east (Germanic basins) and from the southeast (Ligurian Tethys). The transgression was limited by the presence of subaerial structural highs (Armorican Massif, Massif Central and Ardenne) and an elongated gulf developed in close proximity to the Ardenne (see Figures 4.4 and 4.7 for location). Weathering and erosion of subaerial areas provided abundant terrigenous elements to the sediment. With the exception of a shallow platform to the southeast (Burgundy) where biogenic carbonates expanded, siliciclastics were dominant in most of the Paris Basin: sands with minor layers of clay and marl to the north near the Ardenne, grading to dominant clay deposited in shallow-water environments in the remaining of the Early Liassic Gulf.

The transgression accelerated in the Middle Liassic (Figure 4.10), and progressed over the Armorican Massif and the Massif Central particularly. The London–Paris Basin became a seaway between the Ligurian Tethys and the Arctic seas. Coarse siliciclastics dominated at the periphery of the Paris Basin with local occurrences of sandy carbonates and oolites, grading to terrigenous clays in the central parts of the basin. However, a major change occurred at the end of the Middle Liassic when coarse siliciclastics expanded in the northern and eastern parts of the basin, grading to clays and marls to the south, during a further step of subsidence and reactivation of ancient faults.

This event immediately preceded a shift of the area of maximum subsidence from the eastern to the central part of the basin, early in the Late Liassic. When the transgression resumed, shallow-marine environments rapidly reached their maximum extension (Early Toarcian maximum of transgression). The sediments were largely dominated by terrigenous clays and fine silts with rare layers of coarse silt and silt, which increased in frequency in the latest Liassic when a regression began.

The main characteristic of the Liassic sediments of the Paris Basin (and beyond) is the preservation of abundant organic matter, especially when associated to fine siliciclastics which present a black-shale facies. It is likely that the development of narrow sub-basins separated by shallow sills in the Early Liassic restricted the ventilation of the basin and created topographic and oceanographic conditions favorable to anoxia. Organic matter of continental origin (pollen, vegetal debris, lignin, cellulose, etc.) is present in significant amounts in most Liassic sediments. The sediment facies indicate that temperature and precipitation were probably favorable to intense chemical weathering and dense vegetation cover in subaerial areas adjacent to the Paris Basin, as well as to erosion and transport by running waters. This background of continental organic matter is overprinted early in the Late Liassic (Early Toarcian) by a dominance of aquatic organic matter of phytoplanktonic origin, indicative of an exceptional marine productivity. The diversity of the phytoplankton was low and limited to species of low salinity affinity, suggesting that planktonic proliferation was probably of climatic origin: running waters may have transported nutrients from subaerial areas and favored some degree of water stratification in an already poorly ventilated basin. The association of fine siliciclastics and abundant phytoplanktonic organic matter produced a black-shale sedimentary facies, which was widespread in Western Europe (Figure 4.11). Early Toarcian black shales are a potential parent rock for hydrocarbons, and effectively yielded small quantities of oil and gas in areas of significant burial (central part of the Paris Basin) and/or heat flow (Rhine Graben). The black-shale facies characterized the maximum of transgression and ceased in the latest Liassic (Late Toarcian) when increased quantities of silt and sand mark the beginning of a regression.

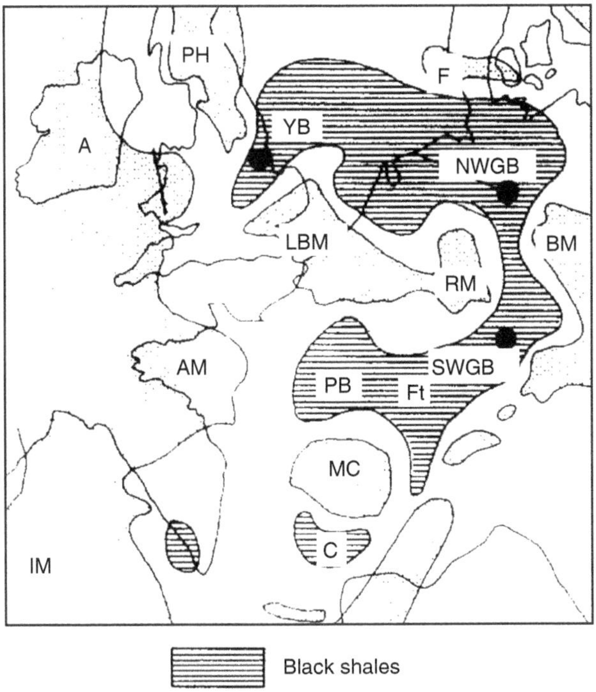

Figure 4.11 Distribution of Late Liassic black-shale facies: C, Causses; Ft, Fecocourt; NWGB, Northwestern Germanic Basin; PB, Paris Basin; SWGB, Southwestern Germanic Basin; YB, Yorkshire Basin. Emerged areas: A, Atlantis; AM, Armorican Massif; BM, Bohemian Massif; F, Funen Plateau; Ft, Fecocourt; IM, Iberian Massif; LBM, London–Brabant Massif; MC, Massif Central; PH, Pennines Plateau; RM, Rhenish Massif. Modified from Disnar J.R., Le Strat, P., Farjanel, G., Fikri, A., 1996. Organic matter sedimentation in the northeast of the Paris Basin: Consequences on the deposition of the lower Toarcian black shales. Chemical Geology, 131, 15–35.

Early Jurassic siliciclastics accumulated mainly during the transgressive part of a major sedimentary cycle (Figure 4.10). The variety of facies illustrates a step-by-step progression of the subsidence with some faulting activity, relative variations of sea level and related changes of siliciclastic input which are compatible with minor sedimentary cycles. The variety of facies also highlights the major role of climate and morphology in the nature and composition of Early Jurassic sediments.

4.3.3. Middle Jurassic Carbonate Platforms of the Paris Basin

After compression led to regression and local emersion near the transition from the Early to the Middle Jurassic (Mid-Cimmerian unconformity), the transgression resumed in the Middle Jurassic. Middle Jurassic sediments extend beyond the Liassic deposits and locally rest directly on the basement, indicating that this transgression was more important than the Early Jurassic one. This is attributable to a progressing thermal subsidence, but also to higher (although fluctuating) sea level.

The lowermost transgressive sediments consist of marls which were rapidly replaced by carbonates deposited in open sea and low energy environments, before early platform sediments (ooliths, crinoids, etc.) developed in the Bajocian and massive carbonate platforms (including corals and oncoids) expanded during the Bathonian when the subsidence increased. The carbonate platforms developed in nearshore areas of the subaerial structural highs of the Armorican Massif, Massif Central and Ardennes as well as on a large and shallow ridge in the central part of the basin, between the Burgundy Platform and the Ardenne (Figure 4.12). The carbonate platforms were separated by open sea marls, which increased in importance to the north in the London Basin, and to the east.

A succession and geographical distribution of carbonate platform environments is reconstructed from the diversity of sedimentary facies observed in the central region (Figure 4.13). They include:

- open sea environments of low energy where marls, ammonitic marls and marly limestones mostly include planktonic biogenic elements together with minor bioclasts;
- fore-reef environments of medium energy where ooliths alternate with bioclastic limestones including fragments of echinoderms, bryozoans and brachiopods, and algal crusts;
- reef environments of high energy which mostly include corals, crinoids and ooliths, with some echinoderms and bryozoans;
- back-reef environments of medium energy where gravelly and bioclastic limestones containing benthic foraminifers and fragments of echinoderms alternate with ooids and oncoliths; and
- lagoonal environments of low energy which produced pelletal and sublithographic limestones with local occurrences of oncoliths.

The distribution of the facies shows that the carbonate platforms prograded over the marls during several time intervals of the Middle Jurassic. The distribution of the facies also shows that the sequence described above is rarely complete. Either the

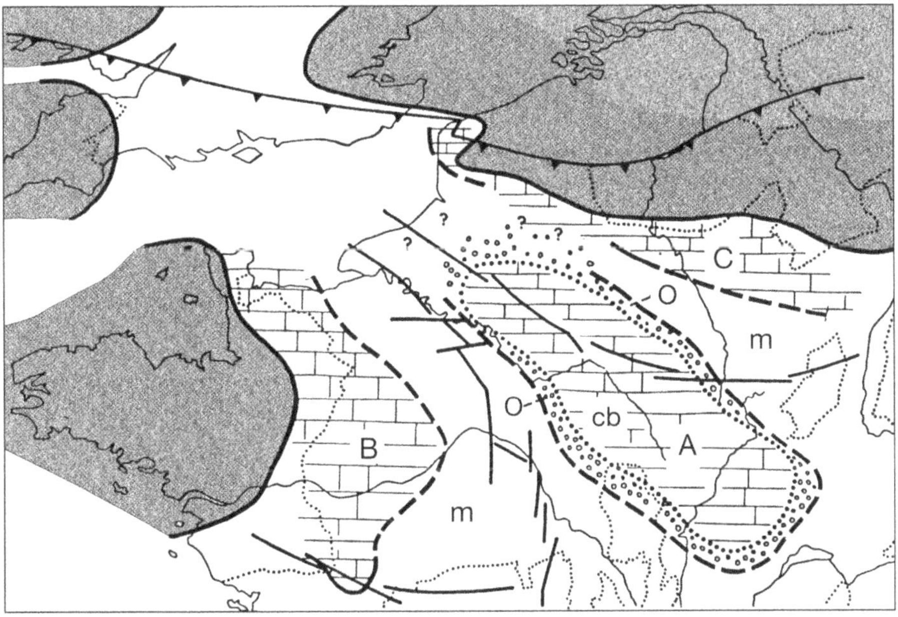

Figure 4.12 Distribution of Middle Jurassic carbonate platforms in the Paris Basin. A, central (Burgundy) Platform; B, Armorican Platform; C, Ardennes Platform; m, open sea marls; o, oolite (reef facies); cb, comblanchien (lagoonal facies). Modified from Pomerol, C., 1989. L'évolution du bassin parisien, in Dynamique et méthodes d'étude des bassins sédimentaires, Technip, Paris.

lower or upper facies are often missing and unconformities are emphasized by channelized, encrusted and fossil-rich hardgrounds.

The Middle Jurassic carbonate platform unit of the Paris Basin includes three intervals of platform development bounded by regional unconformities and is divided into three subunits (Figure 4.14):

- Subunit 1 is largely dominated by marls, indicative of significant erosion from adjacent subaerial areas during an interval of accelerated subsidence. The marls persisted during the whole Aalenian to early Bajocian interval in deeper areas, but marly limestones and minor bioclastic limestones and oolites developed locally in shallower areas at the periphery of the basin and on the central platform. This subunit is capped by an unconformity.
- The lower part of Subunit 2 is also dominated by marls indicative of increased continental erosion as subsidence resumed after an interval of compression. The marls graded to marly limestones later in the Bajocian. This was followed by the extension of bioclastic limestones and significant reef systems in the Late Bajocian. The reefs developed at the periphery of the basin and especially on the central platform (Great Oolite). This subunit is capped by an unconformity.
- The lowermost part of Subunit 3 consists of marly limestones principally with local occurrences of marls which extended to the entire basin during a further interval of increased subsidence of Early Bathonian Age. Subunit 3 is

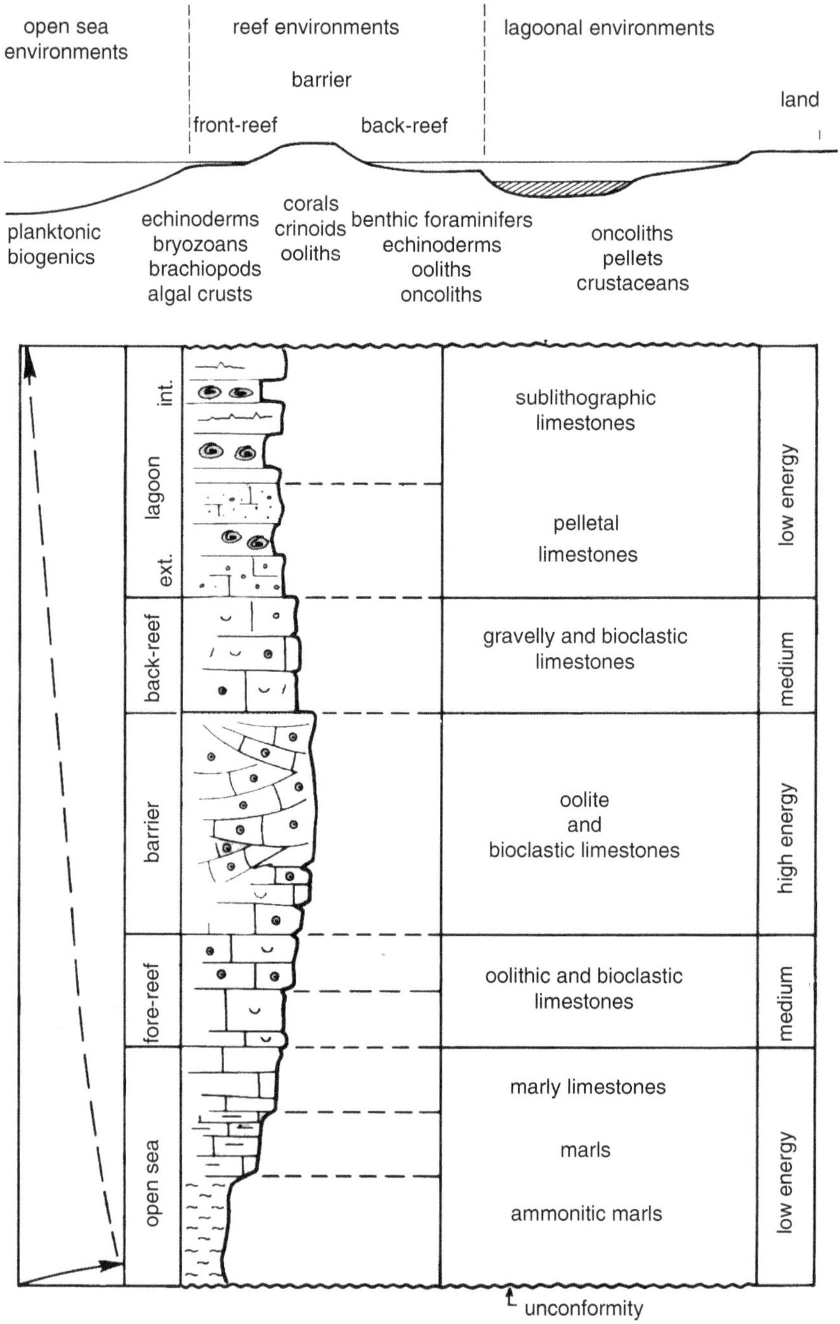

Figure 4.13 Reconstruction of Middle Jurassic carbonate platform environments of the Paris Basin, and succession of facies through time. Note that the complete succession of facies forms a virtual sequence, and that sequences are separated by unconformities. Modified from Mégnien, C. (Editor), 1980. Synthèse Géologique du Bassin de Paris, BRGM, 101. BRGM, Orléans.

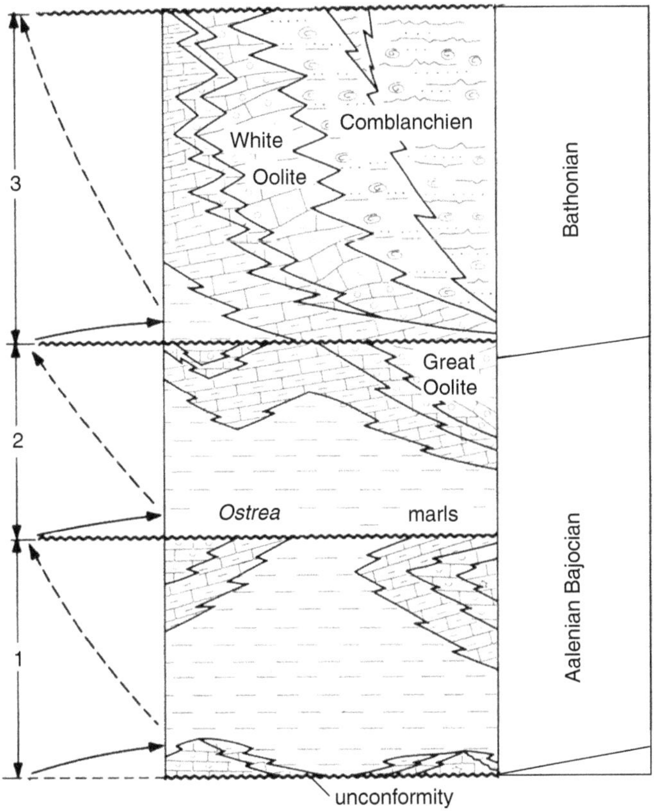

Figure 4.14 Middle Jurassic sedimentary sequences on the central (Burgundy) platform of the Paris Basin, as deduced from the succession of facies. Note that the major Middle Jurassic (Aalenian to Bathonian) cycle is divided into three main sequences, separated by unconformities. Modified from Mégnien, C. (Editor), 1980. Synthèse Géologique du Bassin de Paris, BRGM, 101. BRGM, Orléans.

characterized by a rapid progression of the reef systems, especially on the central platform which prograded to reach maximum extension in the Late Bathonian. The reef environments mostly produced ooliths (White Oolite) and isolated a lagoon which increased in size through time as the reef system prograded. The lagoonal environment produced fine and massive limestones with traces of local emersion, including frequent oncoids, and variable amounts of fine siliciclastics and pellets (Comblanchien facies). This subunit is capped by a major unconformity with traces of hardgrounds.

The Middle Jurassic sediments form a major sedimentary cycle (Figure 4.10) which is predominantly calcareous biogenic in nature, with marly facies dominating in the lower, transgressive part of the cycle and carbonate platforms dominating in the upper, regressive part of the cycle (Great Oolite, White Oolite and Comblanchien). This major cycle is divided into three sequences which all show higher amounts of fine siliciclastics (marls and marly limestones) during the transgressive stage, and variable extension of carbonate platform facies during the regressive stage. The nature and succession of the facies is driven by several factors:

- The formation of structural platforms, especially in the central part of the basin, occurred early in the Middle Jurassic as a consequence of thermal subsidence and intraplate deformation. These uplifted areas provided adequate substrates for the development of carbonate platforms.
- The extension of subaerial areas adjacent to the Paris Basin diminished during the Middle Jurassic because of a higher relative sea level. The production and accumulation of terrigenous siliciclastics, which are unfavorable to the development of reef systems, were limited as a result.
- The Paris Basin was then located in tropical areas of warm water, which are favorable to carbonate production.

The combined long-term effects of thermal subsidence, intraplate stress, sediment accumulation and eustatism maintained the Paris Basin at shallow water depth during the entire Middle Jurassic sedimentary cycle. However, the succession of facies appears mostly forced by an evolving relative sea level, and records subtle variations of eustatism and intraplate stress.

4.3.4. Late Eocene Evaporites of the Paris Basin

Because of increased compression, intraplate deformation and general uplift, which are consequences of collision events and tectonic activity at the plate boundaries of the Alps and Pyrenees, the Paris Basin has been partly continental for most of the Cenozoic. Sporadic marine incursions were limited to the central part of the basin. Most of them occurred in the Paleogene, the most important transgression being of Oligocene Age. At a million-year scale, Paleogene marine incursions in the Paris Basin show significant correlation with intervals of highest eustatic level. The sedimentary cycles are often truncated and combined with lateral variations of facies and alteration, making the Cenozoic record of the Paris Basin very complex.

One of the most characteristic succession of sediments accumulated during the Late Eocene. The lowermost part of the succession consists of shallow-marine siliciclastics of Bartonian Age, which cover a complex erosion surface of Middle Eocene Age. The transgression came from the northwest (the actual English Channel) and progressed in the central part of the basin. The siliciclastics grade to lacustrine carbonates and paleosoils (calcretes and silcretes) at the periphery of the basin. However, marine environments did not persist for a long time. The decreased importance of shallow-water siliciclastics in overlying sediments where evaporites, lacustrine carbonates and paleosoils increase, is an indication that nearshore and continental environments progressed in the basin during the Late Bartonian when anticline forms developed because of compression. This interval of intraplate deformation occurred at a time of intensified tectonic activity on the northern margins of the Tethys Ocean.

The upper part of the succession begins with the accumulation of marls which extended in the central part of the basin during a brief interval of transgression

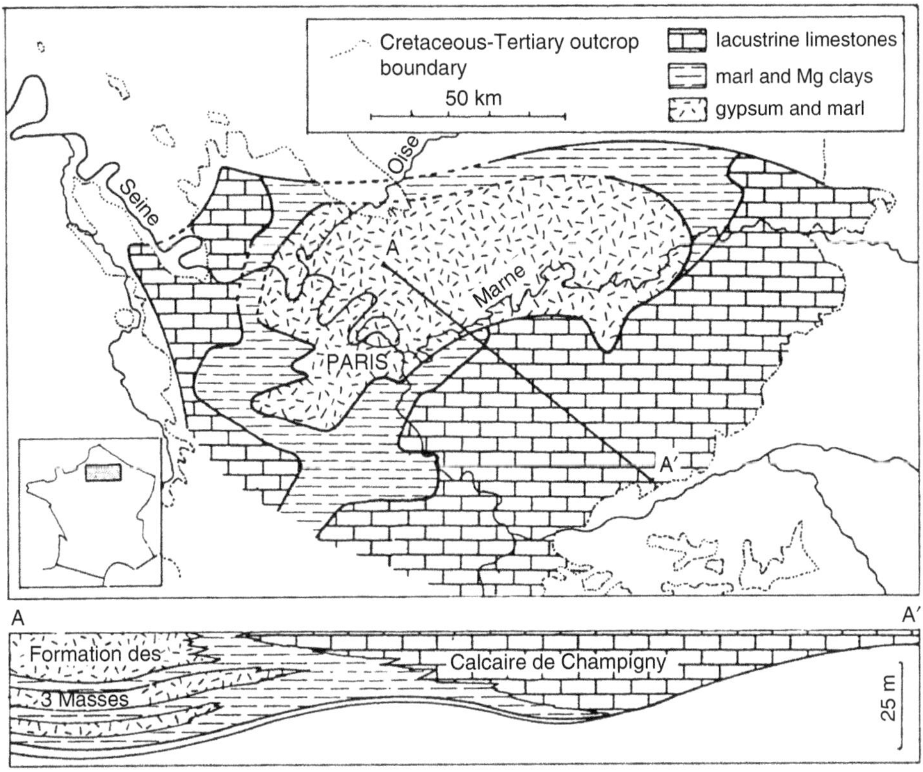

Figure 4.15 Schematic distribution and variations of Late Eocene sedimentary facies in the Paris basin. Note the progression of the lacustrine limestones through time, and the succession of evaporitic facies separated by marly intervals. Reprinted from Thiry, M., 1989. Geochemical evolution and paleoenvironments of the Eocene continental deposits in the Paris Basin. Palaeogeography, Palaeoclimatology, Palaeoecology, 70, 153–163.

of earliest Priabonian Age (Figure 4.15). The marls locally contain biological elements which are very tolerant to low salinity like the mollusk *Pholadomya ludensis*. Lacustrine carbonates persisted at the periphery of the basin, especially to the south where they contain abundant nodular and pseudobreccia intraclasts and nodular chert. Evaporitic environments developed in the central part of the basin during the Early Priabonian when the transgressive trend ceased rapidly and communication with the open sea decreased in importance.

Priabonian evaporites mostly consist of saccharoidal gypsum with abundant traces of hydrodynamism such as ripple marks and flute casts. Oxygen and sulfur isotopes highlight the importance of continental waters in the formation of gypsum. The evaporites are generally separated from the lacustrine carbonates by marls, where terrigenous clays are associated with authigenic clay minerals. Authigenic clays include smectites, palygorskite and sepiolite and their distribution is controlled by the changing fluxes of siliciclastics and chemical elements like magnesium and silica. Where the marls are absent, occurrences of dolomite mark the transition from

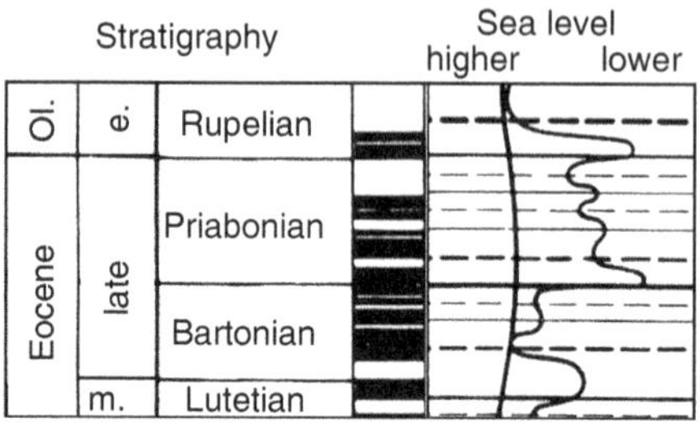

Figure 4.16 Major variations of sea level from the latest Middle Eocene to the earliest Oligocene. Modified from Einsele G., Ricken, W., Seilacher, W., (Editor), 1991. Cycles and events in Stratigraphy. Springer, Berlin.

the lacustrine carbonates to the evaporites. The Priabonian evaporitic environments developed in areas of low relief, isolated from the open sea. Chemical weathering was probably active on adjacent subaerial areas which provided chemical elements but were protected from erosion because of the presence of resistant pedogenetic silcretes. Comparison with modern areas of evaporite formation suggests that average precipitation was relatively low, below 250 mm/year.

Lacustrine environments (Calcaire de Champigny) progressed to the center of the basin through the Priabonian at the expense of the marls (Figure 4.15), probably because of poor erosion and terrigenous fluxes from adjacent subaerial areas. In contrast, evaporitic environments developed during three time intervals which produced three distinct masses of gypsum, separated by marly intervals (Formation des 3 masses). The evaporites are capped by a further interval of marls which are rather similar to the lower *Pholadomya* marls: they accumulated in a shallow environment of brackish affinity, and contain remains of terrestrial mammals. They are truncated by an unconformity of earliest Oligocene Age, with traces of emersion and local paleosoils. Overall, the Priabonian sediments of the Paris Basin show a succession of facies with transgressive (marls) and regressive (evaporites) aspects. It is not clear whether intraplate deformation, climate or eustatism exerted a major control on this succession of facies. However, it is of interest to note that the earliest Oligocene unconformity coincides with an interval of low sea level caused by the development of a transient ice sheet of continental proportion on East Antarctica, and that occurrences of evaporites may coincide with three Priabonian intervals of sea-level fall (Figure 4.16).

FURTHER READING

Association des Sédimentologistes Français, 1989. Dynamique et méthodes d'étude des bassins sédimentaires. Technip, Paris.

Boillot, G., Coulon, C., 1998. La déchirure continentale et l'ouverture océanique. Overseas Publishers Association, Amsterdam.

Boillot, G., Huchon, P., Lagabrielle, Y., 2003. Introduction à la géologie: La dynamique de la lithosphere. Dunod, Paris.

Einsele, G., 1992. Sedimentary basins, evolution, facies, and sediment budget. Springer, Berlin.

Olsen, K.H., 1995. Continental rifts: Evolution, structure, tectonics. Elsevier, Amsterdam.
Ziegler, P.A. (Editor), 1992. Geodynamics of rifting, volume 1: Case history studies on rifts, Europe and Asia. Tectonophysics, 208: 1–363.

Other references used in this chapter

Bourquin, S., Guillocheau, F., 1996. Keuper stratigraphic cycles in the Paris Basin and comparison with cycles in other Peritethyan basins (German Basin and Bresse-Jura Basin). Sedimentary Geology, 105: 159–182.
Disnar, J.R., Le Strat, P., Farjanel, G., Fikri, A., 1996. Organic matter sedimentation in the northeast of the Paris Basin: consequences on the deposition of the lower Toarcian black shales. Chemical Geology, 131: 15–35.
Guillocheau, F., Robin, C., Allemand, P., Bourquin, S., Brault, N., Dromart, G., Friedenberg, R., Garcia, J.-P., Gaulier, J.-M., Gaumet, F., Grosdoy, B., Hanot, F., Le Strat, P., Mettraux, M., Nalpas, T., Prijac, C., Rigollet, C., Serrano, O., Grandjean, G., 2000. Meso-Cenozoic geodynamic evolution of the Paris Basin: 3D stratigraphic constraints. Geodinamica Acta, 13: 189–246.
Mégnien, C. (Editor), 1980. Synthèse géologique du bassin de Paris. Mémoires BRGM, 101. BRGM, Orléans.
Thiry, M., 1989. Geochemical evolution and paleoenvironments of the Eocene continental deposits in the Paris Basin. Palaeogeography, Palaeoclimatology, Palaeoecology, 70: 153–163.

CHAPTER FIVE

Crustal Fissure Systems

5.1. Structure, Tectonics and Sedimentation of Crustal Fissure Systems

The crustal fissure stage, or proto-oceanic trough, or young ocean stage, is characterized by the production of oceanic crust which separates two passive continental margins, as a result of advanced lithosphere thinning and magmatic activity. Because of important tectonic and thermal subsidence at this stage, the structure is generally below sea level (Figure 5.1). The crustal fissure stage is not commonly observed in modern oceans, with the exceptions of the Red Sea and the Gulf of California. However, crustal fissure structures and sediments are incorporated in passive continental margins, where they are deeply buried below continental shelves.

During the crustal fissure stage, the alkaline lavas and/or transitional to tholeiitic (slightly enriched in Si) basalts of the rift stage are replaced by typical Mid-Oceanic Ridge Basalts (MORB), which are derived from the anhydrous fusion of the asthenosphere. MORB basalts often solidify as pillow-lavas on the seafloor and iron-rich minerals preserve the direction of the Earth's magnetic field as they cool below the Curie point which is around 600°C. Therefore, the earliest magnetic anomaly of the seafloor is produced during the crustal fissure stage (Figure 5.2). As sea-floor spreading progresses, the band of oceanic crust increases in size and records the inversions of the Earth's magnetic field. The transition from continental to oceanic lithosphere is often characterized by high seismic wave velocities in the lower continental crust, which are very close to those recorded in typically oceanic crusts and interpreted as indicating the presence of gabbros and dolerites. There, the rising magma crystallizes to form intrusions (dykes, sills, etc.) of gabbros and dolerites, probably because it is still blocked by the thinned continental crust of low density. Such accumulations of gabbros and dolerites underplating thinned continental lithospheres are present in areas of active and passive rifting (Figure 5.3) and may reach up to 15 km in thickness in areas of very active magmatism like the Rockall Plateau area of the North Atlantic.

In areas of active rifting where magmatic processes dominate, the transition from continental to oceanic crust is mostly known from the geophysical investigation of passive volcanic margins and generally occurs on short distances of 20–25 km. At the surface, active rifting led to the accumulation of lava flows which commonly form volcanic trapps. This, together with underplatings of gabbros and dolerites, keeps the crust relatively thick. Continental volcanics grade seaward to buried acoustic sequences characterized by dipping reflectors. These Seaward Dipping Reflector Sequences (SDRS) support magnetic anomalies and result from progressive flexing of lava flows as subsidence increased. They grade seaward to normal oceanic crust.

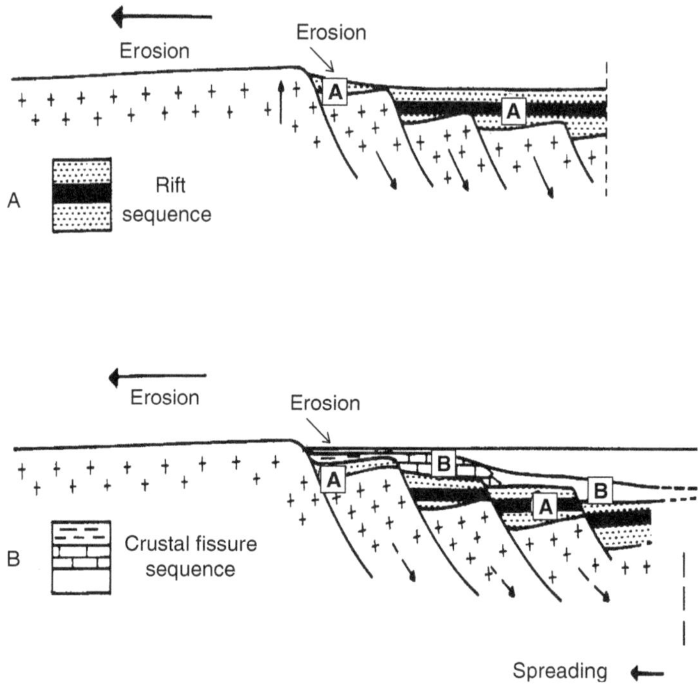

Figure 5.1 General organization of rift (A) and crustal fissure (B) sedimentary sequences. Rift sequences principally include coarse to medium siliciclastics and evaporites. Crustal fissure sequences principally include medium to fine siliciclastics, bioclastic carbonates, carbonate platforms and hemipelagic sediments. Note the subsidence of the structure below sea level during the transition from rift to crustal fissure. Modified from Boillot, G., 1983. Géologie des marges continentales, Masson, Paris.

Such volcanic series are present in the southernmost Red Sea, on the Brazil Margin of the South Atlantic and on the Greenland and conjugate European margins of the North Atlantic (Figure 5.4) where they have been extensively studied during DSDP Legs 38 and 81 and ODP Legs 104, 152 and 163 (Figure 5.5). Among other major results, the investigations have demonstrated that most lavas upwelled in subaerial environments and later subsided below sea level. Also, the SDRS sequences overlay both the thinned continental and the oldest oceanic crust, and are therefore typical of the transitional, early crustal fissure stage. The composition of the basalts shows affinities with abnormal mantle plume lavas near Iceland, and with common MORB basalts in more distal areas.

In areas of passive rifting, tectonic processes dominate and magmatic activity is generally of secondary importance. Transitions from passive rifting to crustal fissures are observed in many areas of the North and South Atlantic oceans, and have been especially investigated on the Galicia Margin of the North Atlantic during ODP Legs 103, 149 and 173. There, advanced extension and lithosphere thinning may result in the complete disparition of the ductile lower continental crust in the central part of the ancient rift, and intrusions of gabbros and dolerites may underplate the tilted blocks of fragile, upper continental crust. As extension progresses, upper mantle

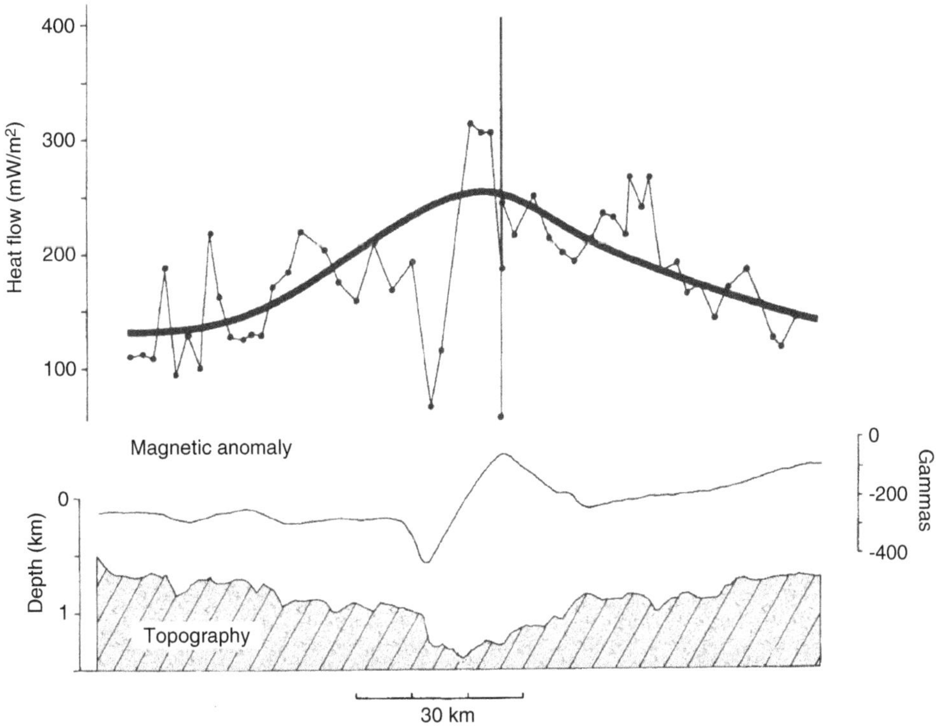

Figure 5.2 Bathymetric, magnetic and heat flow profiles across the Conrad deep, northern Red Sea. The axial trough, where oceanic crust is being produced, coincides with maximum heat flow and the first magnetic anomaly. Modified from Cochran, J.R., Martinez, F., 1988. Evidence from the northern Red Sea on the transition from continental to oceanic rifting. Tectonophysics, 153, 25–53.

peridotites may reach the seafloor, at the transition between the tilted blocks of continental crust and the oceanic crust (Figure 5.6). Therefore, passive rifting may lead to the tectonic exhumation of terranes previously located below the thick continental crust. Because of persisting extension of the lithosphere, shear zones may develop and the gabbros and peridotites are often brecciated and faulted (Figure 5.7). This favors the hydrothermal circulation of seawater, and in turn the metamorphose of gabbros in the greenschist facies and the serpentinization of the upper peridotites, at low temperature ($<600°C$) and shallow depth (<7 km). Such terranes are not associated to magnetic anomalies, commonly extend over distances of several tens of kilometers, and may constitute most of the magnetic quiet zones of the passive continental margins. Magnetic anomalies appear only locally, where the peridotites are covered with basalts because of increased magmatic activity. Persistent stretching of the lithospheric mantle favors the uplift of the asthenosphere, adiabatic melting and the production of oceanic crust (MORB basalts). These processes are especially characteristic of areas of very slow extension.

In areas of fast passive rifting where uplift and adiabatic melting of the asthenospheric mantle take place more rapidly, the exhumation of the lithospheric

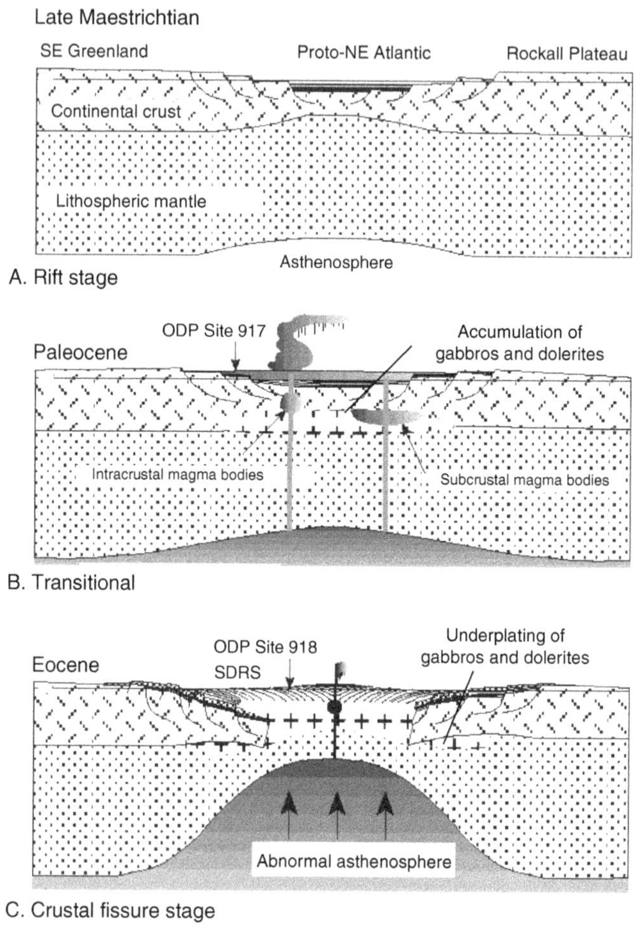

Figure 5.3 Accumulation of gabbros and dolerites underplating thinned continental crust during the transition from rift to crustal fissure. Example from the North Atlantic. SDRS: seaward-dipping reflectors sequences. Modified from Duncan, R.A., Larsen, H.C., Allan, J.F., Brooks, K. (Editors) 1999. Proceedings of the Ocean Drilling Program, Scientific Results, 163. Ocean Drilling Program, College Station, TX.

mantle is of minor importance or even does not occur. There, intrusions of gabbros and dolerites increase in density as the thinning of the lithosphere progresses, and oceanic crust is produced when basaltic lavas upwell and accumulate above mafic intrusions. The newly formed oceanic crust is composed of gabbros in deeper areas of the crust and MORB basalts at the surface. In those areas, the transition from continental to oceanic crust is relatively abrupt. It is probable that the oceanic crust first shows on the floor of axial deeps which also contain metal-rich hot brines and muds, and increases in density through time. The new oceanic crust rapidly expands to form a straight and almost continuous axial band which is only interrupted locally by minor offsets. This type of transition is observed in the Red Sea, where a

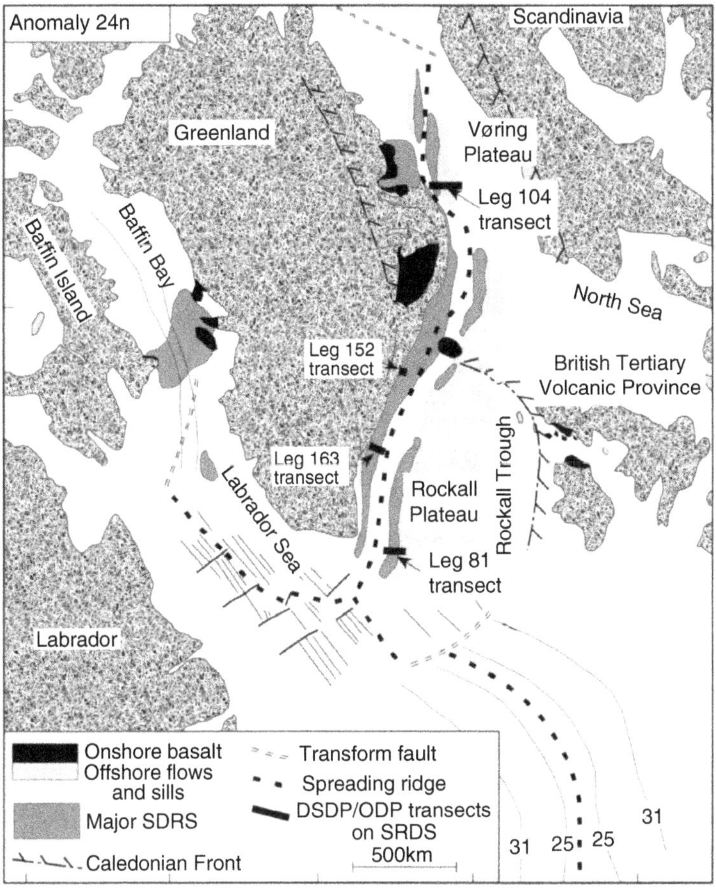

Figure 5.4 Paleoreconstruction of the North Atlantic at the time of Anomaly 24, and location of the ODP Legs 152 and 163 transect shown on Figure 5.5. Note the extension of the North Atlantic Large Igneous Province of active rifting, and related seaward dipping reflectors sequences. Reprinted from Duncan, R.A., Larsen, H.C., Allan, J.F., Brooks, K. (Editors) 1999. Proceedings of the Ocean Drilling Program, Scientific Results, 163. Ocean Drilling Program, College Station, TX.

continuous band of oceanic crust as old as 5 Ma is replaced to the north by a succession of deeps of oceanic crust not older than 2 Ma, suggesting that a northward propagation of sea-floor spreading progressively separates the African and Arabian continental margins (Figure 5.8). However, in some areas a significant volume of basaltic lava may erupt and flow landward over syn-rift deposits: in these transition zones, magnetic anomalies may extend over areas of thinned continental crust, complicating the identification of genuine oceanic crust. This type of transition is observed in the Southern Ocean off the Great Australian Bight, where the commencement of seafloor spreading (anomaly 33, 83 Ma) could be about 15 Myr younger than previously thought (anomaly 34).

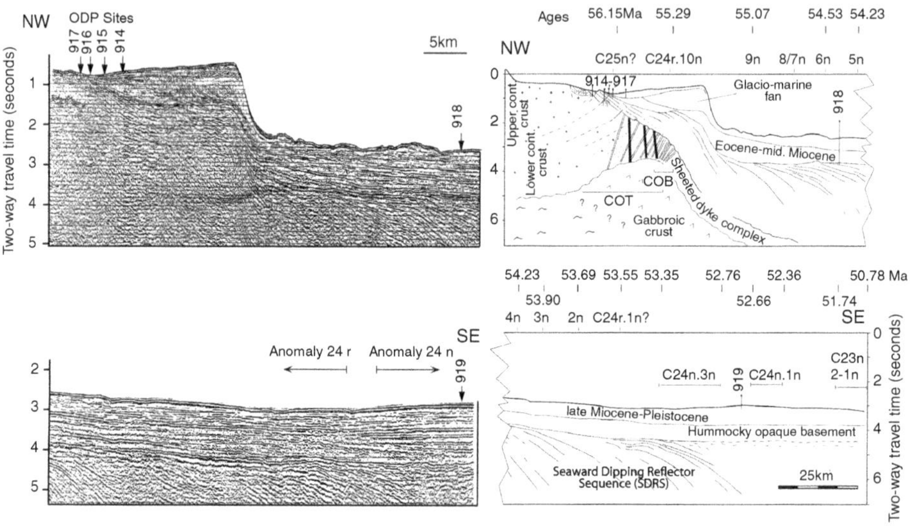

Figure 5.5 Seismic transect across the southeast Greenland margin, and its interpretation. Note the geographic extension of seaward dipping reflectors, the transition to oceanic crust supporting magnetic anomaly 24, and the underplating of thinned continental crust. COB: Continent-Ocean Boundary; COT: Continent-Ocean Transition. Modified from Duncan, R.A., Larsen, H.C., Allan, J.F., Brooks, K., 1999. Proceedings of the Ocean Drilling Program, Scientific Results, 163. Ocean Drilling Program, College Station, TX.

Active and passive continental rifting may ultimately lead to the formation of oceanic crust via a transition zone of variable importance. Similar processes lead to the formation of oceanic crust in pull-apart basins, where the newly formed oceanic crust and the continental crust are in contact along transform segments (Figure 5.9). In these areas, the oceanic crust fills a succession of small basins where hot brines may accumulate, and which increase in size through time. Because of differences in the physical properties of both types of crust, the thermal history of transform areas is relatively complex. The continental crust adjacent to areas of maximum heat flow (where oceanic crust is produced) is uplifted, and subsides later as pull-apart basins increase in size. However, thermal exchanges between the newly formed oceanic crust and the continental crust are limited: the importance of uplift and deformation decreases by comparison with the rift stage, and the thermal alteration of the sediments is moderate.

The morphology of crustal fissures is very close to those of the rifts, and the rift shoulders persist on both sides of the structure (Figure 5.1). The mechanisms responsible for crustal thinning decrease in importance but continue, and the laws of tectonic (see Section 3.1) and thermal (see Section 2.1.3) subsidence apply. Water depths around 2,000 m are recorded in the central deeps of the northern Red Sea and 2,500 m in the deep axial zone of the southern Red Sea. Also, paleodepths of 2,000–3,000 m have been estimated from benthic microfossil assemblages for the magnetic quiet zone (peridotites) of the Galicia Margin during the crustal fissure stage. These water depths are in good agreement with the theoretical subsidence

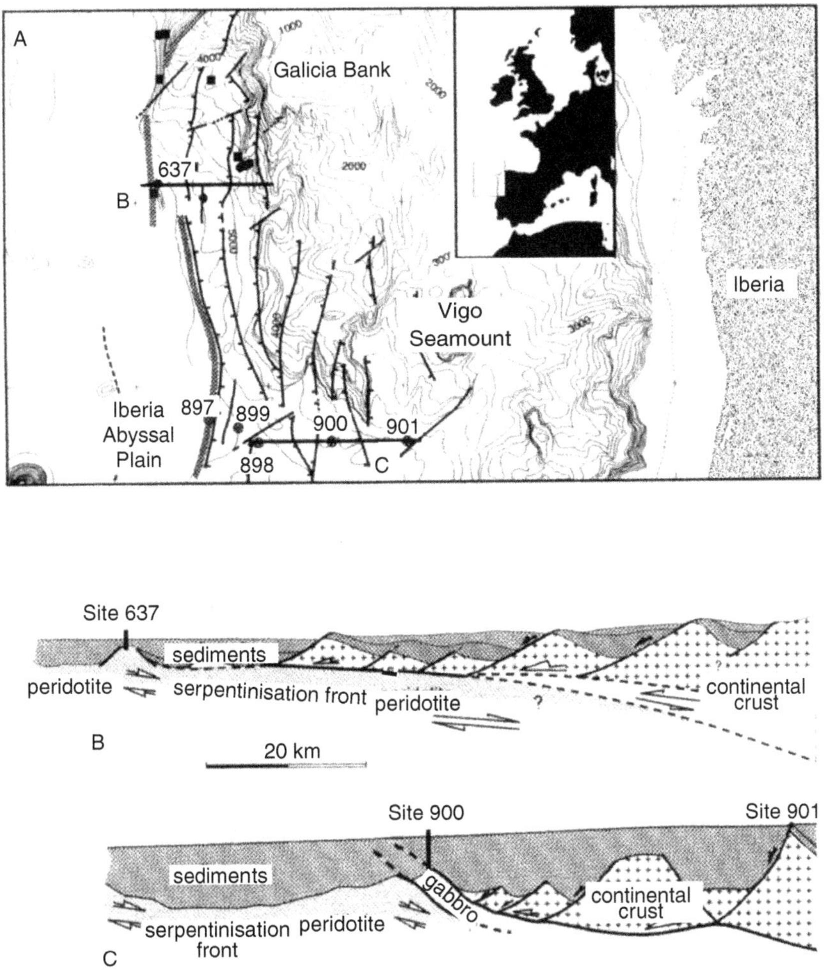

Figure 5.6 Consequences of lithosphere thinning across the Galicia margin of the North Atlantic, as interpreted from seismic profiles and the results of ODP Legs 103 and 149. Note the decreased importance of tilted blocks of continental crust and exhumation of altered upper mantle rocks westward, and the presence of underplated gabbros on the southern Galicia margin. a: location map of transects b and c. Reprinted from Brun, J.P., Beslier, M.-O., 1996. Mantle exhumation at passive margins. Earth and Planetary Science Letters, 142, 161–173.

curves and with the value of 2,500 m which corresponds to the average depth of mid-oceanic ridges where oceanic crust is produced.

Because of the persistence of lithosphere thinning, significant erosion of tectonically rejuvenated continental relief generates important quantities of terrigenous sediments. The rift shoulders move away from the active areas of high heat flow as extension and spreading progress, and decrease in elevation because of thermal subsidence and lower tectonism. As a consequence continental relief and

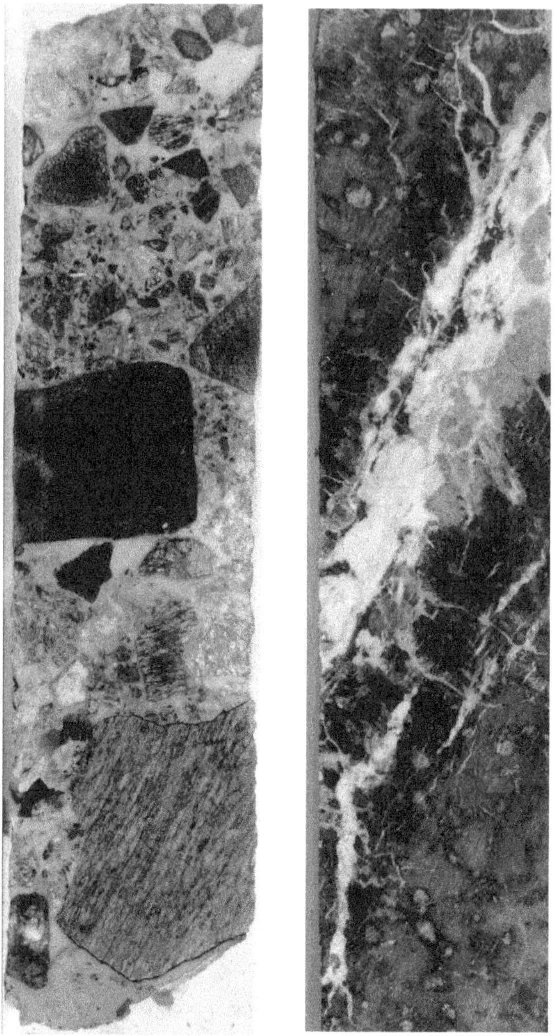

Figure 5.7 Altered gabbro within a peridotite (left) and breccia including elements of gabbro and amphibolite principally, in a calcareous chalk matrix (right). Modified from Whitmarsh, R.B., Beslier, M.-O.,Wallace, P.J., 1998. Proceedings of the Ocean Drilling Program, Initial Reports, 173. Ocean Drilling Program, College Station, TX.

erosion decrease in importance, providing smaller quantities of terrigenous sediment particles of smaller grain size to the basins where the relative importance of particles from other sources, mainly marine biological activity, increases in the sediment. Terrigenous sediments accumulate in coastal areas especially near river mouths while biogenic sediments may dominate in areas of low terrigenous supply, including coral reef formations where climatic and oceanographic conditions are

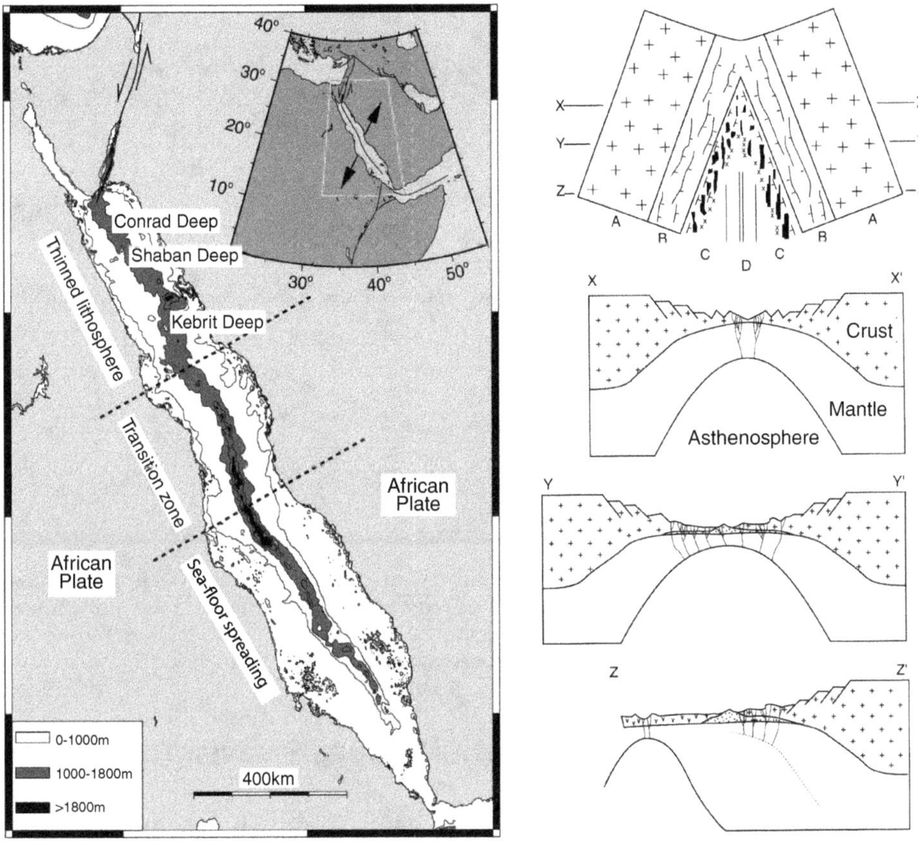

Figure 5.8 Propagation of ocean opening. Left: Division of the Red Sea in three areas of different crustal evolution. Right: Conceptual model for the propagation of ocean opening (top) and theoretical sections across the area of thinned lithosphere (x–x′), the transition zone (y–y′) and the area of seafloor spreading (z–z′). (A) Continental crust; (B) Thinned continental crust; (C) Intermediate crust with intrusions of gabbros and dolerites; (D) Oceanic crust. Modified from Erhardt, A., Hübscher, C., Gajewski, D., 2005. Conrad deep, northern Red Sea: Development of an early stage ocean deep within the axial depression. Tectonophysics, 411, 19–40 and from Larsen, H.C., Saunders, A.D., Wise, S.W., 1998. Proceedings of the Ocean Drilling Program, Scientific Results, 152. Ocean Drilling Program, College Station, TX.

favorable. Because of unstable slope conditions sometimes facilitated by subsidence and tectonism, most terrigenous deposits are rapidly reworked to deeper areas of the basins as gravity flow deposits and turbidites. In more distal areas they mix to calcareous and/or siliceous biogenic particles of planktonic origin principally, to form hemipelagic sediments. The crustal fissure being narrow (Figures 5.4 and 5.8) with significant relief of the seafloor the ventilation of the basins is rather poor, and low oxygen contents locally allow the preservation of organic matter in the sediment. The association of hemipelagic sediments and organic components characterizes the black-shale facies.

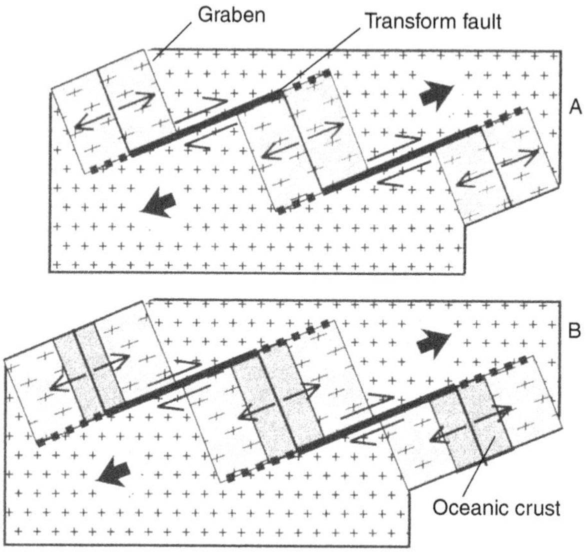

Figure 5.9 Transition from lithosphere thinning (A) to sea-floor spreading (B) in a pull-apart context. Note the sharp contact between oceanic crust and thick continental crust along transform faults. Modified from Boillot, G., Coulon C., 1998. La déchirure continentale et l'ouverture océanique, Overseas Publishers Association, Amsterdam.

5.2. Case Study of a Crustal Fissure System: The Red Sea

The Red Sea is delimited by the Aqaba transform fault to the North, and the Afar/Gulf of Aden triple junction and transform area to the south (Figure 5.10). Like the Gulf of Suez, the Red Sea results from passive rifting associated to tensional stress within the Arabian–African Plate.

5.2.1. Structure, Morphology and Early History of the Red Sea

The initial rifting started during the Oligocene (around 30 Ma) to the South and propagated northwards until the Early Miocene (Aquitanian, 25–20 Ma), according to a rotation pole located in the Eastern Mediterranean. Therefore the degree of structural development of the Red Sea varies regionally, being more advanced to the South, and is used to differentiate three main regions (Figure 5.8):

- A southern region characterized by a continuous, deep (around 2,500 m) axial zone of oceanic crust which shows magnetic anomalies indicating that seafloor spreading has been continuously active there since at least 5 Ma. The oldest seafloor anomaly is separated from the thinned continental crust by a band of intermediate crust heavily intruded by mafic plutons and overlain by lava flows. The narrow (less than 80 km), deep axial zone is flanked by broad shallow areas inherited from the ancient rift, which consist of thick syn-rift sediments overlying the intermediate and continental crusts.

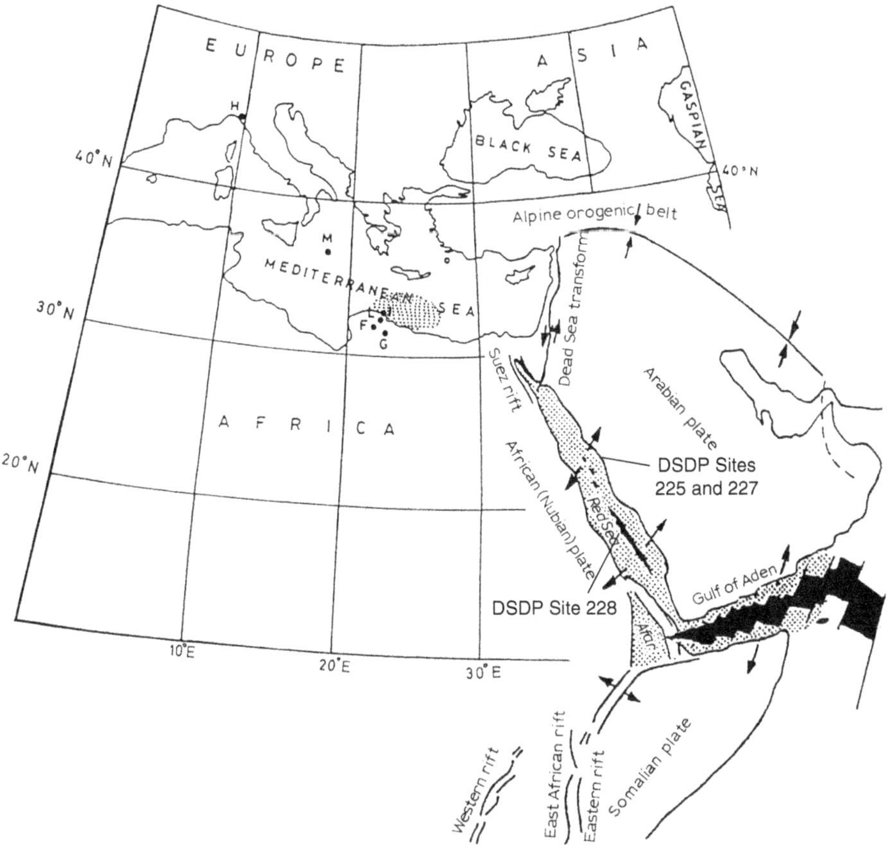

Figure 5.10 The Red Sea in a regional plate tectonic context. Letters indicate the successive poles of rotation of the Arabian plate relative to the African plate. Reprinted from Le Pichon, X., Cochran, J.R. (Editors), 1988. International workshop on the Gulf of Suez and Red Sea rifting. Tectonophysics, 153.

- An intermediate region of relatively shallow depth, where the continuous deep axial zone is replaced by a complex series of axial deeps (around 2,000 m) which decrease in extension and in quantity northwards (Figure 5.11). The floor of the deeps consists of oceanic crust not older than about 2 Ma, is generally overlain by hot brine pools (temperature up to 60°C, salinity up to 300‰) and displays punctual "black smokers". Helium and argon isotope signatures of the brines are typical of active hydrothermal circulation systems and clearly indicate a significant contribution from the mantle. Hydrothermal deposits include sulfides (pyrite, zincite, etc.), sulfates (anhydrite), authigenic carbonates and especially Fe and Mn oxides (for example ferrihydrite and lepidocrocite, which eventually transformed diagenetically into goethite and hematite). Bacterial activity may play a significant role in the formation of some minerals, like authigenic carbonates and pyrite. Hydrothermal deposits are strongly influenced by the chemistry of adjacent evaporitic sediments (anhydrite, gypsum, halite, etc.).

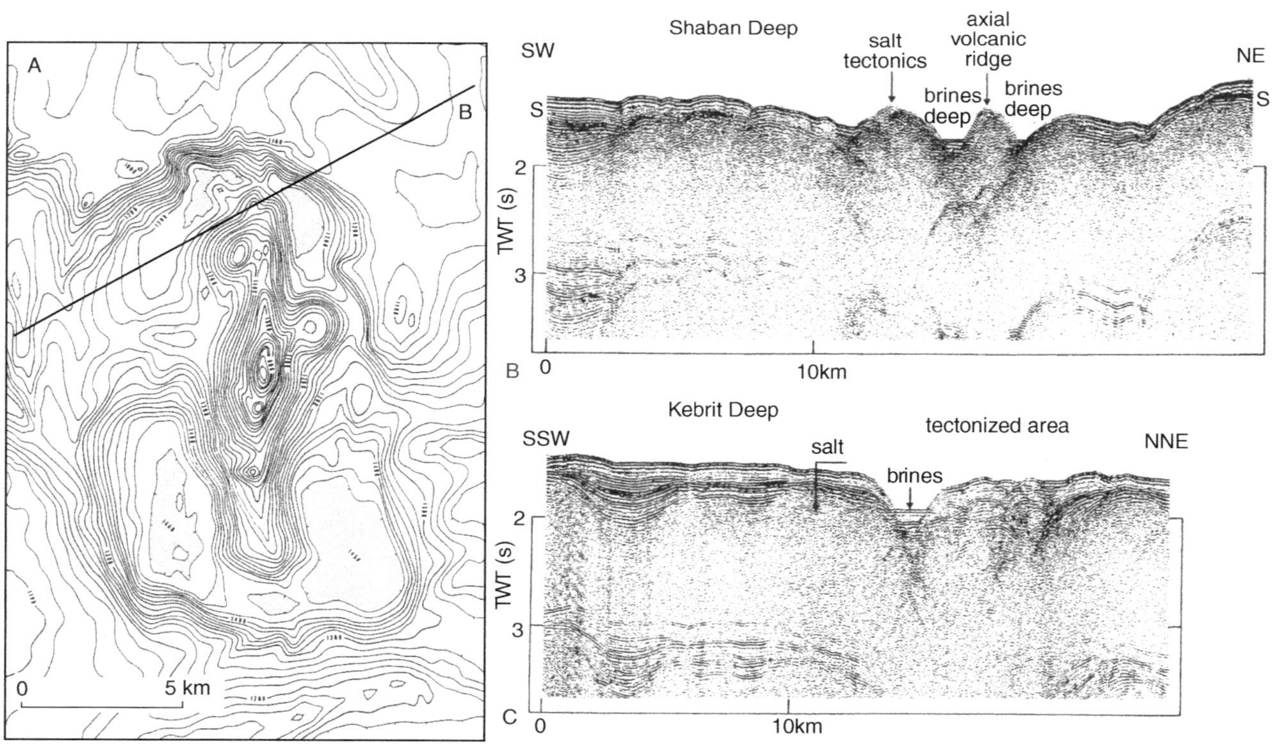

Figure 5.11 Red Sea Deeps. (A) bathymetry of the Shaban Deep (deepest areas are in gray); (B) seismic transect across the Shaban Deep (reflector S represents the top of the evaporites); (C) seismic transect across the Kebrit Deep. See Figure 5.8 for location of Shaban and Kebrit Deeps. Modified from Guennoc, P., Pautot, G., Coutelle, A. 1988. Surficial structures of the northern Red Sea axial valley from 23° N to 28°N: time and space evolution of neo-oceanic structures. Tectonophysics, 153, 1–23 and from Coutelle, A., Pautot, G., Guennoc, P., 1991. The structural setting of the Red Sea axial valley and deeps: implications for crustal thinning processes. Tectonophysics, 198, 395–410.

- A northern region characterized by a subsident but shallow, thinned continental crust attaining maximum water depths around 1,000 m. Thinning processes are here at an advanced stage, since lower continental crust terranes (gneiss, amphibolite, etc.) heavily intruded by gabbros and diabases locally outcrop in the axial zone (Zabargad Island). Deeper water depths of 1,300–1,500 m are only reached in rare, active deeps. Detailed observations of the Conrad Deep, one of the northernmost and most recent (75 Kyr) Red Sea deeps, show an elongated structure oblique to the main extension direction, most probably generated by transtensional processes associated to the emplacement of magmatic bodies nearby. However, investigations of helium and argon isotopes in the Kebrit Deep brines of the same region reveal only a minor contribution of the mantle and suggest a migration from deeper sedimentary or crustal horizons.

It is likely that the Red Sea deeps serve as nucleation points that develop into discrete cells of sea-floor spreading which further grow and coalesce to form a continuous mid-ocean spreading center as extension progresses. Opening processes are therefore more advanced to the South, where extension along the Arabian detachment fault started between 30 and 25 Ma (Figure 5.12). The phase of pure passive rifting lasted for about 6 Myr and produced a divergence of about 20 km before adiabatic decompression generated a rising magma which progressively invaded a relatively thick (20–25 km) but extended continental lithosphere. The magma crystallized as gabbros and diorites, and eventually formed a wide melted plume beneath the rift axis. This early magmatic phase lasted for about 8 Myr and produced a divergence of about 30 km. Intensified activity between 15 and 10 Ma caused a diffuse magmatism and spreading with formation of underplated plutons and probable development of axial deeps. This phase rapidly led to a divergence of about 150 km. Because of increased shear activity along the Aqaba Fault in the latest Miocene and earliest Pliocene, extension strongly accelerated in the Red Sea and highly localized magmatic processes led to the formation of a continuous band of oceanic crust in the axial trough since 7–5 Ma, spreading rates being of about 1.5 cm/year.

5.2.2. Sediments of the Red Sea

Post-evaporite, Pliocene and Pleistocene sediments of the Red Sea accumulated during the crustal fissure stage. They are coeval with the Group D sediments of the Gulf of Suez Rift (see Section 3.4), but major differences exist between both regions. About 1,500 m of Group D sediments amassed in the marine environments of the subsiding central grabens of the Gulf of Suez, whereas a few hundreds of meters of post-evaporite deposits accumulated on an average in the Red Sea. Also, Group D sediments from the Gulf of Suez are essentially terrigenous, whereas post-evaporite sediments from the Red Sea contain a significant, and sometimes dominant, proportion of biogenic components. Lower accumulation rates and proportions of siliciclastic components in the Red Sea probably result from decreased tectonism and erosion in a globally arid context, and dispersion of the

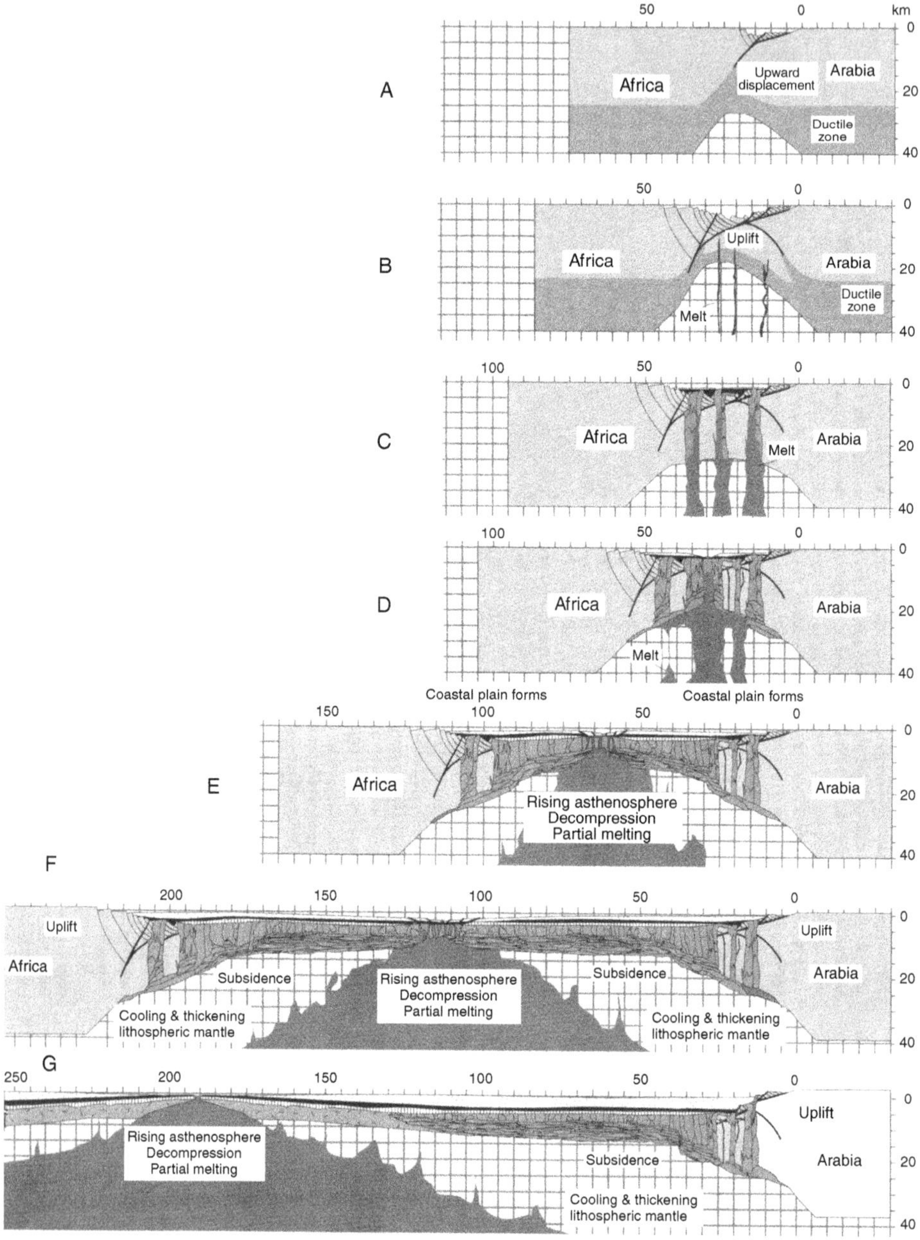

Figure 5.12 History of ocean opening in the southern part of the Red Sea. (A and B) passive rifting phase (Late Oligocene to Early Miocene); (C and D) early magmatic phase (Early Miocene to Late Miocene); (E, F and G) spreading phase (Late Miocene to present). Coordinates are in kilometer. Note progressive extension of the structure and related evolution of the lithosphere and formation of continental margins. Modified from Bohannon, R.G., Eittrem, S.L., 1991. Tectonic development of passive continental margins of the southern and central Red Sea with a comparison to Wilkes Land , Antarctica. Tectonophysics, 198, 129–154.

particles in an extending marine basin (it is noteworthy that accumulation rates significantly increase in the southernmost part of the Red Sea, adjacent to the tectonically active Afar and narrow Bab-el-Mandeb Seaway). In addition, an increasing quantity of biogenic particles was produced in an extending marine environment.

However, terrigenous and biogenic sediments are unevenly distributed in the Red Sea, especially in coastal areas. The uppermost evaporites grade to fine marine siliciclastics intercalated with carbonate muds in subsident areas and especially to the south. The transition occurs via brecciated evaporites and traces of emersion in uplifted areas and especially to the north. These early deposits contain mollusks indicative of restricted conditions and are overlain by open marine sediments including coarse siliciclastics, calcarenites, oolites and patch-reefs. Coarse siliciclastics principally include sand and channelized conglomerates (with lenses of evaporites), which accumulated near river mouths. Marine carbonates accumulated in areas protected from continental discharges, and include species of Indo-Pacific affinity, indicating that the Red Sea became connected to the Gulf of Aden via the Bab-el-Mandeb at that time.

During DSDP Leg 23B, post-evaporite sediments were investigated in the axial zone of the Red Sea: two holes (Sites 225 and 227) were drilled in the intermediate region and one hole (Site 228) was drilled in the southern region (Figure 5.10). The sedimentary series is rather identical in nature in both regions. The uppermost evaporites are associated with stromatolites, oncolites, nodular anhydrite, and provide evidence for a sabkha to very shallow water origin. They are intercalated with siltstones and dolomitic black shales which probably deposited during intervals of brine dilution. The overlying dark gray, organic-rich, dolomitic silty claystone unit (Group D, Unit 3) of latest Miocene to Early Pliocene Age (Figure 5.13) is sometimes in continuity with the evaporites (DSDP Site 225), indicating a gradual change to more open conditions. In other places (DSDP Site 228), the terrigenous unit is separated from the evaporites by an unconformity. Sedimentation rates are highly variable, ranging from about 20 to about 230 m/Myr. Lowest values have been recorded on the eastern edge of the axial zone at DSDP Site 225 whereas highest values have been recorded on the western flank of the axial deep at DSDP Site 228, drilled in close proximity to the Sudanese delta fan. This delta fan is in continuity with an extensive drainage basin which has been active during Pliocene and Pleistocene intervals of accelerated rainfall, providing significant quantities of siliciclastics to the ocean. The dolomite is highly variable in abundance (10%–80%), highest contents being recorded in the deepest part of the axial zone at DSDP Site 227. Higher concentrations of the mineral are associated to relatively coarse beds of silt and sand, suggestive of increased erosion. The dolomite most probably results from early diagenetic alteration of biogenic carbonates within sediments of relatively high porosity and permeability, which accumulated in a poorly oxygenated hypersaline environment of high Mg/Ca ratio.

Beginning in the Early Pliocene at the deeper sites 225 and 227 and in the Late Pliocene at the shallower site 228, the sediment progressively grades to gray calcareous silty claystone to siltstone (Group D, Unit 2). The siliciclastics (mostly

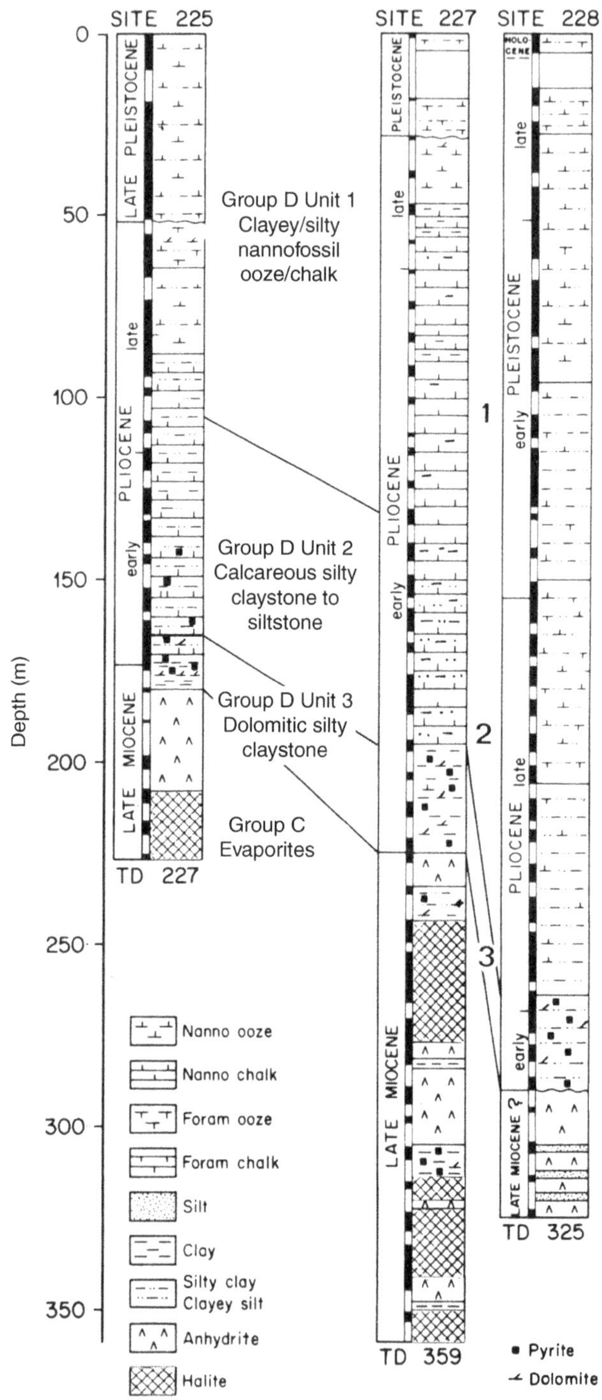

Figure 5.13 Lithostratigraphic summary of DSDP Sites 225, 227 and 228 (Red Sea). Note increase in calcareous biogenics upsection, and increased sedimentation rates at DSDP Site 228 drilled near the Sudanese delta fan. Modified from Whitmarsh, R.B., Weser, O.E., Ross D.E. et al., 1974. Initial Reports of the Deep Sea Drilling Project, volume 23. U.S. Government Printing Office, Washington.

feldspars, but also quartz, micas and clays) largely dominate the sediment (up to 80% at DSDP Site 228). Calcareous components mostly include coccoliths (nannofossils), but also carbonate fragments of unknown origin. Sedimentation rates are roughly similar to those of the underlying unit, and the sediment overall suggests that open marine conditions were firmly established. The proportion of calcareous biogenic components gradually increases from the latest Early Pliocene and the sediment grades to clayey/silty nannofossil ooze and chalk (Group D, Unit 1). Locally the sediment contains significant abundances of foraminifers and pteropods (up to 50%) in probable relation with an increased productivity. The siliciclastics are similar in nature to those of the underlying units, but their abundance decreases below 50%. However, the respective influence of tectonism and climate in the decreased abundance of siliciclastics is not clearly understood, and has to be further investigated.

An important aspect of open marine terrigenous and biogenic sediments is the occurrence of dark layer intervals characterized by high carbon and pyrite contents, indicative of intermittent return to a poorly ventilated environment. The most recent black-shale (or sapropel) deposit accumulated during the last deglaciation and followed an interval of very high salinity (above 50‰) which correlates with the Last Glacial Maximum. Hypersaline conditions led to the disparition of most marine species (especially planktonic forms) and to the precipitation of lithified layers of high aragonite contents. It is probable that severely limited communication with the Indian Ocean via the shallow Bab-el-Mandeb (Figure 5.10), together with an increase in evaporation during an interval of glacial sea level lowering and regional aridity (stronger winter monsoon), led to significant concentration of the Red Sea waters. The end of the hypersaline interval correlates with the deposition of organic-rich sediments in a stagnant environment. The organic matter is mainly derived from terrigenous sources, with only minor contribution from marine sources. The stagnation was probably density driven, triggered by an improved communication with the Indian Ocean as sea level rose after the glacial peak, and an input of freshwater as regional rainfall increased during the deglaciation (stronger summer monsoon). It is possible that significant, relative variations of sea level of eustatic or tectonic origin, associated to variations of the monsoonal regime, have generated most of the black-shale intervals recorded in post-evaporite sediments of the Red Sea.

Gravitational processes are important in the modern Red Sea where they principally occur in the axial trough. They predominantly consist of pelagic carbonate turbidites, with local occurrence of volcaniclastic turbidites. These gravity flow deposits are likely to have been triggered by frequent earthquakes in the axial trough area.

Overall, the evolution of post-evaporite sediments in the Red Sea reflects an increased influence of the marine environment (and correlative decreased influence from continental areas) as the oceanic structure expands and subsides. Evaporitic sediments do not develop anymore and siliciclastic components decrease in size and in abundance. By the same time pelagic productivity develops, and marine biogenic components increase as to dominate the sediment. This global scheme is disrupted by climatic factors such as eustatic changes in sea level which may significantly alter the hydrology of the Red Sea and the monsoonal regime which drives regional precipitation, generating unusual sediment types. However, the impact of climatic factors on the sedimentation is accentuated by some characteristics of crustal fissures, which are elongated, relatively steep and narrow structures where the circulation is constrained by the morphology, and hydrological exchanges limited by the presence of sills and straits.

5.3. Example of Crustal Fissure in a Pull-Apart Context: The Gulf of California

5.3.1. Structure and History of the Gulf of California

Divergence of slab-pull stresses at the west American subduction zones in the Late Oligocene and the Early Miocene led to the fragmentation of the ancient Farallon Plate of the Pacific Ocean in a number of smaller plates. Newly formed subducting microplates were progressively captured by the Pacific Plate as the movement changed to transtensional transform motion, initiating the future San Andreas transform fault system which progressed from North to South. Subduction off Baja California persisted in the Late Miocene till about 12 Ma when transtensional movements started onshore, in coincidence with the emplacement of a triple junction at the mouth of the modern Gulf of California (Figure 5.14). The dextral displacement was relatively minor (a few tens of kilometer) between 12 and 6.3 Ma and was accommodated principally offshore, on the western Baja California Margin. The tensional stress was principally accommodated onshore, generating the Gulf Extensional Province which extended far to the East and consists of N–S oriented grabens of basin and range style (Figure 5.15). The strike-slip motion moved to the East into the modern gulf between 6.3 and 4.7 Ma, when Baja California was captured by the Pacific Plate. Correlation of volcanic tuffs across the gulf indicates that more than 250 km of dextral displacement occurred since 4.7 Ma, at an average rate of 5 cm/year. Transtensional activity led to the formation of pull-apart basins (which mostly consist of half grabens) in the Gulf of California and beyond, in Southern California. Because of regional differences in the intensity of tectonism and subsidence, the sedimentary basins are deeper than 2,000 m but contain less than 1,000 m of sediment in the southern part of the gulf, while they are shallower than 900 m and contain more than 4,000 m of sediments in the northern part of the gulf. Oceanic crust has been identified in some basins but is not older than 3.2 Ma (Middle Pliocene) off the southern tip of the peninsula (Figure 5.16) where about 220 km of oceanic floor has been produced, suggesting average spreading rates around 6 cm/year. The floor of pull-apart basins also consists of oceanic crust in the southern and central parts of the gulf up to the Guaymas Basin, where Pleistocene ages not older than 1.2 Ma have been found.

The reorganization of Pacific lithospheric plates was probably associated with intensified volcanic activity, as lava flows and breccias of Miocene age, interbedded with volcanic sandstones and tephras, overlay a Cretaceous basement in the central part of the Gulf of California. There, the oldest rift deposits consist of continental

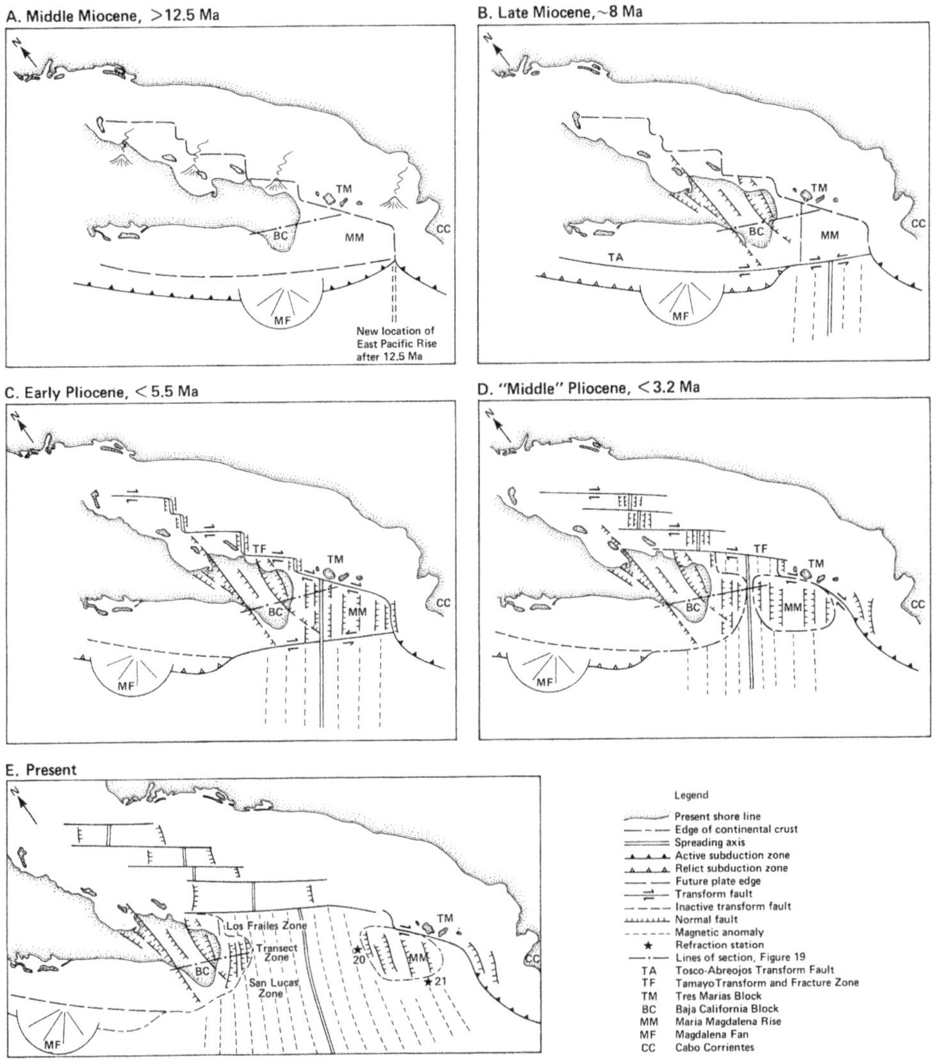

Figure 5.14 Major steps in the Late Neogene evolution of plate boundaries in the Gulf of California area. (A) Subduction of the Cocos plate below North America, prior to 12 Ma; (B) moderate strike-slip activity off Baja California, 12–6.3 Ma; (C) transfer of strike-slip activity within the Gulf Extensional Province to the Northeast, 6.3–4.7 Ma; (D) intensified strike-slip activity and formation of pull-apart basins in the Gulf of California, and seafloor spreading off the southern tip of Baja California, after 4.7 Ma and (E) modern setting. Reprinted from Curray, J.R., Moore, D.G. et al., 1982. Initial Reports of the Deep Sea Drilling Project, volume 64. U.S. Government Printing Office, Washington.

breccias of Late Miocene age (around 9 Ma), uncomformably emplaced on the volcanics. The breccia has been succeeded by ash-flow deposits emplaced in continental to nearshore settings at about 5.7 Ma in the latest Miocene. These initial deposits are overlain by continental and shallow marine conglomerates and

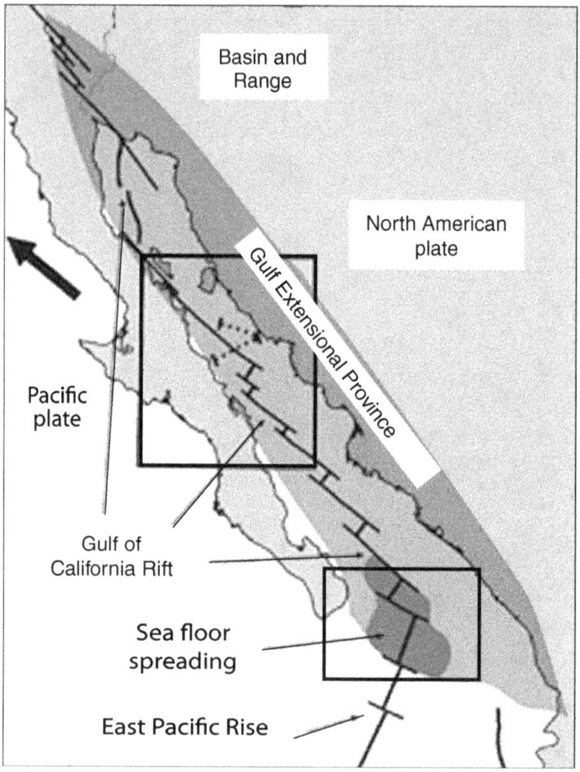

Figure 5.15 Modern setting of the Gulf of California, with approximative location of the initial Gulf Extensional Province and Gulf of California Rift system. NA: North American plate; PA: Pacific plate. Squares represent the areas detailed in the text. Modified from Aragon-Arreola, M., Morandi, M., Martin-Barajas, A., Delgado-Argote, L., Gonzalez-Fernandez, A., 2005. Structure of the rift basins in the central Gulf of California: Kinematic implications for oblique rifting. Tectonophysics, 409, 19–38.

fossiliferous coarse-grained sandstones, which correspond to delta fan deposits, and contain foraminifers indicating Late Miocene to Early Pliocene ages between 6.4 and 4.0 Ma. However, sparse outcrops of marine deposits in the southernmost part of the gulf and near its mouth yielded ages between 8.2 and 7.5 Ma. These data support an early marine incursion from the Pacific Ocean covering a wide area near the modern gateway to the gulf and further extension to the central and northern parts of the gulf at about 6.5 Ma. The transgression occurred in probable relation with an accelerated tectonic subsidence and in close coincidence with the transfer of strike-slip activity and plate boundary there.

5.3.2. Sediments of the Southern Margin of Baja California

The oldest sediments drilled on the Baja California Margin have been recovered from the shallower DSDP Sites 476 and 475 (Figure 5.17). They consist of metamorphic pebbles and cobbles (Unit V) deposited in a subaerial fluvial plain

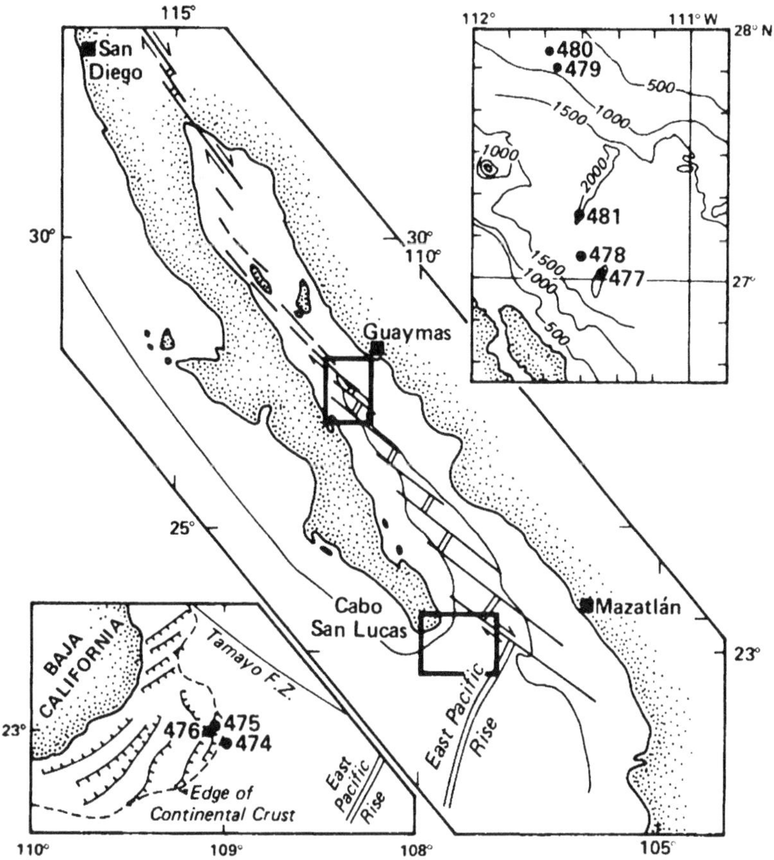

Figure 5.16 Schematic map of the Gulf of California, highlighting the succession of transform segments and pull-apart basins, the morphology of the Guaymas and Yaqui basins (top), and the southern area of oceanic crust (bottom). Reprinted from Curray, J.R., Moore, D.G. et al., 1982. Initial Reports of the Deep Sea Drilling Project, volume 64. U.S. Government Printing Office, Washington.

context, which overlay a deeply weathered granitic terrain (Unit VI). Their age is unknown, but they show similarities with the conglomerate and coarse sandstone unit evidenced further North. These conglomerates are capped by a complex marine unit (Unit IV) which consists of dolomitic claystone at DSDP Site 475, grading to a succession of organic claystone with phosphate rich layers, glauconitic sands and zeolitic clays at DSDP Site 476. Dates on glauconite and an ash layer within the unit yielded ages between 4.4 and 4.0 Ma.

The dolomite has originated through early diagenesis in a poorly oxygenated environment rather than in a nearshore setting. The organic signature of the claystone reveals a mixture of marine algal and terrigenous sources suggestive of shallow offshore, low-energy environments. Glauconite develops at shallow water depths of a few hundreds of meters and consists of sheet silicates which evolve in reduced microenvironments like foraminifers and fecal pellets from pre-existing

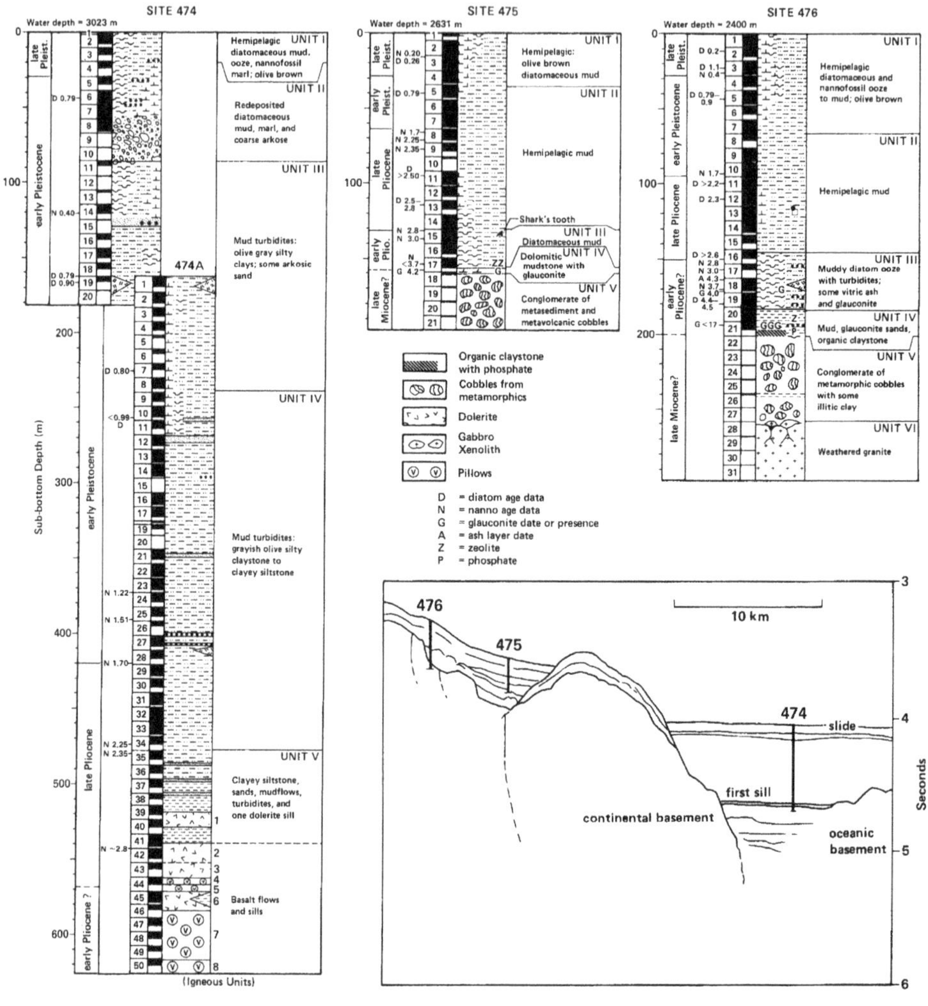

Figure 5.17 Lithostratigraphic summary of Late Neogene sediments drilled during DSDP Leg 64 off the southern tip of Baja California (see Figure 5.15 for location). The schematic NW–SE transect across the margin shows the relative position of the sites, separated by a basement high. Reprinted from Curray, J.R., Moore, D.G. et al., 1982. Initial Reports of the Deep Sea Drilling Project, volume 64. U.S. Government Printing Office, Washington.

clays, by incorporating Fe, K and Mg (see Section 13.2). Its formation requires long exposure to seawater, and glauconite principally develops in areas of low sedimentation rates, hiatuses and active currents. Phosphates are a by-product of biological activity which concentrates phosphorus in fish bones, invertebrate skeletons, guano and fecal pellets (see Section 13.3). They are especially frequent at shallow water depths of the continental slopes where upwelled waters enriched in nutrients including phosphorus support intense biological activity involving plankton, fish and birds. Phosphates are preserved in reduced environments where they are associated with organic matter, and may be concentrated by winnowing. Glauconite and phosphate layers in the Gulf of California reflect higher energy, intense productivity and relatively shallow environments of low sedimentation rates. Combining lithologic evidence, this unit was likely deposited at a depth of 400 m as a maximum, on an offshore bank isolated from major terrigenous influx, within an oxygen-minimum zone beneath an active upwelling. The marine environment was probably unstable, with oxic intervals which allowed formation of zeolitic clays and episodes of winnowing which concentrated glauconitic sands. The overlying Unit III consists of muddy diatom ooze and diatom mud, strongly bioturbated, and is clearly indicative of open marine conditions with significant productivity in the photic zone and decreased importance of terrigenous elements. The sediment also shows frequent subtle grading, indicative of turbiditic sequences and slope instability during the Early Pliocene.

The oldest sediments drilled at the deepest DSDP Site 474 have an age of about 3.0 Ma and overlay alternating basalt flows and dolerite sills of the upper oceanic crust (Figure 5.17). They consist of hard silty claystone and clayey siltstone (Unit V) and include relatively thick (more than 80 cm) mud turbidites and sandy layers. The overlying Unit IV of Late Pliocene and Early Pleistocene Age is very thick (about 80 m), consists of clayey siltstone and silty claystone, and is characterized by frequent mud turbidites and other types of resedimented beds from massive sands to minor debris flows. More than 200 turbidite units have been recognized, which are typically 40–60 cm thick but may locally exceed 320 cm. The thickest turbidites and debris-flow deposits are observed in the lower part of the unit. The average turbidite starts with a sharp contact and a moderately well sorted sand (rarely pebbly sand) which may erode into the substratum. These are succeeded by poorly sorted silty sands which grade rapidly to structureless silts. The lower parts of the turbidites frequently contain shelf carbonates and terrestrial plant remains, whereas the upper part is generally burrowed. The classic Bouma Sequence could not be observed in this unit. The turbidites are rather similar to those generated in lacustrine deltaic regions and indicate an oceanic environment of low energy, as also suggested by occurrences of reduced hemipelagic deposits. This unit is coeval with the hemipelagic muds of the Unit II at the shallower sites, which contain rare microfossils but frequent traces of mud turbidites.

Early to Late Pleistocene deposits are rather similar at all sites (Units III, II and I at DSDP Site 474; Unit I at DSDP Sites 475 and 476) and are characterized by increased contents of siliceous and calcareous microfossils, and decreased mean sedimentation rates. The sediments consist of diatom muds and nannofossil diatom oozes at the shallower Sites 475 and 476. They consist of nannofossil-bearing siliceous clayey silts to silty clays and diatom nannofossil muds grading upwards to nannofossil diatom clayey silts and muddy diatomaceous ooze at the deeper DSDP Site 474, where mud turbidites are still frequent. These units highlight an increased influence of the oceanic environment on the sedimentation, as well as a persistence of gravity processes and slope instability in probable relation with continuous strike-slip activity.

5.3.3. Sediments of the Central Gulf of California: The Guaymas and Yaqui Basins

The Gulf of California is characterized by an active system of pull-apart basins connected by dextral transform faults. In the central part of the gulf, the half-graben Yaqui Basin off Sonora contains about 3.9 km of sediment and is adjacent to the Guaymas Basin off Baja California, which contains about 1.45 km of sediments. Both basins are separated by the Guaymas transform fault (Figure 5.18). The Yaqui Basin is defined by a conjugate system of faults which were especially active during the latest Miocene and the Pliocene but aborted in the Late Pliocene. Since then, the Guaymas and Carmen transform faults accommodate most of the strike-slip strain, and control the development of the Guaymas Basin. Crustal thinning evolved rapidly, and the Guaymas Basin is now a nascent spreading center where DSDP Sites 477, 478 and 481 (Figure 5.16) retrieved tholeiitic and dolerite sills intruding hemipelagic and turbiditic sediments.

Seismic reflection profiles reveal the general organization of sedimentary sequences in the Yaqui and Guaymas basins (Figure 5.18). Irregular reflector configuration in the lowermost Sequence A suggests heterogeneous sedimentation, while the overlying Sequence B looks more uniform. Sequences C, D and E accumulated as tectonism decreased in the Yaqui Basin, leading to progressive

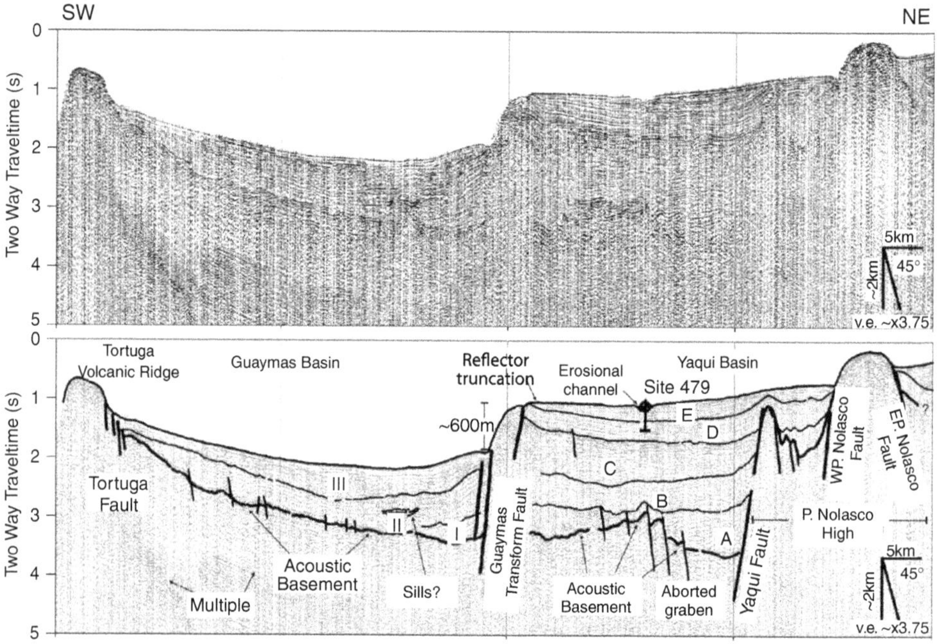

Figure 5.18 Interpretation of a SW–NE seismic transect across the Guaymas and Yaqui basins, highlighting the distribution of sedimentary sequences in both basins and their relations to tectonism. Reprinted from Aragon-Arreola, M., Morandi, M., Martin-Barajas, A., Delgado-Argote, L., Gonzalez-Fernandez, A., 2005. Structure of the rift basins in the central Gulf of California: Kinematic implications for oblique rifting. Tectonophysics, 409, 19–38.

filling and draping of the structures. Sequences E and D have been sampled at DSDP Site 479 (Figure 5.16) where Pleistocene and Late Pliocene sediments, mainly muddy diatom ooze with alternating varved and massive intervals, have been retrieved (Figure 5.19). Sequence D looks dominated by terrigenous elements (diatom mudstone), characterized by significant induration, and separated from Sequence E by a possible hiatus. Although Pleistocene sediments of Sequence E mainly consist of muddy diatom ooze, they contain layers of diatom mud and diatom ooze indicating a variability in productivity and/or continental erosion. The sequence is interrupted by a few sandy intervals and hard dolomitic mudstones of probable diagenetic origin. Varved intervals are controlled by seasonal fluctuations of climate, and minimum oxygen conditions which limit the action of burrowing organisms. The varves include dark laminae which contain more than 60% clay and light laminae which contain more than 60% diatoms (Figure 5.19). The succession of laminae is mainly controlled by the seasonality of climate, i.e., rainfall principally and by wind activity which regulates upwelling conditions. The sediments at DSDP Site 479 document a decreased influence of continental erosion (probably related to decreased tectonism) since the Late Pliocene and a dominant influence of hydrology and climate which control rainfall, upwelling and the ventilation of the basin.

Accumulation of the wedge sequences I and II (Figure 5.18) of the Guaymas half-graben has been driven by the vertical motion of the Guaymas transform fault and subsidence. The accumulation of the hummocky Sequence III and westward shift of the locus of maximum thickness suggest a decrease of tectonic subsidence in the area. The sediments of the Guaymas Basin have been retrieved at DSDP Sites 477 (southern part), 478 (slope) and 481 (northern part, Figure 5.16). Similar sequences of Late Pleistocene age, intruded by dolerite sills, have been recovered at all sites with maximum sedimentation rates of 1,500 m/Myr at DSDP Site 481 drilled along the marginal fault of the northern Guagmas Basin (Figure 5.20). The sediments below the dolerite sill complex are dominated by terrigenous elements and consist of diatom muds which contain some laminated intervals (Units 2 and 3). They are intercalated with mud turbidites in the southern part and mud flows in the northern part. They have been hydrothermally altered and indurated in the presence of exceptionally high heat flows in the southern part, but do not show significant alteration in the northern part where heat flows are weak. The diatom muds are intercalated with dolomitic layers of diagenetic origin at DSDP Site 478 on the basin slope, where they are indurated below 260 mbsf (diatom mudstone, Unit 3). These sedimentary units are principally controlled by tectonism, but have been subsequently modified in areas of intense magmatic activity. The most recent Unit 1 consists of muddy diatom ooze at all sites and is very similar to the uppermost unit of the adjacent Yaqui Basin. The unit includes a number of mud turbidites in the southern part of the basin, grading to sandy turbidites at the more proximal site on the slope. The sediment series also includes some megaturbidites (more than 10 m thick) in the northern part of the basin where DSDP Site 481 has been drilled near an active fault system. The most recent sedimentary unit of the Guaymas Basin reflects an increasing influence of the oceanic environment through an increased proportion of pelagic biogenic elements and a persistence of tectonism and instability via the frequency of turbiditic events.

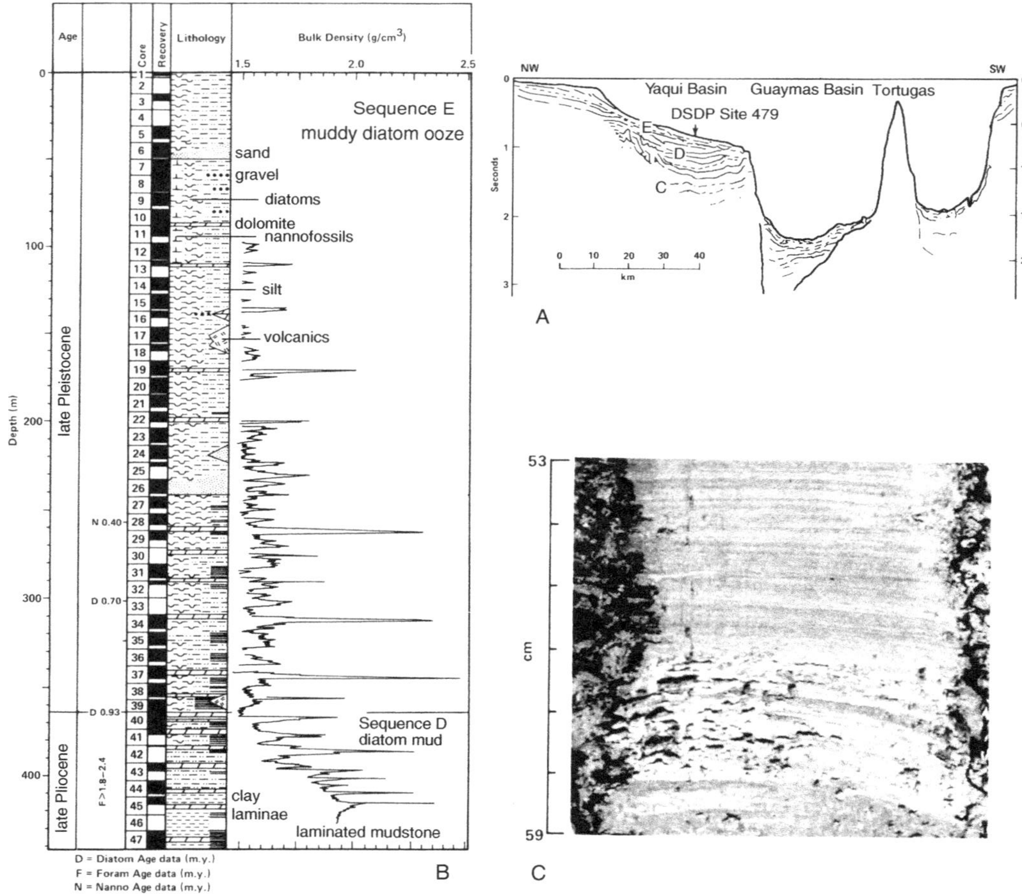

Figure 5.19 The sedimentary sequences of the Yaqui Basin. (A) Location of DSDP Site 479 on a schematic NE–SW transect (letters C to E refer to similar sequences on Figure 5.18); (B) lithostratigraphic summary of DSDP Site 479 (brief increases in density coincide with layers of hard dolomitic mudstone); (C) detail of the late Pliocene laminated diatom mud. Note that a significant increase in density over a 30 m interval coincides with a major seismic reflector separating acoustic sequences D and E and changes in lithology and compaction. Modified from Curray, J.R., Moore, D.G. et al., 1982. Initial Reports of the Deep Sea Drilling Project, volume 64. U.S. Government Printing Office, Washington.

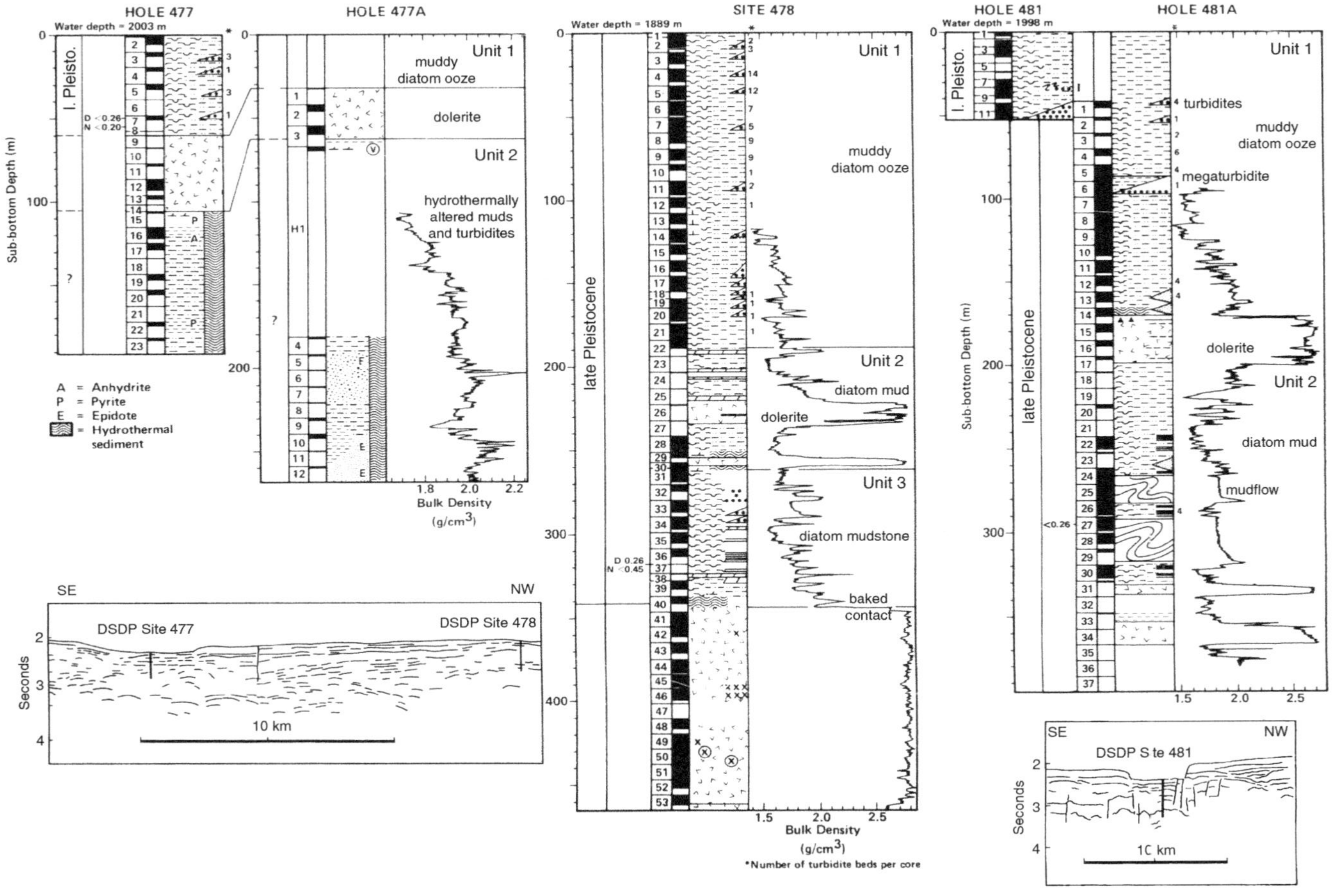

Figure 5.20 Lithostratigraphic summary of Late Pleistocene sediments of the Guaymas basin drilled during DSDP Leg 64 and relative position of the sites (see Figure 5.16 for location). The drill-holes penetrated the upper Sequence III of Figure 5.18. Note coincidence of massive dolerites at DSDP Site 478 with bottom acoustic reflector. Modified from Curray, J.R., Moore, D.G. et al., 1982. Initial Reports of the Deep Sea Drilling Project, volume 64. U.S. Government Printing Office, Washington.

Investigations in the Gulf of California suggest that influences of tectonism and related continental erosion and slope processes are still dominant in a pull-apart context during the crustal fissure stage because of persistent strike-slip activity, whereas these influences decrease through time in rifted areas. However, the influence of the marine environment on the sedimentation is conspicuous in both areas through the remains of the biological activity principally, but also the variable intensity of the hydrodynamism.

5.4. Ancient Crustal Fissures: The Mid-Cretaceous South Atlantic

Paleoreconstructions highlight similarities between the Mid-Cretaceous South Atlantic and the modern Red Sea (Figure 5.21). The oldest magnetic anomalies recorded are of Late Valanginian–Early Hauterivian age (M9, 131 Ma) in the Southern Cape–Argentina Basin, Late Hauterivian age (M4, 127 Ma) in the Northern Cape–Argentina Basin and latest Aptian–earliest Albian age (113 Ma) north of the Rio Grande–Walvis Ridge Barrier in the Angola–Brazil Basin. Therefore the Aptian ocean floor was made of oceanic crust in the southern South Atlantic, and of thinned continental to intermediate crust in the northern South Atlantic.

5.4.1. The Mid-Cretaceous Sediment Facies of the Northern South Atlantic

In most marginal areas of the northern South Atlantic (Brazil, Angola, Gabon), continental syn-rift sediments are separated from transitional sediments of Aptian age by an angular unconformity. The transitional sequence consists of a thick section of siliciclastic and evaporitic sediments principally. The siliciclastics accumulated on a peneplane truncating rift structures and sediments, and result from extensive development of alluvial fans during an interval of intensified tectonism and reactivation of continental erosion. Terrigenous depocenters are separated by very shallow water carbonates, mainly stromatolitic and nodular limestones. In the abandoned rift basins of the continental interior, Aptian siliciclastics are coarse and thin, and locally associated with sabkha deposits. As subsidence progressed, a transgression coming from the South led to the development of a narrow and shallow seaway in the central part of the active structure. The predominance of very restrictive, euxinic and saline conditions controlled the accumulation of as much as 2,000 m of massive halite in the subsiding basins while anhydrite deposited in more stable areas (Figure 5.22). As ocean opening and subsidence progressed, the evaporites were replaced during the earliest Albian by carbonate platforms of high to medium energy where oolitic and oncolitic facies prevailed. These carbonate platforms grade to bioclastic facies toward the axial part of the structure. In some areas like the Cuanza Basin of Angola, the transition is marked by dark gypsiferous clays grading upwards into

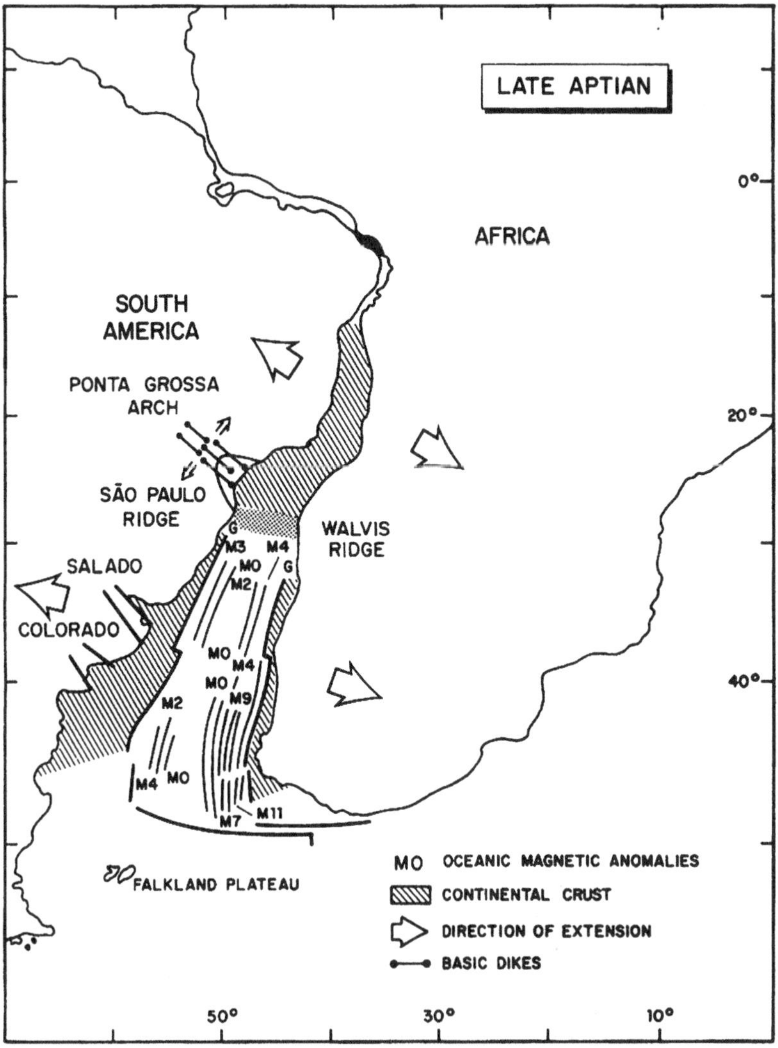

Figure 5.21 Paleoreconstruction of the South Atlantic during the Late Aptian. Note similarities with the modern Red Sea: elongated narrow structure of oceanic crust to the South and thinned continental crust to the North as a result of propagating ocean opening. Reprinted from Chang, H.K., Kowsmann, R.O., Ferreira Figueiredo, A.M., Bender, A.A., 1992. Tectonics and stratigraphy of the East Brazil Rift system: An overview. Tectonophysics, 213, 97–138.

dolomitic and oolitic limestones, indicating that the transgression was associated with a brief resumption of erosion of probable tectonic origin. However, terrigenous accumulations still prevailed in coastal areas of river input, where delta fan systems prograded over the carbonate platforms through time.

Dolomitic limestones also characterize uppermost Aptian sediments (Figure 5.23) in the deepest axial part of the Brazil–Angola Basin at DSDP Site 364 (Unit 7), where high pore-water salinities and seismic reflectors indicate that the drill-hole

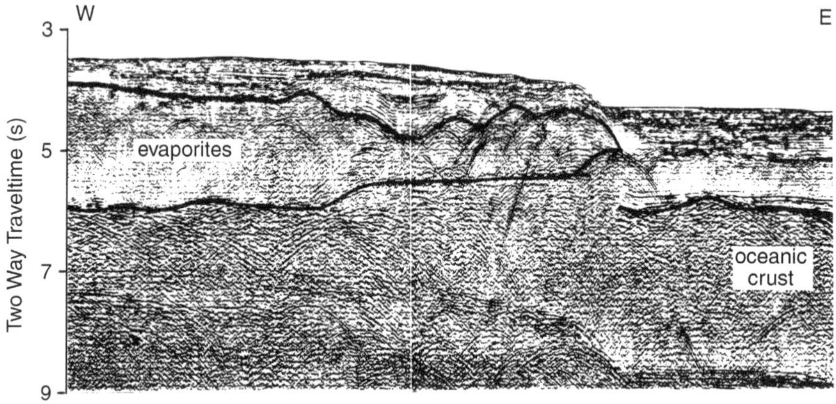

Figure 5.22 Seismic profile across the Brazilian margin. Note the thickness of the evaporitic sequence compared to post-evaporitic sediments, and the coincidence between the limits of the evaporitic sequence and oceanic crust. Modified from Chang, H.K., Kowsmann, R.O., Ferreira Figueiredo, A.M., Bender, A.A., 1992. Tectonics and stratigraphy of the East Brazil Rift system: An overview. Tectonophysics, 213, 97–138.

was terminated a few tens of meters above the Aptian evaporites. The difference in age between axial and marginal areas suggests that the transgression was probably not synchronous over the entire basin, progressing from deeper to shallower areas through time. Sedimentological and geochemical evidence support a diagenetic origin for the dolomite, which replaced calcite in a pore-water environment of high salinity and Mg/Ca ratio. Dolomitic limestones alternate with finely laminated black shales indicative of bottom water conditions stagnant enough to allow preservation of organic elements and hinder benthic life and current-related sedimentary structures. Accumulation of dolomitic limestones and black shales (and related marine environments) persisted until the Middle Albian, when they were replaced by limestones and marly limestones (Unit 6) with slump folds and micro-faults indicative of slope instability, probably related to persisting tectonism and/or thermal subsidence. These facies suggest an improved ventilation of the Brazil–Angola Basin and mark the onset of significant exchanges with the Cape–Argentina Basin at shallow and deep-water depths through the Walvis–Rio Grande Barrier. They accumulated with relatively high sedimentation rates around 55 m/Myr, probably because of improved preservation of calcareous biogenic elements and/or marine productivity. By the end of the Albian, gradual expansion and subsidence of the basin and related marine transgression resulted in the drowning of the coastal carbonate platforms overlain by hemipelagic sediments associating fine siliciclastics to planktonic biogenic elements, and grading to massive limestones in the central part of the basin.

The oldest sediments drilled on the Walvis–Rio Grande Barrier have been retrieved at DSDP Site 363 on an elevated part of the Walvis Ridge where limestones and calcarenites (carbonate sandstone) of Late Aptian age (Unit 3) have been sampled (Figure 5.23). The limestones formed at shallow water depth above 500 m. They contain remains of algae and phosphates, which suggest significant hydrodynamism and productivity. This lithology changed in the Early to Middle Albian to limestones

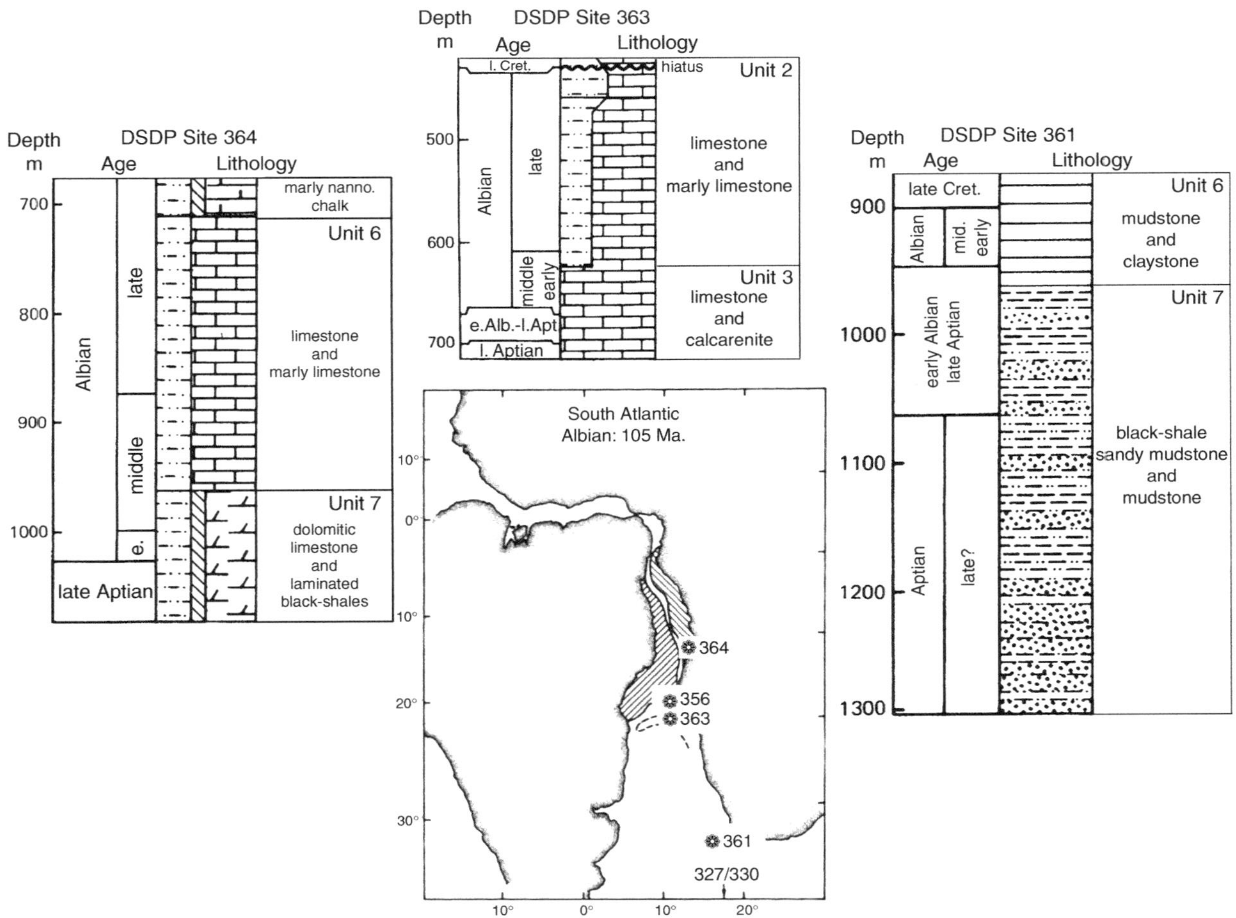

Figure 5.23 Lithostratigraphic summary of Late Aptian and Albian sediments drilled during DSDP Legs 39 and 40 in the South Atlantic. Albian paleoposition of the sites is provided on paleoreconstruction. Note relative homogenization of lithologies through time and higher terrigenous contents at DSDP Site 361 adjacent to a more humid and tectonically more active area. Modified from Bolli, H.M. Ryan, W.B.F. et al., 1978. Initial Reports of the Deep Sea Drilling Project, volume 40. U.S. Government Printing Office, Washington.

with minor layers of marly limestone which increased in importance through time and contain flow and slump structures (Unit 2). At DSDP Site 356 on the Sao Paulo Plateau, which by Mid-Cretaceous time was located north of DSDP Site 363 (Figure 5.23), Albian sediments consist of alternating beds of laminated dolomitic limestones and calcareous mudstones which deposited around 1,000 m water depth. The mudstones contain frequent graded beddings and some glauconite grains, suggesting significant contribution from turbidity flows.

Late Aptian sediments from the Brazil–Angola Basin indicate that a limited marine transgression over a subsiding, peneplaned area of semi-arid climate favored the accumulation of evaporites. Evaporite formation ceased near the Aptian–Albian boundary when the basin deepened in probable relation to an accelerated thermal subsidence and/or increased tectonism. It is remarkable that this transition occurred as the formation of oceanic crust started in the basin. Albian sediments reflect the presence of oxygenated and productive coastal and shallow (neritic) environments, but frequently restricted to anoxic in the deep axial part below 1,000 m water depth. The deeper Albian sediments also contain evidence of slope instability (slumps, turbidites) in probable relation to persisting tectonism and subsidence.

5.4.2. The Mid-Cretaceous Sediment Facies of the Southern South Atlantic

DSDP Site 361 has been drilled on the lower continental rise of Southwest Africa in the Cape–Argentina Basin, in an area where oceanic crust formed at the time of magnetic anomaly M4 (around 127 Ma, during the Hauterivian). The drill-hole penetrated below seismic horizon AII of regional extension, which merges with the acoustic basement interpreted as oceanic crust in the vicinity of the oldest magnetic anomaly (M9, 131 Ma, Hauterivian). There is no seismic nor physical evidence for the presence of evaporites in this part of the South Atlantic. The oldest sediments which have been recovered at DSDP Site 361 consist of alternating black shales, sandy mudstones and sandstones (Unit 7) of Early Aptian to Early Albian Age (Figure 5.23), deposited below 2,000 m water depth. The sandstones increase in frequency in the lowermost cores. Organic components are abundant throughout the unit where only rare, poorly preserved, dissolved calcareous microfossils occur. The abundance of sandstones and organic elements of continental origin (coarse plant debris, pollen grains, etc.) most probably results from the rapid erosion of rugged, marginal block-faulted relief of humid climate and luxuriant plant growth. The sediment grades in the Middle Albian to alternating mudstone and claystone (shales, Unit 6). The mudstones contain abundant parallel and cross-bedded laminae of fine sand and silt typical of distal lower delta fan areas, and abundant plant debris, which highlight the importance of erosion and fluvial transport. The grain size of the siliciclatics at DSDP Site 361 decreased as spreading and thermal subsidence progressed in the Cape Basin. Oxygen-deficient conditions (which had been present for more than 10 Myr) ceased in the Middle Albian when the Falkland Plateau cleared the African Margin, permitting deep-water exchanges with the Southern Ocean.

The succession of sediment facies during the crustal fissure stage of the South Atlantic illustrates the combined influence of tectonism, magmatism, subsidence and

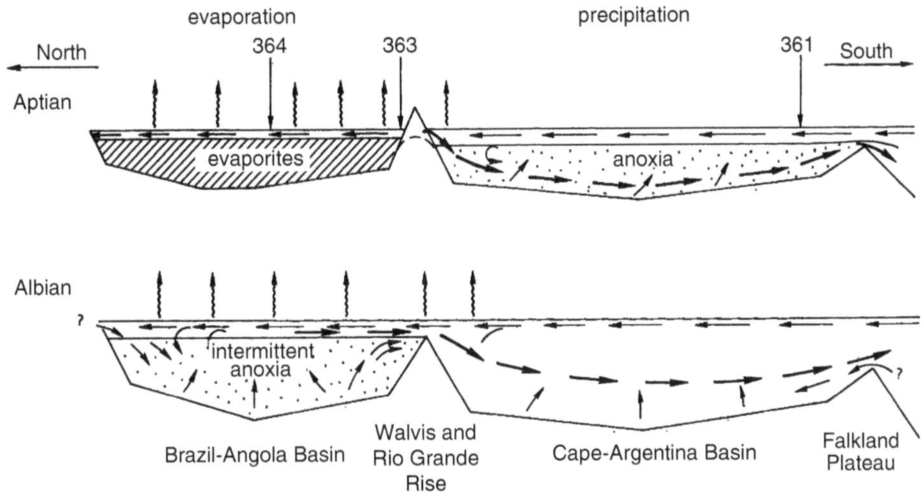

Figure 5.24 Schematic cross-section highlighting major differences in the evolution of the Brazil–Angola and Cape–Argentina basins during the Late Aptian and the Albian. Note the evolution of the oceanic environments as subsidence progresses from South to North. Modified from Bolli H.M. Ryan W.B.F. et al., 1978. Initial Reports of the Deep Sea Drilling Project, volume 40. U.S. Government Printing Office, Washington.

climate on the morphology of the oceanic basins, marine circulation and conditions of sedimentation (Figure 5.24). Tectonism and magmatism controlled the persistence of morphological barriers of the Falkland Plateau and Walvis–Rio Grande areas, respectively. In turn, tectonism and magmatism were involved in the modulation of productivity and water mass exchanges with the Southern Ocean and between the South Atlantic basins. Tectonism and subsidence principally controlled the morphology of the basins, the stability of the slopes and the intensity of continental erosion. Consecutively, tectonism and subsidence initiated the formation of elongated basins favorable to water stagnation, controlled the accumulation of evaporites and siliciclastics and the frequency of gravity flows. Regional climate participated in the control of the sedimentation, facilitating the erosion of siliciclastics and plant remains in humid areas adjacent to the Argentina–Cape Basin, and the formation of evaporites in the semi-arid Brazil–Angola Basin. The influence of the oceanic environment increased through time as spreading progressed in the Mid-Cretaceous South Atlantic like in other crustal fissure areas, leading ultimately to the formation of marine pelagic carbonates in the Angola–Brazil Basin. This is however less obvious in the southern Cape Basin where fine siliciclastics dominated at DSDP Site 361, close to the tectonically active Falkland Plateau transform area.

FURTHER READING

Association des Sédimentologistes Français, 1989. Dynamique et méthodes d'étude des bassins sédimentaires. Technip, Paris.

Banda, E., Torné, M., Talwani, M. (Editors), 1995. Rifted ocean–continent boundaries. Klüwer, Dordrecht.

Boillot, G., Coulon, C., 1998. La déchirure continentale et l'ouverture océanique. Overseas Publishers Association, Amsterdam.

Boillot, G., Huchon, P., Lagabrielle, Y., 2003. Introduction à la géologie: la dynamique de la lithosphère. Dunod, Paris.

Makris, J., Mohr, P., Rihm, R. (Editors), 1991. Red Sea: Birth and early history of a new oceanic basin. Tectonophysics, 198: 129–468.

Purser, B.H., Bosence, D.W.J., 1998. Sedimentation and tectonics in Rift Basins: Red Sea, Gulf of Aden. Chapman and Hall, London.

Other references used in this chapter

Aragon-Arreola, M., Morandi, M., Martin-Barajas, A., Delgado-Argote, L., Gonzalez-Fernandez, A., 2005. Structure of the rift basins in the central Gulf of California: Kinematic implications for oblique rifting. Tectonophysics, 409: 19–38.

Bolli, H.M., Ryan, W.B.F. et al., 1978. Initial Reports of the Deep Sea Drilling Project, volume 40. U.S. Government Printing Office, Washington.

Brun, J.P., Beslier, M.-O., 1996. Mantle exhumation at passive margins. Earth Planetary Science Letters, 142: 161–173.

Chang, H.K., Kowsmann, R.O., Ferreira Figueiredo, A.M., Bender, A.A., 1992. Tectonics and stratigraphy of the East Brazil Rift system: An overview. Tectonophysics, 213: 97–138.

Cochran, J.R., Martinez, F., 1988. Evidence from the northern Red Sea on the transition from continental to oceanic rifting. Tectonophysics, 153: 25–53.

Curray, J.R., Moore, D.G., et al., 1982. Initial Reports of the Deep Sea Drilling Project, volume 64. U.S. Government Printing Office, Washington.

Duncan, R.A., Larsen, H.C., Allan, J.F. et al., 1996. Proceedings of the Ocean Drilling Program, Initial Reports, volume 163. Ocean Drilling Program, College-Station, TX.

Duncan, R.A., Larsen, H.C., Allan, J.F., Brooks, K. (Editors), 1999. Proceedings of the Ocean Drilling Program, Scientific Results, volume 163. Ocean Drilling Program, College-Station, TX.

Ehrardt, A., Hübscher, C., Gajewski, D., 2005. Conrad Deep, northern Red Sea: Development of an early stage ocean deep within the axial depression. Tectonophysics, 411: 19–40.

Hemleben, C., Meischner, D., Zahn, R., Almogi-Labin, A., Erlenkeuser, H., Hiller, B., 1996. Three hundred eighty thousand year long stable isotope and faunal records from the Red Sea: Influence of global sea level change on hydrography. Paleoceanography, 11: 147–156.

Larsen, H.C., Saunders, A.D., Clift, P.D. et al., 1994. Proceedings of the Ocean Drilling Program, Initial Reports, volume 152. Ocean Drilling Program, College-Station, TX.

Larsen, H.C., Saunders, A.D., Wise, S.W. (Editors), 1998. Proceedings of the Ocean Drilling Program, Scientific Results, volume 152. Ocean Drilling Program, College-Station, TX.

Oskin, M., Stock, J., 2003. Marine incursion synchronous with plate-boundary localization in the Gulf of California. Geology, 31: 23–26.

Oskin, M., Stock, J., Martin-Barajas, A., 2001. Rapid localization of Pacific–North America plate motion in the Gulf of California. Geology, 29: 459–462.

Rohling, E.J., 1994. Glacial conditions in the Red Sea. Paleoceanography, 9: 653–660.

Supko, P.R., Perch-Nielsen, K. et al., 1977. Initial Reports of the Deep Sea Drilling Project, volume 39. U.S. Government Printing Office, Washington.

Whitmarsh, R.B., Beslier, M.-O., Wallace, P.J. (Editors), 1998. Proceedings of the Ocean Drilling Program, Scientific Results, volume 173. Ocean Drilling Program, College Station, TX.

Whitmarsh, R.B., Weser, O.E., Ross, D.E. et al., 1974. Initial Reports of the Deep Sea Drilling Project, volume 23. U.S. Government Printing Office, Washington.

Winckler, G., Aeschbach-Hertig, W., Kipfer, R., Botz, R., Rübel, A.P., Bayer, R., Stoffers, P., 2001. Constraints on origin and evolution of Red Sea brines from helium and argon isotopes. Earth and Planetary Science Letters, 184: 671–683.

CHAPTER SIX

Mature Oceans in a Context of Plate Divergence

6.1. Structure, Tectonics and Sedimentation of Mature Divergent Oceans

Mature ocean systems are essentially characterized by the development of mid-oceanic ridges, deep oceanic basins, passive continental margins and continental shelves.

6.1.1. Mid-Oceanic Ridges and Oceanic Basins

As oceanic systems widen, magmatic and tectonic activity focus on a narrow band where new oceanic lithosphere is being continuously produced (Figure 6.1).

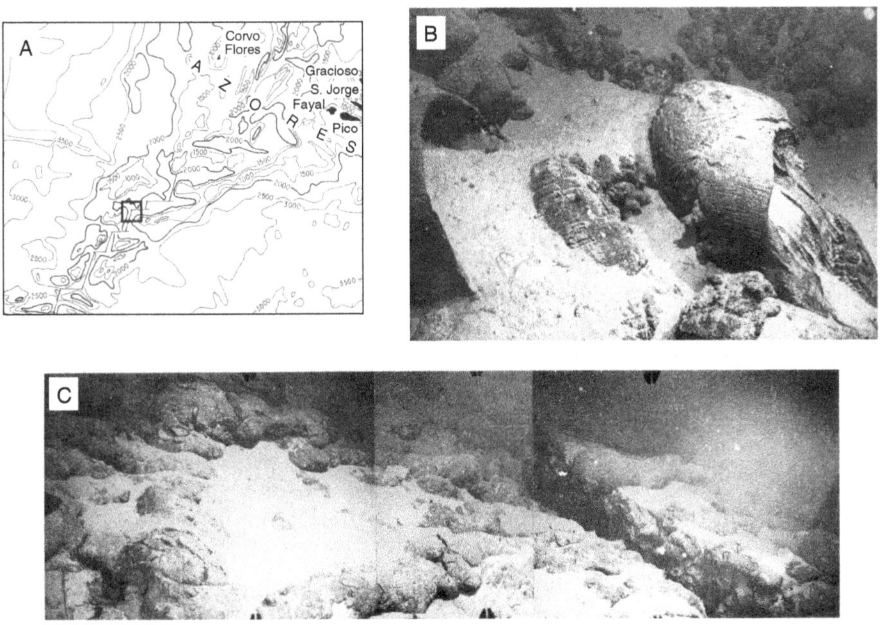

Figure 6.1 Recent oceanic crust and sediment from the active area of the Mid-Atlantic Ridge, FAMOUS (French-American Mid-Oceanic Underwater Survey) area. A, location (square) of the FAMOUS area, SW of the Açores archipelago; B, basaltic pillow-lavas; C, fissure on the floor of the ridge. Note thin cover of biogenic pelagic ooze. Modified from Arcyana, 1978. Atlas FAMOUS, Bordas, Paris.

For isostatic reasons, the depth of this geologically active area varies from 1,500 to 3,000 m with the degree of magmatic activity, the average depth being around 2,500 m (see Section 2.1.3). However, there are some exceptions like the North Atlantic Large Igneous Province where intense activity of the Iceland hot spot produces upper oceanic crust basalts in subaerial conditions. Because of continuous creation of oceanic crust the ocean widens and areas of older oceanic crust move away from the active area, and increase in depth as a result of thermal subsidence. The contrast of density and isostatic response between young and old oceanic lithosphere produces a typical morphology of ocean floors where uplifted, active mid-oceanic ridges of young lithosphere are bounded by deep oceanic basins of ancient lithosphere. This morphological pattern is locally interrupted by the presence of seamounts or aseismic ridges of basaltic substrate. These structures are typically derived from intense magmatism: hot spot activity and/or large igneous provinces. For example, the greater Kerguelen Plateau of the Southern Ocean was attached to Broken Ridge and the Ninetyiest Ridge before separation occurred near 45 Ma when the Southeast Indian Ridge became active (Figure 6.2). The whole structure is probably the largest igneous province known to date. The structure was created near an active spreading center during separation of the Indian plate from the Australia–Antarctica plate. Intensified magmatism started around 120 Ma with the position of the Kerguelen hot spot below young oceanic lithosphere (creating the South Kerguelen Plateau) and lasted till about 35 Ma (in the North Kerguelen Plateau) with peak activity recorded around 119–110 and 105–95 Ma. Kerguelen Plateau is principally made of basalts, which were emitted in subaerial conditions principally when magmatism was the most intense. Assuming that the Southern Kerguelen Plateau subsided at the same rate as adjacent Indian oceanic crust, maximum subaerial elevations of 1,000–2,000 m should have been frequent during the Middle Cretaceous (Figure 6.3).

Because of distance from terrigenous sources the sediments in many mid-oceanic ridge (and aseismic ridge) areas consist of biogenic oozes, the sediment cover being very thin or absent in active areas of recent oceanic accretion (Figure 6.1). The sediments are commonly thermally altered near contact with weathered basalts and locally interact with hydrothermal circulation and massive hydrothermal mineralization. The sediment cover usually increases in age and thickness with distance from active areas of mid-oceanic ridges. The sediment may also change in nature in deeper areas which are under dominant influence of deep water circulation. In the modern context of thermohaline circulation, for example, deep oceanic waters derived from high latitudes are cold and supplemented in carbon dioxide, and may dissolve most or all carbonates which are principally of planktonic origin. As a result the deepest oceanic sediments are often dominated either by fine siliciclastics, metalliferous products or siliceous biogenic elements.

Because deep water circulation is principally constrained by the morphology of the seafloor (including mid-oceanic ridges which act as barriers) and the Coriolis force (which deflects oceanic currents to the right in the northern hemisphere and to the left in the southern hemisphere), the control exerted by deep water masses and currents on the sedimentation may strongly vary between and within deep oceanic basins. For example, the relief of the mid-oceanic ridge in the South Atlantic and the Southern Ocean, together with the Coriolis effect, limit the penetration of Antarctic

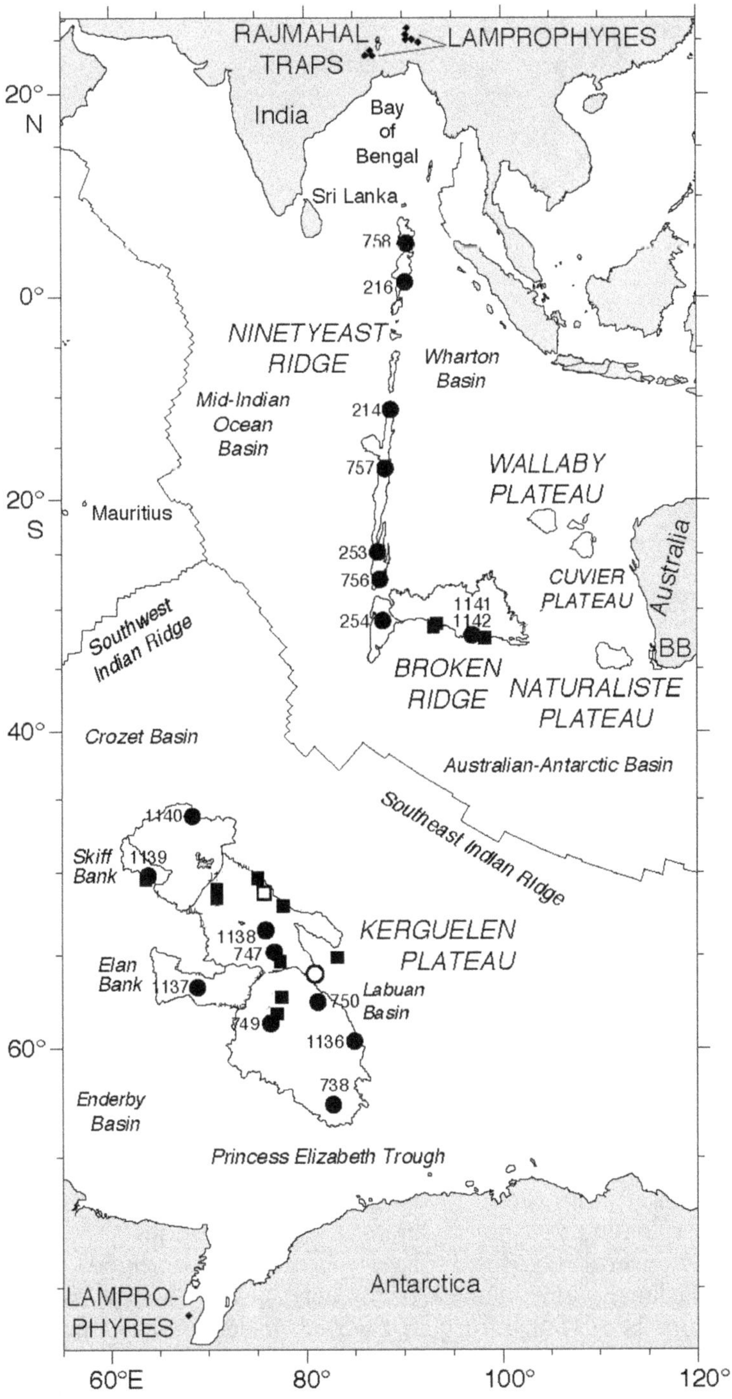

Figure 6.2 Basalt provinces of the eastern Indian Ocean attributed to the Kerguelen plume. Circles represent locations of igneous basement sites drilled or cored, squares represent dredge locations. Reprinted from Coffin, M.F., Frey, F.A., Wallace, P.J., et al., 2000. Proceedings of the Ocean Drilling Program, Initial Reports, volume 183, Ocean Drilling Program, College Station, TX.

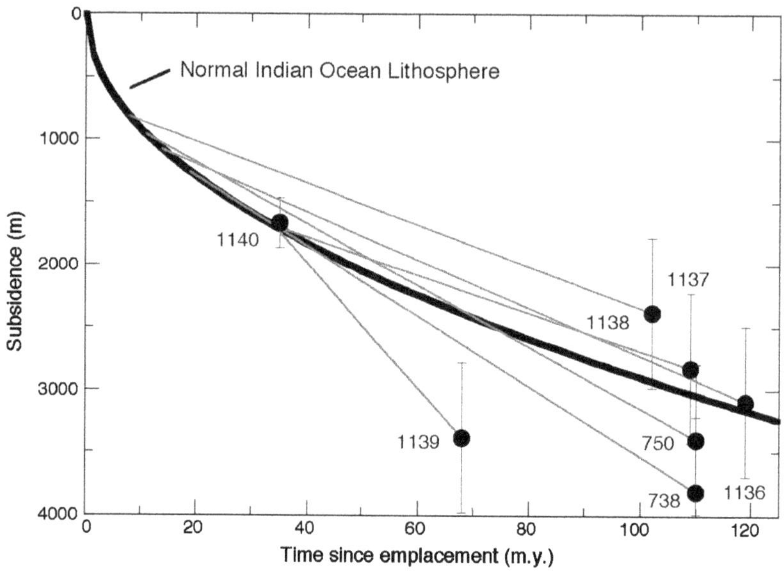

Figure 6.3 Subsidence estimates of ODP sites drilled on Kerguelen Plateau since the eruption of basement basalts. Basalts at all sites but ODP Site 1140 were erupted in subaerial environments. For each site, subsidence from the time of the oldest marine sediment is shown as a thin line. Reprinted from Coffin, M.F., Frey, F.A., Wallace, P.J., et al., 2000. Proceedings of the Ocean Drilling Program, Initial Reports, volume 183, Ocean Drilling Program, College Station, TX.

Bottom Water in the Cape Basin of the South Atlantic. Also, interaction of Northeast Atlantic Deep Water masses with the morphology of Rockall Plateau and the mid-oceanic ridge in the Northeast Atlantic, and the Coriolis effect, focus the path of circulation of Northeast Atlantic Deep Water which forms contour currents and scours surface sediments along channels and furrows (erosional troughs) in some areas. Sediment particles settle and accumulate in areas of maximum turbidity and/or lower current velocity (for example, at the margin of the main flow) where they form sediment drifts which are commonly hundreds of kilometers long and tens of kilometers wide (Figure 6.4). The main characteristics of sediment drifts may vary with the morphology of the seafloor, the velocity of the current, the nature and quantity of the existing sediment, and the time interval of current activity principally. This context determines four major types of sediment drifts: (i) sheet drifts are present in abyssal plains; (ii) elongated drifts occur in slope areas where they vary in size, morphology and structure according to local conditions; (iii) confined drifts are trapped in small basins; (iv) and channel-related drifts develop in areas of momentous current activity (Figure 6.5). Sediment drifts are made of contourites, which consist of a succession of fine sedimentary beds with normal or reversed graded beddings (principally clays and silts) with rare, poorly preserved microfossils. As the path and velocity of deep water circulation may vary through time, migrating and prograding sediment drifts show alternating contourite deposits and hiatuses leading to the formation of lenticular sedimentary units separated by major discontinuities (Figure 6.6).

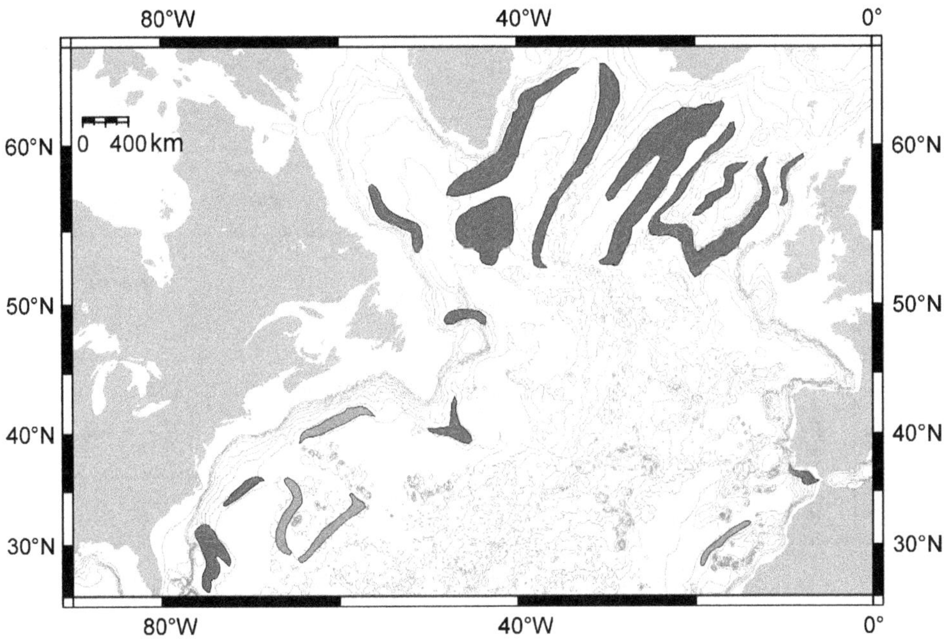

Figure 6.4 Contourite sediment distribution in the North Atlantic. Note the importance of sediment drifts in areas of deep water (North Atlantic Deep Water) formation and circulation. Reprinted from Faugères, J.-C., Stow, D.A.V., Imbert, P., Viana, A., 1999. Seismic features diagnostic of contourite drifts. Marine Geology, 162, 1–38.

6.1.2. Passive Continental Margins and Associated Continental Shelves

As the ocean system widens, areas of transitional and thinned continental lithosphere, as well as the ancient rift shoulders, are exposed to increased thermal subsidence. The importance of the subsidence may vary regionally with the degree of lithosphere thinning and underplating. Because tectonism ceased, erosion smooths continental relief. When leveled, subsident passive continental margins sink below sea level and shallow marine environments progress inland. This early marine transgression is sometimes characterized by neritic, terrigenous lagoonal and/or evaporitic facies according to local conditions of climate and environment, and marks the transition from crustal fissure to the mature stage of ocean evolution (Figure 6.7). Drainage basins progressively expand over passive continental margins and into continental hinterlands, and major river (and glacier) systems carry significant amounts of terrigenous elements eroded from continental interiors to the ocean, especially in tropical areas of high precipitation and in high latitude areas of intense ice erosion. For example, the drainage basins of the Amazon and Congo rivers extend over vast continental areas up to the Andean Cordillera and the East African Rift respectively, and carry to the South Atlantic terrigenous loads among the highest. At southern high latitudes, the Lambert Glacier carries vast quantities of terrigenous elements to the Prydz Bay area of the Southern Ocean, and locally favored the accumulation of more than 1,000 m of Plio-Pleistocene glacial and glacio-marine sediments. Terrigenous elements first

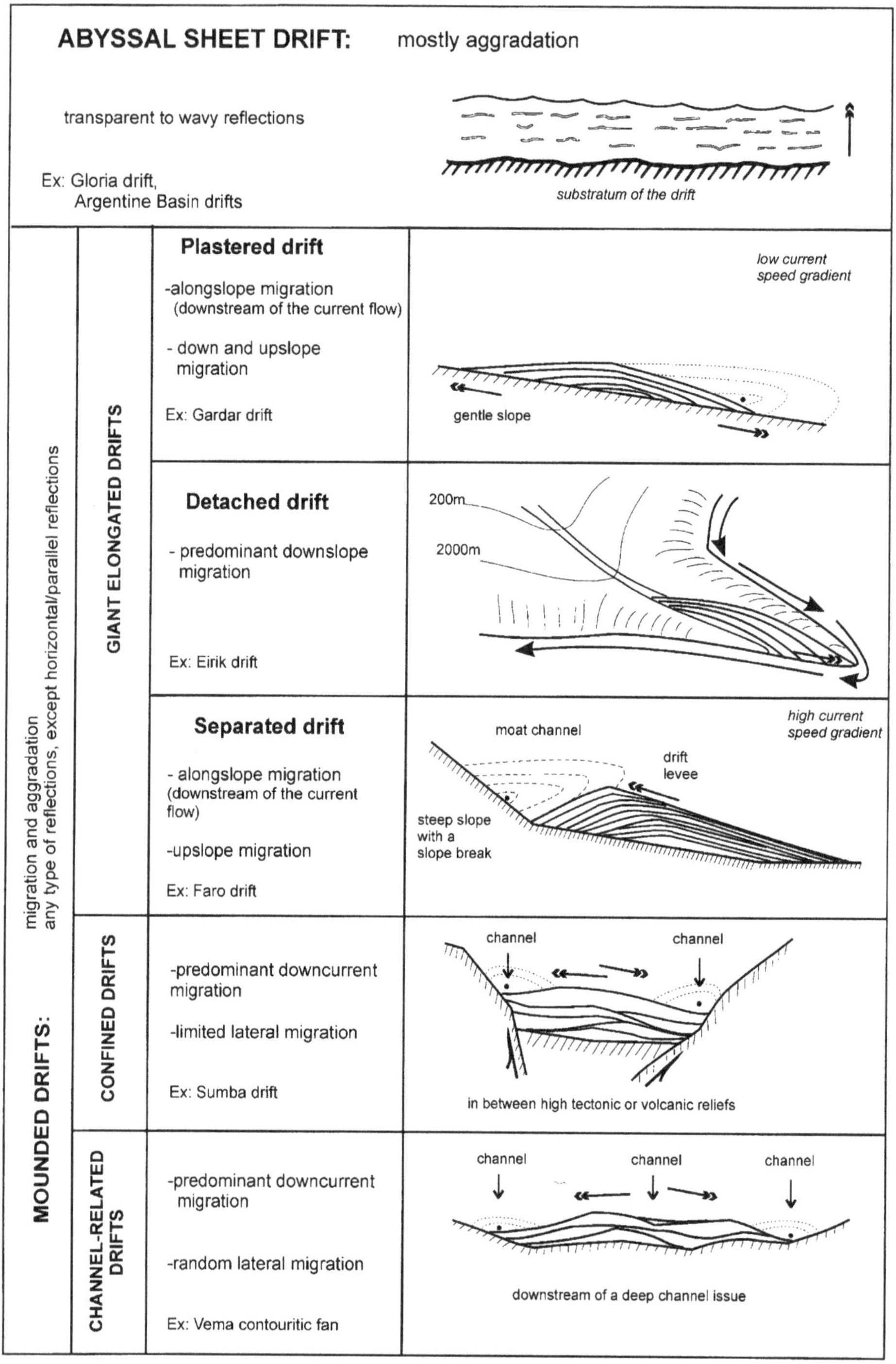

Figure 6.5 Main types of contourite drifts: morphology and characteristics. Reprinted from Faugères, J.-C., Stow, D.A.V., Imbert, P., Viana, A., 1999. Seismic features diagnostic of contourite drifts. Marine Geology, 162, 1–38.

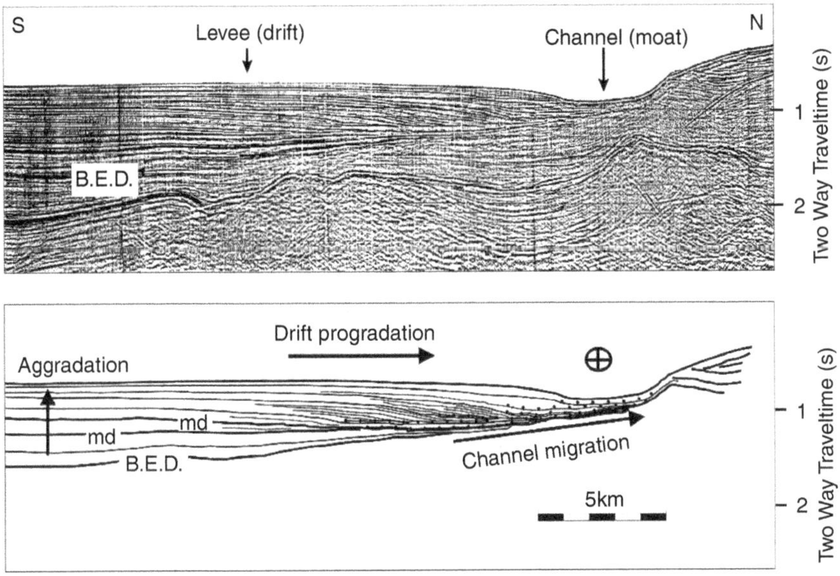

Figure 6.6 Seismic reflection profile across an elongate-mounded drift, the Faro Drift in the Gulf of Cadiz. Note upslope progradation of the sediment drift and migration of the channel through time. B.E.D., basal erosive discontinuity; md, major discontinuity. Reprinted from Faugères, J.-C., Stow, D.A.V., Imbert, P., Viana, A., 1999. Seismic features diagnostic of contourite drifts. Marine Geology, 162, 1–38.

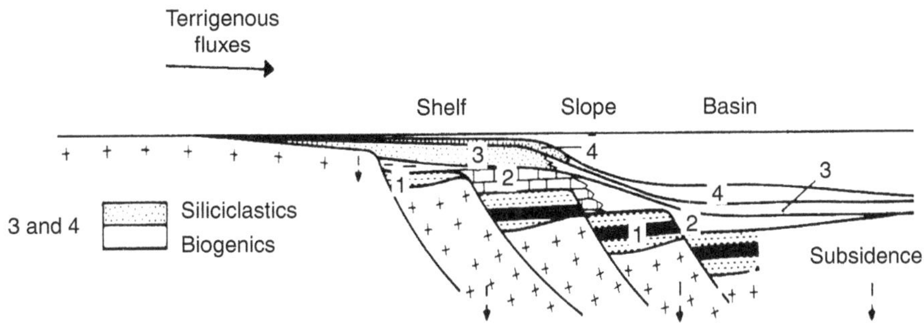

Figure 6.7 Schematic cross section of a divergent passive margin. 1, syn-rift deposits; 2, crustal fissure deposits; 3 and 4, mature ocean phase deposits. Note landward extension of neritic deposits (siliciclastics) basinward extension of the shelf and vertical accumulation in the basin from 3 to 4. Modified from Boillot, G., 1983. Géologie des marges continentales. Masson, paris.

accumulate near river or glacier mouths, prior to their redistribution by nearshore (wave and tide related currents) and gravity processes principally (see Section 2.3).

6.1.2.1. The formation of continental shelves

The terrigenous load of river and glacier systems actively participates in the formation of continental shelves and slopes, which prograde offshore through time over early transgressive deposits of passive continental margins. In highly subsident areas, the progradation of continental shelves can be associated to significant aggradation. Because of shallow water depths generally less than 200 m, continental shelves are highly sensitive to sea level variations, of eustatic or tectonic origin. During intervals of lower relative sea level, the shorelines and river mouths move offshore, and upper continental shelves eventually become subaerial. Continental shelf sediments are more easily reworked by wave and tide related currents to the outer shelf and slope, where they accumulate together with newly transported river loads. An erosion surface (discontinuity) develops on top of continental shelves. Intervals of lower relative sea level favor the transfer of terrigenous loads directly to the slope and the deep oceanic basin. As sea level starts increasing again, sediments first accumulate in shallow water conditions near the shelf edge where they form a prograding lowstand wedge of sigmoidal structure on the erosion surface. As relative sea level increases, a transgression progresses on the erosion surface where sediments accumulate almost horizontally, aggrading the shelf to form a transgressive wedge. During intervals of maximum sea level, a prograding wedge of sigmoidal structure progressively accumulates over the shelf, overlapping the transgressive series (Figure 6.8). Because the accumulation of sediments on continental shelves is primarily controlled by nearshore processes and water depth, vertical accumulation of sediments is limited during intervals of stable sea level (stillstand), and terrigenous loads bypass the shelf to accumulate on the shelf edge and the continental slope. For example, the Holocene stabilization of sea level about 6,500 years ago shaped the morphology of modern coastal and shelf areas. With the exception of river mouth and adjacent areas, the accumulation of sediments in coastal areas is currently limited and an estimated 70% of continental shelf areas is covered by ancient "relict" sediments (toplap) where interaction with biological and chemical processes may locally favor authigenesis of phosphates and glauconite. Some major river systems like the Amazon are even strong enough to allow partial deposition of their terrigenous load directly to the shelf edge and slope.

In warm water areas remote from terrigenous influences like the Florida coast, carbonate mounds made of debris from benthic organisms like bryozoans, algae,

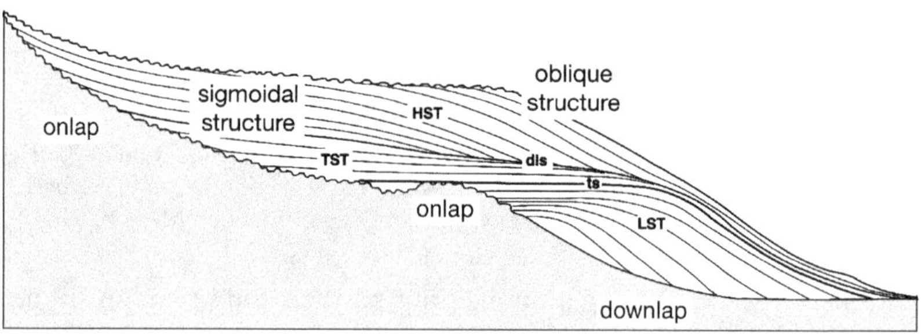

Figure 6.8 Schematic cross section of a typical shelf sequence which accumulated during an interval of sea level rise. The sequence rests on an erosion surface developed during a preceding fall of sea level and has been partly eroded during a subsequent fall of sea level. LST, lowstand system track; ts, transgressive surface; TST, transgressive system track; dis, discontinuity; HST, highstand system track. Modified from Boillot, G., Coulon, C., 1998. La déchirure continentale et l'ouverture océanique. Overseas Publishers Association, Amsterdam.

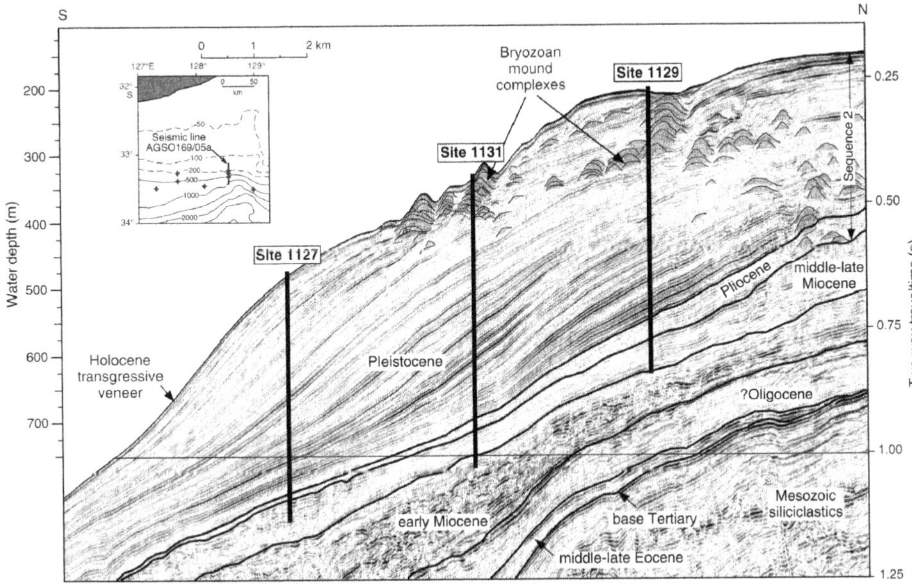

Figure 6.9 Seismic reflection profile across the shelf and upper slope of the southern Australian margin (Great Australian Bight). Note the thick Pleistocene sequence which includes a number of Bryozoan mound complexes near shelf edge. Reprinted from Feary, D.A., Hine, A.C., Malone, M.J., et al., 2000. Proceedings of the Ocean Drilling Program, Initial Reports, volume 182, Ocean Drilling Program, College Station, TX.

phanerogams and other bioclastic elements may accumulate and coral reefs may expand, according to the degree of hydrodynamism. Although typical coral reefs are usually restricted to warm waters of high hydrodynamism, carbonate mounds may extend to areas of lower temperature and/or lower hydrodynamism. For example, carbonate mounds are typical constituents of the Great Australian Bight, on the northern passive margin of the Southern Ocean. There, Pleistocene carbonate mounds developed on the lower shelf during intervals of low sea level, covered by fine blankets of mud during sea level highs (Figure 6.9). Carbonate platforms made of a succession of carbonate mounds and/or coral reefs may accumulate over considerable thicknesses through geologic time in areas distant from persistent, significant sources of terrigenous elements, as to dominate in some continental shelves. This is the case in parts of the Central Atlantic off Morocco and the southern United States, and on fossil passive margins of the Tethys Ocean (see Section 9.3).

6.1.2.2. Progradation of continental shelves and sediment transfer across continental slopes

Continental slopes represent active progradation fronts and transit areas for terrigenous sediment particles, especially in regions of high terrigenous input. Their average inclination is of about 3°, lower values being recorded near the transition to abyssal plains. Accumulations of terrigenous sediments freshly transported to the outer shelf and slope by major river systems and nearshore processes have a high water content and are unstable. They are easily reworked and transported along continental slopes as gravitational bodies (slides, slumps), laminar viscoplastic flows (debris flows, mud flows) and hyperconcentrated flows (hyperpycnal flows, turbidity currents) principally (see Section 2.3), progressively feeding the lower slope and the deepest areas of oceanic basins.

Continental slopes are frequently incised by submarine canyons where most of the downslope sediment transfer occurs, in the form of hyperpycnal flows and turbidity currents principally. Most canyons are located off river mouths, and have been connected to adjacent river systems during intervals of low sea level and related strengthening of downstream erosion. Submarine canyons frequently develop in close proximity and relationship to geological structures of regional importance such as fault systems, suggesting a partial control of their morphology by ancient tectonism. Retrogressive erosion during intervals of lower relative sea level, together with the eroding action of sediment flows, maintain the capacity of submarine canyons to ensure downslope sediment transfer through time. Submarine canyons commonly show a succession of active and abandoned meandering channels associated to flat and stacked terraces and levees, indicative of successive phases of intense activity. It is assumed that spill over from turbidity currents favors the accumulation of sediments on terraces and levees. Additional contribution may come from tributaries, and gravity flows along the canyon walls which are sometimes relatively steep. For example, the Capbreton Canyon is a 300 km long structure located near the North Pyrenean front in the Bay of Biscay (Figure 6.10). The upper part of the canyon was naturally connected to the Adour River which drains the Western Pyrenees until the river mouth was artificially shifted from15 km to the South during the 16th century AD, while its lower part is connected to the Cap-Ferret deep sea fan. The Capbreton Canyon is a unique structure which starts at 30 m water depth on the shelf (about 250 m from the shoreline), has an average slope of 1.3% in its upper part and a maximum width of 32 km. Gravity core studies suggest that sediment discharge from the Adour river and turbidity current activity through the canyon were significantly higher during the last deglaciation than the Holocene, but occurred only sporadically in modern times. One such event resulted in the accumulation of a 30 cm thick turbidite near the main channel of the canyon in December 1999. No earthquake nor flooding event were recorded at that time, and the turbidity current was most probably generated during the "Martin" storm of December 27, which was a centennial event in the Bay of Biscay where wave height reached up to12 m. It is likely that wave and current activity during the storm increased the shear force exerted on unstable surface sediments, triggering the turbidity current.

Continental rises have a low declivity ($<1^\circ$ on average) and make the transition from continental slopes to abyssal plains. They consist of accumulations of terrigenous sediments principally, fed by downslope gravity processes and later reworked by bottom currents. Continental rises show alternating deep sea fans where turbidites dominate, and sediment drifts where contourites dominate. Elongated deep sea fans develop downstream of submarine canyons and major river systems, and are characterized by the partition of the main channel into an increasing number of secondary channels and sedimentary lobes as a function of the decreasing energy and

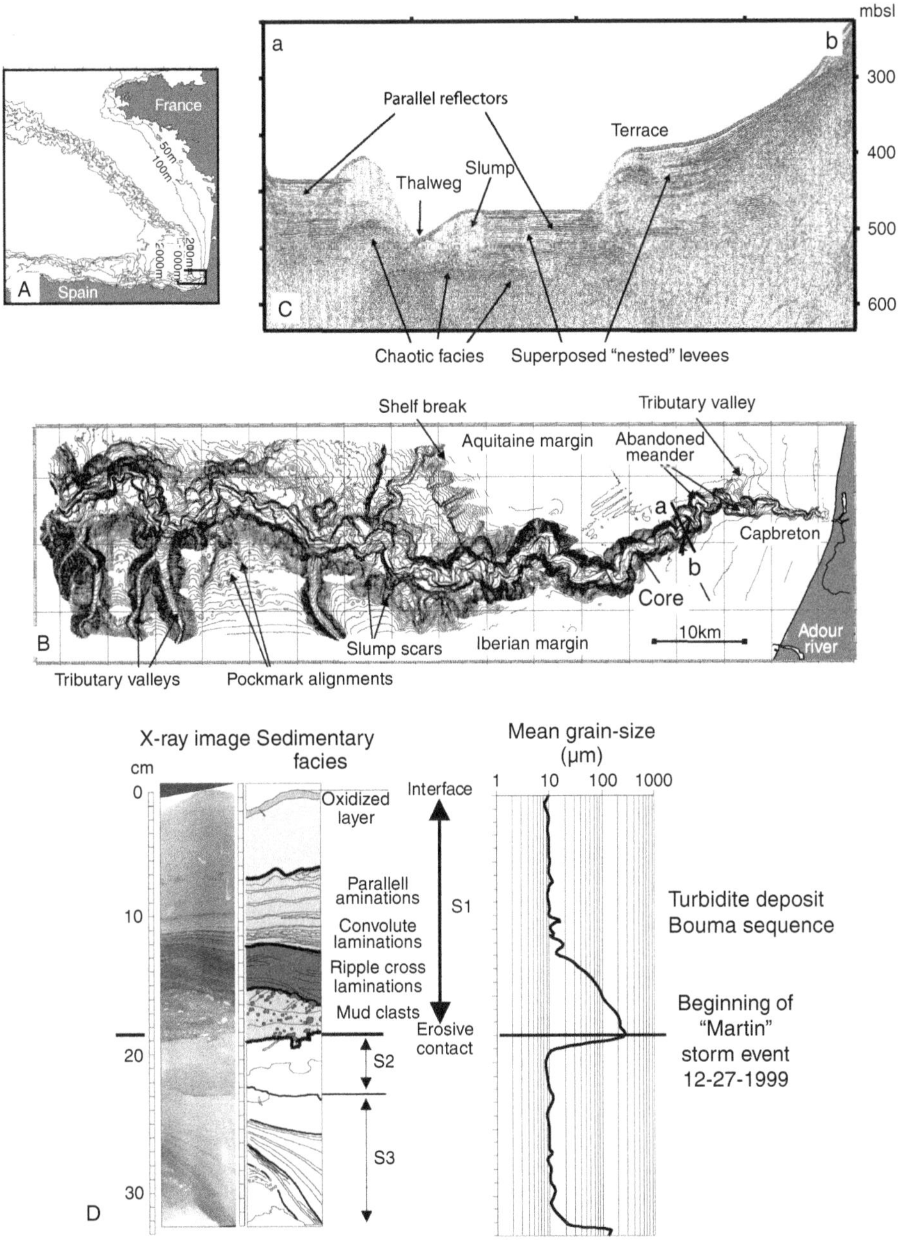

Figure 6.10 The 1999 "Martin" turbidite event in the Capbreton canyon (Bay of Biscay, NE Atlantic Ocean). A, location map; B, bathymetry and main features of the Capbreton canyon, with location of seismic profile and gravity core; C, NW-SE seismic profile across the upper Capbreton canyon; D, sedimentological data from a 30 cm long gravity core taken from a terrace near the main channel of the upper Capbreton canyon soon after the "Martin" storm event of 1999. Note the succession of levees and terraces on seismic profile and the typical Bouma sequence (S1) recovered by coring. Modified from Mulder, T., Cirac, P., Gaudin, M., Bourillet, J.-F., Tranier, J., Normand, A., Weber, O., Griboulard, R., Jouanneau, J.-M., Anschutz, P., Jorissen, F.J., 2004. Understanding continent-ocean sediment transfer. EOS, American Geophysical Union, 85, 257–262.

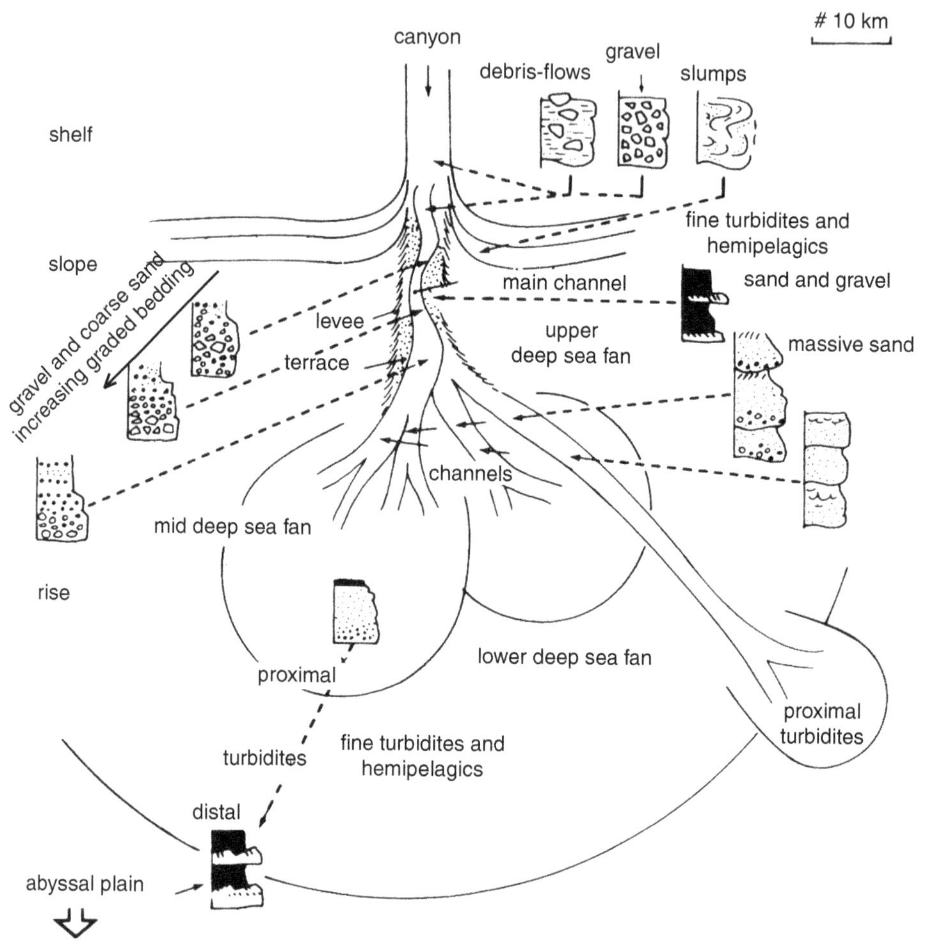

Figure 6.11 Schematic organization of an elongated deep sea fan. Note seaward decrease in grain size and alternations of distal turbidites and hemipelagics in the lower fan. Deep sea fan units are dynamic, and the mid fan may prograde over the lower fan. Modified from Chamley, H., 2000. Bases de Sédimentologie. Dunod, Paris.

velocity of turbidity flows. As a consequence, the mean grain size of the sediment decreases with distance from the feeder canyon. Deep sea fans are divided into three dynamic units, based on sediment structure and facies (Figure 6.11):

- The proximal (upper) deep sea fan is generally characterized by a single, meandering channel in continuity with the feeder submarine canyon, bounded by levees and terraces. The main channel is typically meters to tens of meters deep and hundreds of meters wide, and principally ensures the downslope transit of turbidity currents at relatively high velocities. The main channel therefore only contains structureless to roughly graded very coarse sand and gravel. Turbidity currents frequently spill off the main channel and rapidly loose energy laterally, facilitating the accumulation of coarse to fine turbidite sequences and aggradating lateral levees and terraces. Proximal deep sea fan channels may migrate laterally or avulse through time, leading to the accumulation of stacked lenticular bodies made of coarse structureless sand turbidites, frequently incised by erosion surfaces.

- The mid deep sea fan is characterized by the presence of secondary channels and the formation of sedimentary lobes. The lateral levees decrease in importance and mid deep sea fan channels are generally floored by massive sand deposits. Individual lobes extend over distances of kilometers to tens of kilometers. They are prograding structures made of a succession of turbidites fed by secondary channels, which accumulate until channel migration or avulsion initiates the formation of a new sedimentary lobe. Turbidites from the mid fan are dominated by sands and mostly include the lower units of the Bouma sequence (proximal turbidites). As a result, mid deep sea fan deposits consist of stacked elongated lenses dominated by proximal, sandy turbidites.
- The distal (lower) deep sea fan is characterized by rare, minor channels and the dominance of fine (mud) turbidites. Turbidity currents may eventually reach the distal fan but they have lost most of their energy and have already deposited most of their coarsest load. Turbidites from the distal fan are dominated by silts and clays and mostly include the upper units of the Bouma sequence. Fine turbidites from the distal fan alternate with hemipelagic sediments, which are frequently bioturbated.

Being fed by major river systems and submarine canyons, deep sea fans accumulate terrigenous elements principally, eroded from continental drainage basins and continental shelves. They are therefore very sensitive to variations in continental climate and morphology, as well as to relative changes of sea level. As a consequence, the locus of deep sea fan sedimentation may vary through time with the dynamics of downslope sediment transfer, in the same manner as continental shelf sequences (Figure 6.12). However at geological time scale, a succession of facies from distal to mid and proximal fan deposits and upward coarsening of grain size at a single observational point illustrate the progression of the prograding continental slope through time. A detailed study of the Late Quaternary Amazon

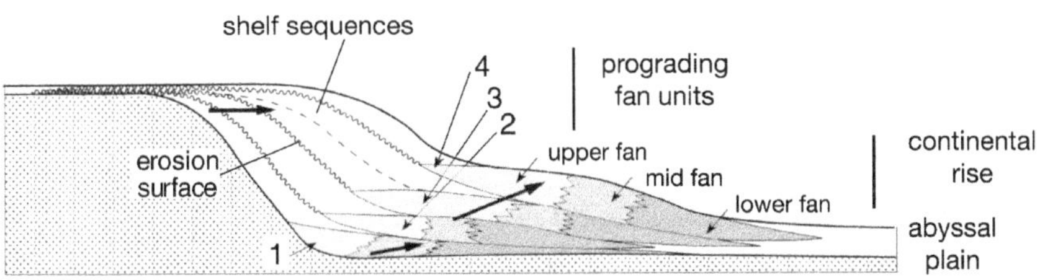

Figure 6.12 Progradation of deep sea fan facies as illustrated by a succession of four sequences and relationships with continental shelf and abyssal plain sediments. Note the coincidence between maximum development of deep sea fan sequences and intervals of shelf erosion/non-deposition which characterize intervals of low and stable sea level. Modified from Boillot, G., Coulon, C., 1998. La déchirure continentale et l'ouverture océanique. Overseas Publishers Association, Amsterdam.

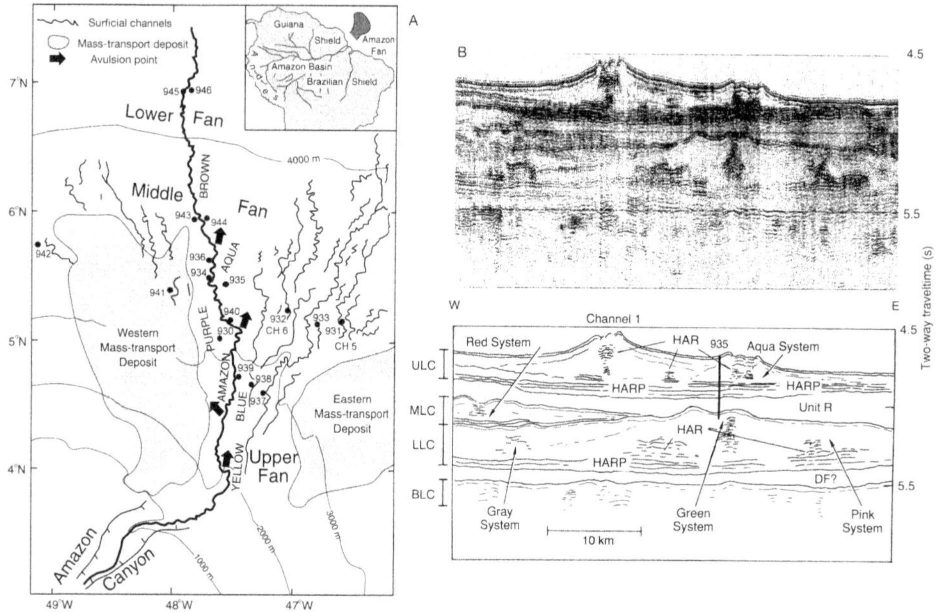

Figure 6.13 The Amazon Fan. A, map of the Amazon fan showing the location of ODP Leg 155 sites and names of the channels of the Upper Levee Complex. Note that the currently active Amazon Channel is represented by a thick line. B, W-E seismic profile across the Middle Amazon Fan in the vicinity of ODP Site 935, highlighting a succession of Late Pleistocene (last 250 kyr) channels and levees. ULC, Upper Levee Complex; MLC, Middle Levee Complex; LLC, Lower Levee Complex; BLC, Bottom Levee Complex; DF and Unit R, Debris flows; HAR, High amplitude reflections; HARP, High amplitude reflection packets. Note that ULC accumulated during isotope stages 2–4, MLC during isotope stage 6, LLC during the early stage 7 and BLC during stage 8. Reprinted from Flood, R.D., Piper, D.J.W., Klaus, A., et al., 1995. Proceedings of the Ocean Drilling Program, Initial Reports, volume 155. Ocean Drilling Program, College Station, TX.

deep sea fan has been conducted during ODP Leg 155 and brings additional information on the dynamics of deep sea fans (Figure 6.13):

- aggradation of the Amazon deep sea fan mostly progresses during lowstands of sea level, whereas the accumulation of sediments switches to new depocenters (lobes) after highstands;
- deep sea fan sediments are frequently replaced by mass-transport deposits during major intervals of falling sea level;
- only one distributary channel is active at any given time, avulsion being related to local conditions (sediment fluxes, levee wall failure, ...); and
- average sedimentation rates may reach up to 25 m/kyr on levees of active channels which may transport coarse elements down to the lower fan on exceptional occasions.

6.1.2.3. The evolving morphology of passive continental margins

The lithostatic pressure exerted by continental shelves and slopes on passive continental margins of thinned and intermediate lithosphere reinforces the consequences of thermal subsidence. For isostatic reasons, the sediment load increases the subsidence of the lithosphere below continental shelves and slopes, facilitating further accumulation of sediments and progradation (Figure 6.14). To compensate, adjacent portions of thicker lithosphere slightly uplift, facilitating further erosion of continental surfaces. It is assumed that an initial escarpment is formed by vertical displacement of normal faults inherited from the rift and/or crustal fissure stages and retreats inland as erosion progresses, maximum denudation rates being recorded downstream of the retreating scarp. Isostatic adjustment can progressively lead to an upwarp relief parallel to the coast. In many cases, the presence of rivers and drainage divides leads to rapid incision of streams and development of escarpments near the initial position of drainage divides (Figure 6.15). In fact, these conditions produce a downwearing pattern of denudation and significant erosion. Estimated denudation rates are of the order of a few tens of meters per million years. Continuous uplift and denudation of coastal relief and sedimentation and subsidence of shelf and slope areas progressively produce a bending of the passive continental margin. In turn, the coastal morphology of passive continental margins (including the extension of coastal plains and position of the shoreline) is partly controlled by

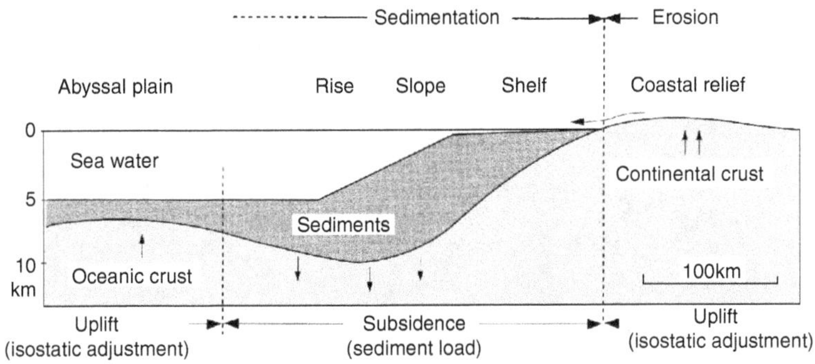

Figure 6.14 Influence of sediment load on the morphology and structure of passive margins. Note the formation of a bulge of oceanic crust and the persistence of coastal relief and erosion as continental shelf and slope progress. Modified from Boillot, G., Coulon, C., 1998. La déchirure continentale et l'ouverture océanique. Overseas Publishers Association, Amsterdam.

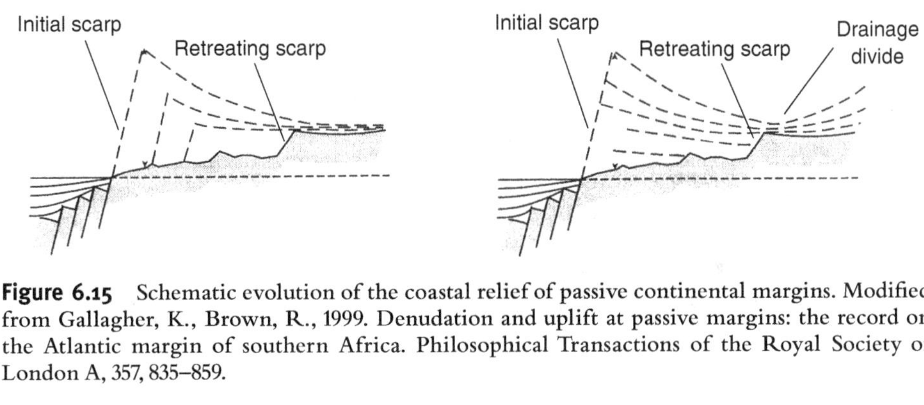

Figure 6.15 Schematic evolution of the coastal relief of passive continental margins. Modified from Gallagher, K., Brown, R., 1999. Denudation and uplift at passive margins: the record on the Atlantic margin of southern Africa. Philosophical Transactions of the Royal Society of London A, 357, 835–859.

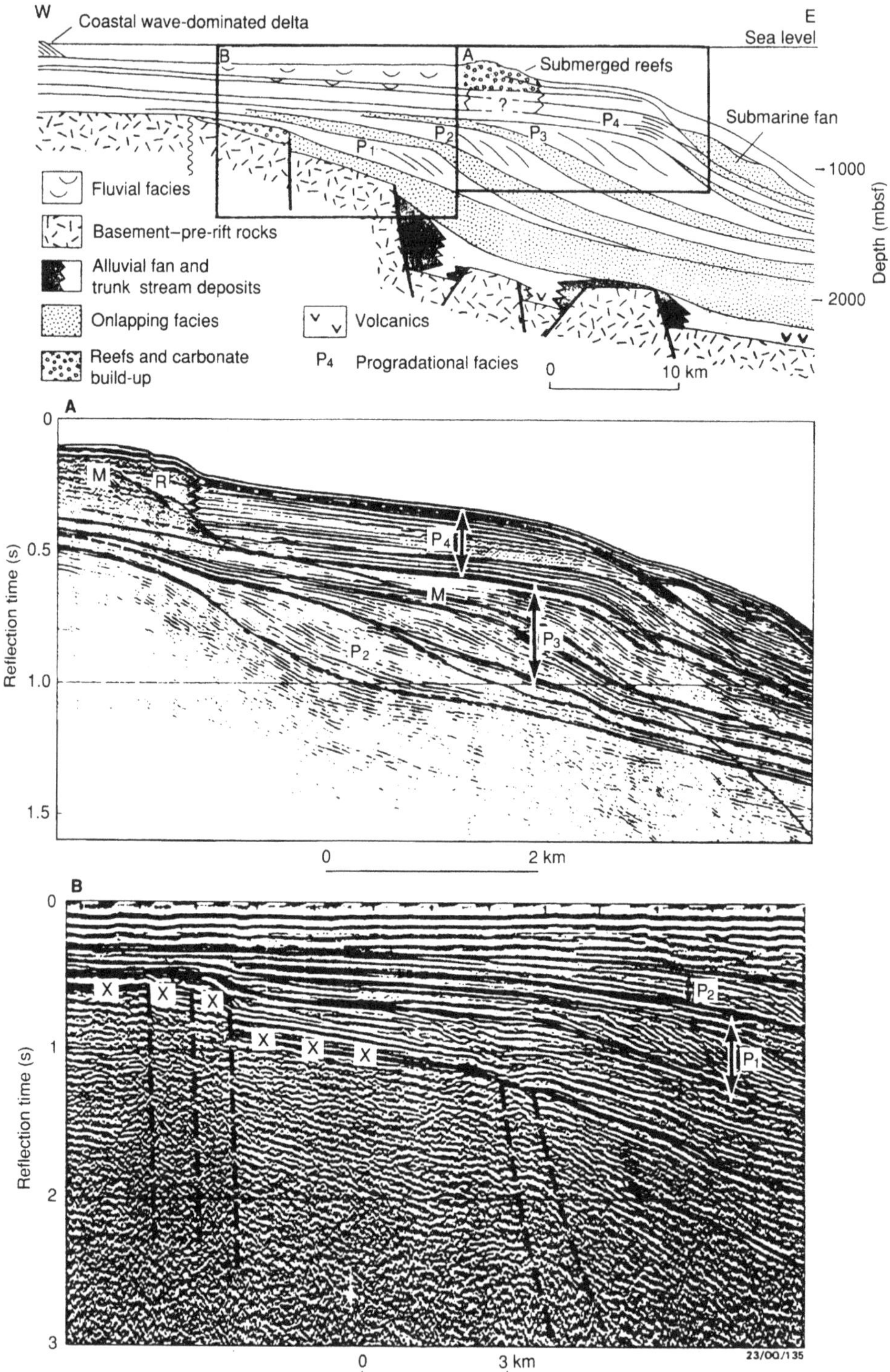

W
Coastal wave-dominated delta
E
Sea level
B
A
Submerged reefs
?
P4
P3
P2
P1
Submarine fan
1000
2000
Depth (mbsf)
Fluvial facies
Basement–pre-rift rocks
Alluvial fan and trunk stream deposits
Onlapping facies
Volcanics
Reefs and carbonate build-up
P4 Progradational facies
0
10 km
A
M
R
P4
M
P3
P2
Reflection time (s)
0
0.5
1.0
1.5
0
2 km
B
X
P2
P1
Reflection time (s)
0
1
2
3
0
3 km
23/OO/135

isostasy, and isostatic adjustment progressively increases the inclination of fragile areas of thinned and intermediate lithosphere.

However, the morphology of passive continental margins is principally controlled by tectonic traits inherited from the rift and crustal fissure stages, as well as by the nature and quantity of sediments which accumulate through time. Passive continental margins commonly consist of predominantly siliciclastic sediments which cover areas of thinned continental, intermediate and sometimes old oceanic lithosphere. For example, the Northeast Australian margin of the Coral Sea NE of Cairns is dominated by 1,000–1,500 m of prograding fluvio-deltaic and onlapping slope siliciclastics (Figure 6.16). There, only 100–150 m of reef carbonates accumulated during the Pliocene and the Pleistocene as the Great Barrier extended southward.

In areas where the absence of significant continental drainage basin and/or persistently dry climatic conditions result in low siliciclastic accumulation rates, carbonate platforms made of coral reefs and/or carbonate mound complexes (bryozoans, foraminifers, algae, ...) may develop, providing that surface seawater temperatures and hydrodynamism are important enough. This is currently the case in many areas of low latitudes, where these conditions remained roughly the same since the Miocene. For example, carbonate platforms developed as early as the Early and Middle Miocene in some areas of the Northeast Australian margin: the distal Queensland Plateau NE of Cairns, and the Marion Plateau SE of Townsville, where terrigenous supply is low (Figure 6.17).

Where seawater conditions are unfavorable to the development of carbonate formations, a relatively thin blanket of predominantly siliciclastic sediments covers areas of thinned and intermediate lithosphere. Locally, tilted blocks of continental crust and upper mantle rocks may even outcrop on the ocean floor. These points are characteristic of starved margins, which are more frequent in middle to high latitude areas. For example, the European margin of the North Atlantic is a typical starved margin (Figure 6.18): a significant proportion of continental areas is submerged by shallow seas (Celtic Sea, English Channel, ...), the continental drainage basins are relatively short, and most terrigenous elements eroded from Western Europe have been trapped into adjacent geological structures like the North Sea, the London-Paris basin and the Mediterranean Sea. As a consequence, less than 1,000 m of post-rift sediments accumulated in many areas since the Albian, and the Hercynian basement still outcrops as escarpments of NW-SE direction over distances of several tens of kilometers. Starved continental margins have been extensively studied because they provide easy access to the early stages of ocean opening and crustal processes.

In regions of important continental drainage basins and/or persistently high rainfall, thick accumulations of terrigenous sediments (commonly several kilometers

Figure 6.16 Seismic profiles across the Australian passive margin of the Coral Sea off Cairns (A and B), and their interpretation. M: multiple; P1 to P4, siliciclastic prograding units; R, submerged reef; X, basement structure. Note the predominance of thick prograding terrigenous facies on the continental shelf and slope, while occurrences of coral reefs are limited to the most recent sequences in association with thin terrigenous units. Reprinted from Davies, P.J., McKenzie, J.A., Palmer-Julson, A.A., et al., 1991. Proceedings of the Ocean Drilling Program, Initial Reports, volume 133. Ocean Drilling Program, College Station, TX.

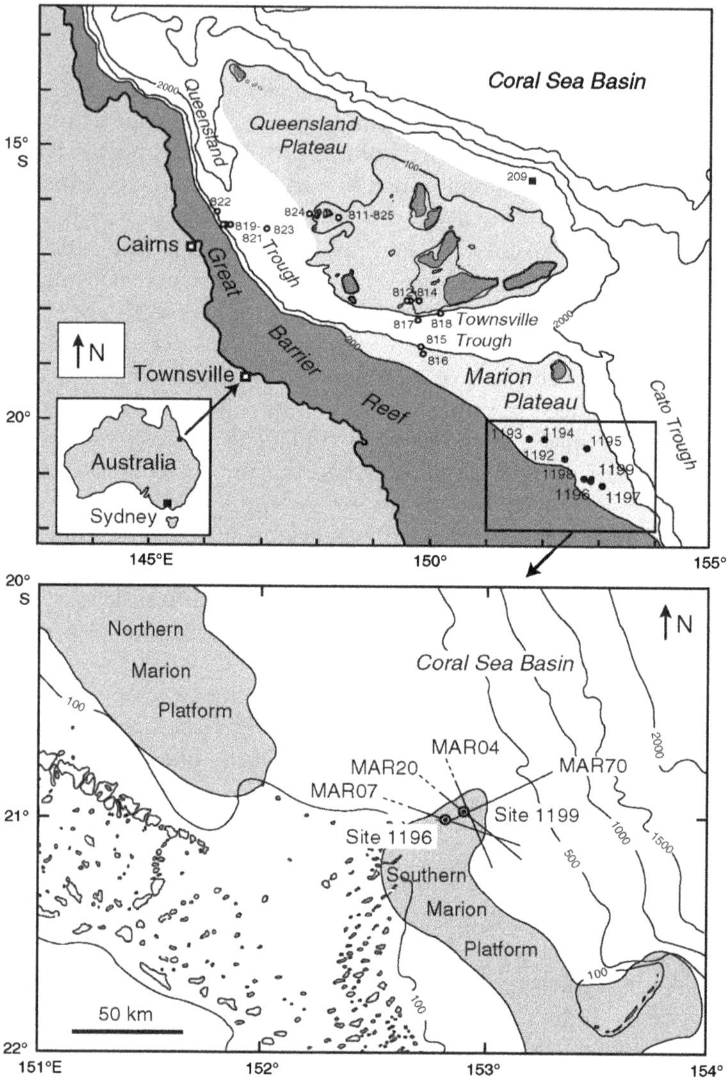

Figure 6.17 Map of the Australian passive margin of the Coral Sea showing the modern distribution of coral reef (dark gray) and bioclastic (light gray) facies and the location of ODP sites and seismic profiles. Note the geographical extension of both facies at a time of relatively low terrigenous fluxes. Reprinted from Conesa, G.A.R., et al., 2004. In: Anselmetti, F.S., Isern, A.R., Blum, P., Betzler, C. (Editors), Proceedings of the Ocean Drilling, Scientific Results, volume 194. Ocean Drilling Program, College Station, TX.

Figure 6.18 A typical starved margin, the passive European margin of the North Atlantic. A, location map; thick lines, seismic profiles; numbers, DSDP sites. Note the importance of epicontinental seas and intraplate basins in Western Europe. B and C, typical seismic profile and interpretation of a seismic survey of the Goban Spur area. Note outcrops of tilted blocks of Hercynian basement rocks and thin Cretaceous and Cenozoic sediment cover. Modified from de Graciansky, P.C., Poag, C.W., et al., 1985. Initial Reports of the Deep Sea Drilling Project, volume 80. U.S. Gov. Print. Office, Washington, DC.

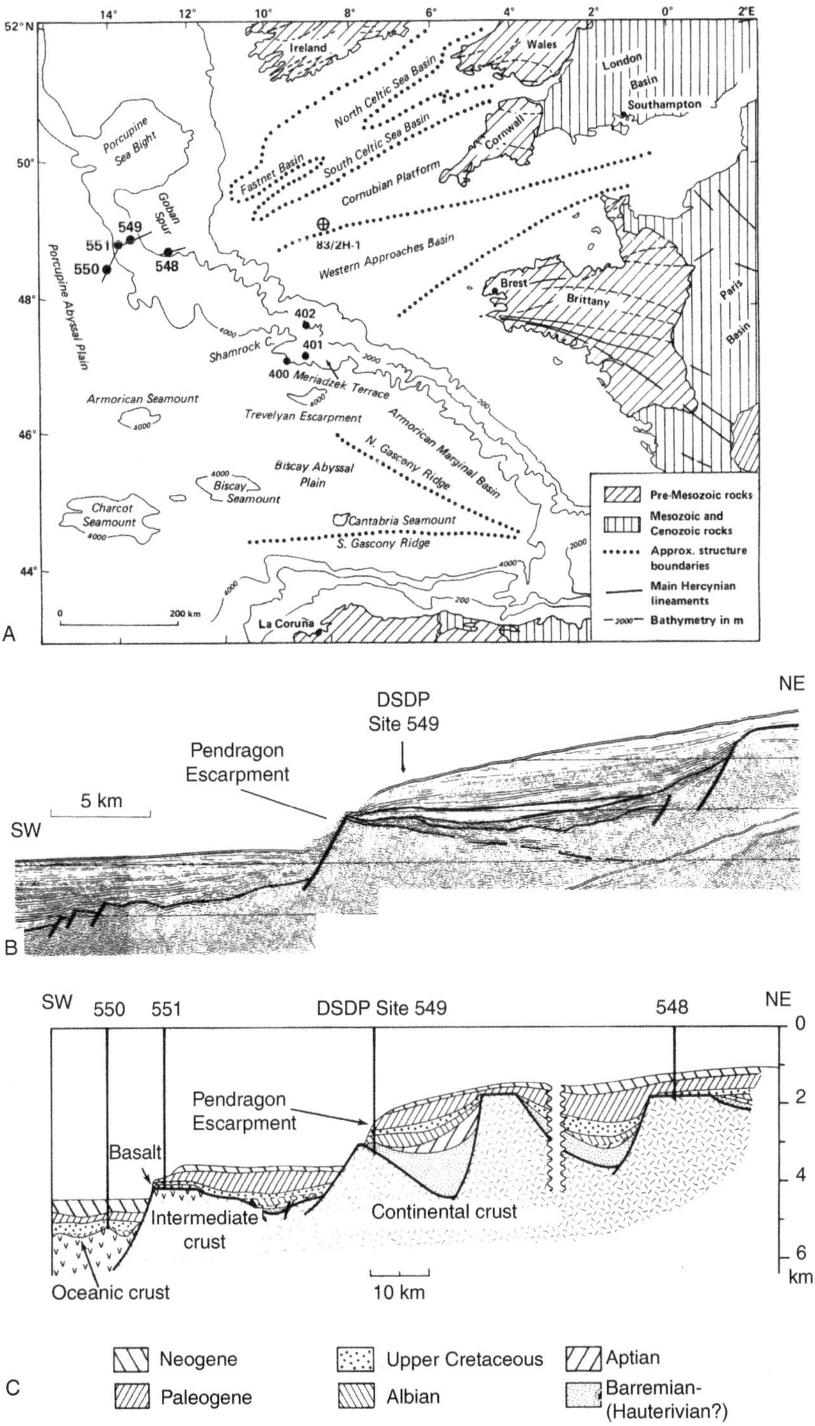
A
52°N
50°
48°
46°
44°
14°
12°
10°
8°
6°
4°
2°
0°
2°E
Ireland
Wales
London Basin
Southampton
North Celtic Sea Basin
South Celtic Sea Basin
Fastnet Basin
Cornwall
Cornubian Platform
Porcupine Sea Bight
Goban Spur
549
551
550
548
83/2H-1
Western Approaches Basin
Brest
Brittany
Paris Basin
Porcupine Abyssal Plain
402
401
400
Shamrock C.
Meriadzek Terrace
Armorican Seamount
Trevelyan Escarpment
Armorican Marginal Basin
N. Gascony Ridge
Biscay Abyssal Plain
Biscay Seamount
Charcot Seamount
Cantabria Seamount
S. Gascony Ridge
La Coruña
200 km
Pre-Mesozoic rocks
Mesozoic and Cenozoic rocks
Approx. structure boundaries
Main Hercynian lineaments
Bathymetry in m
B
SW
NE
5 km
Pendragon Escarpment
DSDP Site 549
C
SW
550
551
DSDP Site 549
548
NE
0
2
4
6
km
Pendragon Escarpment
Basalt
Intermediate crust
Continental crust
Oceanic crust
10 km
Neogene
Paleogene
Upper Cretaceous
Albian
Aptian
Barremian-(Hauterivian?)

thick) drape areas of thinned and intermediate lithosphere (Figure 6.19). As a result of sediment surcharge, some deformation (syn-sedimentary faulting and folding) may affect parts of the sediment cover. In areas where evaporites formed during earlier stages of ocean opening, the sediment load may initiate and sustain intense diapirism. These points are characteristic of fat margins, which are more frequent in low latitude areas. For example, fat margins are typical of the equatorial South Atlantic where major river systems such as the Amazon, the Niger and the Congo drain vast continental areas of high rainfall, significant relief and intense erosion. Fat margins are also present in glaciated high latitude areas, because of the capacity of glaciers to erode and transport significant quantities of terrigenous elements to coastal areas. This is the case in many Antarctic areas, as for example in the Ross Sea and on the Wilkes Land margin of the Southern Ocean, where sediment accumulation reaches up to 2,500 m although significant erosion marked by unconformities of regional importance occurred during intervals of glacial progression on the continental shelf and related sea level fall (Figure 6.20). Fat margins may also develop in the absence of significant terrigenous fluxes, in areas of high carbonate productivity. This is the case on the Florida and African margins of the Central Atlantic where about 10 km of carbonate platform sediments accumulated since the Jurassic (Figure 6.21), and on many fossil passive margins of the Tethys Ocean.

In regions of transform and pull-apart activity, the mature ocean stage begins when the active area of the mid-oceanic ridge clears the adjacent continental crust

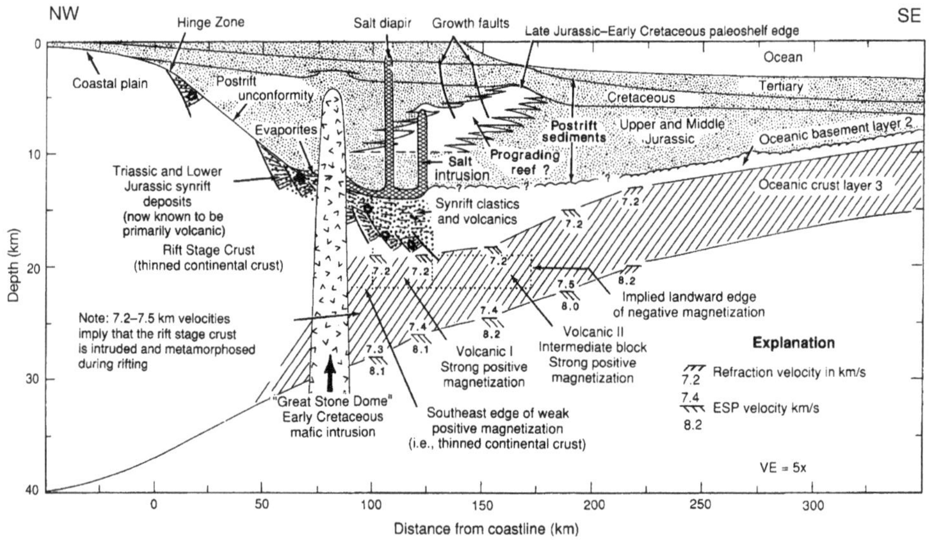

Figure 6.19 Schematic cross section of a typical fat margin, the passive New-Jersey margin of the North Atlantic, as interpreted from reflection and refraction seismic surveys. Note the thickness (up to 15 km) of post-rift sediments, increased subsidence below the shelf (compare to Figure 6.14), formation of salt diapirs and faults facilitated by high sediment load, and Late Jurassic/Cretaceous transgression over continental basement. Reprinted from Austin, J.A., Christie-Blick, N., Malone, M.J., et al., 1998. Proceedings of the Ocean Drilling Program, Initial Reports, volume 174A. Ocean Drilling Program, College Station, TX.

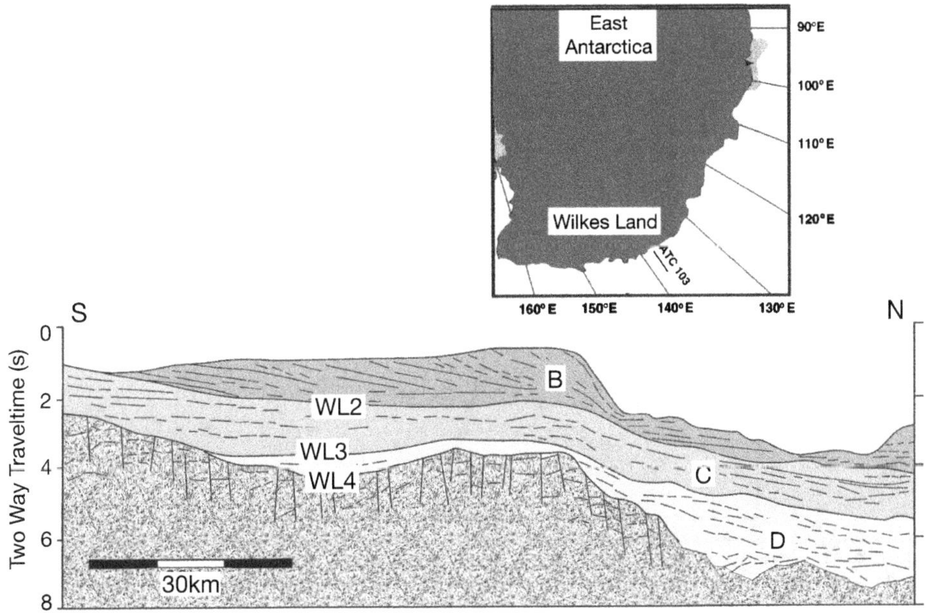

Figure 6.20 Schematic interpretation of a seismic profile across the fat passive margin of the Wilkes Land of East Antarctica. B to D, major siliciclastic sequences; WL2 to WL4, major discontinuities. Note the importance of erosional features in a shelf environment dominated by ice processes. Modified from Anderson, J.B., 1999. Antarctic Marine Geology. Cambridge University Press, Cambridge.

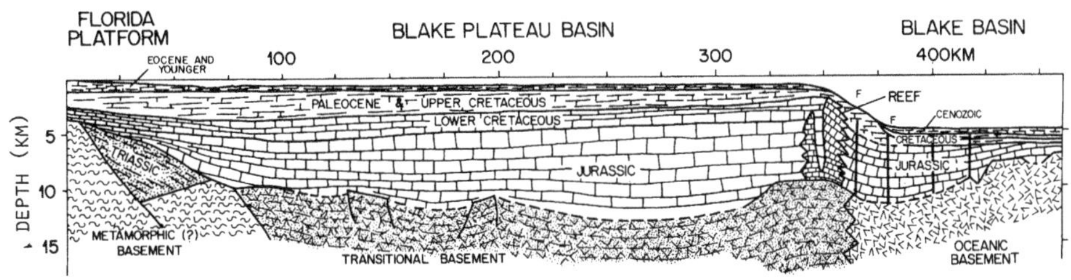

Figure 6.21 Schematic interpretation of a seismic survey of the Blake Plateau of the western Central Atlantic. Note the accumulation of reef and carbonate platform facies of Jurassic and Cretaceous age, when this region was part of the northern passive margin of the Tethys Ocean. Reprinted from Talwani, M., Hay, W.W., Ryan, W.B.F. (Editors), 1979. Deep Drilling Results in the Atlantic Ocean: continental margins and paleoenvironment, American Geophysical Union, Washington.

(Figure 6.22). Strike-slip activity does not involve portions of continental and oceanic lithosphere anymore but two segments of oceanic lithosphere, and thermal subsidence accelerates. Transform margins such as the Ghana-Ivory Coast margin of the South Atlantic are characterized by an abrupt transition from continental to oceanic lithosphere (over distances of 10–15 km), where thick continental crust is separated from thin oceanic crust by a narrow fracture zone (Figure 6.23). Transform margins are predominantly in a straight line, which is locally interrupted

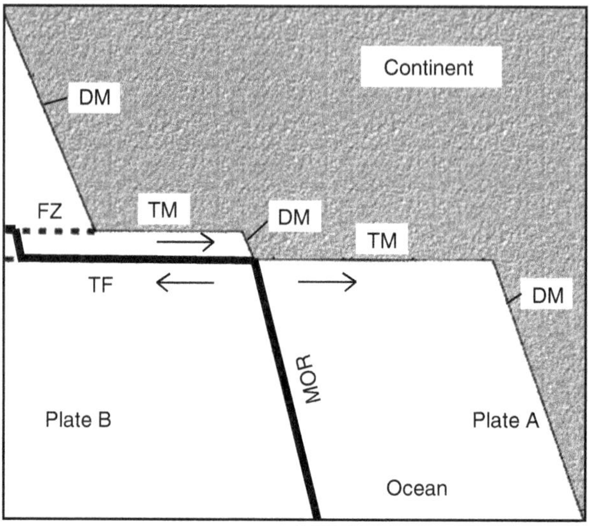

Figure 6.22 Schematic model of a transform margin. Thick line represents active plate boundary. DM, divergent margin; FZ, fracture zone; MOR, mid-oceanic ridge; TF, transform fault; TM, transform margin. Note that the transform margin is interrupted by a segment of divergent margin, and that the active mid-oceanic ridge clears the continental crust. Modified from Boillot, G., Coulon, C., 1998. La déchirure continentale et l'ouverture océanique. Overseas Publishers Association, Amsterdam.

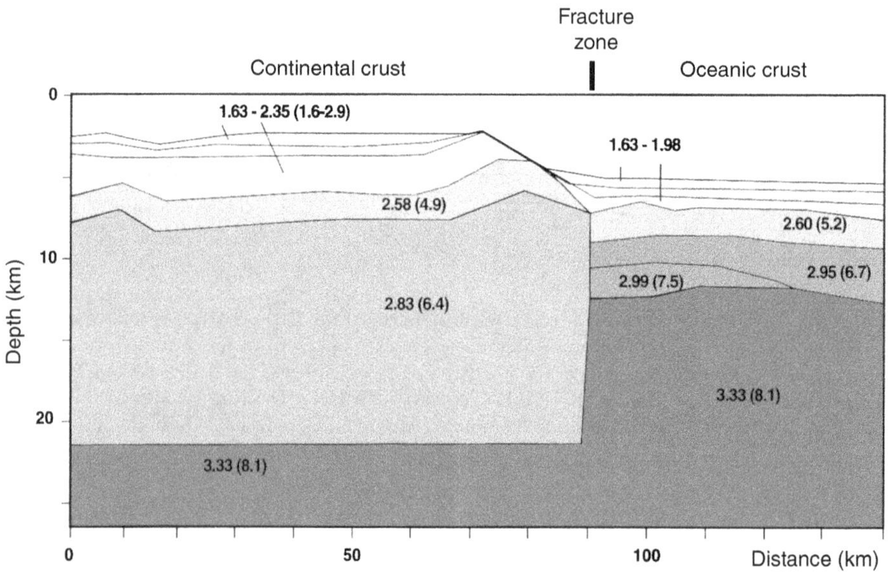

Figure 6.23 Crustal structure of a typical transform margin, the Ghana-Ivory Coast margin of the South Atlantic, as deduced from seismic refraction data. Densities and seismic velocities (in brackets) are indicated. Note the abrupt transition from continental to oceanic crust. Modified from Boillot, G., Coulon, C., 1998. La déchirure continentale et l'ouverture océanique. Overseas Publishers Association, Amsterdam.

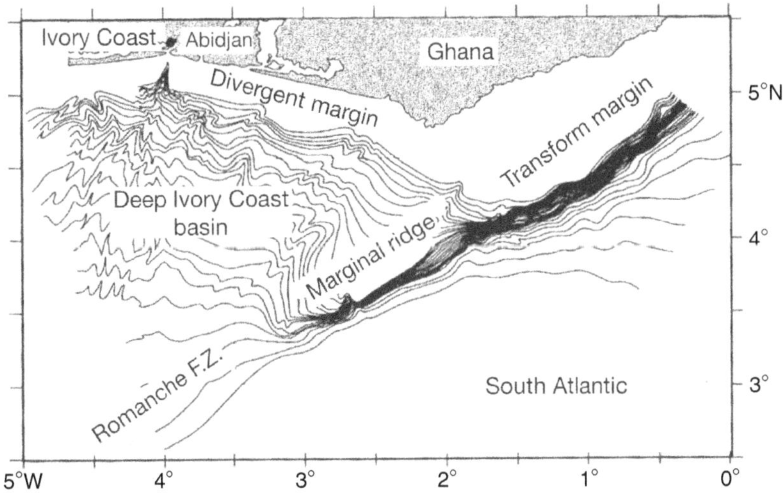

Figure 6.24 Bathymetry of the Ivory Coast and Ghana margin of the South Atlantic. Note the transition from transform to divergent margin and the presence of an elongated marginal ridge in the direction of the strike-slip activity. The Deep Ivory Coast Basin is located on thinned continental lithosphere and bounded by a fracture zone and a divergent margin, and represents part of a pull-apart basin of NE-SW extension. Modified from Boillot, G., Coulon, C., 1998. La déchirure continentale et l'ouverture océanique. Overseas Publishers Association, Amsterdam.

by segments of divergent passive margins where the continental lithosphere has been thinned during the rift and crustal fissure stages (Figure 6.22). The transition from transform to divergent margins may be associated to a marginal ridge, made of syn-rift deposits. Marginal ridges may result either from differential subsidence (lower subsidence in areas of thicker continental crust) or from deformation and uplift during intervals of transpression (Figure 6.24).

6.2. Example of a Starved Passive Margin: The Goban Spur Area of the Celtic Continental Margin

6.2.1. Structure and Early History of the Goban Spur Area

The starved Celtic passive margin of Western Europe extends from Rockall Plateau, which is part of the North Atlantic Igneous Province, to the Bay of Biscay. A segment of this margin has been extensively investigated in the Goban Spur area (Figure 6.18), during DSDP Leg 80 and submersible expeditions. Goban Spur is a marginal plateau bounded to the North by the Porcupine Sea Bight which is a marginal basin, and to the South by the Jean Charcot Escarpment and a series of submarine canyons. Goban Spur deepens abruptly to the West along two escarpments to the Porcupine Abyssal Plain below 4,400 m water depth (Figure 6.25). The Jean Charcot escarpment follows a hercynian direction, whereas the western escarpments have the same variscan orientation as major regional faults and basement highs inherited from the rifting stage of the North Atlantic.

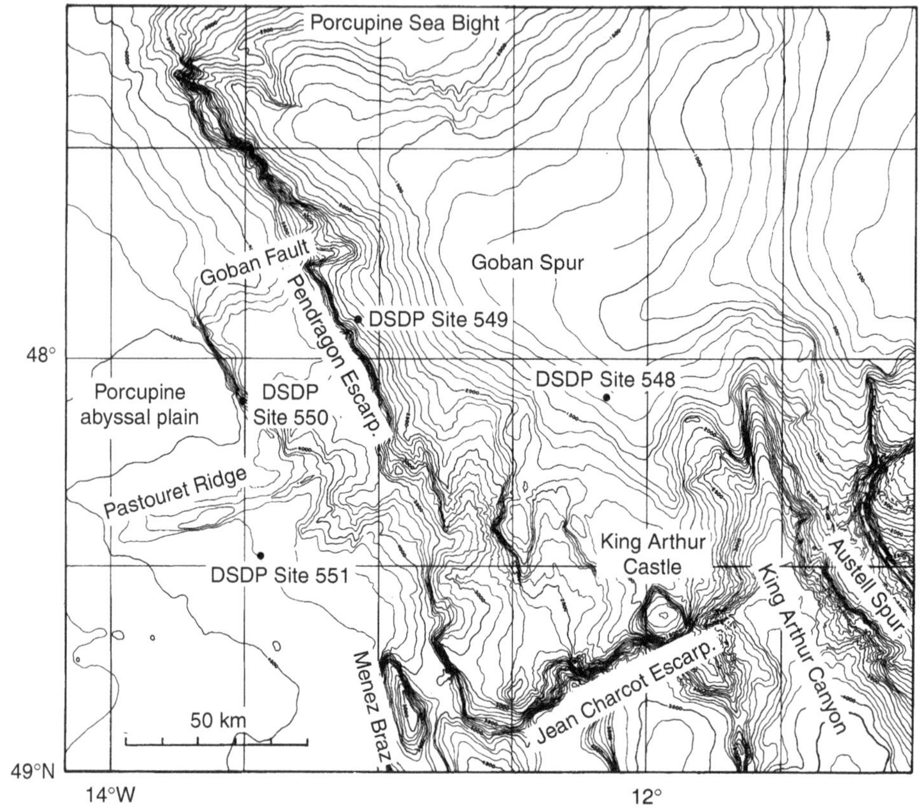

Figure 6.25 Detailed bathymetric map of the Goban Spur area of the starved Celtic passive margin of the North Atlantic. Equidistance of isobaths is 100 m. See Figure 6.18 for location. Note that Goban Spur is bounded by escarpments inherited from rifting, of NW-SE (variscan) and SW-NE (hercynian) directions. Reprinted from de Graciansky, P.C., Poag, C.W., et al., 1985. Initial Reports of the Deep Sea Drilling Project, volume 80. U.S. Gov. Print. Office, Washington, DC.

Seismic profiles indicate that the Goban Spur margin is characterized by a series of tilted fault-blocks, horsts and grabens. The tilted basement blocks outcrop along the escarpments (Figure 6.26). Basement rocks have been sampled at DSDP Sites 548 and 549 where they consist of Devonian quartzites and black-shales which have been metamorphosed during the Hercynian orogeny, and during *Cyana* dives on Pendragon Escarpment where they consist of Paleozoic schists and gneisses. Permo-Triassic sequences exist only farther East on the margin and in the English Channel, and upper Jurassic shallow water limestones of peri-reefal environment cap basement highs farther South in the Meriadzek Terrace area (see Section 4.3). In the Goban Spur area, syn-rift sediments are of Barremian to Early Albian age. Above the basement, seismic profiles define three major units. The lower syn-rift unit (Figure 6.26) fills wedge-shaped, fault-bounded basins, and shows successions of divergent reflectors indicating that deposition took place during rotation of the basement blocks and chaotic reflectors suggesting intense syn-sedimentary faulting. Syn-rift sediments drilled at DSDP Sites 548 and 549 (Figure 6.27) mainly consist of

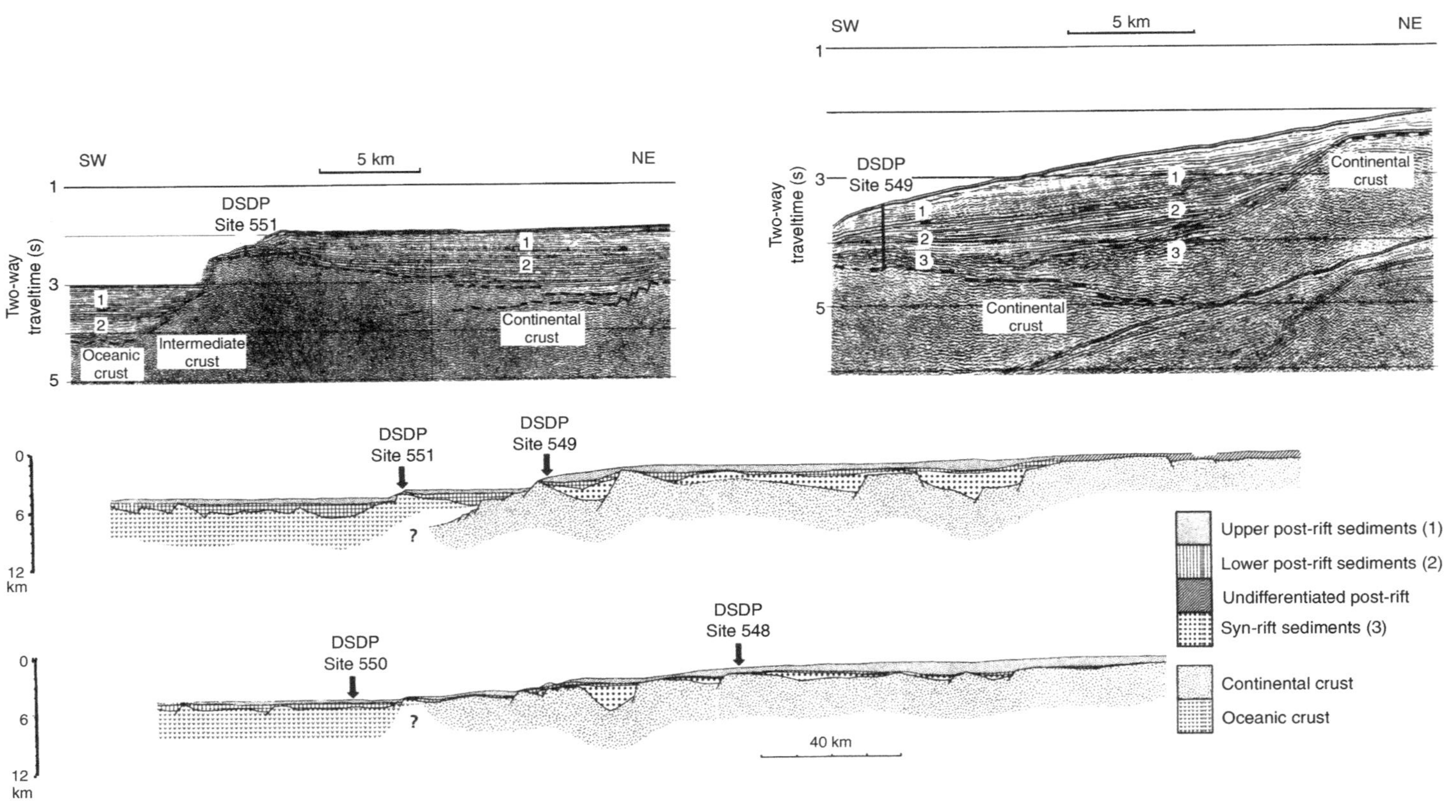

Figure 6.26 Structure and major sedimentary sequences of the starved Celtic passive margin (Goban Spur area) of the North Atlantic, as evidenced from seismic surveys and DSDP drill holes. Top: segments of seismic reflection profiles across the Goban Spur area. Bottom: Interpretative cross sections of the Goban Spur area. See Figure 6.25 for location of DSDP sites. Note minor subsidence of the basement, thin sediment cover and presence of escarpments in coincidence with basement highs. Modified from de Graciansky, P.C., Poag, C.W., et al., 1985. Initial Reports of the Deep Sea Drilling Project, volume 80. U.S. Gov. Print. Office, Washington, DC.

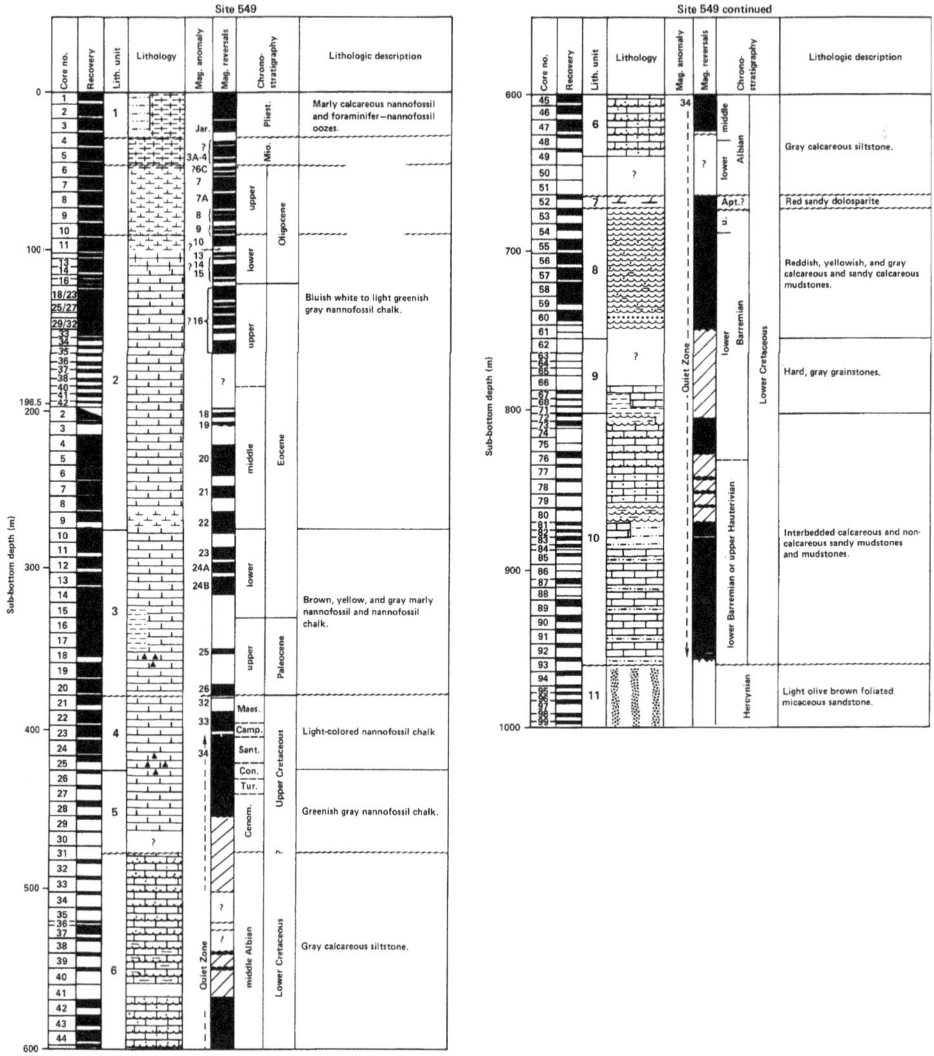

Figure 6.27 The sedimentary column at DSDP Site 549 on Goban Spur. Syn-rift sediments extend from the Barremian to the Early Albian and are dominated by siliciclastics, but note the local importance of carbonates. Post-rift sediments are Middle Albian to Pleistocene in age and consist of biogenic carbonates. Note occurrences of siliciclastics in the Paleocene/Eocene and in the Pleistocene. Reprinted from de Graciansky, P.C., Poag, C.W., et al., 1985. Initial Reports of the Deep Sea Drilling Project, volume 80. U.S. Gov. Print. Office, Washington, DC.

shallow water siliciclastics commonly mixed with continental organic remains, sandy dolomitic limestones and shallow marine bioclastic siltstones. The nature of the sediments clearly indicates that the expansion of the thermal bulge and related tectonics and erosion were limited (see Section 3.4). The accumulation of syn-rift sediments extended from the Barremian to the Early Albian, an interval of uplift and erosion in the adjacent London and Paris Basin (see Section 4.3).

6.2.2. Post-Rift Sediments of the Goban Spur Area

Post-rift sediments of the Goban Spur area are separated from syn-rift sediments by an unconformity (Post-rift unconformity) and can be divided into two major sedimentary sequences, based on their respective acoustic properties and geographic distribution (Figure 6.26). The lower post-rift sequence is characterized by alternating transparent and layered units. The thickness of this sedimentary sequence is highly variable, being absent on top of some basement highs but reaching up to 1,000 m in some intervening basins. The lower post-rift sequence lies directly over the oldest oceanic crust in the Porcupine Abyssal Plain and does not show evidence for significant tectonic control, and therefore probably corresponds to the crustal fissure stage of the North Atlantic at least for its lower part. The oldest sediments overlying tholeiitic basalts drilled off the western escarpments at DSDP Site 550 are of Late Albian age, providing a constraint for the commencement of the crustal fissure stage. Assemblages of benthic microfossils indicate that DSDP Sites 549 and 550 were already at middle bathyal and lower bathyal depths, respectively, during the Late Albian. However, subsidence estimates indicate that some basement highs may have reached altitudes of about 1,000 m, the deepest parts of the oceanic basin reaching water depths of about 2,000 m. This is very consistent with current observations of the Red Sea deeps and many mid-oceanic ridges, where oceanic accretion takes place at 2,000–2,500 m water depth.

The lower part of the lower post-rift sequence at DSDP Leg 80 sites is made of highly condensed Late Cretaceous sediments and is characterized by several unconformities and hiatuses which do not show a clear correlation between sites. The relative northward motion of Africa relative to Eurasia started in the Middle Cretaceous and the related northwestward motion of Iberia and compression facilitated faulting and uplift in Western Europe. This is a possible origin for many unconformities and hiatuses observed on the European margin of the North Atlantic.

From the Albian to the Santonian, sediments and environmental conditions varied considerably across Goban Spur because the basins were still isolated from each other by basement highs (Figure 6.28). The most characteristic Late Cretaceous facies consists of white to greenish-gray nannofossil chalks (Figure 6.27) which accumulated in well-oxygenated environments and highlight the paucity of terrigenous fluxes in this area adjacent to intraplate basins and epicontinental seas. Late Albian to Turonian nannofossil chalks alternate with carbonaceous silty mudstones (black-shale facies) containing organic matter of oceanic origin principally, which is indicative of oxygen-depleted environments of poor ventilation. Several of these black-shale intervals are correlated with similar events in the Central Atlantic and Tethys oceans. They are intimately linked to global climatic events involving the atmosphere and ocean, and can hardly be only attributed to a poor ventilation of the North Atlantic during the crustal fissure stage.

Because of low terrigenous fluxes which limit the progression of the continental shelf, anoxic intervals at ocean scale and a number of unconformities and hiatuses,

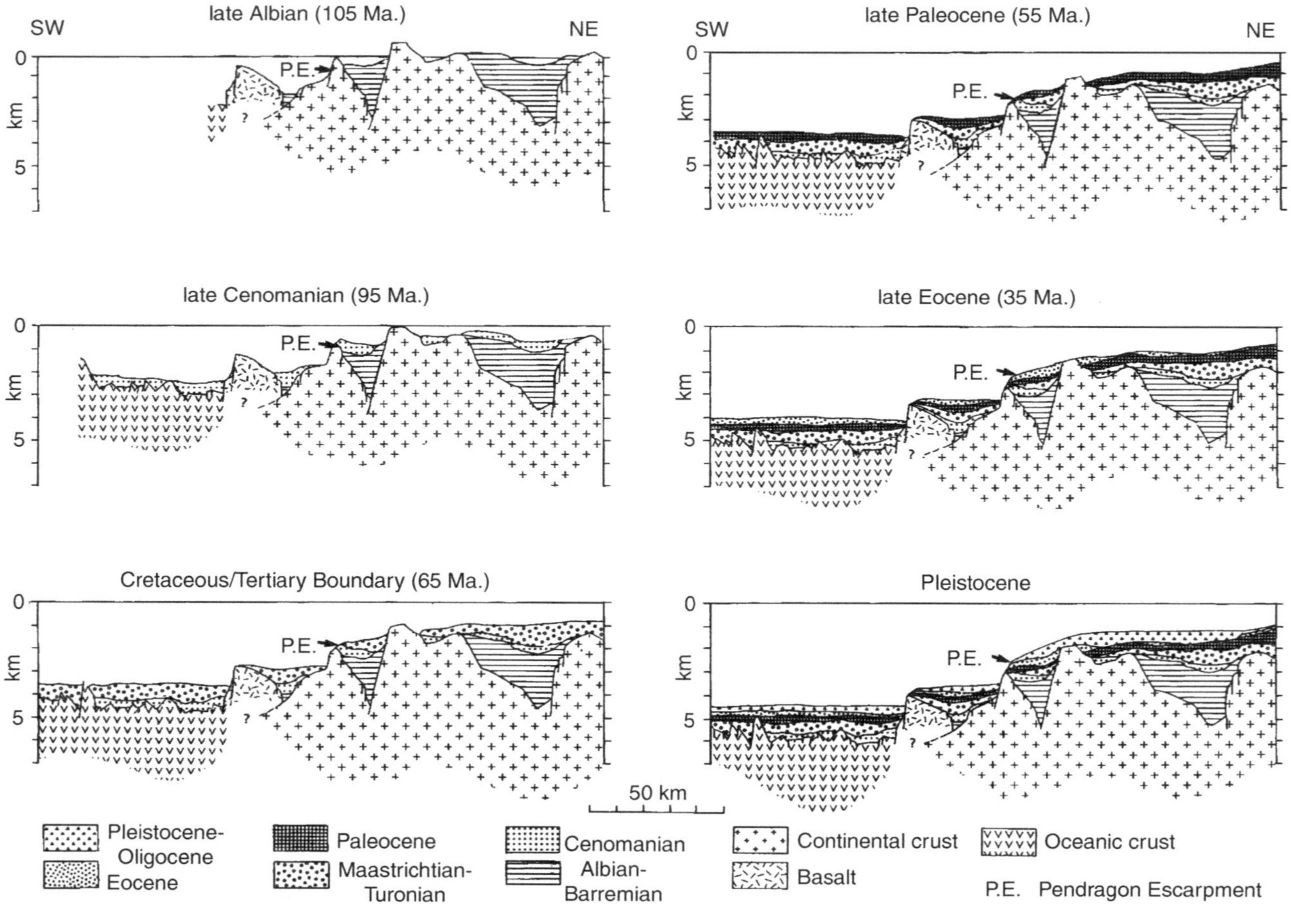

Figure 6.28 Post-rift evolution of the starved Celtic margin of the North Atlantic. Note the persistence of isolated basins during the Late Albian and Cenomanian, limited subsidence and persistence of basement highs through time, and absence of significant changes in morphology since the Late Eocene. Modified from de Graciansky, P.C., Poag, C.W., et al., 1985. Initial Reports of the Deep Sea Drilling Project, volume 80. U.S. Gov. Print. Office, Washington, DC.

it is difficult to estimate the age of the transition from crustal fissure to mature ocean stages on the starved Celtic passive margin area of the North Atlantic:

- A major marine transgression occurred during the Turonian but coincides with an interval of high sea level of global extension, anoxia and limited terrigenous fluxes, and therefore is not related to the thermal subsidence of the passive margin.
- The accumulation of marine sediments started at the shallowest DSDP Site 548 during the latest Campanian, at a paleodepth of about 500 m. The latest Campanian was also the first time that almost identical marine sediments (white to light-colored nannofossil chalks to foraminifer-nannofossil chalks, Figure 6.29) were uniformly and widely distributed across the entire Goban Spur region. This is compatible with an important step in the progression of the thermal subsidence of the passive Western European margin and the beginning of the mature ocean stage in the North Atlantic. However, the seismic profiles do not provide evidence for a major Late Cretaceous transgression on the inner margin, where post-rift sediments look generally thin and indifferentiated.
- In some areas an upper post-rift sequence clearly overlaps the lower post-rift sequence and hercynian basement. Comparison of seismic profiles and DSDP sites indicates that the lower and upper post-rift sequences are locally separated by an unconformity, and that the upper post-rift sequence started accumulating in the Middle Eocene. However, the relative northward motion of Africa reached a maximum in the Eocene, leading to intraplate deformation before collision along the northern Tethys margin initiated the Pyrenean tectonic phase during the Late Eocene. Traces of Eocene compression (faulting, folding, slumping) are visible in the Goban Spur and Trevelyan Escarpment areas of the Celtic margin.

The timing of the transition from crustal fissure to mature ocean stages in the Celtic area of the North Atlantic is therefore poorly constrained, but sediments of the upper post-rift sequence clearly accumulated during the mature ocean stage. The upper post-rift sequence at DSDP Sites 548 and 549 (Figure 6.27), drilled at 1,256 and 2,533 m water depth respectively, principally consists of nannofossil to foraminifer-nannofossil oozes and chalks, associated to minor clayey nannofossil and foraminifer-nannofossil oozes and chalks. The large dominance of biogenic carbonates clearly indicates that the sedimentation of the starved Celtic passive margin was principally controlled by planktonic productivity and carbonate preservation. The turbiditic facies are rare, and principally consist of carbonate debris. The upper post-rift sequence at DSDP Site 549 is interrupted by several hiatuses which encompass parts of the Middle to Late Eocene, Middle Oligocene, latest Oligocene to earliest Miocene, Middle to Late Miocene and Pliocene. The number of hiatuses, which overlap major intervals of climate change and/or paleoceanographic shifts, illustrates the influence of climate driven deep currents on the accumulation of sediments. Fine siliciclastic particles are only a minor component of the sediments which accumulated on the starved Celtic passive margin during the mature ocean stage, because of the importance of epicontinental seas and intraplate basins in the hinterland. However, the production of siliciclastics locally increased during the Paleocene and Eocene interval of intensified compression and related erosion on adjacent continents, but most terrigenous

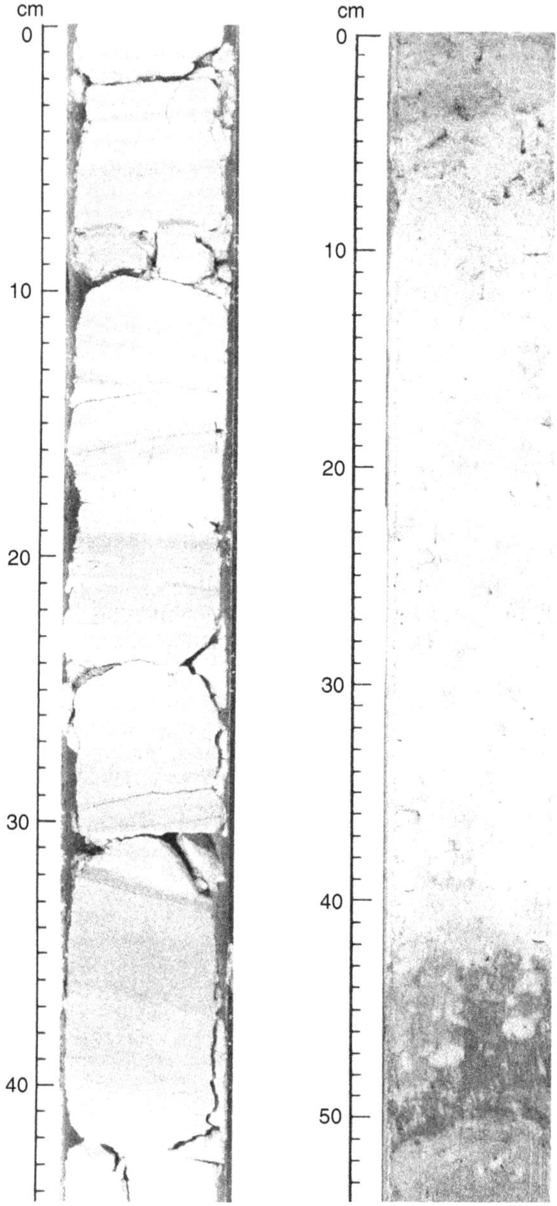

Figure 6.29 Left: laminated Late Cretaceous nannofossil chalk, DSDP Site 549. Right: Pleistocene successions of nannofossil ooze and clayey nannofossil ooze, DSDP Site 549. Darker colors indicate higher contents in siliciclastics. Note bioturbation of Pleistocene sediments. Reprinted from de Graciansky, P.C., Poag, C.W. et al., 1985. Initial Reports of the Deep Sea Drilling Project, volume 80. U.S. Gov. Print. Office, Washington, DC.

sediments have been trapped in the Celtic Sea basins where they alternate with pelagic deposits. Later on, sporadic increases of the proportion of fine siliciclastics in the Late Pliocene and Pleistocene (Figure 6.29) probably result from lower sea level, emergence of many European epicontinental seas and correlative extension of the drainage basins during intervals of polar ice expansion.

To summarize, the development of the starved Celtic passive margin of the North Atlantic most probably results from the association of several characteristics of crustal and environmental origin:

- a relatively low, passive rift activity (and related tectonism and volcanism) might explain the limited accumulation of syn-rift deposits and small regional subsidence;
- the presence of intraplate basins, short drainage basins and large areas of epicontinental seas on the adjacent European plate probably limited the terrigenous fluxes to the Celtic margin; and
- vigorous deep oceanic currents and related erosion sporadically limited the accumulation of post-rift sediments, especially during key intervals of climate history.

As a result, the starved Celtic passive margin lacks clearly identified shelf sequences and its depth and morphology did not change significantly since the Late Eocene.

6.3. Example of a Fat Passive Margin: The New-Jersey Area of the North American Atlantic Margin

The New-Jersey area of the fat passive margin of North America extends from the Georges Bank area off Cape Cod to the North to Cape Hatteras to the South and is bounded by the Piedmont Province of the Appalachians to the West (Figure 6.30). Up to 18 km of sediments accumulated on the passive margin of New-Jersey, building a large coastal plain and continental shelf which prograded over distances of 100–150 km offshore (Figure 6.19). The sediment load increased the subsidence of the lithosphere facilitating further accumulation, leading to the formation of an elongated sag basin below the coastal plain and shelf, the Baltimore Canyon Trough. The upper and middle continental slope is locally incised by canyons which die out on the lower slope (Figure 6.31). Otherwise, the slope shows traces of erosion by gravity processes, especially the lower slope where Eocene sediments outcrop. The transition from slope to rise is marked by a significant decrease in seafloor gradient, from 4.5° to 1.5° and the sedimentary sequences of the upper rise onlap the erosion surface of the lower slope (Figure 6.31).

6.3.1. Early History of the New-Jersey Margin

Rifting started during the Triassic in the New-Jersey area of the North American margin (Figure 6.19), which then was part of the future Western Tethys Ocean. Conglomerates, sandstones, reddish mudstones, gray lacustrine shales associated to volcanic products outcrop in grabens of Triassic age onland, and are presumed to fill the oldest depocenters of the Baltimore Canyon Trough. Syn-rift deposits are capped by

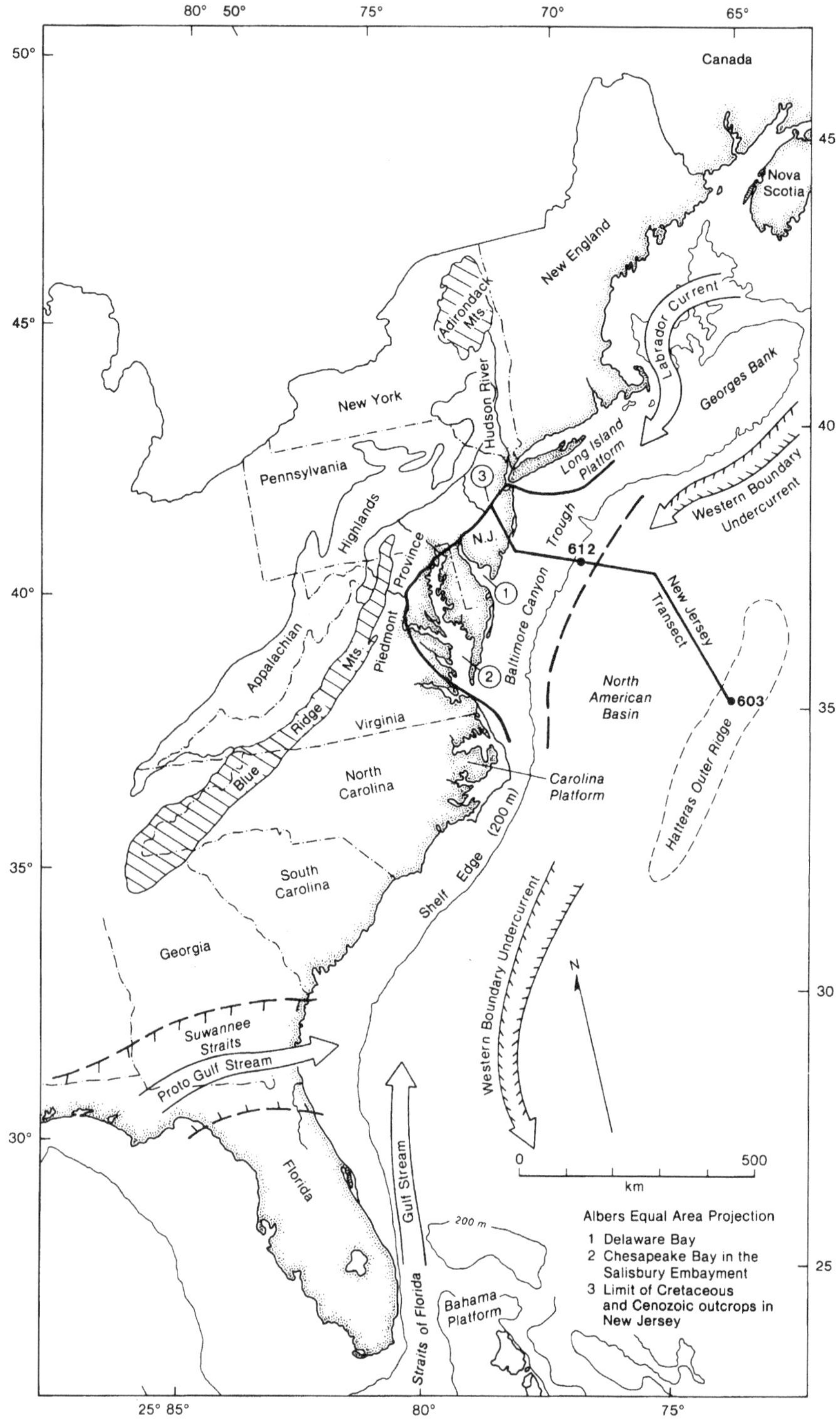

Figure 6.30 The passive continental margin of New-Jersey and its morphologic and oceanographic context. Note the position of DSDP Site 603, on the lower rise. Reprinted from Poag, C.W., Watts, A.B., et al., 1987. Initial Reports of the Deep Sea Drilling Project, volume 95. U.S. Gov. Print. Office, Washington, DC.

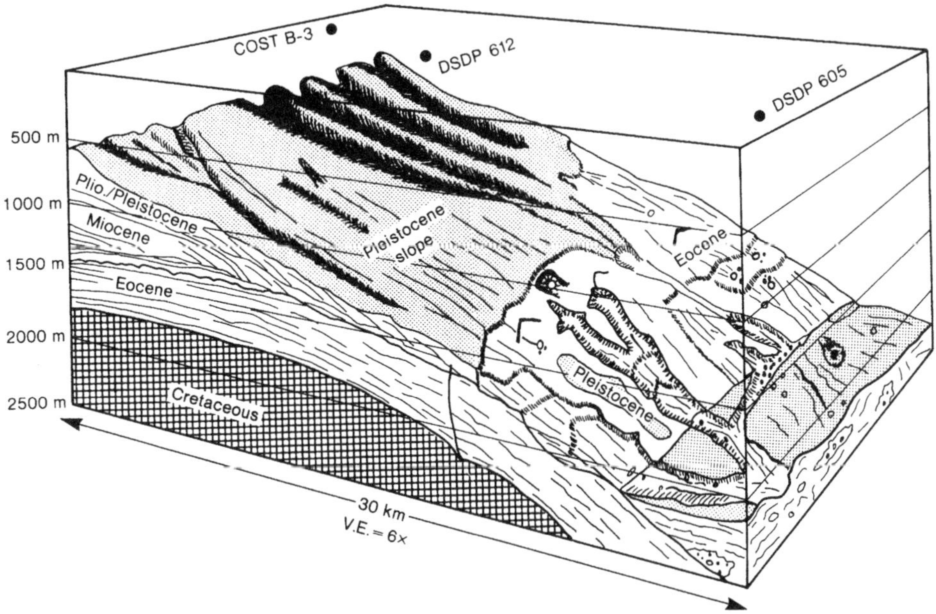

Figure 6.31 3D representation of the fat New-Jersey margin. Note the presence of canyons and the outcrop of Eocene sediments on the continental slope. Reprinted from Poag, C.W., Watts, A.B., et al., 1987. Initial Reports of the Deep Sea Drilling Project, volume 95. U.S. Gov. Print. Office, Washington, DC.

evaporites, later exposed to halocinetic deformation as post-rift sediments accumulated. The crustal fissure stage presumably began in the Middle Jurassic and was characterized by the accumulation of 8–10 km of shallow water limestones (including a barrier reef complex) in the New-Jersey area, grading to shales offshore (Figure 6.32). However, clay to sand-sized siliciclastics locally persisted nearshore during the Late Jurassic in relation to deltaic environments. The oldest sediments drilled on the continental rise at DSDP Sites 105 and 603 are of latest Jurassic to earliest Cretaceous (Berriasian) age. They principally consist of light gray, bioturbated nannofossil limestones which deposited at bathyal depth. The pelagic limestones alternate with gray, laminated nannofossil claystones which suggest a periodic anoxia of bottom conditions, most probably related to changes in the deep oceanic circulation.

Beginning in the Hauterivian, two different types of turbidites interfered with the pelagic sedimentation. Clay turbidites, enriched in organic matter and coccoliths with minor quartz silt, probably originated from an oxygen minimum zone of the continental slope. From the Late Hauterivian, they were progressively replaced by silt and sand turbidites (Figure 6.33) enriched in quartz, feldspars, micas, plant debris and glauconite reworked from shallow water areas (DSDP Site 603 was located about 500 km off the shelf edge during the Hauterivian). A peak in silt and sand turbidity current activity occurred in the Barremian, which is disconformably capped by about 30 m of sand. The sand is associated to a deep sea fan complex which prograded during an Early Aptian interval of low relative sea level. Deposition of massive sands alternating with fine siliciclastics persisted in shallow

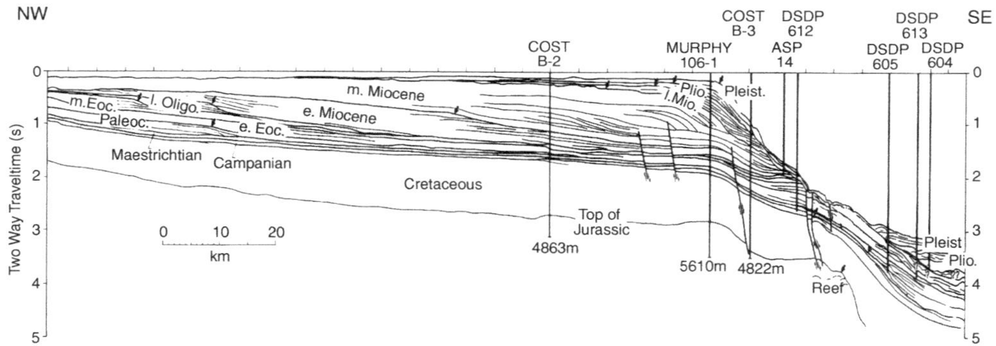

Figure 6.32 Sedimentary structure of the continental shelf, slope and rise of New-Jersey, as interpreted from seismic reflection profiles. Note changes in seafloor gradient, outcrop of Eocene sediments on the slope, and onlap of Neogene sediments of the upper rise. Reprinted from Poag, C.W., Watts, A.B., et al., 1987. Initial Reports of the Deep Sea Drilling Project, volume 95. U.S. Gov. Print. Office, Washington, DC.

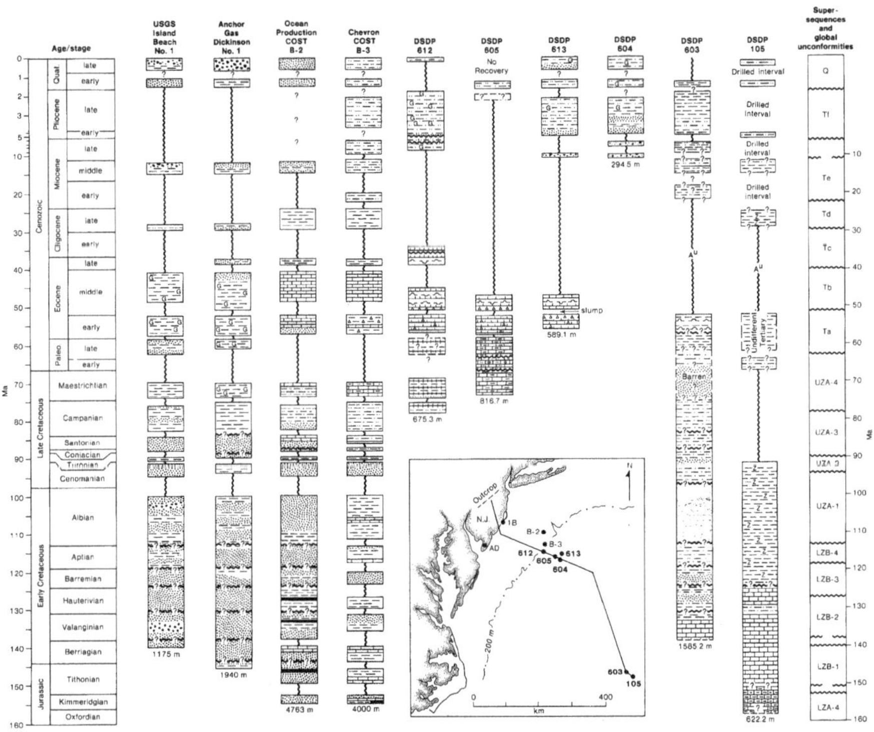

Figure 6.33 Lithologic summary and location map of the deepest holes drilled on the fat New-Jersey passive margin, from the coastal plain to the abyssal plain. Dots, dashes and black represent terrigenous sediments whereas closed and open bricks represent calcareous biogenic sediments. Note the respective contributions of siliciclastic and biogenic sediments. Reprinted from Poag, C.W., Watts, A.B., et al., 1987. Initial Reports of the Deep Sea Drilling Project, volume 95. U.S. Gov. Print. Office, Washington, DC.

water areas during the whole Early Cretaceous interval, progressively burying the reef system. It is noteworthy that terrigenous elements and turbidite activity in the New-Jersey area increased during the Early Cretaceous when rift activity began further North, progressively separating Iberia from Newfoundland.

Massive sands dominated till the Santonian in most coastal areas of the New-Jersey passive margin, when they were replaced by fine siliciclastics. At DSDP Sites 105 and 603, the Late Aptian to Turonian interval of sedimentation is characterized by dark claystones (black-shales) containing up to 15% of organic matter which includes elements of oceanic and continental origins (type II kerogen and plant debris, see Section 12.3). The black-shale facies (Figure 6.34) include intervals of homogeneous claystones which correspond to mud turbidites, and intervals of faintly laminated claystones. The black-shale facies alternate with intervals of variegated (greenish gray to reddish brown) claystones. Mid-Cretaceous intervals of black-shale deposition have been observed throughout the North Atlantic and Western Tethys oceans and represent anoxic events of global extension. In most areas, anoxia has been attributed to enhanced productivity caused by more

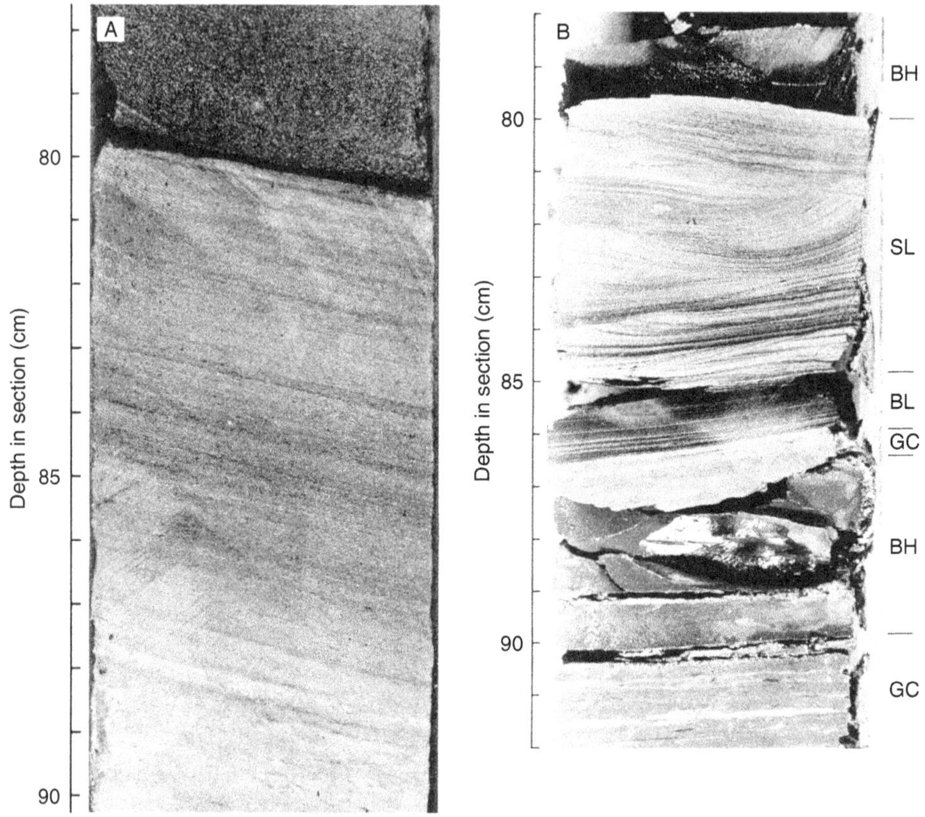

Figure 6.34 Typical Early Cretaceous facies of the New-Jersey continental rise, DSDP Site 603. A, transition from nannofossil limestone to laminated nannofossil claystone, topped by a silt and sand turbidite, Barremian; B, interval of black-shale facies, Aptian-Albian; BH, homogeneous black claystone (mud turbidite); SL, parallel and cross-laminated siltstone; BL, laminated black claystone; GC, faintly laminated to slightly bioturbated variegated claystone. Modified from Van Hinte, J.E., Wise, S.W., et al., 1987. Initial Reports of the Deep Sea Drilling Project, volume 93. U.S. Gov. Print. Office, Washington, DC.

rapid turnover rates of surface waters during warmer intervals, and increased stratification of deep waters due to increased production and circulation of warm saline deep waters. The variegated claystones became predominant in the Santonian where they alternate with rare layers of siltstone, but some degree of anoxia may have persisted as suggested by the presence of pyrite and organic remains, and the absence of microfossils. The Late Aptian to Santonian interval is also characterized by a number of hiatuses (which are often marked by glauconitic, shelly sands and/or poorly preserved microfossils) in nearshore as in more distal areas, and corresponds to a limited expansion of the continental shelf (see Section 6.2).

Besides a number of seismic reflection profiles, the deep latest Jurassic to Late Cretaceous sediments of the continental margin of New-Jersey are known from a few commercial boreholes drilled in the coastal area of the Baltimore Canyon Trough and from a couple of DSDP sites drilled on the continental slope and rise

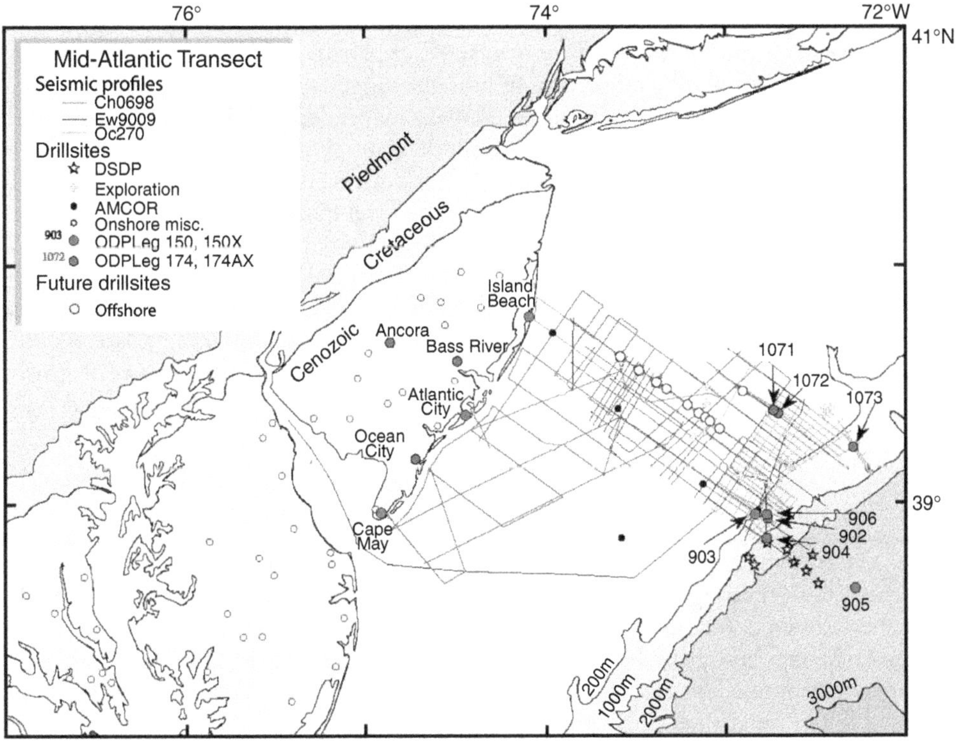

Figure 6.35 Location map of DSDP, ODP and commercial holes drilled on the fat passive margin of New-Jersey. DSDP Site 603 is southeast of ODP Site 905. Note the extension of the continental shelf and coastal plain where Cenozoic and Cretaceous sediments outcrop due to the bending of the passive margin and relatively low Pleistocene sea level. Reprinted from McHugh, C.M.G., Damuth, J.E., Mountain, G.S., 2002. Cenozoic mass-trasport facies and their correlation with relative sea-level change. Marine Geology, 184, 295–334.

(Figure 6.35). Post-Santonian and especially Cenozoic sediments have been further detailed from ODP sites drilled on the shelf, slope and rise (ODP Legs 150 and 174A) and in the coastal plain (ODP Legs 150X and 174AX). The combined results of these investigations show that at least 12 major Late Cretaceous and Cenozoic sedimentary sequences are bounded by major unconformities of regional extension. Near the modern shoreline (Island Beach, Figure 6.35) Campanian sediments consist of dark fossiliferous, lignitic and pyritic clays and silty clays, which contain microfossil assemblages indicating deposition in sublittoral environments of 100–200 m water depth. Similar Campanian sediments outcrop in the inner coastal plain (Figure 6.35) where microfauna indicate paleodepths of 50–100 m. On the modern outer shelf at Sites B2 and B3 (Figure 6.33), dark siltstones and claystones contain benthic microfossil assemblages which are typical of outer sublittoral and upper bathyal environments of 200–300 and 300–350 m water depth, respectively. Thicknesses of Campanian sediments vary from about 95 to 165 m nearshore, but increase to about 200 m on the upper Campanian slope in the vicinity of DSDP Site 612 where dark gray to black chalks, shales and mudstones have been sampled.

Further East on the rise, thin variegated claystones and sandstones separate Paleocene from undifferentiated Coniacian–Campanian sediments.

The distribution of Campanian sediments suggests that abundant siliciclastics eroded from adjacent continental areas (Appalachians) deposited in shallow, oxygen-depleted marine environments, and built a continental shelf which extended approximately from the Piedmont to the modern shelf break. The buried Late Jurassic reef system probably still partly controlled the transit of terrigenous elements to the deep ocean. On the ocean side, fine siliciclastics sporadically increased the slope where an oxygen minimum zone was still present at upper bathyal depth. A stratification of water masses persisted in the deep ocean, where oxygen depletion favored the dissolution of biogenic carbonates and helped maintaining relatively low sedimentation rates. The Campanian sequence is capped by a major unconformity that extends from the coastal plain to the rise, suggesting that it was caused by a major drop of relative sea level.

6.3.2. Cenozoic Intervals of Progradation and Erosion

The Maestrichtian, Paleocene and Early Eocene sequences form thin blankets of sediments separated by major unconformities (Figure 6.32). Maestrichtian and Early Eocene sediments (Figure 6.33) cover the entire continental shelf where they range from sandstones and glauconitic and/or calcareous claystones to limestones, whereas Paleocene dark fossiliferous, calcareous, pyritic and glauconitic clays and sandy silts are limited to the inner shelf only. The shelf edge migrated about 10 km seaward from its Campanian position during the Maestrichtian, and benthic microfauna suggest paleodepths of 100–150 m on the inner shelf and 100–200 m on the outer shelf, slightly deepening in the Early Eocene. On the upper slope at DSDP Site 612, thicker Maestrichtian and Early Eocene sections include foraminifer-nannofossil chalks and limestones with minor siliciclastics which deposited at upper bathyal paleodepths (200–500 m), sediments of Paleocene age being absent. The Maestrichtian and Paleocene sections increase in thickness downslope. At DSDP Site 605, Maestrichtian to Early Eocene sediments consist of foraminifer and clayey/silty nannofossil chalks and limestones with sporadic occurrences of nannofossil claystones. These sediments grade to zeolitic and radiolarian claystones on the rise at DSDP Site 603. The continental shelf slightly prograded especially during the Maestrichtian, but fine siliciclastics and the dominance of calcareous biogenics on the slope suggest a decrease of continental erosion and terrigenous fluxes to a still highly productive and oxygen-depleted marine environment of stratified water masses. However, reworking of shelf sediments to the slope and rise was still active.

The relative sea level reached its maximum Cenozoic height during the Middle Eocene, and entrained significant changes in the depositional patterns of the New-Jersey margin (Figures 6.32 and 6.33). A second shelf break developed near the modern shoreline, separating a prograded wedge of fine siliciclastics locally topped by thin evaporites on the inner shelf, from dominant clayey biogenic carbonates on the outer shelf where bathyal microfossil assemblages indicate paleodepths of 500–600 m. The carbonates grade to biosiliceous nannofossil chalks on the upper slope, where the Middle Eocene section thickens to about 150 m. Middle Eocene sediments remain identical in composition and thickness downslope, but are absent from the continental rise. The importance of biogenic carbonates on the outer shelf and slope may partly result from low terrigenous fluxes and accumulation of fine siliciclastics on the hinter shelf. However, high pelagic productivity confirmed by the presence of biosiliceous remains and good preservation in oxygenated warm waters probably played a key role in the accumulation of biogenic carbonates. The sequence is capped by a major erosional surface of regional extension which is exposed on the lower slope and coalesces with the Early Eocene/Middle Miocene contact on the upper rise.

Sediments of Late Eocene to Early Miocene age are poorly developed on the New-Jersey margin where this interval is principally characterized by significant erosional unconformities (Figures 6.32 and 6.36). Small prograded wedges of siliciclastics locally accumulated at sublittoral paleodepths on the inner shelf during the Late Eocene and Oligocene (Figure 6.37), and the depocenter migrated to the East in the Early Miocene when the hinter shelf break prograded by about 50 km. Late Eocene sediments are missing on the foreshelf. When present, Late Oligocene and Early Miocene sediments of the foreshelf principally consist of claystones, siltstones and silty/glauconitic sands which deposited at upper bathyal paleodepths

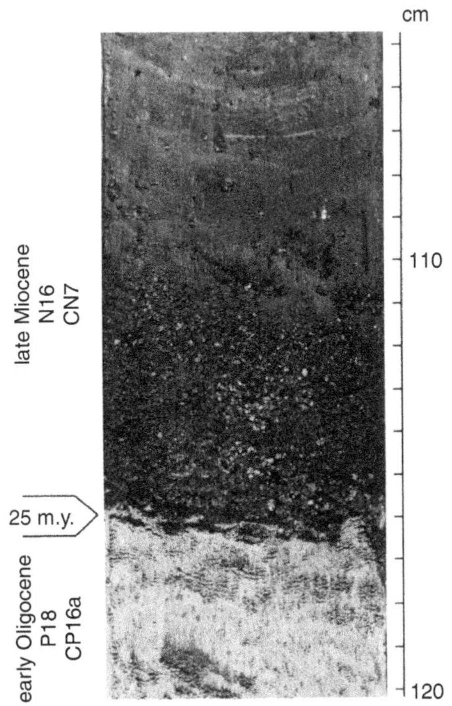

Figure 6.36 Sedimentological pattern of a significant unconformity at DSDP Site 612 on the upper continental slope of New-Jersey. Early Oligocene light gray, slightly burrowed biosiliceous and foraminifer-nannofossil ooze is separated from Late Miocene, dark olive gray homogeneous claystone by a 5 cm thick section of dark glauconitic sand. Reprinted from Poag, C.W., Watts, A.B., et al., 1987. Initial Reports of the Deep Sea Drilling Project, volume 95. U.S. Gov. Print. Office, Washington, DC.

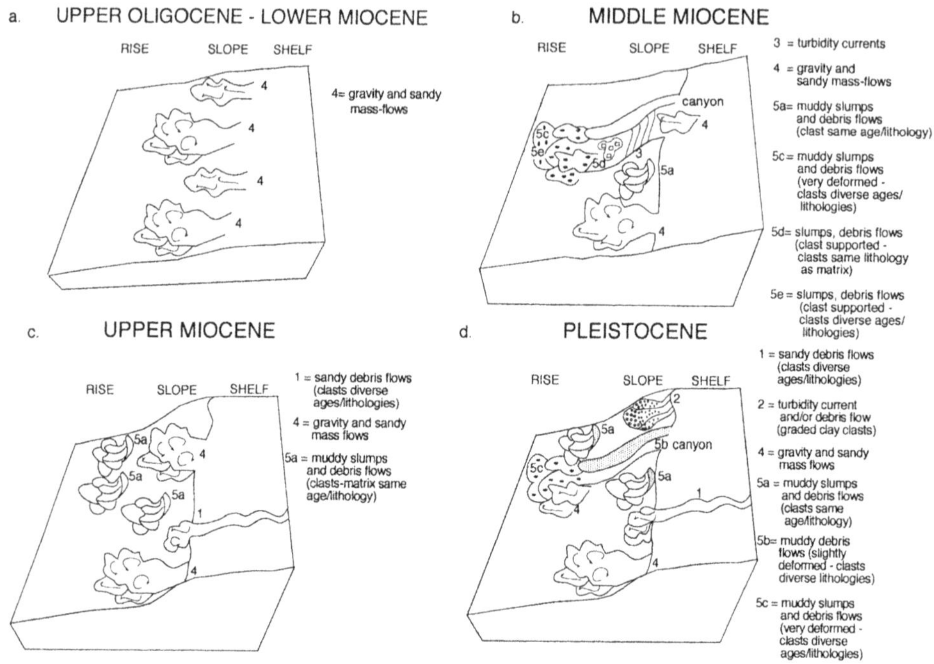

Figure 6.37 Schematic outline of mass-transport deposition on the fat continental margin of New-Jersey. Note the role of canyons during the Middle Miocene and Pleistocene, and increasing importance of slope processes during the Neogene. Reprinted from McHugh, C.M.G., Damuth, J.E., Mountain, G.S., 2002. Cenozoic mass-transport facies and their correlation with relative sea-level change. Marine Geology, 184, 295–334.

(Figure 6.33). Thin strata of biosiliceous nannofossil chalks still deposited during the Late Eocene on the upper slope in mid-bathyal environments (*ca.* 1,000 m water depth), where they are locally capped by Late Oligocene fine siliciclastics. However, most Oligocene and Early Miocene sediments have been eroded on the continental slope (Figure 6.38). On the rise, a thin section of Early Miocene claystone rests on Early Eocene sediments.

Major changes in sedimentation occurred during the Late Eocene on the shelf and during the Oligocene on the slope, when dominant carbonate biogenics are being replaced by dominant siliciclastics. The Late Eocene and Oligocene interval coincides with a reorganization in the relative motion of the North American and European plates (cessation of seafloor spreading in the Labrador Sea, switch to East-West spreading between Greenland and Europe and related regional subsidence), which improved the communication between the North Atlantic and the Arctic oceans. The Late Eocene and Oligocene interval also coincides with significant cooling trends of both regional (from pollen assemblages) and global (from oxygen isotopes) climate, and an expansion of the thermohaline circulation. Cooler surface water temperatures in the New-Jersey area may have inhibited carbonate production and preservation, whereas lower sea level may have reinforced the erosion in the drainage basins and nearshore areas. By the same time, invigorated intermediate and

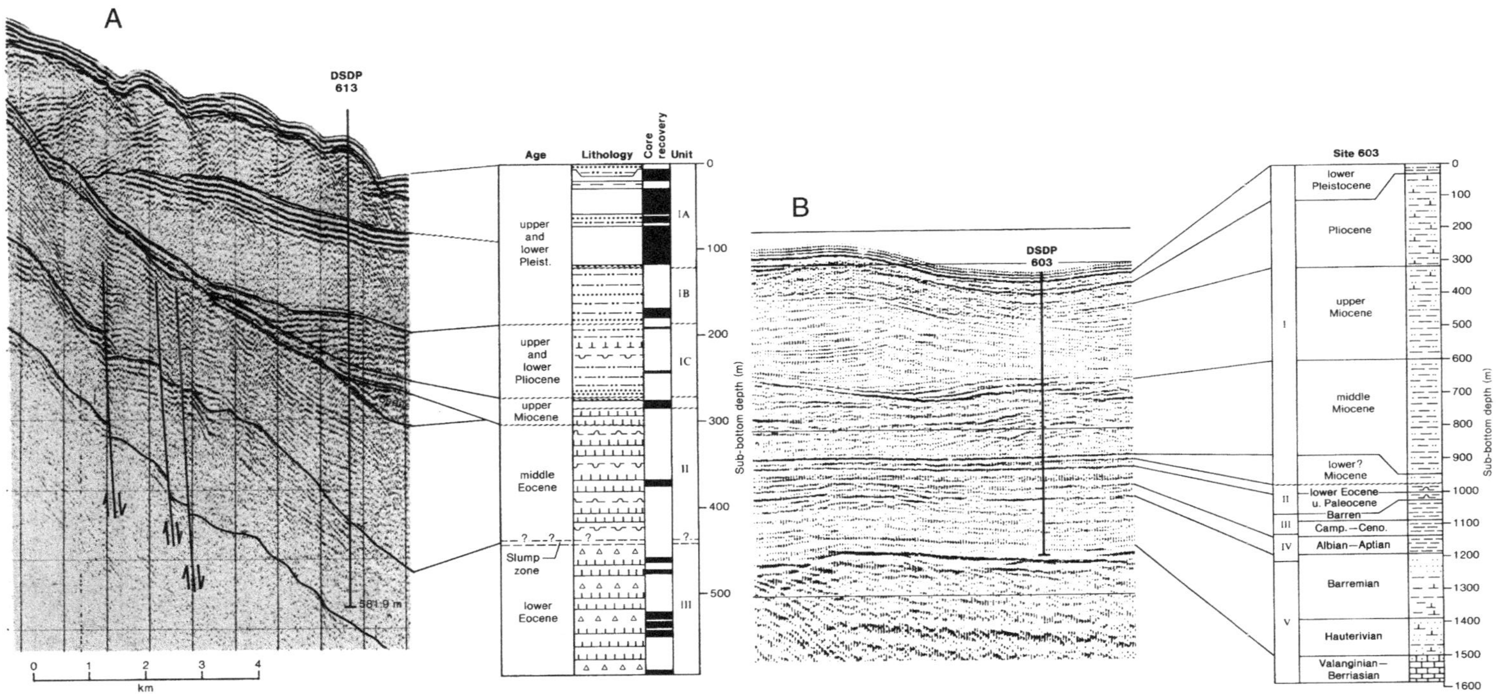

Figure 6.38 Comparison of NW-SE seismic profiles across the fat passive margin of New-Jersey and stratigraphic columns for DSDP Site 613 on the lower slope (A) and DSDP Site 603 on the lower rise (B). Dots and dashes represent siliciclastics. Note the importance of the unconformity on top of Eocene sediments, the onlap of Neogene sequences on the lower slope, and the thickness of Middle Miocene sediments on the rise. Modified from Poag, C.W., Watts, A.B., et al., 1987. Initial Reports of the Deep Sea Drilling Project, volume 95. U.S. Gov. Print. Office, Washington, DC.

deep oceanic currents may have removed significant quantities of sediment from the foreshelf, slope and rise, producing a number of hiatuses and unconformities.

During the Middle Miocene, the accumulation of siliciclastics increased to approximately 240 m/Myr in prograding deltaic environments of the continental shelf (Figure 6.37). Locally, up to 1,200 m of Middle Miocene sediments are present. They consist of shelly, diatomaceous sands, glauconitic and micaceous sands, micaceous and lignitic silty clays and clays, which accumulated in paralic to sublittoral environments (Figure 6.33). Terrigenous fluxes increased around 13–14 Ma. As a result the continental shelf had prograded by about 30–60 km seaward at the end of the Middle Miocene and was again characterized by a single shelf break. Seismic profiles suggest that Middle Miocene deposits initially extended from the shelf to the rise, but have been subsequently eroded on most of the continental slope (Figure 6.38). The Middle Miocene section decreases in thickness to about 200 m near the current shelf break where glauconitic, micaceous and organic silty clays accumulated in bathyal environments of 1,000–1,500 m water depth. Silty clays dominate in a 300 m thick Middle Miocene section on the continental rise (Figure 6.38), which is characterized by distal turbidite facies. Middle Miocene sediments also contain a proportion of organic matter, derived from both upwelling activity and continental detritus. Middle Miocene deposits can be divided into three separate sequences on the shelf and on the rise, which are capped by a widespread erosional surface (Figures 6.32 and 6.38). The transition from a sediment-starved regime during the Late Eocene–Early Miocene interval to a prograding regime of high sedimentation rates during the Middle Miocene is most probably the consequence of concurrent events of lithospheric and climatic origins:

- isostatic compensation, intraplate deformation and uplift of the Appalachian hinterland;
- sea level fall and canyon cutting subsequent to the development of Antarctic ice around 13–14 Ma and
- intensified precipitation as the warm western boundary surface system responded to increased gradients of temperature.

Beginning in the Late Miocene, shelf deposition was minimal: a thin wedge of siliciclastics formed at the shelf break during the Pliocene and thinned rapidly shoreward, whereas Pleistocene gravel and sand deposited in paralic environments in the coastal plain (Figure 6.33). Terrigenous fluxes were still significant during most of the Late Miocene–Pleistocene interval, decreasing in the Pliocene, but the sediments were principally channeled to the rise (Figure 6.38) where they were partly reworked by bottom currents to feed sediment drifts. Minor gravel and sand of Late Miocene and Pliocene age, typical of channel fills, were locally sampled from the slope. Late Miocene to Pleistocene sediments are organized in four sequences which increase in size from the continental slope to the rise and are separated by marked erosional unconformities. A prism made of fine siliciclastics enriched in diatoms and organic matter developed in the upper slope area, especially during the Pleistocene. Because of the bending of the fat continental margin and low sea level during Pleistocene glacial intervals, proximal sources of siliciclastics in the coastal plain area began to dominante over those in the Appalachians. On the upper rise, thin Late Miocene sediments still consist of reworked gravel, glauconitic sand and pieces of Eocene chalk which are typical of channel facies. Thicker Pliocene and Pleistocene sequences onlap the Eocene strata and grade to clays and nannofossil clays principally. However, layers of gravel, sand and reworked Eocene chalk are still present, indicating the persistence of significant downslope processes such as turbidity currents and debris flows. The Pliocene and Pleistocene sequences also onlap the Eocene strata of the lower slope (Figures 6.32 and 6.38), but the thinness of onlapping sediments suggests that the Western Boundary Undercurrent (North Atlantic Deep Water) was an effective agent of sediment dispersion. Pliocene and Pleistocene sediments consist of turbiditic silts and clays with a few reworked, glauconitic sand layers, which alternate with nannofossil or diatomaceous hemipelagic clays. Late Miocene sediments principally accumulated on the middle and lower rise where more than 300 m of claystones have been drilled at DSDP Site 603 (Figure 6.38). The Pliocene and Pleistocene sequences have been locally eroded by bottom circulation on the middle rise, but reach maximum thicknesses on the lower rise (up to 300 m). They principally consist of distal turbidites (Figure 6.37) dominated by silty clays, and hemipelagic clays. The presence of turbiditic facies and the thickness of the sedimentary sequences on the rise mark a seaward migration of the main prograding wedge during the Late Neogene.

6.3.3. A Step-by-Step Development of the Fat Passive Margin of New-Jersey

Several concurrent geological and environmental conditions facilitated the development of a fat passive margin in the New-Jersey area of the Atlantic Ocean:

- This region of the Atlantic Ocean was formerly part of the Western Tethys Ocean and did open early in the Jurassic. Therefore, sediments accumulated there for about 200 Myr.
- Erosion of the adjacent hinterland was significant during many time intervals of the mature oceanic stage of the New-Jersey margin because of the persistence of regional continental relief for lithospheric reasons (ocean opening processes in adjacent North Atlantic areas, intraplate deformation, isostatic readjustment and related bending of the margin) and significant precipitation (location of the New-Jersey in a Western Boundary context of warm surface waters and ocean to continent heat transfer).
- During several intervals of low continental erosion, carbonate productivity and preservation in the ocean maintained relatively high sedimentation rates.
- A number of erosional unconformities highlight the role of oceanic currents and sea level variations on the sedimentary regime of the margin. However, the thickest terrigenous sequences represent relatively short time intervals, suggesting that brief periods of intense continental erosion are sufficient to build a fat continental margin.

The thickest sedimentary sequences formed during the Cenozoic, especially during the Neogene. They provide an opportunity to detail the role of sea level change on continental shelf sedimentation, in relation to global climate. For example, detailed investigation of Late Paleogene and Neogene sediments from the slope led to the identification of five major facies: (1) disorganized sand; (2) organized sand; (3) laminated to thin-bedded organized silt and sand (Figure 6.39); (4) mud with disseminated silt and sand; (5) matrix or clast supported chaotic mud. These facies represent three major environments of deposition (intercanyon regions of the slope, canyons, upper continental rise). Changes of facies often coincide with strata surfaces

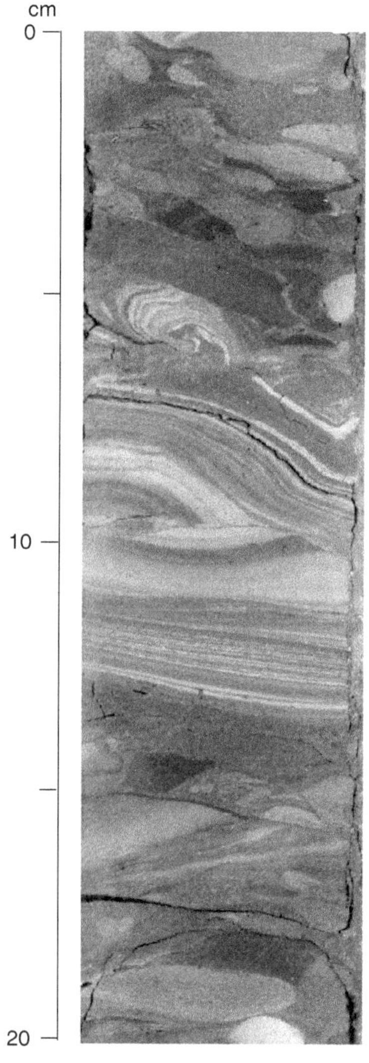

Figure 6.39 Example of a typical facies of mass-transport deposit, and related soft-sediment deformation. Laminated to thin-bedded organized silts and sands have been folded and faulted. Note the presence of mud clasts. Modified from McHugh, C.M.G., Damuth, J.E., Mountain, G.S., 2002. Cenozoic mass-transport facies and their correlation with relative sea-level change. Marine Geology, 184, 295–334.

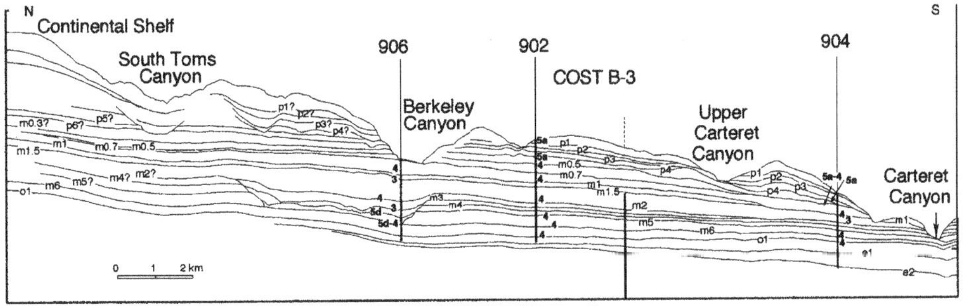

Figure 6.40 Interpretation of a high-resolution seismic profile across the New-Jersey passive margin. Bold numbers represent the main facies identified at ODP sites. Note the number of Neogene unconformities and sedimentary sequences. Reprinted from McHugh, C.M.G., Damuth, J.E., Mountain, G.S., 2002. Cenozoic mass-transport facies and their correlation with relative sea-level change. Marine Geology, 184, 295–334.

and sequence boundaries identified from high-resolution seismic profiles (Figure 6.40), suggesting some degree of correlation between mass-transport deposits (slumps, debris flows, turbidity currents) and sea level variations: gravity flow activity increases as sea level drops, leading to an intensified erosion of the continental slope and to the formation of major erosional unconformities and sequence boundaries.

Detailed investigations also highlight significant differences between Middle Miocene and younger sedimentary sequences of the slope and rise. Middle Miocene sequences show a thick onlapping lowstand wedge overlain by thin (less than 50 m) regressive highstand deposits, and lasted for about 1–2 Myr (Figure 6.41). The younger Late Miocene to Pliocene sequence lacks a distinct lowstand wedge but present thick transgressive deposits on the shelf, thick regressive deposits seaward, and lasted for about 7 Myr (Figure 6.41). Dissimilarities between both types of sequences most probably result from changes in the long-term evolution of climate, ice volume and relative variations of sea level. High frequency eustatic fluctuations during a Middle Miocene long-term fall of relative sea level exposed the shelf to periodic erosion and produced a succession of short and highly progradational sequences. In contrast, a long-term rise of relative sea level of Late Miocene to Pliocene age produced an overall increase in accommodation space and continuous accumulation of siliciclastics. As a result, high frequency eustatic fluctuations did not produce distinct sequence boundaries but could be recorded as parasequences.

6.4. Example of Sedimentation in Active and Ancient Areas of Seafloor Spreading: The Mid-Atlantic Ridge

6.4.1. Sediments at the Contact of Oceanic Crust

Oceanic basalts being emitted at water depths of 1,500–3,000 m for isostatic reasons, mid-oceanic ridges are generally isolated from gravity-driven mass-transport

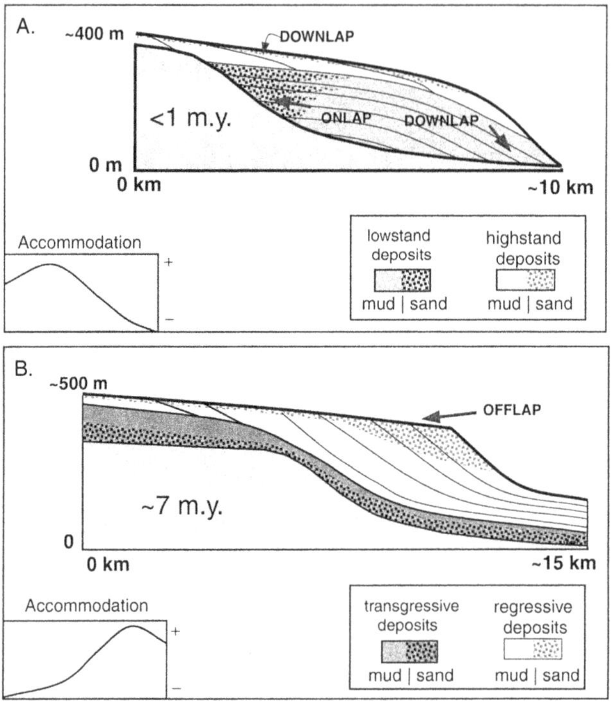

Figure 6.41 Typical Neogene sedimentary sequences from the fat passive New-Jersey margin. A, Middle Miocene sequence, characterized by a thick onlapping lowstand wedge; B, Late Miocene–Pliocene sequence, characterized by thick regressive deposits. Reprinted from Metzger, J.M., Flemings, P.B., Christie-Blick, N., Mountain, G.S., Austin, J.A., Hesselbo, S.P., 2000. Late Miocene to Pleistocene sequences at the New-Jersey outer continental shelf (ODP Leg 174A, Sites 1071 and 1072). Sedimentary Geology, 134, 149–180.

deposits of siliciclastics. Sediment accumulations on mid-oceanic ridges mainly include biogenic elements derived from planktonic production in the photic zone. Biogenic oozes may locally rest on devitrified palagonite, a basaltic glass altered to clay minerals principally (smectite, corrensite, celadonite, ...), in areas where large quantities of seawater could move through the cooling young basalt. In some places, biogenic oozes may cover basaltic sand and breccia. In other places, successive lava flows may have diagenetically altered interbedded layers and reworked clasts of biogenic ooze. However in most areas, biogenic oozes rest directly on fresh basalt as in the FAMOUS (French-American Mid-Oceanic Underwater Survey) area of the North Atlantic (Figure 6.1) and increase in thickness with distance from the active ridge. In addition, the dominant lithology may change as the seafloor drifts to areas of different oceanic environments. At DSDP Site 417 in the Hatteras abyssal plain of the North Atlantic, an oceanic crust of Mid-Cretaceous age is overlain by predominantly biogenic sediments which consist of nannofossil chalk (Figure 6.42), the concentration of carbonates being about 85%. There, the sediment grades to organic claystone and chalk and to clay in

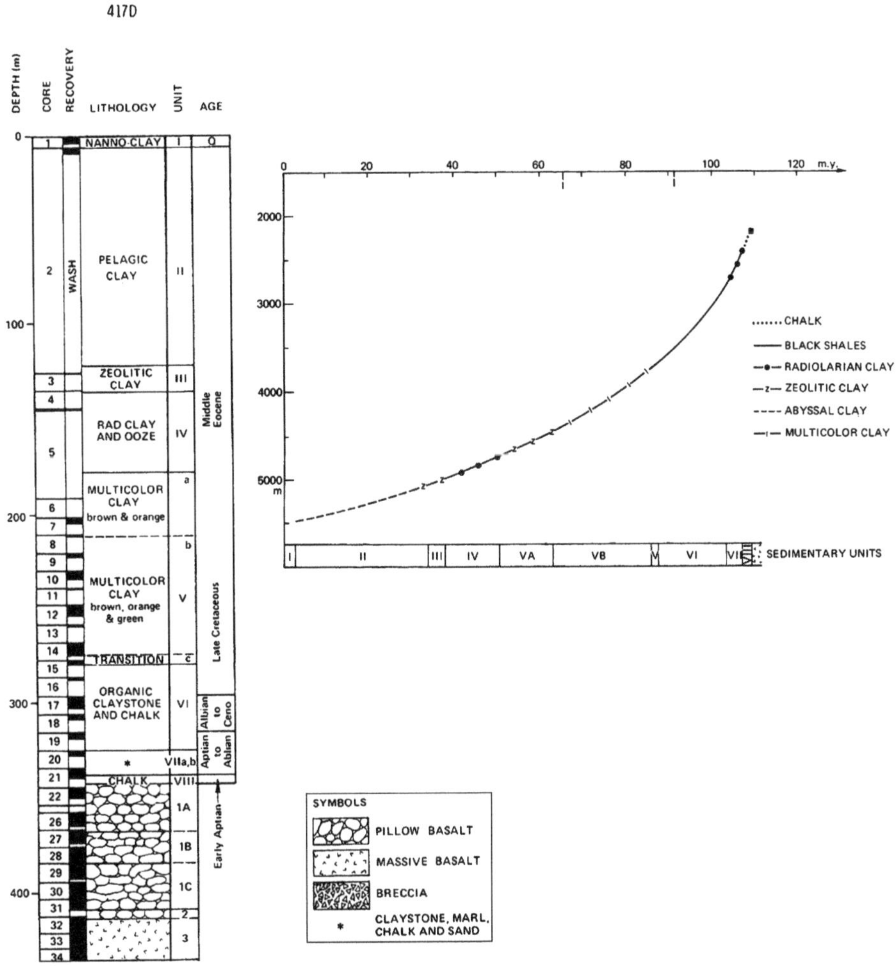

Figure 6.42 Lithostratigraphy (left) and subsidence curve (right) for DSDP Site 417, drilled in the Hatteras abyssal plain of the North Atlantic. Note changes in sedimentation as subsidence progressed. Modified from Donnelly, T., Francheteau, J., Ryan, W., Robinson, P., Flower, M., Salisbury, M. et al., 1979. Initial Reports of the Deep Sea Drilling Project, volumes 51, 52, 53. U.S. Gov. Print. Office, Washington, DC.

the Late Cretaceous, the contents of carbonates decreasing to 65% and then to sporadic minor amounts. The decrease in biogenic carbonate contents coincides with the subsidence of the ocean floor to greater depths of increasingly aggressive waters supplemented in carbon dioxide. High contents of carbon dioxide in deep waters are facilitated either by low temperatures (for example, in the Neogene and modern ocean of thermohaline circulation) or by water stratification and high organic contents (for example, during Cretaceous intervals of high productivity and/or halothermal circulation).

6.4.2. Hydrothermal Deposits of Active Mid-Oceanic Ridges

Accumulation of biogenic oozes on young basaltic crust may also interfer with hydrothermal activity, which in some cases may lead to massive hydrothermal mineralization. This is the case in the Trans-Atlantic Geotraverse (TAG) area East of the axis of the Mid-Atlantic Ridge at about 26°N (Figure 6.43) which includes, for example, more than 5 million metric tons of sulfides. The active segment of the Mid-Atlantic Ridge in the TAG area consists of an asymmetric median valley which is floored by overlapping volcanic edifices and bounded by fault controlled walls, the eastern wall being higher and steeper. Micro-seismic activity is currently limited to the axial part of the valley, where relatively low average spreading rates around 24 mm/year have been recorded for the past 10 Myr. The TAG hydrothermal field is located on the floor of the rift valley at 3,670 m water depth, 2.4 km east of the axis and 1 km from the east wall, in an area where fissures and faults parallel to the active ridge segment intersect a series of obliquely oriented faults which are likely to provide conduits for the circulation of the hydrothermal fluids. This portion of the Mid-Atlantic Ridge also includes a number of fossil (Late Pleistocene) and currently active hydrothermal fields which have produced significant accumulations of sulfides and oxides. The TAG mound is circular in shape with a diameter of about 200 m, culminates at about 40–50 m above seafloor, and has been investigated in detail during ODP Leg 155.

6.4.2.1. The nature and compostion of hydrothermal deposits

The TAG mound comprises two superposed platforms that may represent different phases of active growth (Figure 6.43). The larger lower platform shows evidence of mass wasting and supports a complex of white smokers venting fluids at temperatures of 260–300°C. The narrower upper platform is characterized by a black smoker complex of high temperature (363°C) fluid discharge. The white smoker fluids (which principally precipitate low Fe sphalerite, with minor marcasite and amorphous silica) are believed to be derived from the black smoker fluids (which principally precipitate pyrite, chalcopyrite and anhydrite) as a result of cooling, mixing with seawater and precipitation of sulfides within the mound. The flanks of the mound are relatively steep and in part covered by aprons of sulfide debris and fine-grained oxides eroded from the weathered slopes which grade outward into biogenic oozes. The sediments are locally silicified (red chert). It is hypothesized that the growth and collapse of sulfide chimneys progressively built accumulations of brecciated sulfide deposits during the Late Pleistocene and the Holocene. Continuous flow of hydrothermal fluids as well as diffuse venting through the growing accumulation of sulfide rubble caused recrystallization, annealing and cementation within the mound.

Seventeen holes were drilled at ODP Site 957 on the active TAG mound (Figure 6.43). The holes are distributed between four locations in the black smoker area and one location in the white smoker area. Six main zones of hydrothermal deposit stratigraphy which vary in thickness are recognized in the black smoker area,

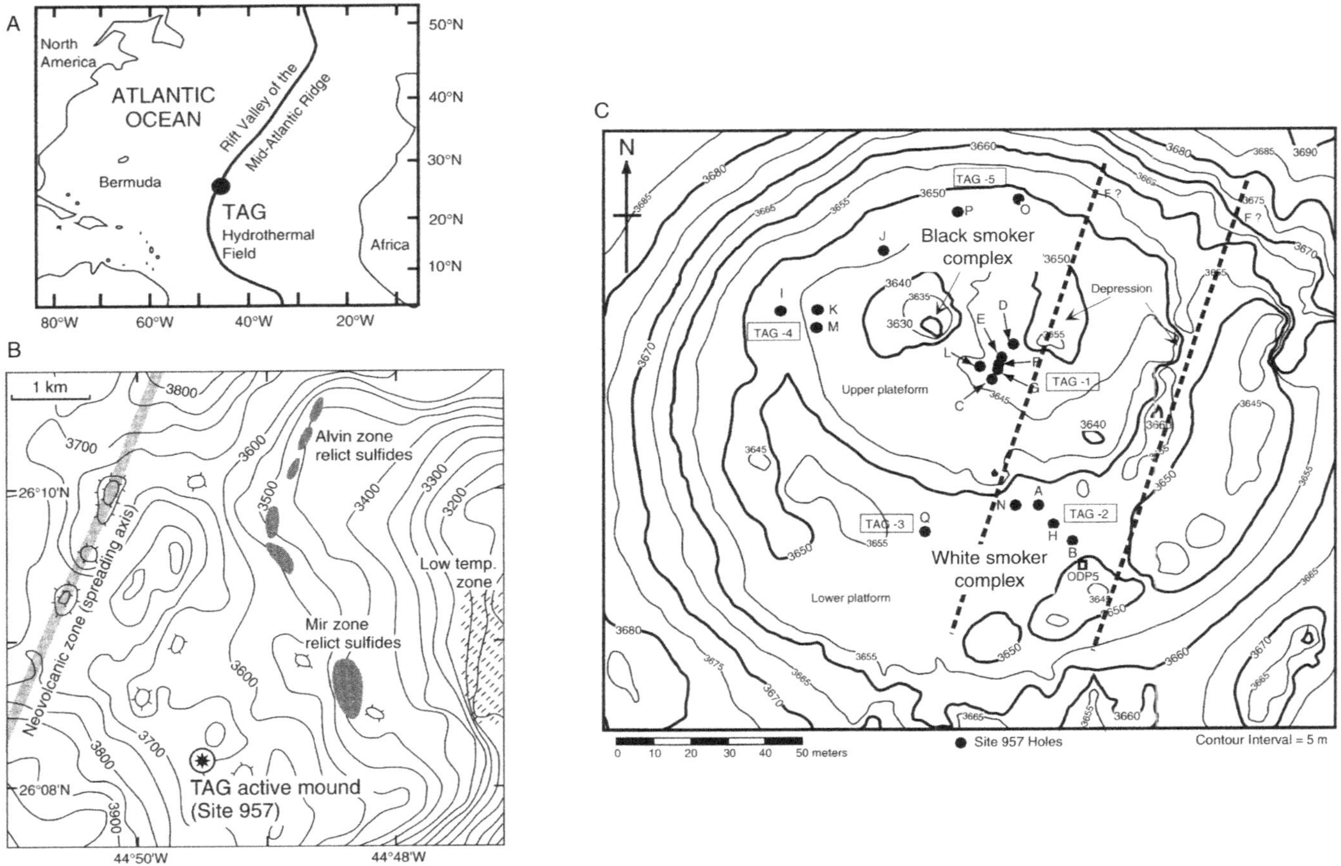

Figure 6.43 Location and morphology of the TAG (Trans-Atlantic Geotraverse) hydrothermal field of the Mid-Atlantic Ridge. A, location map; B, bathymetry of the TAG ridge segment with location of active and relict hydrothermal fields; C, detailed bathymetry of the active TAG mound. Dashed lines represent active faults. Note the location of the hydrothermal fields in the axial valley of the ridge close to the axis, the circular morphology of the TAG mound and the platforms sustaining the smoker complexes. Modified from Herzig, P.M., Humphris, S.E., Miller, D.J., Zierenberg, R.A. (Editors), 1998. Proceedings of the Ocean Drilling Program, Scientific Results, volume 158. Ocean Drilling Program, College Station, TX.

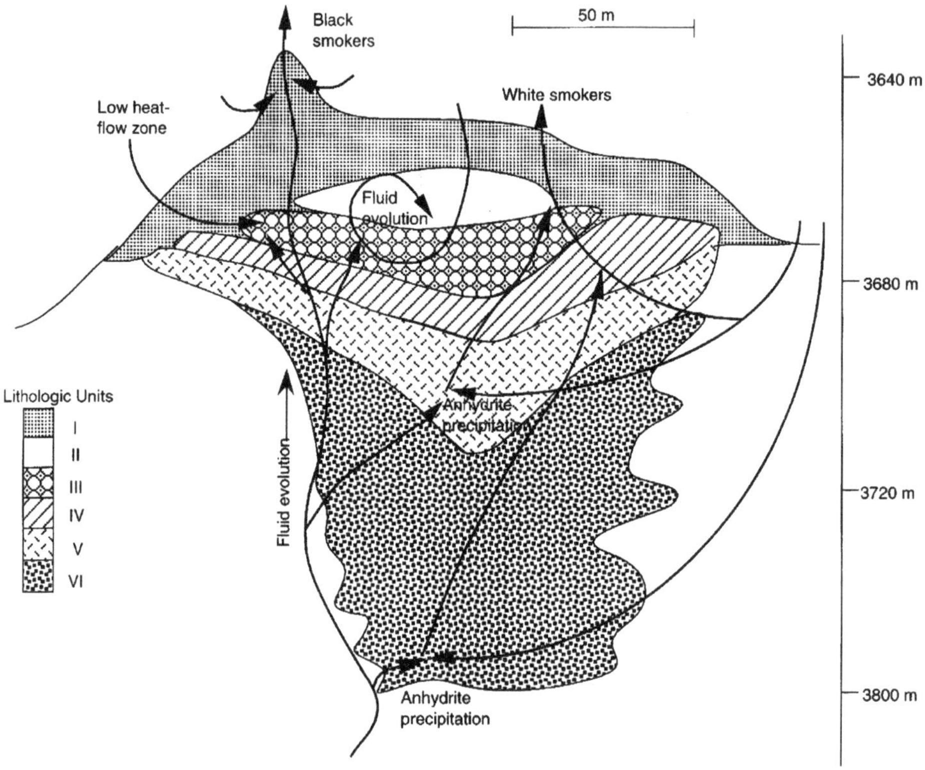

Figure 6.44 Schematic cross-section of the TAG mound. I, massive sulfides; II, pyrite-anhydrite breccia; III, pyrite-silica-anhydrite breccia; IV, pyrite-silica breccia; V, silicified basalt breccia; VI, chloritized basalt breccia. Arrows represent the paths of seawater and hydrothermal fluids. Note the extension of the upper unit and the deep alteration of the basalt below the active hydrothermal vents, decreasing at the periphery. Reprinted from Herzig, P.M., Humphris, S.E., Miller, D.J., Zierenberg, R.A. (Editors), 1998. Proceedings of the Ocean Drilling Program, Scientific Results, volume 158. Ocean Drilling Program, College Station, TX.

from the seafloor to the fresh basalt (Figure 6.44):

- The upper unit consists of massive sulfides which grade from porous Fe-Cu sulfide near surface to massive pyrite breccia downwards. Locally the porous Fe-Cu sulfide is substituted by porous Fe-Zn sulfide, silica, and Si-Fe oxyhydroxide. The pyrite breccia contains clasts that probably developed from recrystallization of porous sulfide.
- A gradual increase in the proportion of anhydrite in the matrix marks the transition to the pyrite-anhydrite breccia (Figure 6.45). This unit is characterized by further increases in the proportions of anhydrite in the matrix, and anhydrite veins which increase in number and thickness downsection.
- The pyrite-anhydrite breccia is succeeded by a pyrite-silica-anhydrite breccia which is characterized by pyrite clasts, a quartz matrix and anhydrite veins that

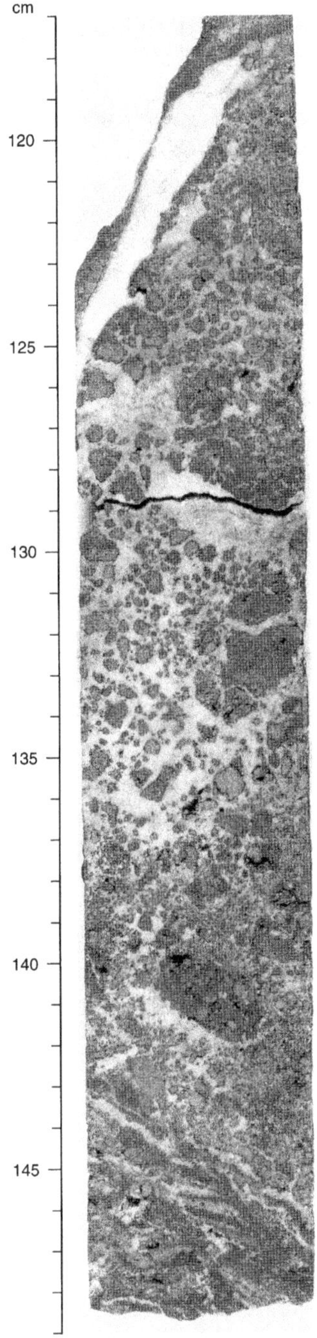

Figure 6.45 Typical pyrite-anhydrite breccia drilled at ODP Site 957, TAG 1 area. Pyrite in dark and anhydrite in light colors. Note the presence of an anhydrite vein upsection. Reprinted from Herzig, P.M., Humphris, S.E., Miller, D.J., et al., 1996. Proceedings of the Ocean Drilling Program, Initial Reports, volume 158. Ocean Drilling Program, College Station, TX.

occur in fractures postdating quartz deposition. Both pyrite-anhydrite and pyrite-silica-anhydrite breccias are locally missing at the periphery of the TAG mound.

- The pyrite-silica breccia is characterized by pyrite and pyrite-quartz clasts and a matrix of cryptocrystalline quartz, and contains large pieces of brecciated, silicified basalt. This unit underlies the massive sulfides at the periphery of the mound and its upper limit probably coincides with the location of the initial seafloor surface.
- A silicified basalt breccia is characterized by the replacement of all primary basaltic minerals by paragonite, quartz and pyrite.
- A chloritized basalt breccia where most primary basaltic minerals are replaced by chlorite, pyrite and quartz, is present in the deepest part of the stockwork. It is remarkable that chloritization is reduced to millimeters to centimeters thick halos parallel to the surface of basalt fragments at the periphery of the hydrothermal mound, and that some anhydrite veins have been observed in the three lower units.

In the white smoker area, the upper unit of massive sulfides mainly consist of Fe-Zn sulfide associated to minor chert and Fe-oxyhydroxide. Below, the lithologic units are roughly similar to those in the black smoker area. However, the pyrite-anhydrite zone and anhydrite veining are poorly developed, and overall the anhydrite is less abundant.

6.4.2.2. Evolution of hydrothermal structures and sediments

The succession of lithologies and deposition textures provide information on the growth and evolution of the TAG hydrothermal mound. Fine grain size, primary deposition textures and absence of oxidation indicate that the upper porous sulfides developed recently at the mound surface from diffuse hydrothermal flow. Initial precipitation of pyrite at temperatures between 250°C and 300°C was followed by insulation and channeling of the hydrothermal fluid into newly created interconnected pore spaces. Higher fluid temperatures of 300°C to more than 350°C led to the formation of well-crystallized pyrite and chalcopyrite principally. Lower temperatures and entrainment of seawater led to a different lithology in areas of Fe-Zn sulfide deposition and in the white smoker area where sphalerite, marcasite and silica are present. Collapse and mass wasting of the porous sulfide during burial were followed by recrystallization of most minerals into pyrite. There is textural evidence that the anhydrite matrix in the pyrite-anhydrite breccias did not replace a pre-existing matrix and did not fill pre-existing cavities. It is probable that *in situ* precipitation of anhydrite was made possible because of high fluid pressure compensating for burial and fluid percolation through displaced sulfide clasts. This process probably led to significant internal expansion of the TAG mound. Variations of hydrothermal activity during the Late Pleistocene and the Holocene probably resulted in a succession of collapses (anhydrite dissolution) and reactivations (anhydrite precipitation) leading to disruption of the breccias and uplift of older fragments of breccia and altered basalt. However, anhydrite veins seem to have formed through filling of fissures and fractures which developed within the mound and acted as pathways for hydrothermal fluids. The pyrite-silica breccia evolved from pre-existing pyrite-anhydrite breccia through progressive replacement of the anhydrite matrix by quartz aggregates. It is probable that initial deposition of amorphous silica and/or cryptocrystalline quartz was followed by hydrothermal recrystallization and related minor decrease in volume. The relationships between pyrite and quartz are confused by mutual intergrowth and replacement. The upper pyrite-silica breccia corresponds to the inferred depth of the oceanic crust, and the presence of altered basaltic clasts increasing with depth suggests that for a large part this unit was initially oceanic basalt.

The hydrothermal alteration of the oceanic crust started with the chloritization of the basalts through interaction with Mg-bearing hydrothermal solutions. Chloritization was especially intense below the currently active hydrothermal mound, decreasing in importance at the periphery. As alteration progressed, most of the previously altered and relict minerals were replaced by paragonite, quartz and pyrite. The presence of pyrite crystals with quartz corona and zoned quartz with pyrite veins suggest that recurrent intervals of silicification and pyritization deeply altered the basaltic stockwork. Anhydrite locally precipitated as veins in open spaces, as a consequence of significant infiltration of seawater in the hydrothermal system.

The development of hydrothermal deposits seems to result from the combination of surface precipitation of porous crusts of sulfides and oxyhydroxydes followed by collapse, dissolution and recrystallization, hydrothermal mineralization of sulfides, anhydrite and silica within the mound, and progressive alteration of the underlying oceanic basalt. Hydrothermal activity decreases with distance from active areas of mid-oceanic ridges but produces ore deposits which are progressively buried below variable thicknesses of pelagic or hemipelagic sediments as the oceans become wider.

6.5. Example of Sedimentation in a Transform Passive Margin Area: The Ivory Coast and Ghana Margin of the South Atlantic

6.5.1. Structure and Early History of the Ivory Coast and Ghana Margin

The transform passive margin which bounds the South Atlantic to the North extends from the Benue Trough of Nigeria to the East to the Saint-Paul and Romanche fracture zones to the West (Figure 6.46). The first detailed investigation of a transform passive margin associated geophysical surveys, submersible and drilling expeditions in selected areas off Ivory Coast and Ghana. There, the deep Ivorian Basin is an extensional basin of probable pull-apart origin where half-grabens are filled with thick syn-rift deposits. The basin is bounded by a segment of rifted margin to the East and by oceanic crust to the West. Its northern limit is the Ivory Coast nearshore escarpment which is in continuity with the Saint-Paul Fracture Zone. Its southern limit is the Ivory Coast-Ghana Marginal Ridge which is in continuity with the steep Ghana shear margin to the East and the Romanche Fracture Zone to the West (see Section 2.1.4). The investigations focused on the

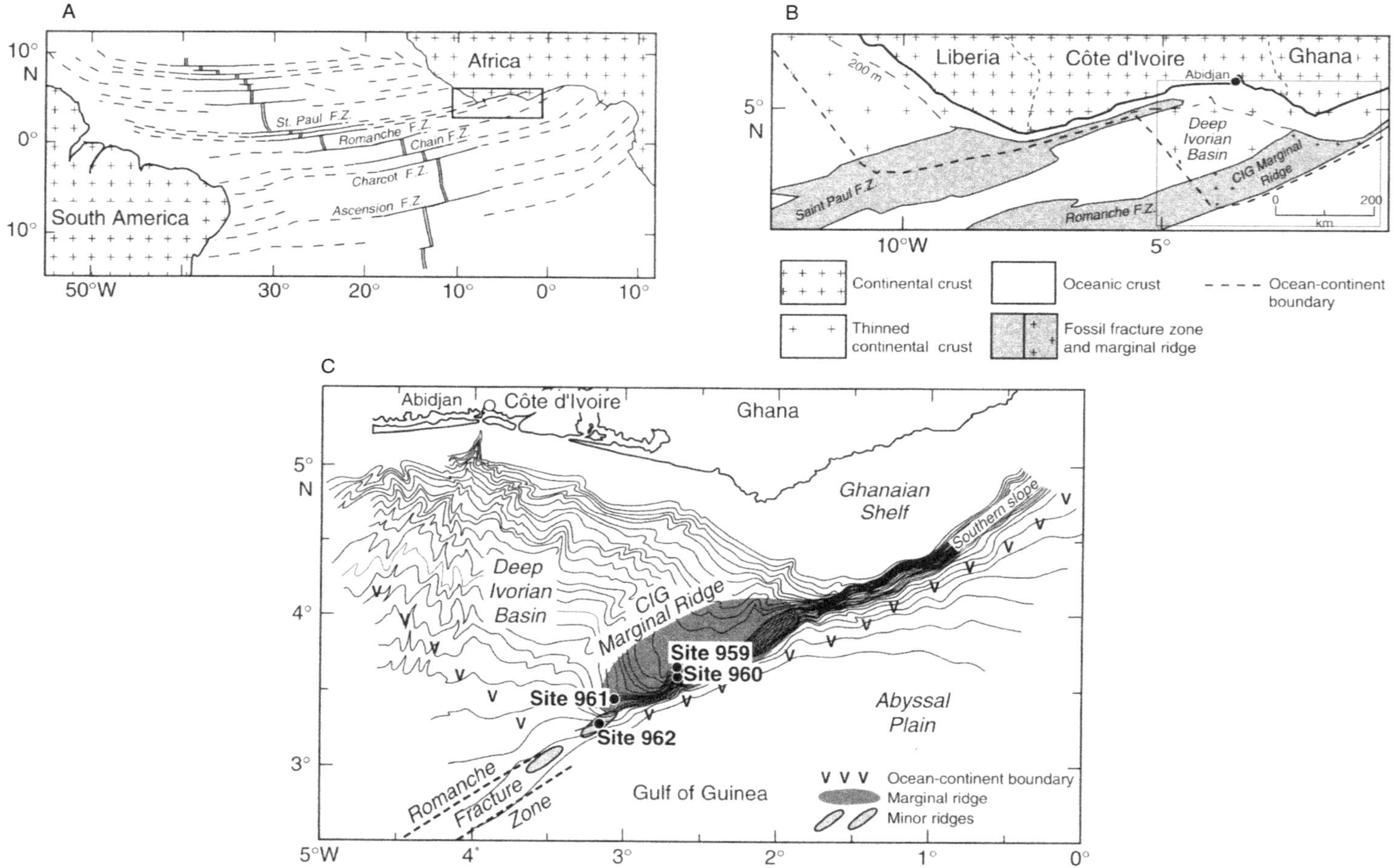

Figure 6.46 The Ivory Coast-Ghana transform margin. A, plate tectonic framework. B, main regional traits. Note the alternance of divergent segments of thinned continental crust and strike-slip segments in continuity with oceanic fracture zones. C, detailed bathymetry of the Ivory Coast-Ghana area and location of ODP sites drilled during Leg 159. Note the extension of the marginal ridge. Modified from Mascle, J., Lohman, G.P., Clift, P.D. et al., 1996. Proceedings of the Ocean Drilling Program, Initial Reports, volume 159. Ocean Drilling Program, College Station, TX.

Ivory Coast Marginal Ridge and its relations to the Ghana margin and Romanche Fracture Zone (Figure 6.46). The Ivory Coast-Ghana Marginal Ridge is a 135 km long and 25 km wide structure which corresponds to the transition from thinned continental crust of the deep Ivorian Basin to the oceanic crust of the Angola Basin. The ridge culminates at 2,000 m water depth (about 1,300 m above the deep Ivorian Basin and 3,000 m above the abyssal plain) but deepens to the West where it is progressively buried below oceanic sediments. The shear margin and marginal ridge represent very steep (up to 20°) and narrow (less than 40 km) transition zones from continental to oceanic crusts.

Africa and South America were in contact during the Early Cretaceous, before rifting and strike-slip activity slowly separated the deep Ivorian Basin from its conjugate Bareirinhos Basin of Brazil. The uppermost part of thick syn-rift deposits, which form a distinct and often chaotic seismic unit (Unit A, Figure 6.47), has been sampled during ODP Leg 159 at Sites 959, 960 and 961. The sediments (Figure 6.48) principally consist of siliciclastics ranging from silty sandstones to silty claystones (sometimes laminated) with minor breccias, and are of Aptian-Albian age. Laminated fine siliciclastics probably deposited in lacustrine environments, whereas massive to convoluted sandy successions suggest gravity-driven deposition in deltaic environments. Lacustrine sediments are overlain by storm influenced, cross-bedded sandstones of shallow marine environments. The siliciclastics alternate with biogenic, partly pelagic carbonates at ODP Site 962, which could represent a deeper and relatively sediment-starved environment.

Microfaults, folds, pop-up structures and slumps affected syn-rift sediments after deposition. Soft-sediment deformation was followed by hydrothermal mineralization of calcite, pyrite, quartz and barite principally, as irregular patches or within microfaults and fluid-escape structures. Syn-rift deposits were heated to temperatures of 120–180°C, as deduced from the maturation of organic elements and clay minerals, and fluid inclusions. The semi-lithified syn-rift unit was later tilted and faulted. It is remarkable that thermal event and deformation progressed from East to West, from the Late Aptian–Early Albian at ODP Sites 959 and 960 to the Latest Albian at ODP Site 962. Seismic reflection data suggest that deformed syn-rift deposits constitute most of the marginal ridge sedimentary pile (Figure 6.47). Syn-rift sediments outcrop on the southern flank of the marginal ridge, where they have been observed and sampled during submersible expeditions (Figure 6.49). At all sites, the syn-rift unit has been deformed and heated before an erosional phase associated to a jump of the main transform activity, which moved from North to South of the marginal ridge.

6.5.2. Sedimentation During the Cretaceous Interval of Continent/Ocean Strike-Slip Activity

The creation of the oldest oceanic crust during the Latest Albian to Cenomanian marks the beginning of continent/ocean strike-slip activity (equivalent to the crustal fissure stage, see Sections 5.1 and 5.3). In the study area of the marginal ridge, African continental crust was in contact with thinned South American continental crust until the Coniacian. The transform motion involved African continental crust

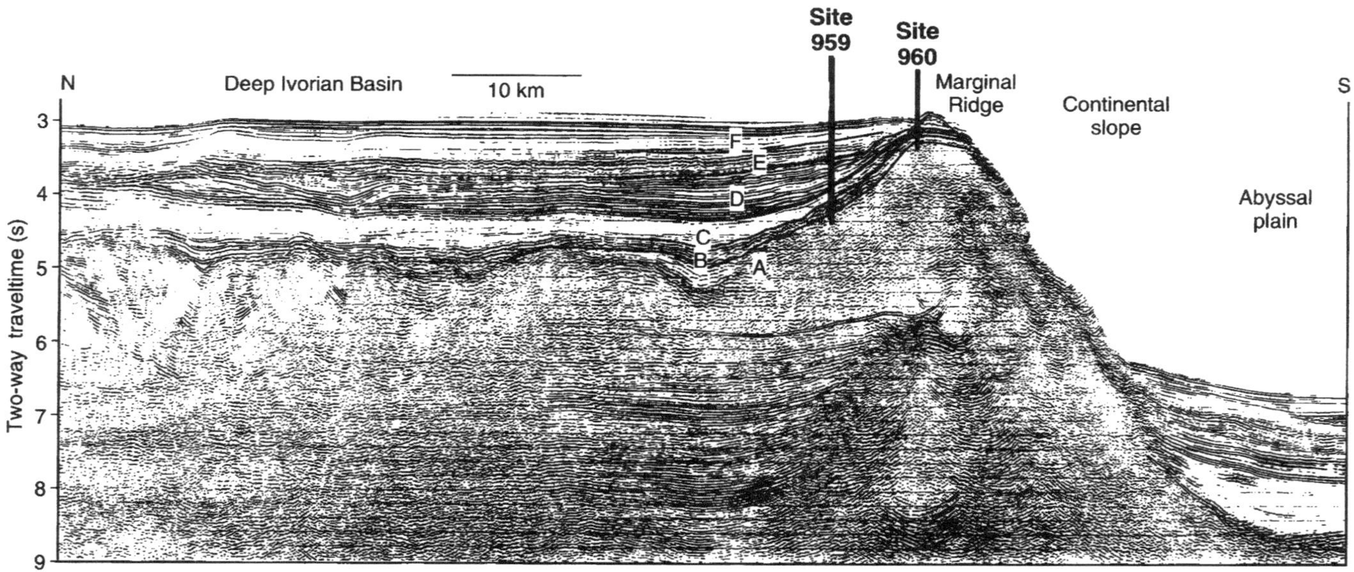

Figure 6.47 Seismic reflection profile across the Deep Ivorian Basin, the marginal ridge and the abyssal plain. Note the pile of syn-rift deposits below the marginal ridge, the steep southern flank where syn-rift deposits outcrop, and the succession of parallel post-rift units in the subsiding Deep Ivorian Basin which are pinched up toward the crest of the marginal ridge. Reprinted from Mascle, J., Lohman, G.P., Clift, P.D. et al., 1996. Proceedings of the Ocean Drilling Program, Initial Reports, volume 159. Ocean Drilling Program, College Station, TX.

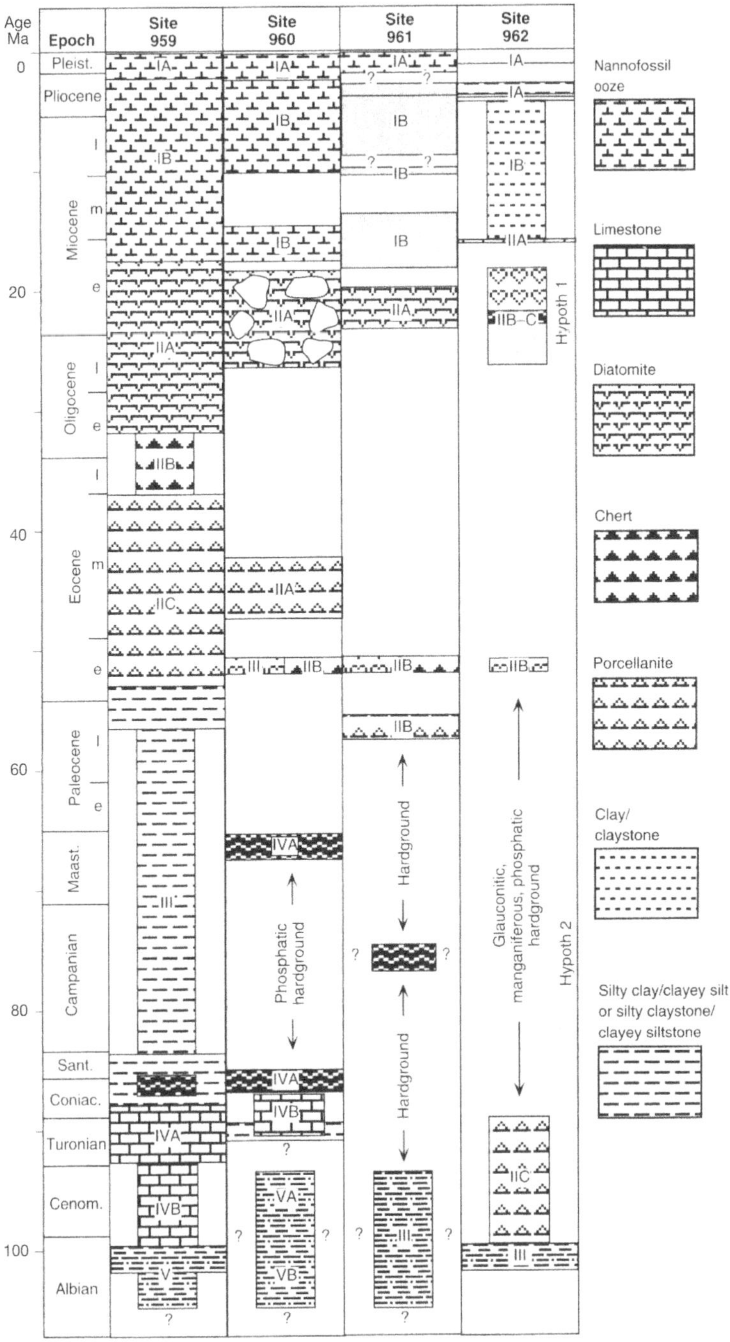

Figure 6.48 Lithostratigraphic summary of ODP Holes drilled during Leg 159. Correlation between lithostratigraphic units at ODP Site 959 and acoustic units from Figure 6.47: Unit V correlates with Unit A, Unit IV with Unit B, Unit III with Unit C, Subunits IIC and IIB with Unit D, Subunit IIA with Unit E, Unit I with Unit F. Note the dominance of siliciclastic facies from the Mid-Cretaceous to the Paleocene, biosiliceous facies from the Eocene to the Early Miocene and biocalcareous facies since the Early Miocene, the frequency of hardgrounds and condensed series at the shallower Sites 960 and 961, and the importance of Neogene siliciclastics at the western sites within the reach of direct continental input. Reprinted from Mascle, J., Lohman, G.P., Clift, P.D. et al., 1996. Proceedings of the Ocean Drilling Program, Initial Reports, volume 159. Ocean Drilling Program, College Station, TX.

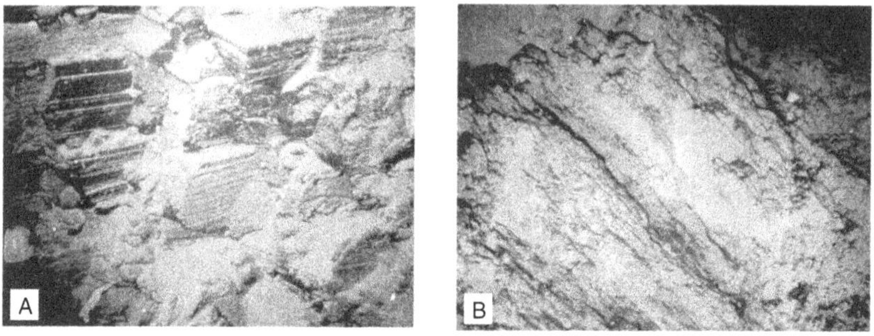

Figure 6.49 Syn-rift sediments outcropping on the southern flank of the marginal ridge, as observed during the Equanaute submersible expedition. A, fine-grained sandstones to siltstones cut by orthogonal brittle joints, 4,000 m water depth; B, alternating claystones and siltstones tilted toward the southeast, 4,525 m water depth. Reprinted from Mascle, J., Lohman, G.P., Clift, P.D. et al., 1996. Proceedings of the Ocean Drilling Program, Initial Reports, volume 159. Ocean Drilling Program, College Station, TX.

and oceanic crust from the Coniacian to the Paleocene when the spreading axis cleared the marginal ridge and strike-slip activity ceased. These two phases roughly coincide with deposition of acoustic Units B and C respectively, which are separated by an unconformity in the marginal ridge area (Figure 6.47). The migration of the spreading axis along the margin and marginal ridge did not produce a significant thermal uplift of the continental lithosphere, probably because strike-slip activity involved a highly fractured zone of vertical hydrothermal circulation with limited horizontal conduction. However, contrasts in thickness and physical properties of the two adjacent lithospheres probably produced isostatic readjustments and related changes in relief and erosion during the entire continent/ocean transform stage. It is remarkable that the tilting phase of the marginal ridge is correlated to the active transform contact with the thinned Brazilian crust, and that the subsidence of the marginal ridge increased in the Early Coniacian together with the difference in depth between the continental margin and the oceanic basin, producing slope instabilities and significant erosion. Above the post-rift unconformity, Turonian to Early Coniacian sediments at ODP Sites 959 and 960 principally consist of limestones (Figure 6.48) with skeletal fragments, red algae and mollusks, grading to sandy limestone, sandy dolomite and calcareous sandstone. They are interpreted as periplatform deposits transported as grain flows and debris flows. Higher thicknesses at the shallower Site 960 and northward transport direction suggest that carbonate platforms and/or reefs colonized shallow areas of the marginal ridge which was probably significantly uplifted during the post-rift erosional event. Further West, siliciclastics still dominated at ODP Site 961 in the deep Ivorian Basin, whereas cherts and porcellanites at the shallower Site 962 (Figure 6.48) could derive from biogenic elements and high pelagic productivity.

Carbonate sedimentation persisted on the marginal ridge until the Early Coniacian when they were replaced by calcareous claystones, chalks, phosphates

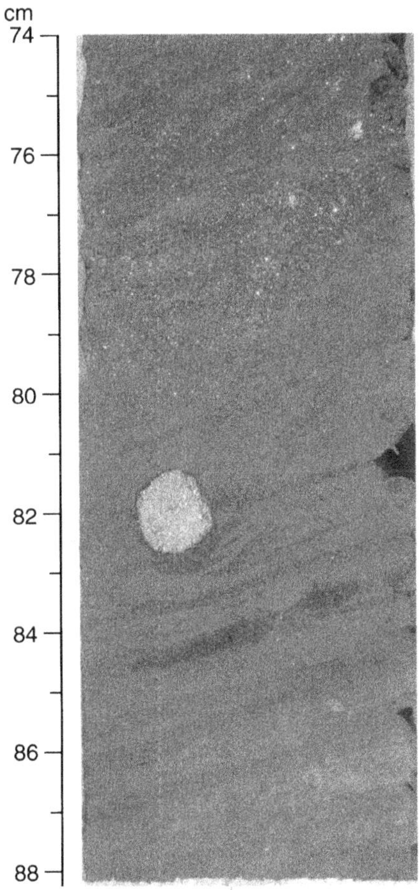

Figure 6.50 Maestrichtian black-shale facies (black claystone) sampled at ODP Site 959 (Section 159-959D-54R-1). Note pyrite nodule, scatterred pyrite crystals in the upper part of the section, and faint bioturbation in the lower part of the section. Reprinted from Mascle, J., Lohman, G.P., Clift, P.D. et al., 1996. Proceedings of the Ocean Drilling Program, Initial Reports, volume 159. Ocean Drilling Program, College Station, TX.

and glauconite at the shallower Site 960. The change in sedimentation indicates that carbonate platforms did recede because of increased subsidence of the marginal ridge: the crest of the ridge was still isolated from terrigenous input as indicated by low proportions of fine siliciclastics, but remained at shallow water depths of strong circulation and high productivity as attested by occurrences of glauconite and phosphates. Early Coniacian to Late Paleocene sediments at the deeper Site 959 principally consist of faintly laminated to massive dark organic claystone (Figure 6.48) with disseminated pyrite and occasional occurrences of barite and glauconite. Interbedded lighter intervals contain low proportions of carbonates (mainly coccoliths) and are slightly bioturbated. This typical black-shale facies (Figure 6.50) is indicative of dysoxic to anoxic environments which developed in a context of water stratification. The presence of phosphatic nodules and glauconite in black-shale deposits suggests some reworking from the upper marginal ridge where oxygenated environments persisted at the shallower Site 960.

It is remarkable that the inception of black-shale facies off Ivory Coast and Ghana in the Early Coniacian significantly postdates the Mid-Cretaceous Anoxic Events, which extended throughout the Atlantic and Tethys oceans from the Aptian to the Turonian, and resulted from the combined influence of high productivity, production of warm saline deep water and stratification of the water column. The black-shale facies off Ivory Coast and Ghana encompass an interval of oxygenated and ventilated environments in adjacent regions of the Atlantic Ocean. They do not reflect global conditions, but are most probably consequences of local environmental conditions.

The black-shale facies started when the transform motion involved an African continental crust and oceanic crust in the Early Coniacian, and lasted until the spreading ridge cleared the marginal ridge in the Paleocene. Therefore, the black-shale facies characterize an interval of maximum contrast in density and other physical properties between adjacent lithospheres, and related isostatic readjustments. This is illustrated by the dip of the bedding surfaces which varies from 40° at the base, to 15° at the top of the black-shale unit, indicating a progressive northwestward tilting of the marginal ridge which subsided much more slowly than the adjacent deep Ivorian Basin. It is likely that the marginal ridge acted as a barrier to water exchanges and limited the ventilation of the deep Ivorian Basin. Moreover, the Early Coniacian to Late Paleocene was an interval of high productivity, as attested by the presence of phosphates at the shallower Site 960 and dominance of lithologies derived from biogenic remains farther West at ODP Site 962 (Figure 6.48). The combination of high productivity and poor ventilation probably sustained dysoxic to anoxic conditions in the deep Ivorian Basin. This in turn facilitated the preservation of organic matter (of both terrestrial and marine origins) and carbonate dissolution, and the persistence of the black-shale facies for about 30 Myr.

6.5.3. The Cenozoic Passive Margin

Biogenic sedimentation resumed at ODP Site 959 when the black-shales were succeeded by light greenish gray, moderately bioturbated nannofossil claystone of Late Paleocene age. Sedimentation rates, which were very low (7–7.5 m/Myr) in the black-shale unit increased to 15–22 m/Myr in the Cenozoic because of the preservation of biogenic elements principally. It is probable that ventilation of the deep Ivorian Basin improved when the Mid-Atlantic Ridge cleared the African margin. Strike-slip activity ceased as a result, and the entire transform margin of the Gulf of Guinea became passive. The transition was diachronous, and may have locally occurred as soon as the Campanian. Acoustic units D to F (Figure 6.47), which clearly overlap the crest of the marginal ridge, deposited during the passive margin stage. The dip of Eocene bedding surfaces is below 10° at ODP Site 959 where it is clearly compatible with differential compaction, but increases strongly at ODP Site 960 where sediments are highly fractured is some places and subjected to plastic flows in others.

With the exception of a 10 m thick Early Eocene section at the shallower Site 960 (and some intervals at ODP Site 961) where palygorskite claystones containing barite nodules deposited in probable relation to local environmental conditions, Eocene to Miocene sediments at all sites clearly reflect the importance of pelagic productivity (Figure 6.48). At the shallowest locations, cherts and porcellanites alternate with glauconitic and zeolitic claystones at ODP Site 960, whereas nannofossil chalks and porcellanites alternate with palygorskite and zeolitic claystones at ODP Site 961. At the deeper Site 959, diatomites, cherts and porcellanites alternate with nannofossil chalks and clays. Silica dominates at all sites, as in modern areas of coastal upwelling and oceanic divergence, suggesting that such processes developed or intensified in the Early Eocene. Biosiliceous elements have been diagenetically altered at these sites where highly condensed series, frequent hiatuses and glauconitic hardgrounds suggest periodic erosion by active oceanic circulation. Diagenetic alteration might have been favored by increased temperatures in this area where traces of hydrothermal mineralization are commonly found in Eocene sediments. The association of calcareous and siliceous biogenics, siliciclastics, authigenic and diagenetic elements reflects the complexity of oceanographic and environmental processes in the equatorial Atlantic Ocean during the Paleogene.

Calcareous biogenic elements increased in the late Early Miocene (Figure 6.48), the sediment grading to nannofossil and foraminifer-nannofossil oozes and chalks principally at ODP Sites 959 and 960. The proportion of siliciclastics increases farther West, where intervals of nannofossil clays are recorded at the shallower Site 961. Clays and silty clays containing plant debris alternate with nannofossil clays at the deeper Site 962 which is the most easily reached by terrigenous input from the adjacent African continent of tropical climate. The drastic decrease in siliceous biogenics observed in the Early Miocene might reflect either a decrease or southward migration of the high productivity divergence zone, relative to the northward migrating African plate. Biogenic elements in Neogene pelagic and hemipelagic sediments of the marginal ridge essentially consist of carbonates, making the Ivory Coast-Ghana passive transform margin very comparable to other passive margins of tropical and temperate latitudes. The sedimentary sequence of the marginal ridge presents lower accumulation rates of carbonates and significant hiatuses of Middle Miocene age, followed by increased accumulation rates of carbonates in the Late Miocene and Early Pliocene which are also observed in many other areas of the Atlantic and Indian oceans. This succession indicates that Neogene sediments from the Ivory Coast-Ghana passive transform margin contain a record of global oceanographic events.

FURTHER READING

Anderson, J.B., 1999. Antarctic marine geology. Cambridge University Press, Cambridge.

Banda, E., Torné, M., Talwani, M. (Editors), 1995. Rifted ocean-continent boundaries. Klüwer, Dordrecht.

Boillot, G., Coulon, C., 1998. La déchirure continentale et l'ouverture océanique. Overseas Publishers Association, Amsterdam.
Boillot, G., Huchon, P., Lagabrielle, Y., 2003. Introduction à la géologie: La dynamique de la lithosphère. Dunod, Paris.
Einsele, G., 1992. Sedimentary basins, evolution, facies, and sediment budget. Springer, Berlin.
Kennett, J.P., 1982. Marine geology. Prentice-Hall, Englewood Cliffs, NJ.

Other references used in this chapter

Austin, J.A., Christie-Blick, N., Malone, M.J., et al., 1998. Proceedings of the Ocean Drilling Program, Initial Reports, volume 174. Ocean Drilling Program, College Station, TX.
Brun, J.P., Beslier, M.-O., 1996. Mantle exhumation at passive margins. Earth and Planetary Science Letters, 142: 161–173.
Duncan, R.A., Larsen, H.C., Allan, J.F., et al., 1996. Proceedings of the Ocean Drilling Program, Initial Reports, volume 163. Ocean Drilling Program, College Station, TX.
Duncan, R.A., Larsen, H.C., Allan, J.F., Brooks, K. (Editors), 1999. Proceedings of the Ocean Drilling Program, Scientific Results, volume 163. Ocean Drilling Program, College Station, TX.
Faugères, J.-C., Stow, D.A.V., Imbert, P., Viana, A., 1999. Seismic features diagnostic of contourite drifts. Marine Geology, 162: 1–38.
Flood, R.D., Piper, D.J.W., Klaus, A., et al, 1995. Proceedings of the Ocean Drilling Program, Initial Reports, volume 155. Ocean Drilling Program, College Station, TX.
Gallagher, K., Brown, R., 1999. Denudation and uplift at passive margins: the record on the Atlantic margin of southern Africa. Philosophical Transactions of the Royal Society of London A, 357: 835–859.
de Graciansky, P.C., Poag, C.W., et al., 1985. Initial Reports of the Deep Sea Drilling Project, volume 80. U.S. Gov. Print. Office, Washington, DC.
Herzig, P.M., Humphris, S.E., Miller, D.J., et al., 1996. Proceedings of the Ocean Drilling Program, Initial Reports, volume 158. Ocean Drilling Program, College Station, TX.
Herzig, P.M., Humphris, S.E., Miller, D.J., Zierenberg, R.A. (Editors), 1998. Proceedings of the Ocean Drilling Program, Scientific Results, volume 158. Ocean Drilling Program, College Station, TX.
Larsen, H.C., Saunders, A.D., Clift, P.D., et al., 1994. Proceedings of the Ocean Drilling Program, Initial Reports, volume 152. Ocean Drilling Program, College Station, TX.
Larsen, H.C., Saunders, A.D., Wise, S.W. (Editors), 1998. Proceedings of the Ocean Drilling Program, Scientific Results, volume 152. Ocean Drilling Program, College Station, TX.
McHugh, C.M.G., Damuth, J.E., Mountain, G.S., 2002. Cenozoic mass-transport facies and their correlation with relative sea-level change, New-Jersey continental margin. Marine Geology, 184: 295–334.
Mascle, J., Lohman, G.P., Clift, P.D., et al., 1996. Proceedings of the Ocean Drilling Program, Initial Reports, volume 159. Ocean Drilling Program, College Station, TX.
Mascle, J., Lohman, G.P. Moullade, M. (Editors), 1998. Proceedings of the Ocean Drilling Program, Scientific Results, volume 159. Ocean Drilling Program, College Station, TX.
Metzger, J.M., Flemings, P.B., Christie-Blick, N., Mountain, G.S., Austin, J.A., Hesselbo, S.P., 2000. Late Miocene to Pleistocenesequences at the New-Jersey outer continental shelf (ODP Leg 174A, sites 1071 and 1072). Sedimentary Geology, 134: 149–180.
Miller, K.G., Sugarman, P.J., Browning, J.V., et al., 1998. Proceedings of the Ocean Drilling Program, Initial Reports, volume 174AX. Ocean Drilling Program, College Station, TX.
Mountain, G.S., Miller, K.G., Blum, P., et al., 1994. Proceedings of the Ocean Drilling Program, Initial Reports, volume 150. Ocean Drilling Program, College Station, TX.
Mulder, T., Cirac, P., Gaudin, M., Bourillet, J.-F., Tranier, J., Normand, A., Weber, O., Griboulard, R., Jouanneau, J.-M., Anschutz, P., Jorissen, F.J., 2004. Understanding continent-ocean sediment transfer. EOS, American Geophysical Union, 85: 257–262.

Poag, C.W., Watts, A.B., et al., 1987. Initial Reports of the Deep Sea Drilling Project, volume 95. U.S. Gov. Print. Office, Washington, DC.

Sawyer, D.S., Whitmarsh, R.B., Klaus, A. (Editors), 1994. Proceedings of the Ocean Drilling Program, Scientific Results, volume 149. Ocean Drilling Program, College Station, TX.

Scrutton, R.A., Stocker, M.S., Shimmield, G.B., Tudhope, A.W. (Editors), 1995. The tectonics, sedimentation and paleoceanography of the North Atlantic region. Special Publication 90, Geological Society, London.

Whitmarsh, R.B., Beslier, M.-O., Wallace, P.J. (Editors), 1998. Proceedings of the Ocean Drilling Program, Scientific Results, volume 173. Ocean Drilling Program, College Station, TX.

CHAPTER SEVEN

AULACOGENS

7.1. STRUCTURE, TECTONICS AND SEDIMENTATION OF AULACOGENS

Aulacogens are paleorifts that have been reactivated by compression at some stage of their history. Although oceanic crust is not necessarily involved, aulacogens have recorded most sequences of the Wilson cycle of ocean evolution. Like many failed rift branches, aulacogens originate in hot-spot areas where a mantle plume is associated to rising magma and high heat flow. Reinforced thermal and isostatic processes produce a thermal doming of the continental lithosphere of high amplitude (up to 3–4 km above sea level) and regional importance, and the brittle continental crust may eventually fracture along a series of three rift valleys radiating away from the center of the dome, which becomes a triple junction (Figure 7.1). Intense magmatism rapidly leads to gabbro intrusions and basaltic lava flows. Intense heat may locally melt the continental crust, leading to granite intrusions and rhyolitic lavas. The Afar Triangle of East Africa is an example of an evolving triple junction. The Afar Triangle is a seismically and tectonically active area of recent and intense volcanism, at the intersection of three major structures: the oceanic crust of the southern Red Sea, the Gulf of Tadjura which is in continuity with the Carlsberg Ridge of the Indian Ocean and the Ethiopian section of the East African Rift (see Section 5.2).

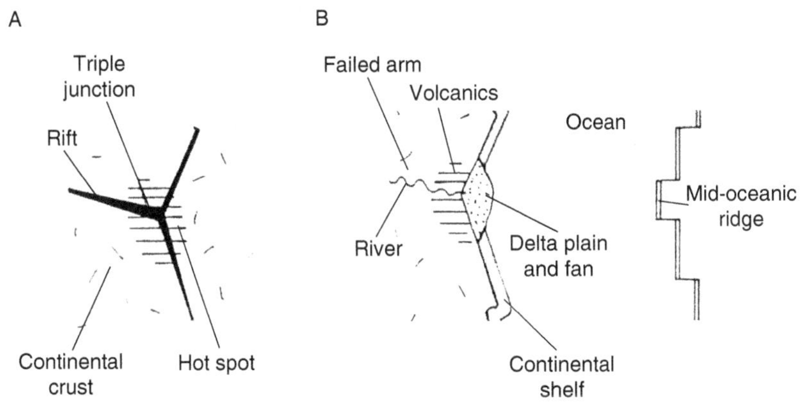

Figure 7.1 Most common evolution of a triple junction. (A) Magmatic activity and related thermal doming initiate three rift valleys radiating from the triple junction at the center of the dome. (B) Ocean opening processes focus on two rift branches and the failed arm is incorporated into the passive margin as an elongated coastal basin where river systems converge. Modified from Prothero, D., Schwab, F., 1996. Sedimentary geology. Freeman, New-york.

In some cases, all rift branches radiating from the triple junction may evolve into active spreading ridges. This has been the case in the Azores area of the North Atlantic during the Cretaceous, at the intersection of the African, European and North American plates. The Azores are volcanic islands which have been created by hot-spot activity. In most cases however, geological activity concentrates on two rift branches only, which progressively separate two passive continental margins, and the third branch becomes inactive (Figure 7.1). Because of lower heat flow, the density of the lithosphere increases and its volume decreases. As a result the inactive area subsides, and running waters converge toward the topographic low, which may eventually become a conduit for major river systems.

When activity ceases during the early rift stages, the failed arm is rapidly incorporated into the passive margin as an elongated coastal basin extending into the continental hinterland where it gradually disappears. For example, the formation of the Tucano Basin of Brazil may have been facilitated by an interval of mantle upwelling during the Neocomian (see Section 3.5). However, lithospheric extension may persist sufficiently to produce extensively thinned continental lithosphere or intermediate to oceanic lithosphere, before activity ceases and the structure is incorporated into a passive continental margin. Those areas of weakened lithosphere are bounded by sections of thick continental lithosphere and likely to accommodate internal stress transmitted from plate boundaries, for example during changes in the rotation and velocity of lithospheric plates, and collision events. They may record a complex history of tensional and compressional events through a succession of continental and marine environments, faulting and folding episodes and intervals of metamorphism, which are a major characteristic of aulacogens. Examples of aulacogens include the Aquitaine Basin/Pyrenees Mountains of the European North Atlantic margin, and the Benue Trough of the African South Atlantic margin.

Aulacogens may contain thick accumulations of primarily siliciclastic sediments. Several kilometers of volcaniclastics, lava flows, coarse to fine siliciclastics and lacustrine sediments commonly accumulate during the period of active rifting, like in other rift systems. As magmatic and tectonic activity cease, continental relief and erosion decrease in importance, and the structure may subside below sea level (see Section 4.1). Smaller quantities of siliciclastics of lower grain size are transported to the basin where they accumulate in coastal areas, principally near river mouths. By the same time, the relative importance of particles derived from marine biological activity increases in the sediment. As the accommodation of intraplate compressional stress is generally associated to regional uplift, shallow to deep marine sediments are commonly overlain by siliciclastics, eroded from adjacent areas of increased topography, which accumulate in continental environments. If compressional stress persists, parts of the sedimentary column are faulted and folded, and metamorphosed in high-pressure/low-temperature conditions. Intraplate stress is generally accomodated on one side of the aulacogen principally, where deformation and metamorphism culminate. Sediment records suggest that aulacogens may experience successive intervals of extension and compression through geological time.

7.2. Example of Sedimentation in an Aulacogen: The Benue Trough of Nigeria

The Benue Trough extends northeastward from the Niger Delta of the equatorial Atlantic into the continental hinterland of Nigeria (Figure 7.2). The structure is approximately 1,000 km long and 120–150 km wide, and its northeastern section is separated into two branches, which are in continuity with other basins and shear zones of the West and Central African Rift System. This rift system formed during the early separation of Africa and South America in the latest

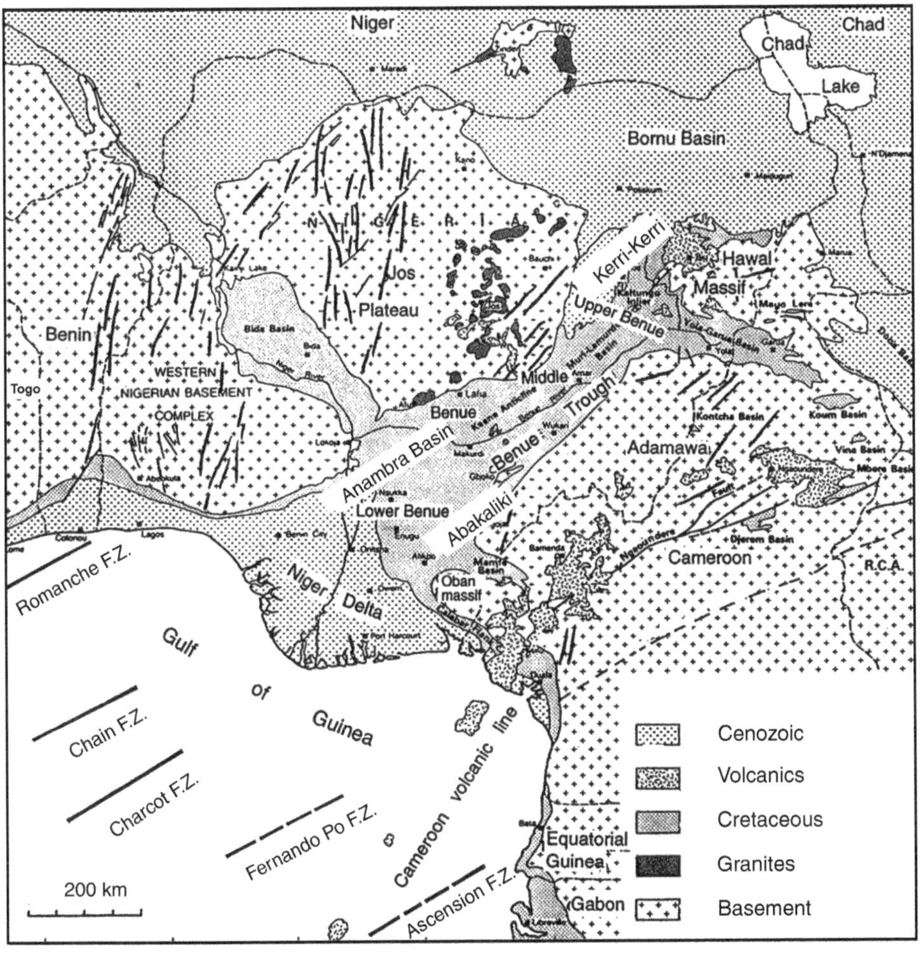

Figure 7.2 Location map of the Benue Trough. Note that major river systems converge into the aulacogen. Modified from Mascle, J., Lohman, G.P., Moullade, M. (Editors), 1998. Proceedings of the Ocean Drilling Program, Scientific Results, volume 159. Ocean Drilling Program, College Station, TX.

Jurassic and Early Cretaceous. The trough is bounded by Precambrian to Paleozoic terranes of the Pan-African basement and looks in continuity with the Chain and Charcot fracture zones of the Equatorial Atlantic (Figure 7.3). No oceanic lithosphere has been identified in the Benue Trough, but a thinned continental lithosphere containing a number of intrusive bodies, the degree of thinning decreasing from the southwest to the northeast. The Benue Trough consists in a succession of pull-apart basins filled with 5–6 km of terrigenous sediments principally. The sediments were subsequently folded and uplifted, and locally eroded. Large folds, locally reversed, characterize most areas, grading to tight folds with sub-vertical axial plane cleavage to the southeast (in the Abakaliki anticlinorium), and conjugate strike-slip faults.

7.2.1. Opening and Subsidence of the Benue Trough

The opening of the South Atlantic propagated from the southernmost parts of the African-South American block northward in the Late Jurassic (see Section 3.5), and early rift activity in the Benue Trough started during the latest Jurassic to Early Cretaceous interval as indicated by the oldest volcanic emissions, which consist of alkaline basalts and hyperalkaline rhyolites. It is probable that the St Helen hot spot was then active beneath the modern Niger Delta area. The related triple junction may have facilitated the formation of transtensional basins in the Equatorial Atlantic and Benue Trough areas, and extensional rift basins in the northern South Atlantic. However, mantle plume activity was probably limited and did not lead to massive lava flows, and most basins in the Benue Trough area result from a combination of strike-slip and extensional stress (pull-apart basins) in a passive rifting context.

Early rift deposits outcrop in the upper Benue where they consist of conglomerates and debris flows interbedded with basaltic and rhyolitic lava flows of latest Jurassic age (Figure 7.4). Beginning in the Neocomian, an increased tectonism and related erosion led to the accumulation of thick series of sandstones interbedded with conglomerates (Bima Formation). The environments of sedimentation were continental, and range from alluvial fans to braided river systems principally. They fill asymmetric half-grabens and are cut by normal faults which extend into the basement, and constitute typical syn-rift deposits. The accumulation of fluviatile sediments ceased during the Late Aptian in the lower to middle Benue (Figure 7.5). They were replaced by alternating sandstones, siltstones and mudstones (including black-shale facies) with fossilized ammonites, which are locally interbedded in the middle Benue with limestones containing gastropod remains and foraminifers (Asu River formation). These principally terrigenous facies mark the first marine transgression in the Benue Trough (Figure 7.6). Their distribution indicates that shallow marine environments flooded most of the lower and middle Benue, grading to lagoonal and nearshore environments at the periphery. The accumulation of sandstones in fluviatile environments resumed in the upper to middle Benue above an erosional unconformity, and persisted throughout the Albian. The latest Aptian/Albian transgression is the probable consequence of decreased tectonism and magmatism during an interval of thermal subsidence in the Benue Trough, which may coincide with the creation of oceanic

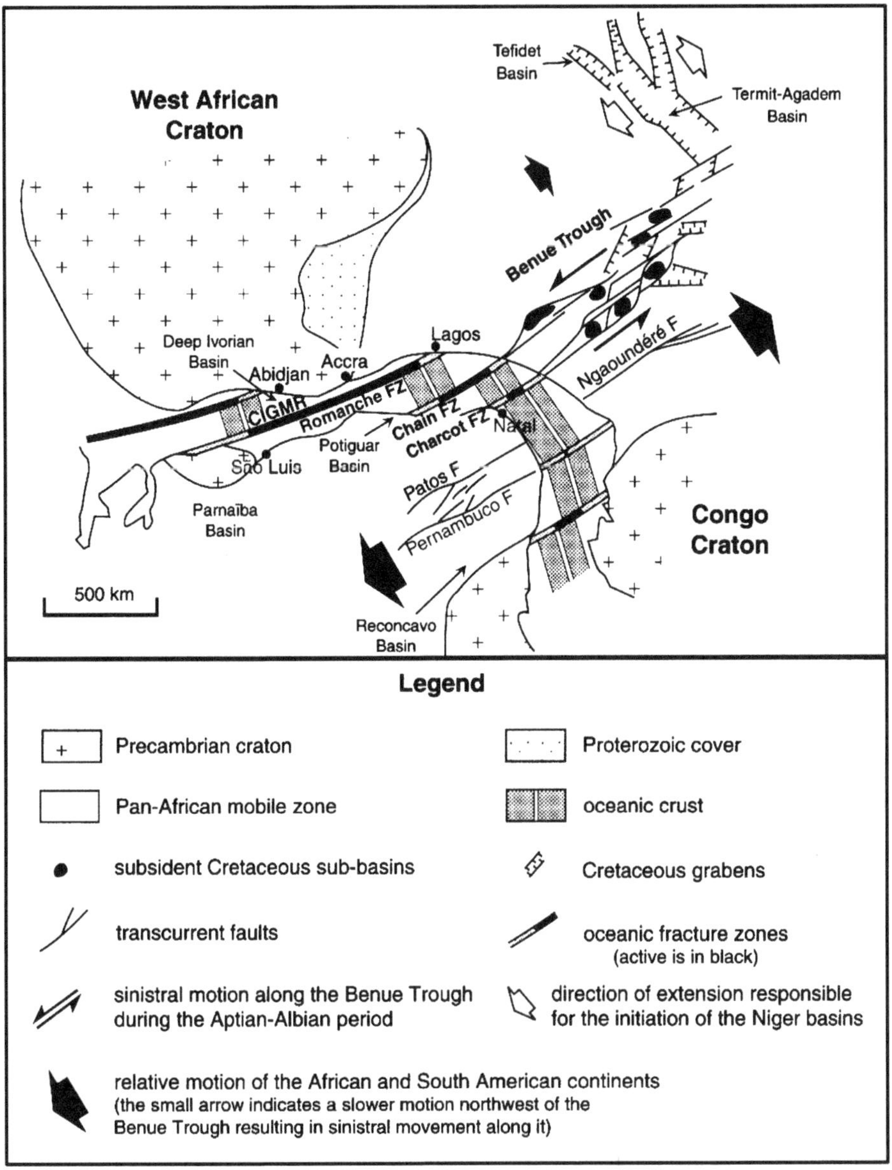

Figure 7.3 Reconstruction of the Equatorial Atlantic in the Albian, highlighting the structure of the Benue Trough and its relations with the Atlantic fracture zones. Reprinted from Mascle, J., Lohman, G.P., Moullade, M. (Editors), 1998. Proceedings of the Ocean Drilling Program, Scientific Results, volume 159. Ocean Drilling Program, College Station, TX.

crust and the inception of the crustal fissure stage in the Brazil-Angola Basin of the South Atlantic (see Section 5.4). It is possible that extensional rifting ceased at that time. Strike-slip activity persisted, as well as related tectonism and erosion in the upper Benue.

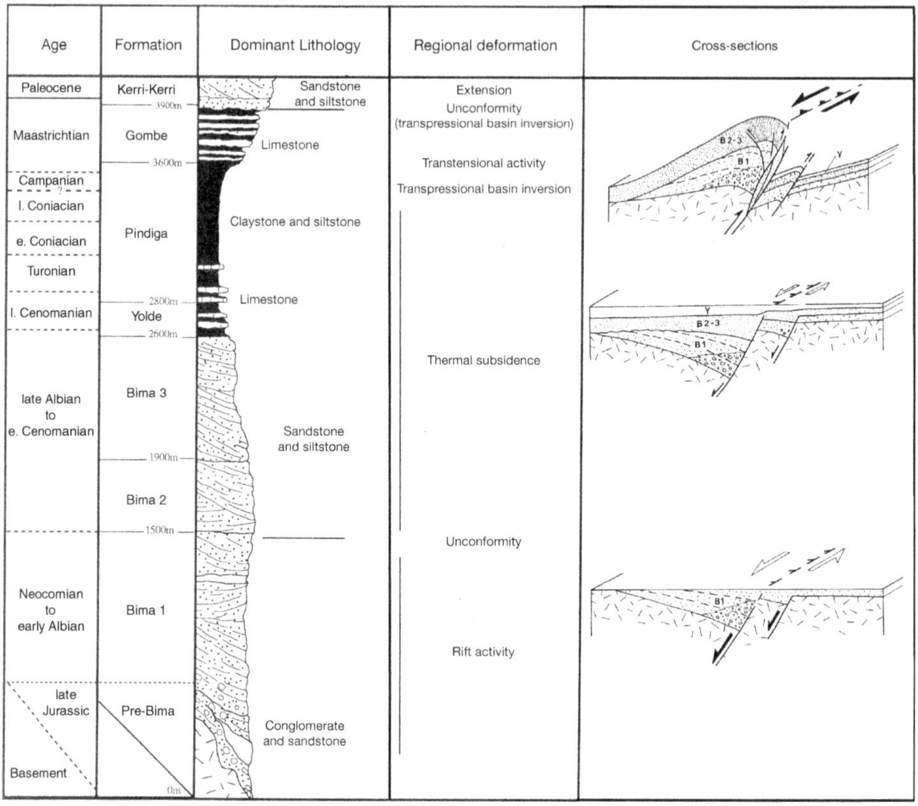

Figure 7.4 Schematic lithostratigraphy and tectono-sedimentary evolution of the upper Benue Trough. Note the large predominance of siliciclastics and the succession of tensional and compressional intervals. Modified from Guiraud, M., 1993. Late Jurassic-Early Cretaceous rifting and Late Cretaceous transpressional inversion in the upper Benue Basin (NE Nigeria). Bulletin Centres Recherches Exploration-Production Elf-Aquitaine, 17, 371–383.

The subsidence of the structure was probably not uniform. For example, uplift and erosion may have locally occurred during the Cenomanian (Figure 7.5), as suggested by the presence of unconformities and the scarcity of Cenomanian sediments in many areas. However, marine environments persisted in other areas of the lower and middle Benue where fossiliferous shales and sandstones accumulated. Thermal subsidence accelerated in the Late Cenomanian and a marine transgression progressively extended over the entire Benue Trough to a Late Turonian maximum (Figure 7.6), and persisted until the Santonian (Figure 7.5). By this time, seawater probably connected the South Atlantic to the Tethys via parts of the West and Central African Rift System, and the Niger and Chad basins. Transgressive sedimentary series accumulated above unconformities in many areas. They principally consist of biogenic carbonates and mudstones, frequently of black-shale facies, grading locally to sandstones of alluvial fan environment.

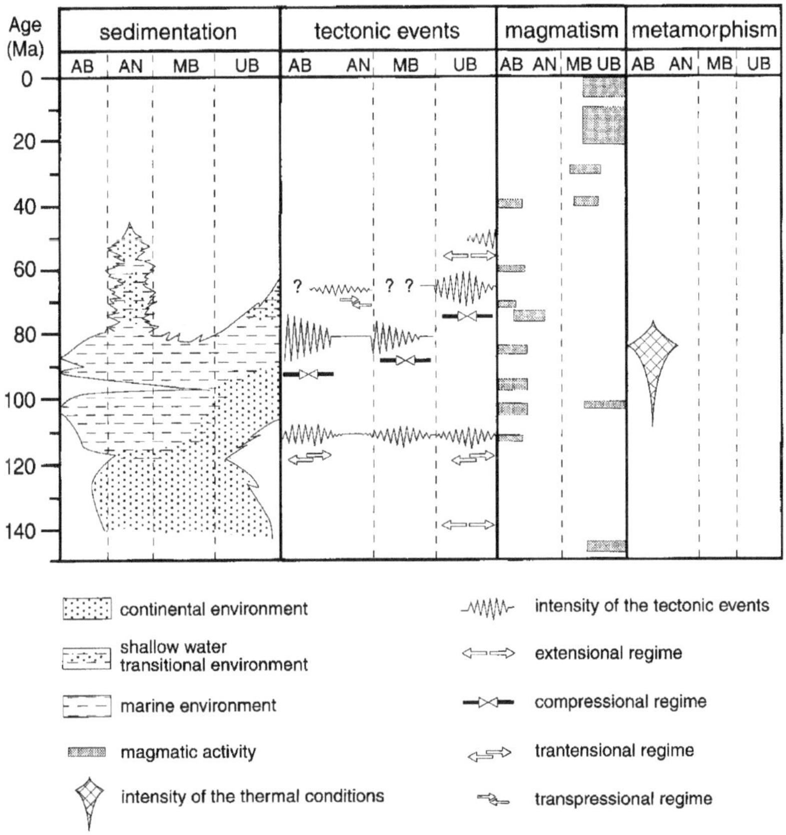

Figure 7.5 Summary of major geological events in the Benue Trough. AB, Abakaliki area of the lower Benue; AN, Anambra Basin of the lower Benue; MB, middle Benue; UB, upper Benue. Note the succession of extensional and compressional regimes, transgressions and regressions and magmatic events. Reprinted from Mascle, J., Lohman, G.P., Moullade, M. (Editors), 1998. Proceedings of the Ocean Drilling Program, Scientific Results, volume 159. Ocean Drilling Program, College Station, TX.

7.2.2. Compression and Closure of the Benue Trough

A regression started in the Coniacian with an increase in the contribution of siliciclastics to the sediment. Also, basaltic alkaline pyroclastics accumulated locally, especially in the Abakaliki area. No significant accumulation of sediments occurred in the Santonian, with the exception of coarse to fine-grained sandstones along the northwestern flank of the middle Benue Trough.

The emersion of most of the Benue Trough was followed by an interval of intense folding and faulting, which involved the entire sedimentary series that filled the grabens as well as the structures that developed during the rift stage (Figure 7.7). The tectonic inversion was especially important along the southeastern flank of the Benue Trough in the Abakaliki area, where the Santonian compressional event generated tight folding and reversed faulting, locally associated to the formation of

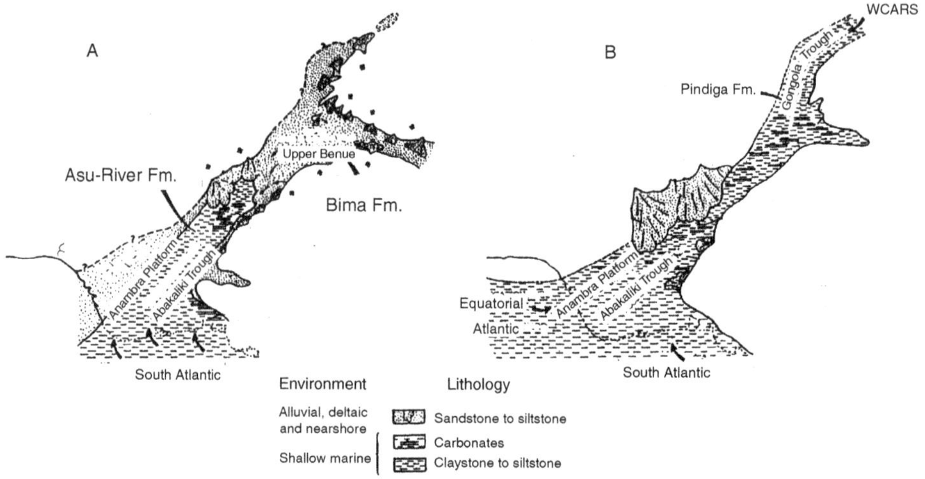

Figure 7.6 Paleogeographic reconstruction of the Benue Trough. (A) Extension of the Albian transgression. Note the persistence of continental environments in the upper Benue. (B) Turonian maximum of the transgression. Note the marine connection with other basins of the West and Central African Rift System (WCARS). Modified from Benkhelil, J., 1989. The origin and evolution of the Cretaceous Benue Trough (Nigeria). Journal of African Earth Sciences, 8, 251–282.

mylonites. The intensity of the compressional event is also deduced from the presence of sediment microstructures such as cleavage planes. Local occurrences of mica crystals along cleavage planes indicate that the Abakaliki area also experienced an interval of low-grade metamorphism (Figure 7.5). The Santonian compressional event, which peaked at 83–85 Ma, has been observed beyond the Benue Trough in other areas of the West and Central African Rift System. The Santonian compressional event coincides with major plate tectonic events centered on magnetic anomaly 34, around 84 Ma. Among them, a change in the rotation of the African plate increased its collisional coupling with the European plate. Other consequences of this major event in the Atlantic domain include the cessation of seafloor spreading in the Bay of Biscay and its commencement in the Labrador Sea.

The Santonian compressive events were followed by further intervals of passive rifting and subsidence, in response to renewed extensional and transtensional stress. The Cameroon volcanic line initiated in the Late Campanian. Thick series of sandstones, siltstones and claystones interbedded with coal seams accumulated from the Campanian to the Late Maestrichtian in paralic and shallow marine environments of the Anambra Basin (Figure 7.5), which developed west of the Abakaliki folded area in the lower Benue (Figure 7.7), as well as along the western margin of the upper Benue where sandstones dominate (Figure 7.4). The whole Benue was uplifted in the Late Maestrichtian, when N–S compression of the African plate during a major change in the rotation of lithospheric plates initiated a transpressional regime at regional scale. The latest Maestrichtian tectonic inversion

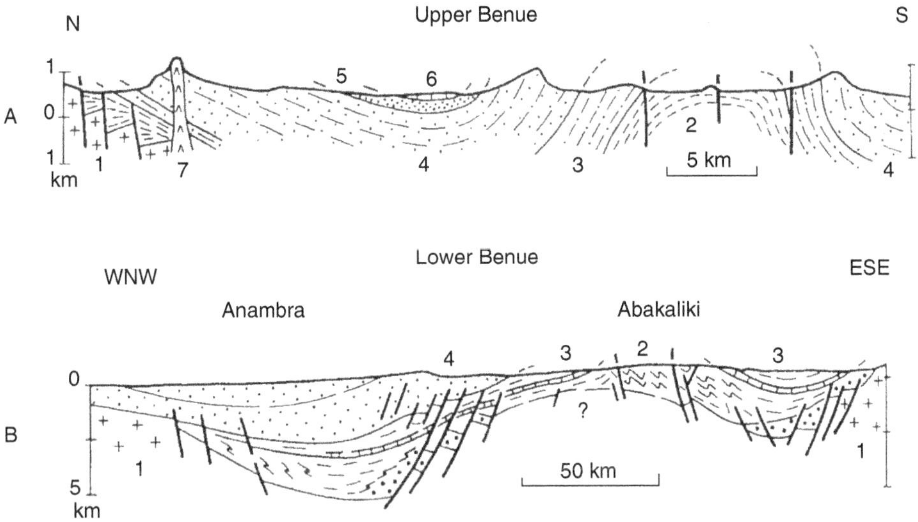

Figure 7.7 Schematic cross sections of the Benue Trough. (A) Upper Benue. 1, Basement; 2,3,4, Neocomian to Cenomanian continental sandstones (Bima formation); 5,6, Cenomanian to Coniacian marine siliciclastics; 7, volcanic intrusion. (B) Lower Benue. 1, Basement; 2, Neocomian to Albian continental siliciclastics; 3, Albian to Coniacian marine limestones and fine siliciclastics; 4, post-Santonian continental and marine sandstones. Note post-Santonian accumulations in the Anambra Basin. Modified from Guiraud, R., Bosworth, W., 1997. Senonian basin inversion and rejuvenation of rifting in Africa and Arabia: synthesis and implications to plate-scale tectonics. Tectonophysics, 282, 39–82.

was especially important in the upper Benue where the Santonian event was limited. Minor extension in the Paleocene produced the fault-bounded Kerri-Kerri Basin of the upper Benue, filled with sandstones, siltstones and claystones. Sedimentation resumed in the Anambra Basin of the lower Benue, where a subsequent interval of subsidence and related marine transgression in the Late Paleocene and Early Eocene was followed by the construction of the Niger Delta. Although younger compressional events have been identified, the fault systems principally controlled the location of volcanic activity in the Benue area for the remaining of the Cenozoic.

FURTHER READING

Banda, E., Torné, M., Talwani, M. (Editors), 1995. Rifted ocean-continent boundaries. Klüwer, Dordrecht.

Benkhelil, J., 1989. The origin and evolution of the Cretaceous Benue Trough (Nigeria). Journal of African Earth Sciences, 8: 251–282.

Ofoegbu, C.O. (Editor), 1990. The Benue Trough: Structure and evolution. Vieweg, Wiesbaden.

Ziegler, P.A. (Editor), 1992. Geodynamics of rifting, volume 2: case history studies on rifts, North and South America and Africa. Tectonophysics, 213: 1–284.

Other references used in this chapter

Chang, H.K., Kowsmann, R.O., Ferreira Figueiredo, A.M., Bender, A.A., 1992. Tectonics and stratigraphy of the East Brazil Rift System: an overview. Tectonophysics, 213: 97–138.

Guiraud, M., 1993. Late Jurassic-Early Cretaceous rifting and Late Cretaceous transpressional inversion in the upper Benue Basin (NE Nigeria). Bulletin Centres Recherches Exploration-Production Elf-Aquitaine, 17: 371–383.

Guiraud, R., Bosworth, W., 1997. Senonian basin inversion and rejuvenation of rifting in Africa and Arabia: synthesis and implications to plate-scale tectonics. Tectonophysics, 282: 39–82.

Guiraud, R., Maurin, J.C., 1991. Le rifting en Afrique au Crétacé inférieur: synthèse structurale, mise en évidence de deux étapes dans la genèse des bassins, relations avec les ouvertures océaniques péri-africaines. Bulletin Société Géologique de France, 162: 811–823.

Mascle, J., Lohman, G.P., Moullade, M. (Editors), 1998. Proceedings of the Ocean Drilling Program, Scientific Results, Vol. 159. College Station, TX, Ocean Drilling Program.

CHAPTER EIGHT

Oceans in a Context of Plate Convergence

8.1. Structure, Tectonics and Sedimentation of Convergent Oceans

Convergent ocean systems are essentially characterized by the development of active continental margins and/or active island arcs. Both continental and oceanic lithospheres are involved in active continental margins, whereas active island arcs implicate two oceanic lithospheres. The morphological elements of such active areas include oceanic trenches, accretionary wedges, volcanic arcs and backarc systems (see Section 2.1.5).

8.1.1. The Consequences of Subduction on the Activity of Convergent Margins

The accretion and relative motion of lithospheric plates generate internal stress which is transmitted within the lithosphere to plate boundaries. Where internal stress converges, minor compression is accommodated via intraplate deformation, whereas important compression may cause the rupture of the lithospheric plate in areas of mechanical weakness and the subduction of the lithosphere of highest density. Such areas of mechanical weakness include for example transition zones from oceanic to continental lithosphere and transform areas where the contrast in thickness and/or physical properties of adjacent lithospheres is important, and mid-oceanic ridges where oceanic lithosphere is very thin. However, it is generally assumed that old oceanic lithospheres are easily separated from adjacent continental lithospheres and entrained in the asthenosphere because of high contrasts in density.

Active continental margins and island arcs are generally concave toward the overriding plate and their curvature can be approximated using the formula:

$$r = R\theta$$

where r is the radius of the curvature and R the radius of the Earth. 2θ is the angle subtended by the curvature of the active system at the center of the Earth, and also the angle of dip of the subducted plate (Figure 8.1). This provides a partial explanation for the curvature of active continental margins and island arcs which is especially visible in areas of high dip angle of the subducted plate, for example in the northwest Pacific Ocean. The degree of thermal evolution and the velocity of the subducted lithosphere are important factors for controlling the angle of dip (see Section 2.1.5), together with convection processes within the asthenosphere. However, the curvature of active margins and island arcs may change at regional scale with the characteristics of the convergent plates. For example, major oceanic

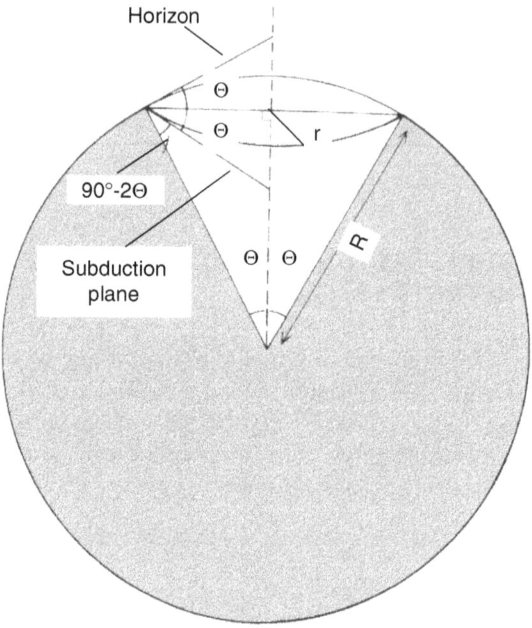

Figure 8.1 Relationships between the curvature of active margins and island arcs, and the angle of dip of the subducted plate. Modified from Fowler, C., 2000. The solid Earth: an introduction to global geophysics. American Geophysical Union, Washington.

relief may resist subduction, the persistent convergence creating a convex curvature toward the overriding plate. In some cases the relief is finally subducted, but it may also collide with the overriding plate. This is the case in the Izu Peninsula of Japan where an active island arc located at the eastern periphery of the Philippine plate, generated by the subduction of the Pacific plate beneath the Philippine plate, penetrates significantly into the active margin of Japan.

The distribution and focal mechanisms of subduction zone earthquakes indicate that they principally concern the brittle parts of the subducted lithosphere and that major stress direction follows the slope of the Wadati–Benioff zone. However, a number of earthquakes with hypocenters shallower than 300–400 km have focal mechanisms with down-dip extension, especially when they are located within the brittle lithospheric mantle of the subducted slab (Figure 8.2). They suggest that the subducted slab, made of relatively high-density oceanic lithosphere principally, is entrained by gravity principally in the asthenosphere of lower density. However, most earthquakes with hypocenters deeper than 300–400 km have focal mechanisms with down-dip compression, suggesting that a physical barrier around 650–700 km opposes the penetration of the subducted slab. The thermal structure of active margins and island arcs provides an explanation for the observed variations of the subduction processes. Heat flows below average are recorded in the oceanic trench and accretionary wedge areas, whereas heat flows higher than average increase to very high values from the backarc system to the volcanic arc (Figure 8.2). Modelling studies include the effects of frictional heating, mineralogical phase changes and

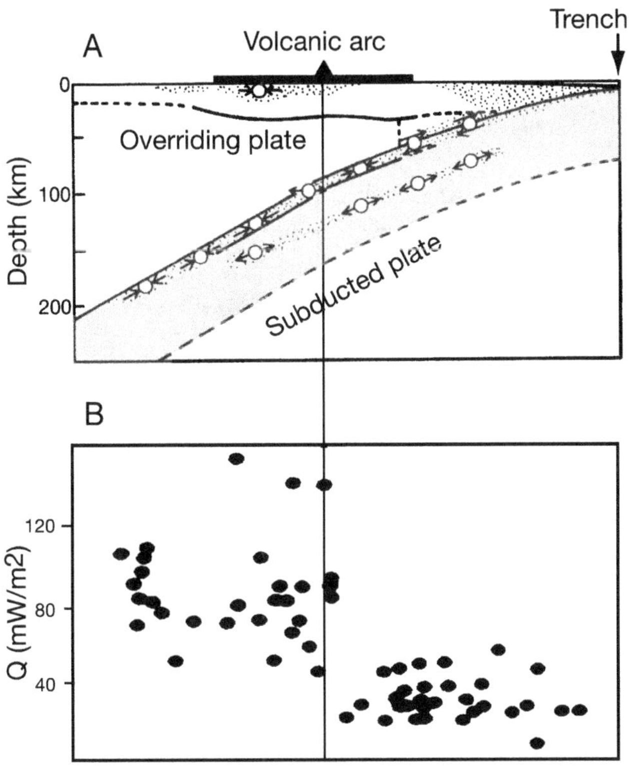

Figure 8.2 (A) Focal mechanisms of shallow to intermediate earthquakes within a subducted plate. Note down-dip compression within the brittle subducted crust and down-dip extension within the brittle lithospheric mantle. (B) Variability of heat flow from the trench to the back-arc system. Note low heat flows in the trench area and high heat flow in the volcanic arc area (average is around 40 mW/m^2). Modified from Brahic, A., Hoffert, M., Schaaf, A., Tardy, M., 1999. Sciences de la Terre et de l'Univers, Vuibert, Paris.

temperature gradients for the oceanic lithosphere in conductive thermal models. They indicate that the subduction of cold oceanic lithosphere at velocities of a few centimeter/year is sufficient to maintain low heat flows in the oceanic trench and accretionary wedge areas, as well as a significant contrast of temperature between the subducted slab and the asthenosphere.

The resulting thermodynamical context generates significant alteration of the rocks and sediments of the subducted slab. The dehydration and metamorphose of the sediments and rocks (basalts and gabbros principally) of the subducted oceanic crust in the blueschist, greenschist, amphibolite and eclogite facies start early and are generally completed before the slab reaches a depth of 100–150 km (Figure 8.3). The loss of interstitial fluids from the subducted oceanic crust is sufficient to lower melting temperatures considerably but generates an explosive volcanism. A magma forms from partial melting of the subducted sediments and oceanic crust as well as of the overriding mantle, and rises within the overriding lithosphere where it fractionates and produces additional melting. Therefore, the nature and

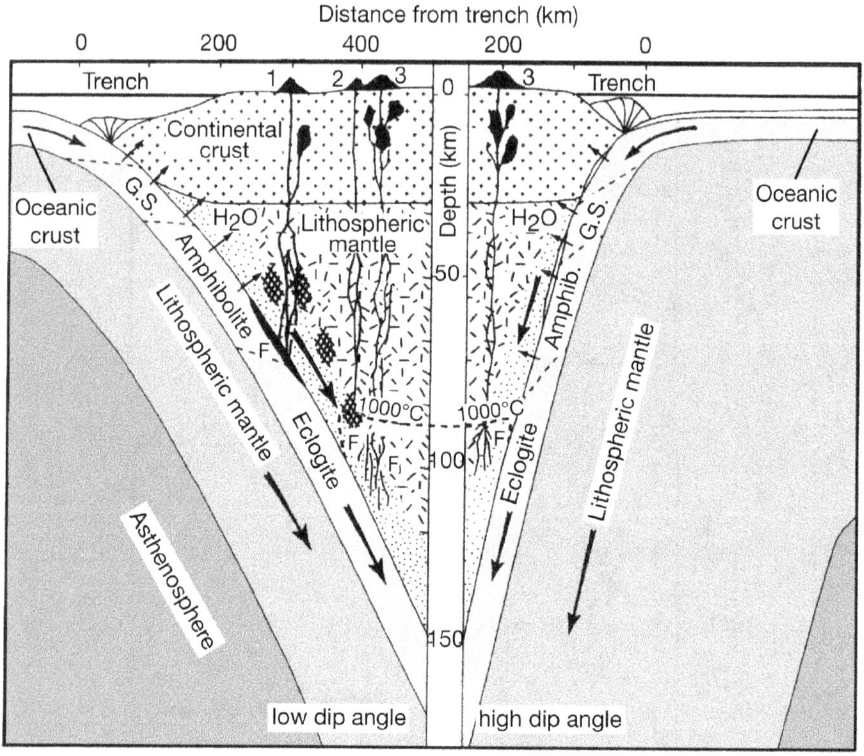

Figure 8.3 Alteration of subducted slabs and related volcanism. G.S, Greenschist facies; F, main sites of partial fusion of the lithosphere; 1, Si-rich basalt; 2, Ca-K-rich basalts; 3, andesites. Note early dehydration and progressive metamorphose of subducted slabs, and the variability of the volcanism in relation to the dip angle of the subducted slab. Modified from Brahic, A., Hoffert, M., Schaaf, A., Tardy, M., 1999. Sciences de la Terre et de l'Univers, Vuibert, Paris.

composition of volcanic products in active margins and island arcs are directly related to those of the lithospheres involved: the contents of silica, alkaline elements and potassium increase from areas of dominant oceanic crust to areas of dominant continental crust. As a result, a basaltic volcanism dominates in active island arc areas, whereas dominant andesitic and sometimes rhyolitic volcanisms occur in active margin areas. The focus of volcanism generally occurs in areas where the subducted lithosphere reaches a depth of 100–150 km.

A further step in the alteration of the subducted slab begins around 300 km depth. There, abundant olivine of the oceanic lithosphere is metamorphosed in spinel. The transformation releases energy and is associated to an increase in density. The phase change occurs at lower pressure (shallower depth) in the subducted slab (300 km) of lower temperatures than in the adjacent asthenosphere (400 km) of higher temperatures (Figure 8.4). The resulting increase in the contrast of density may account for down-dip extension observed around 300–400 km and supports the role of gravity in subduction processes. A further phase change from spinel to perovskite (postspinel) begins around 650 km in the subducted lithosphere.

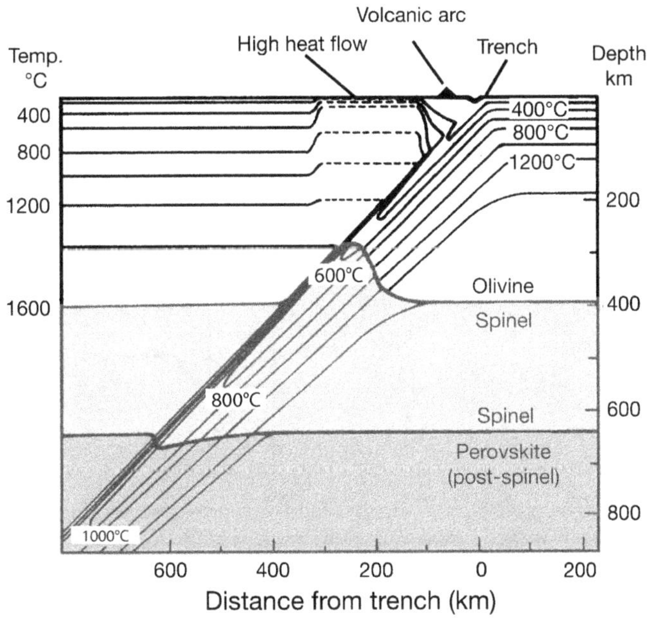

Figure 8.4 Variability of temperatures in the subduction area and deep metamorphose of the subducted lithosphere. Note lower temperatures within the subducted slab and related changes in the depth of phase changes. Modified from Brahic, A., Hoffert, M., Schaaf, A., Tardy, M., 1999. Sciences de la Terre et de l'Univers, Vuibert, Paris.

The transformation absorbs energy and is associated to a decrease in density. The phase change occurs at higher pressure (deeper depth) in the subducted slab of lower temperatures than in the adjacent asthenosphere of higher temperatures (Figure 8.4). The resulting decrease in the contrast of density may account for down-dip compression observed around 650–700 km and related physical barrier to subduction. However, subducted slabs may either further penetrate into the Earth's mantle or follow the convective cells of the asthenosphere.

8.1.2. The Evolving Structure and Morphology of Active Convergent Margins

Active margins and island arcs are also characterized by a series of gravity anomalies (Figure 8.5). The bulge of the subducted plate is associated to a minor positive anomaly due to local uplift of the oceanic lithosphere of higher density. The oceanic trench and accretionary wedge areas are associated to a strong negative anomaly of gravity which results from local importance of low-density elements such as seawater and oceanic sediments above a sinking oceanic lithosphere. However, the volcanic arc is associated to a positive anomaly of gravity which principally results from the presence of intrusions and volcanic elements partially derived from the lithospheric mantle within an uplifted overriding plate, and mineralogical phase changes within the subducted plate. The contrast between negative and positive

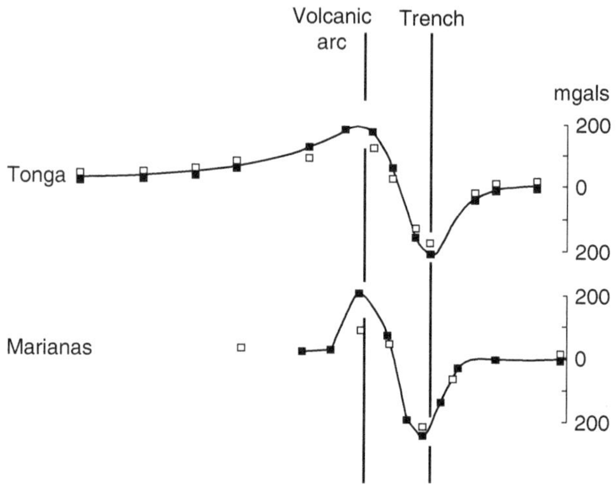

Figure 8.5 Examples of gravity anomalies in subduction areas. Note negative anomalies in the trench and accretionary wedge areas, and positive anomalies in the volcanic area. Modified from Boillot, G., 1983. Géologie des marges continentales, Masson, Paris.

anomalies may commonly reach 400 milligals and clearly results from the dynamic consequences of the subduction which oppose any isostatic readjustment. Therefore, it is likely that any variation in plate convergence and subduction velocities is associated to significant changes in the morphology of active margins and island arcs. For example, a decrease in velocity or a cessation of subduction should result in significant uplift of the oceanic trench and accretionary wedge areas, and subsidence of the volcanic arc.

The lithospheric plates are not uniformly smooth but include a number of submarine morphological elements such as seamounts and other relief which interfere with subduction processes as they reach the oceanic trench area. Important shear stress may separate at least part of the submarine oceanic relief from the subducted plate, the detached block of oceanic crust being incorporated into the accretionary wedge. For example, in the Japan and Kuril Trench of the Northwest Pacific, the Erimo Seamount is being subducted together with the Pacific plate (Figure 8.6). In some cases, accretionary wedges may include a succession of such pieces of ocean floor (ophiolites) alternating with oceanic sediments. This is especially apparent in areas of low sediment fluxes and poorly developed accretionary wedge such as active island arcs. However, minor pieces of relief are generally entrained together with the subducted slab. Shear stress generated by the friction between the subducted and overriding plates may also separate deformed blocks of sediment from the base of the accretionary wedge, the detached blocks of sediments being entrained in the subduction (tectonic erosion). In some cases however the entrainment of low-density sediments could be limited, the sediments being later incorporated at the base of the overriding plate (underplating). Tectonic erosion principally occurs in areas of low angle subduction where the coupling between the subducted and overriding plates is important, such as active margin areas of South America (Figure 8.7). For example, the active margin of Peru shows a

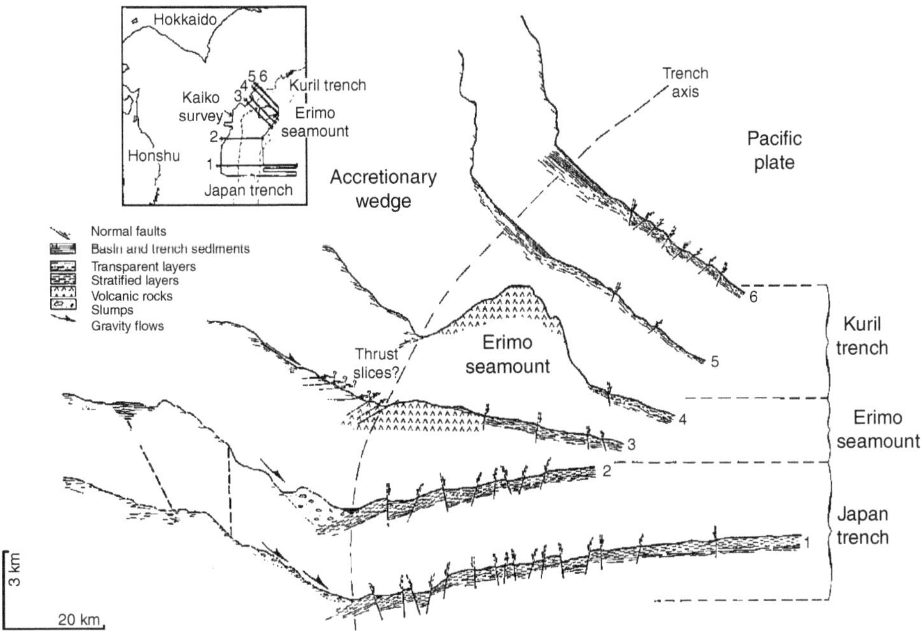

Figure 8.6 Cross-sections of the Japan and Kuril trench area, Kaiko Project. Note ongoing subduction of the Erimo seamount, and probable incorporation of seamount fragments to the accretionary wedge. Modified from Cadet, J.P. et al., 1987. The Japan Trench and its juncture with the Kuril Trench: Cruise results of the Kaiko Project, Leg 3. Earth and Planetary Science Letters, 83, 267–284.

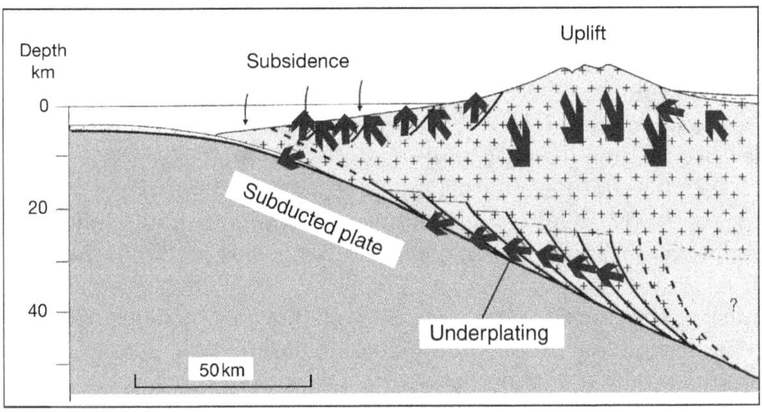

Figure 8.7 Schematic representation of tectonic erosion. Note the detachment of blocks from the base of the accretionary wedge and their incorporation at the base of the overriding continental crust. Modified from Boillot, 1983. Géologie des marges continentales, Masson, Paris.

small accretionary wedge despite important relief and erosion in the continental hinterland of the Andes Cordillera. Moreover, the accretionary wedge is characterized by normal faulting and environments of sedimentation which increase in depth through time, suggesting a collapse of the active margin. Related underplating might partly account for the thickness of the overriding continental crust, but this needs to be further demonstrated.

8.1.3. The Sediments of Active, Convergent Margin Areas

8.1.3.1. Sediments of the incoming lithospheric plate

In convergent oceanic systems, the subducted plate progressively brings pelagic sediments to the trench. In the Pacific Ocean, the nature of those sediments varies with the age of the lithosphere. Because of an increasing density, old oceanic lithospheres subside and form deep oceanic basins (see Section 2.1). Deep basins are filled with water masses of high density which are also enriched in CO_2 because of low temperatures and/or decay of organic elements, and are generally aggressive for biogenic carbonates which are dissolved. Convergent margins involving old oceanic lithosphere generally include large backarc basins (see Section 2.1), where most siliciclastics eroded by runoff from the continental hinterland of the overriding plate are trapped. Only minor quantities of fine, principally windblown siliciclastics can reach the deep oceanic basin. Therefore, deep Pacific basins of old oceanic lithosphere generally contain thin sedimentary series of low sedimentation rates where biogenic silica, volcaniclastic and authigenic elements dominate.

For example, ODP Site 801 has been drilled at 5,685 m water depth in the Pigafetta Basin east of the Marianas Trench where Pacific oceanic lithosphere of Jurassic age is being subducted. There, only 460 m of sediments accumulated since the Bathonian and the series is interrupted by many hiatuses indicative of active bottom circulation. Most Jurassic and Cretaceous sediments consist of radiolarites (Figure 8.8) which accumulated as the site was near the equator, suggesting noticeable productivity and a permanence of the equatorial divergence. Yet, a pulse of volcanic activity in the Middle Cretaceous led to massive redeposition of volcanic material from adjacent seamounts, in the form of turbidites and debris flows. Beginning in the Campanian, the sediment grades to brown clay which is principally windblown. However, important authigenesis of metal oxides and hydroxides (see Section 13.1), zeolites and clays within the sedimentary column probably took place during intervals of active bottom circulation.

Because of lower density and subsidence, young oceanic lithosphere form oceanic basins of shallower water depth than old oceanic lithosphere. With the exception of areas of high productivity, upper water masses generally have lower CO_2 contents than lower water masses, and biogenic carbonates are generally preserved in Pacific basins of young oceanic lithosphere. Convergent margins involving young oceanic lithosphere generally include cordilleras of high relief (see Section 2.1) where intense erosion and run-off transfer important quantities of siliciclastics and volcaniclastics to the ocean. Most of the clastic elements fill the trench, but part of them are dispersed by the currents. Therefore, Pacific basins of young oceanic lithosphere generally contain thick sedimentary series of relatively high sedimentation rates where biogenic carbonates dominate, sometimes associated to biogenic silica and/or minor fine siliciclastics.

For example, DSDP Site 495 has been drilled at 4,150 m water depth west of the Middle America Trench where Cocos oceanic lithosphere of Early Miocene age is

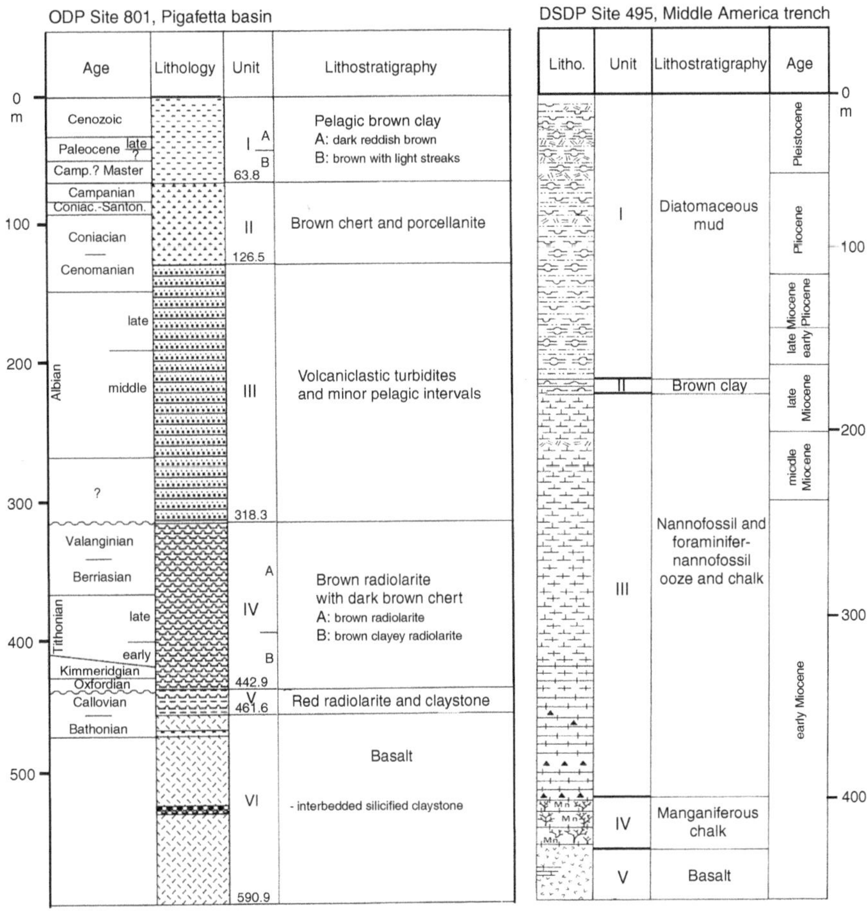

Figure 8.8 Lithologic summary of sedimentary columns covering the old oceanic lithosphere of the Pacific plate (ODP Site 801, Pigafetta Basin) and the young oceanic lithosphere of the young Cocos plate (DSDP Site 495, Middle America Trench). Note differences in the nature of sediments and sedimentation rates. Modified from Lancelot, Y., Larson, R., Fisher, A. et al., 1990. Proceedings of the Ocean Drilling Program, Initial Reports, volume 129. Ocean Drilling program, College Station, TX and from Von Huene, R., Aubouin, J. et al., 1985. Initial Reports of the Deep Sea Drilling Project, volume 84, U.S. Gov. Print. Office, Washington.

being subducted. There, 425 m of sediments accumulated since the Early Miocene. Early to Middle Miocene sediments consist of nannofossil chalk and foraminifer-nannofossil chalk which rest on the basalts of the subducted plate (Figure 8.8). Late Miocene to Pleistocene sediments consist of diatomaceous mud which reflect greater proximity of the site to both the coastal upwelling and continental landmass.

8.1.3.2. Trench deposits

The pelagic and hemipelagic sediments carried by the subducted lithosphere are overlain by clastic elements eroded from the active margin or island arc as they reach

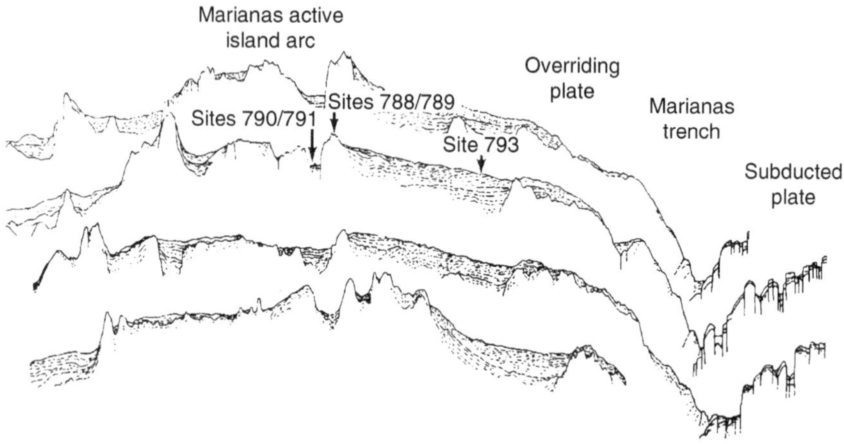

Figure 8.9 Schematic cross sections of the Marianas active island arc, from seismic reflection profiles. Note the presence of fault-controlled basins on the forearc and outcrops of basement lithosphere on the trench walls. Modified from Fryer, P., Coleman, P., Pearce, J.A., Stokking, L.B. (Editors), 1992. Proceedings of the Ocean Drilling Program, Scientific Results, volume 125, Ocean Drilling program, College Station, TX.

the trench area. In active island arcs like the Marianas and the Bonins where old, sediment-starved Pacific lithosphere is being easily subducted, only minor quantities of sediment accumulate in the trench and the accretionary wedge is of limited extension or missing. As a consequence, maximum water depths of up to 11 km are recorded in those areas. Seismic surveys conducted in the Marianas suggest that the basaltic substratum of the overriding plate forms the inner wall of the trench (Figure 8.9). ODP Sites 788 and 790 drilled on basement highs near the trench area recovered thin series of vitric and pumiceous sands and conglomerates interbedded with nannofossil clay of Pliocene and Pleistocene ages, and bottomed in massive basalts of the overriding lithosphere.

In active margin areas, the erosion of the overriding plate provides significant quantities of terrigenous particles which accumulate in the trench above the sediment blanket of the subducted plate. As a consequence, water depths of 4–6 km are commonly found in those areas. Erosion and accumulation are especially important in active margin areas bounded by a cordillera, where young lithosphere is being subducted at low dip angle. At ODP Sites 1039 and 1043 drilled in the Middle America Trench off Costa Rica (Figure 8.10), the sediment series carried by the subducted lithosphere are identical to those recovered at DSDP Site 495 (see earlier), and consist of biogenic carbonates grading upwards to biogenic siliceous oozes and muds. At ODP Site 1039 in the outer trench, uppermost sediments of Late Pleistocene age consist of a few meters of diatom ooze interbedded with sand to silt layers thinning upwards (Unit U1a, Figure 8.10), which contain glauconite and shell fragments indicative of reworking from shallow water areas of the active margin. Further east in the deformed area of the inner trench, ODP Site 1043 recovered about 150 m of silty clay and breccia (Unit T1, Figure 8.10) and abundant turbidites.

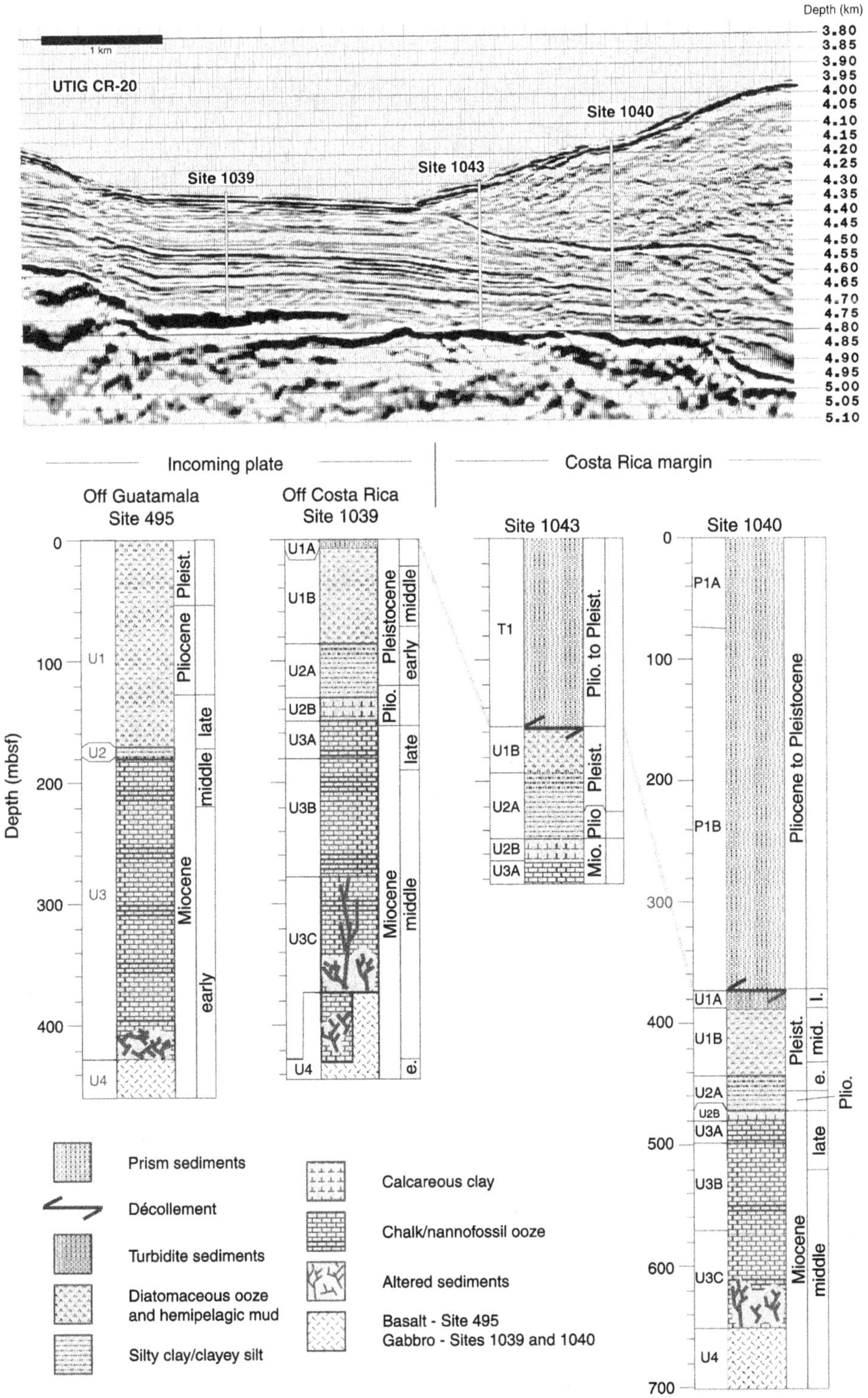

Figure 8.10 Top, Seismic profile across the Middle America Trench and Costa Rica active margin and location of ODP Leg 170 sites; bottom, lithologic summary of incoming sediments and trench and prism sediments. Note contrasts in lithology. Modified from Silver, E.A., Kimura, G., Shipley, T.H. (Editors), 2001. Proceedings of the Ocean Drilling Program, Scientific Results, volume 170, Ocean Drilling program, College Station, TX.

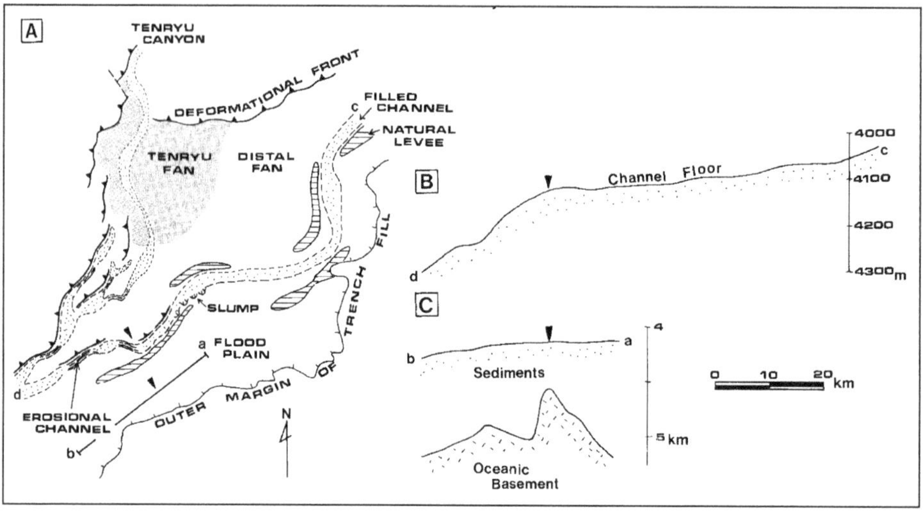

Figure 8.11 The Nankai Trough off Japan. (A) Distribution of major sedimentological traits; (B and C), cross sections of the trench floor. Note the importance of canyons, deep-sea fans and submarine channels in the distribution of terrigenous sediments which overlay those of the subducted plate and the variable relief of the trench floor and basement. Reprinted from Le Pichon, X. (Editor), 1987. The Kaiko project. Earth and Planetary Science Letters, 83, 181–375.

Gravity flows such as slumps and debris flows play a significant role in the transport of terrigenous sediments from the active margin to the trench in areas where continental erosion is poor because of limited tectonic activity and/or arid conditions. The role of turbidity currents increases in areas of active tectonism and/or runoff where continental erosion is high. Turbidity currents flow through deep-sea canyons cutting the active margin and through channels along the trench, where they form deep-sea fans (Figure 8.11). The sediments are then further reworked along axial channels in areas of active bottom circulation. Axial channels of the Nankai Trough off Japan are sometimes about 3 km wide, meandering and bounded by well-developed levees, but sometimes narrow (1 km wide), deep (about 100 m) and erosional. However, the transfer of sediment along the trench is also controlled by submarine relief (such as a basement highs) which may act as a dam.

8.1.3.3. The accretionary wedge

Because of plate convergence, the sediments filling the trench are progressively displaced toward the inner trench, where they are deformed before being either subducted or incorporated to the accretionary wedge. The residence time (t) of the sediment in the trench is a function of the trench width (w) and convergence velocity (v):

$$t = w/v$$

For example, in the southern Chili trench which is about 30 km wide and where the convergence velocity is about 10 cm/yr, the sediments eroded from the

Andes Cordillera may accumulate in the trench over those of the subducted plate for about 300 kyr before being incorporated to the accretionary wedge. Deformation starts with proto-thrusts of small offset which precede large offset faults cutting the section from top to a surface of decollement, beginning at the deformation front. The decollement separates the upper part of the sedimentary series where maximum compressive stress is horizontal and which is incorporated to the accretionary wedge, from the lower part where the principal stress is vertical and which is subducted (Figure 8.12).

The location of the decollement within the sedimentary series is highly variable at regional scale: the decollement may occur near the volcanic basement of the subducted plate, within the incoming sedimentary series, at the transition with or within the trench fill section. In fact, the position of the decollement is controlled by the physical properties of the sediment. Porosity data indicate that sediments below the decollement have a higher porosity than those above the decollement. Such contrasts in porosity may result either from differences in the nature of the sediment or in diagenetic processes. For example, the decollement at ODP Site 808 in the Nankai Trough off Japan separates underconsolidated sediments from overconsolidated sediments within a 600 m thick section of hemipelagic mudstone of Miocene to Early Pleistocene age. In contrast, the decollement at ODP Site 1043 in the Middle America Trench off Costa Rica occurs at the transition from biogenic siliceous oozes and muds (higher porosity) of the incoming plate to alternances of silty clays and turbidites (lower porosity) of the trench fill. In the Caribbean, the decollement in the Barbados Trench occurs within an interval of radiolarian mudstone of high porosity compared to overlying and underlying mudstones.

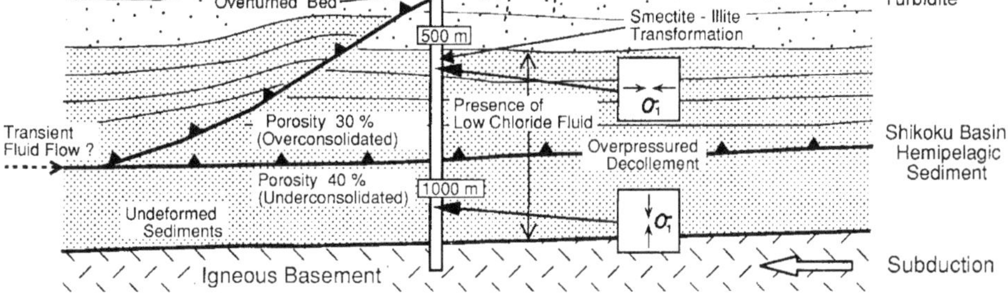

Figure 8.12 Summary figure showing the principal results of ODP Site 808 which cut the frontal and decollement thrusts in the Nankai Trough off Japan. Note changes in porosity and stress in the decollement zone. Low chloride fluids originate from the deep subduction zone and smectite/illite transformation may result from high pressure. Reprinted from Taira, A., Hill, I., Firth, J. et al., 1992. Sediment deformation and hydrogeology of the Nankai Trough accretionary prism: Synthesis of shipboard results of ODP Leg 131. Earth and Planetary Science Letters, 109, 431–450.

Beginning at the deformation front the sediments are folded and faulted, forming sometimes sedimentary ridges (Figure 8.13) before their incorporation to the accretionary wedge. The accretion process consists in a shortening and vertical thickening of the sediment section above the decollement, associated to thrust faulting which emplaces surface sediments of high porosity to greater depth. Subsequent expulsion of pore fluids and consolidation tend to reestablish an equilibrium of the porosity versus depth profile. By comparison, the sediments below the decollement can be either deformed or undeformed.

Surface manifestations of fluid outflows are visible in many accretionary wedge areas where they often sustain deep-sea communities including clams (*Calyptogena*), sea anemone, sea cucumbers, tube worms and/or bacterial mats, depending on

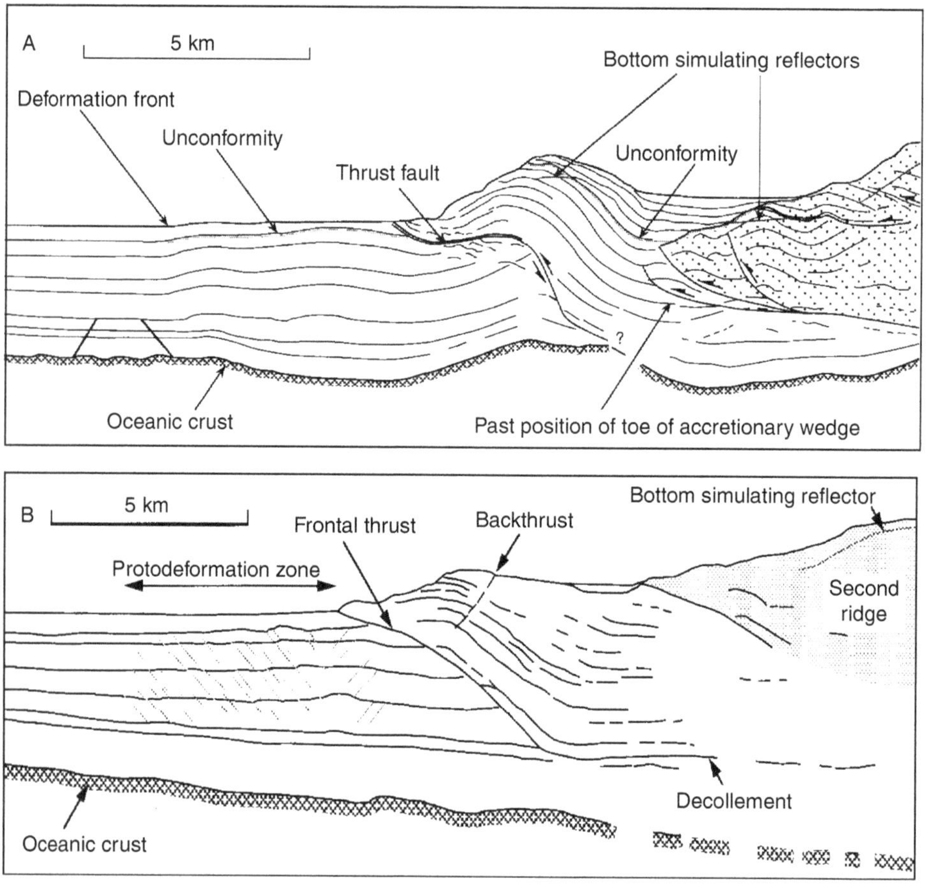

Figure 8.13 Schematic cross sections of the Cascadia Trench. (A) Off Vancouver Island. Note early deformation of trench sediments, probable decollement near oceanic crust, and formation of a single sedimentary ridge. (B) Off Oregon. Note formation of proto-thrust faults, decollement within the incoming sediment and succession of sedimentary ridges. Modified from Westbrook, G.K., Carson, B., Musgrave, R. et al., 1994, Proceedings of the Ocean Drilling Program, Initial Reports, volume 146, Ocean Drilling Program, College Station, TX.

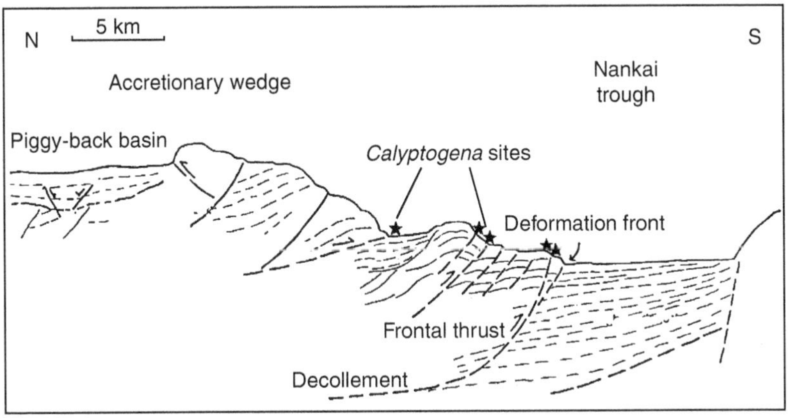

Figure 8.14 Schematic cross section of the Nankai Trough. Note coincidence between *Calyptogena* sites (indicated by a star) and main thrust faults which act as conduits for methane and other fluids. Modified from Kobayashi, K., Ashi, J., Boulegue, J., Cambray, H., Chamot-Rooke, N. et al., 1992. Deep-tow survey in the KAIKO-Nankai cold seepage areas. Earth and Planetary Science Letters, 109, 347–354.

the chemistry of the fluids (Figure 8.14). The fluids are principally derived from sediment compaction within the accretionary wedge, the primary source of energy for deep-sea ecosystems being methane of biogenic origin. However, part of the fluids formed at temperatures of 100–150°C which are found as deep as 10–15 km, and are probably derived from diagenetic dewatering and metamorphose of sediments.

Active expulsion of fluids in accretionary wedges is favored by horizontal compression and the presence of pervasive conduits. Besides vertical diffusion to the seafloor which occurs principally near surface, most fluids are channeled within permeable sediments to thrust faults, which are active fluid conduits. Most seepage sites on the seafloor coincide with fault outcrops. However, bottom-simulating reflectors (BSR) highlight the presence of methane hydrates (see Section 12.3.2) in many accretionary wedge areas (Figure 8.15). The strongest reflectors often coincide with coarse-grained series and anticlinal structures. It is probable that biogenic methane migrates through the sediment until pressure and temperature reach the conditions for gas hydrate stability. These conditions being fulfilled, the formation of methane hydrates within the gas hydrate stability field is associated to the accumulation of free methane below, and only minor quantities of the gas seep to the seafloor, together with other fluids. In addition, continuous deformation of sediment series causes sporadic dissociation of gas hydrates within underthrusted sediments, and related recycling of methane gases.

8.1.3.4. The forearc area

The forearc corresponds to the edge of the overriding lithosphere, which is bounded by the volcanic arc on the inner side, and by the active accretionary wedge on the outer side. Forearc areas may therefore include a prism of accreted sediment which developed at the active accretionary wedge during earlier intervals of plate

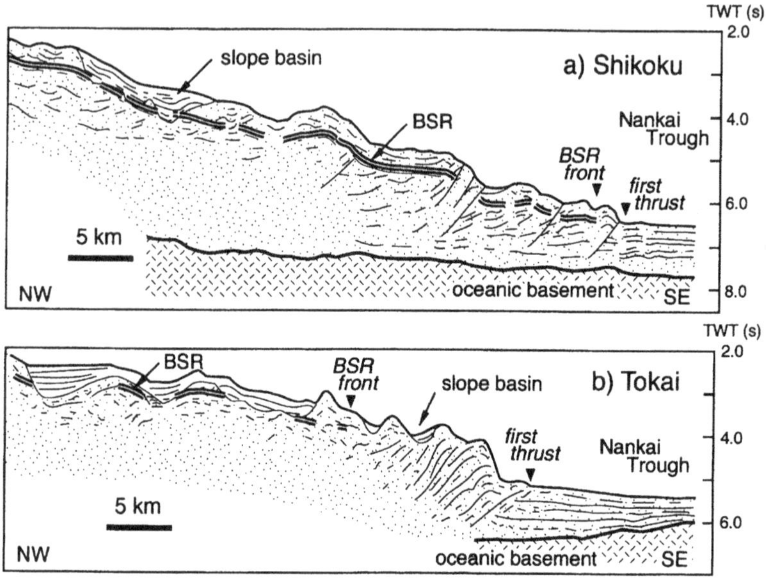

Figure 8.15 Schematic cross sections of the Nankai Trough off Japan, highlighting the distribution of bottom simulating reflectors (BSR) and related methane hydrates. The Shikoku and Tokai areas correspond to the southern and northern parts of the Nankai Trough, respectively. Note absence of bottom simulating reflectors in areas of major thrust and slope (piggy-back) basins. Reprinted from Ashi, J., Tokuyama, H., Taira, A., 2002. Distribution of methane hydrate BSRs and its implication for the prism growth in the Nankai Trough. Marine Geology, 187, 177–191.

convergence. For example, the forearc of the Bonin–Marianas active island arc shows outcrops of overriding oceanic lithosphere, whereas the forearc of the active margin of Japan consists of deformed, nearly vertical sediment strata of Jurassic to Cenozoic age which increase in age landward where they have been metamorphosed and uplifted (Shimanto belt). The forearc is generally faulted and deformed (Figure 8.9), because of compressive stress and subduction related processes such as underplating and magmatic activity. As a consequence, a variety of sedimentary basins may develop in forearc areas, which are filled with sediments eroded or reworked from adjacent relief.

In areas of limited accretion where basement rocks outcrop, elongated fault controlled rift basins may develop in front of the active volcanic arc. For example, the Bonin/Marianas forearc basins of the Philippine plate formed during initial development of the active island arc in the Eocene and Early Oligocene, and remained active till the Miocene. They contain variable thicknesses of sediments, up to about 1,000 m. Because of the active island arc context, the sediments are principally volcanogenic in nature. They principally consist of pyroclastics, pumice and volcanic breccias and conglomerates, often in the form of turbidites and gravity flows. These sediments are interbedded with siltstones, claystones, and calcareous/nannofossil claystones. In areas of tectonic erosion and underplating where the active accretionary wedge is of limited extension and the slope of the active margin is important, a slope apron may develop.

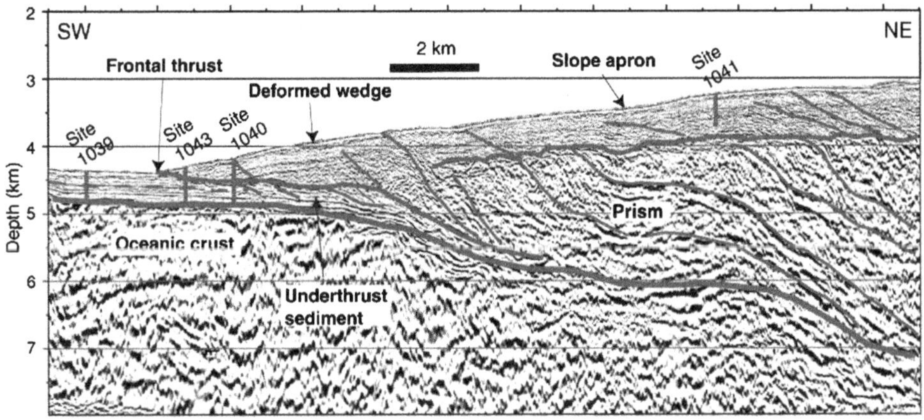

Figure 8.16 Seismic profile across the Costa Rica active margin. The slope apron covers the entire forearc area. Reprinted from Silver, E.A., Kimura, G., Shipley, T.H. (Editors), 2001. Proceedings of the Ocean Drilling Program, Scientific Results, volume 170. Ocean Drilling Program, College Station, TX.

The slope apron is sometimes cut by submarine canyons and channels, as it is on the path of gravity flows to the trench. The slope apron of the Costa Rica margin (Figure 8.16) has been drilled at ODP Site 1041, where it consists of 420 m of silty clay/claystone and siltstone grading to sandy siltstone and silty sandstone downward. The sediment increases in compaction with depth, and the base of the apron is marked by an unconformity capped by a breccia containing fragments of sandstone and basalt. In areas of important active accretionary wedge, continuous deformation of sediments produces a progressive tilting of strata and uplift of anticlinal structures. As deformation progresses, sediments accumulate at the back of submarine relief in piggy-back basins which are gradually incorporated to the prism (Figure 8.15). Like the trench, piggy-back basins are fed by gravity flows principally. Piggy-back basins are often associated to mud volcanoes which grow in adjacent anticlinal areas principally. It is assumed that continuous loading by the piggy-back basins creates overpressured zones within the deformed accretionary wedge, which facilitate the destabilization of methane hydrates and the development of mud volcanoes. For example, piggy-back basins of the Barbados accretionary wedge (Figure 8.17) are bounded by anticlines which support important mud volcanoes and do not show the typical BSR produced by the accumulation of methane hydrates.

8.1.3.5. Backarc systems

Backarc (or marginal) basins are produced by extensional tectonism within the overriding plate, at the back of the volcanic arc. In active margin areas bounded by a cordillera, backarc basins are generally of limited extension and consist of grabens which develop in areas of high elevation, like the Altiplano of the Andes Cordillera. In other areas of active continental margins and active island arcs, extensional tectonism may produce significantly thinned lithosphere and related rift basins, sometimes isolating part of the volcanic basement as a remnant arc. In areas of

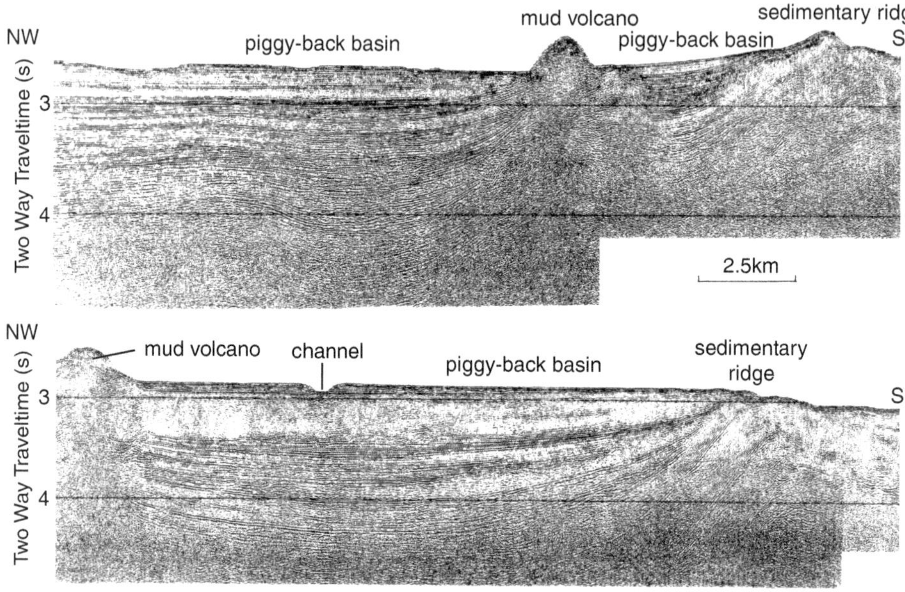

Figure 8.17 Seismic profiles across the Barbados Ridge accretionary complex of the Caribbean. Two large piggy-back basins develop to the west of growing anticlinal sedimentary ridges. The mud volcano of the lower section is located on top of an anticlinal structure, whereas that of the upper section is isolated in the basin. Modified from Mascle, A., Moore, J.C., Taylor, E., Underwood, M.B. (Editors), 1990. Proceedings of the Ocean Drilling Program, Scientific Results, volume 110. Ocean Drilling Program, College Station, TX.

continuous backarc extension, the stage of lithosphere thinning may be followed by the creation of new oceanic lithosphere. In such cases, backarc basins are small oceanic basins, bounded by one or several passive continental margins. However, most basins are asymmetrical in their structure and sediment series.

The volcanic arc margin is a relatively steep area of active tectonism and magmatism which produces high quantities of terrigenous and volcanic sediments, and may eventually experience compressive events. The inner margins are typical, subsident passive margins where terrigenous elements derived from the continental drainage basins progressively build prograding continental shelves, whereas hemipelagic and biogenic sediments dominate in the center parts of the basin. Such backarc basins generally evolve rapidly but remain relatively shallow because the spreading center ceases activity as distance from the volcanic arc increases. However, persistent extensional tectonism in the volcanic arc area may result in the formation of further backarc basins. A typical example is the Marianas island arc area where the Marianas Trough, a young basin in its initial opening stages, formed by intra-arc rifting (Figure 8.18). The Marianas Trough is floored by pillow basalts and filled with turbiditic sequences of volcaniclastics derived from the Marianas Arc which alternate with hemipelagic and biogenic intervals, all sediments being of Pleistocene age. The Marianas Trough is separated from the Parece-Vela Basin by the West Marianas Ridge, an arc basement complex of Late Miocene age where

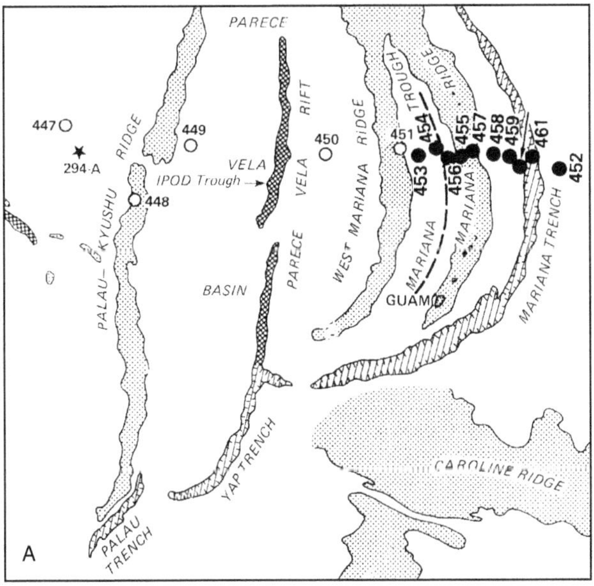

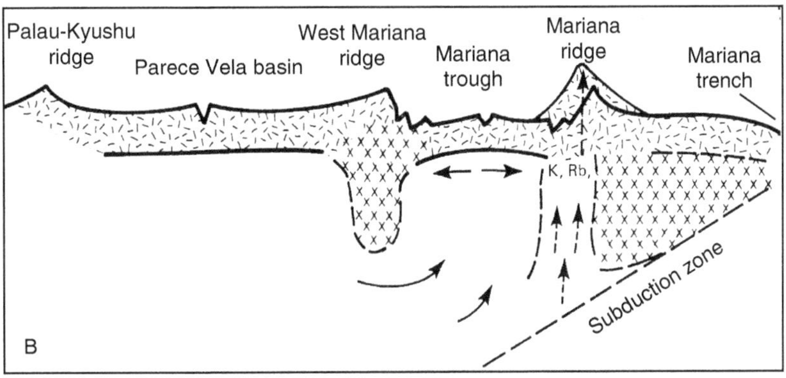

Figure 8.18 The Marianas active island arc and backarc complex. (A) Schematic map highlighting a succession of backarc basins (Marianas Trough and Parece Vela Basin) separated by a remnant volcanic arc (West Marianas Ridge). Black and white circles show the position of DSDP sites. (B) Schematic cross section highlighting the succession of basins (note the presence of spreading centers) and ridges. Modified from Hussong, D.M., Uyeda, S. et al., 1982. Initial Reports of the Deep Sea Drilling Project, volume 60, U.S. Gov Print Office, Washington.

activity ceased 5 Myr ago. There, DSDP Site 451 retrieved a thick sequence of Late Miocene volcaniclastics (breccias, tuffs, conglomerates, ash) capped by a few meters of Pliocene and Pleistocene nannofossil ooze. Beyond the West Marianas Ridge, the Parece-Vela Basin formed by backarc spreading during the Early Miocene. On the inner side of the basin, a latest Oligocene to Pleistocene sequence of nannofossil ooze, radiolarian ooze and clay overlie a 24 Ma old basement at DSDP Site 449. Above an 18 Ma old basement on the arc side of the basin, thick (257 m) Middle

Miocene volcaniclastics are capped by a thin (83 m) section of Middle Miocene to Pleistocene clays at DSDP Site 450.

In fact, backarc basins are often complex geological structures. For example, the Japan Sea is a complex backarc system which initially formed because of tensional and translational stresses within the proto-Japan arc, and related crustal thinning (Figure 8.19). The Japan Sea expanded along the eastern edge of the Asian plate from the Late Eocene to the Late Miocene but sea-floor spreading occurred to the northwest in the Japan Basin only, the Yamato Basin to the southeast being floored by thinned continental lithosphere. Both basins have experienced compressional stress since the Late Miocene. The sediments are especially thick (up to 3,000–5,000 m) along the margin of Japan. They consist of sandstones and siltstones which accumulated rapidly in deltaic environments, capped by biocalcareous and/or biosiliceous claystones.

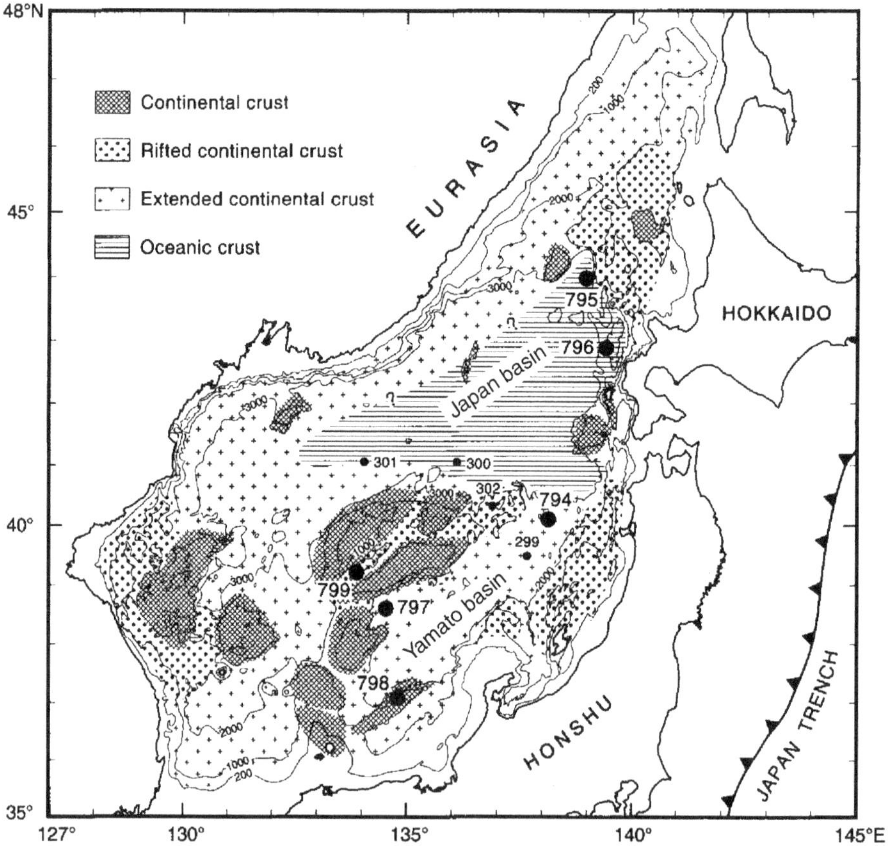

Figure 8.19 The main structural traits of the Japan Sea backarc basins. Note the presence of oceanic crust in the northern Japan Basin and extended continental crust in the southern Yamato Basin. Reprinted from Tamaki, K., Suyehiro, K., Allan, J., Mc Williams, M. (Editors), 1992. Proceedings of the Ocean Drilling Program, Scientific Results, volume 127/128. Ocean Drilling Program, College Station, TX.

8.2. Example of Active Island Arc System: The Tonga Trench–Lau Basin System

8.2.1. Structure and History of the Tonga Trench–Lau Basin System

The active Tonga Island arc marks the subduction of the Pacific plate beneath the Indo-Australian plate, which involves two oceanic lithospheres. The island arc system includes the Tonga Trench and Ridge, the active Tofua arc, the Lau backarc basin and the Lau Ridge which is a remnant arc (Figure 8.20). High subduction rates of 16–23 cm/yr result from the combined effects of plate convergence and backarc spreading. The oldest evidence for subduction-related processes in this region consists of arc-tholeiitic basalts of Early to Middle Eocene age which outcrop on the Tonga Ridge, and in the Fiji Islands at the northwestern extremity of the Lau Ridge. It is likely that the Tonga Ridge and Fiji area once formed a single active volcanic arc and that the South Fiji Basin evolved as a backarc basin in the Eocene and Oligocene. The current position of Eocene volcanic products at about 5,200 m water depth in the forearc of the Tonga Ridge indicates great subsidence since their production. At DSDP Site 205 drilled west of the Lau Ridge in the South Fiji Basin, oceanic basalts are capped by mid-Oligocene sediments. There and at DSDP Site 285 in the center part of the basin, dominant siliciclastics grade to nannofossil ooze enriched in volcanic ash and glass in the Middle Miocene and to a variety of biogenic oozes later in the Miocene. Activity in the South Fiji backarc basin ceased early in the Late Miocene around 12–13 Ma.

However, most of the Fiji Islands and Lau Ridge consist of volcanic arc rocks of Neogene age as does most of the Tonga Ridge, whereas the youngest volcanic rocks of Late Pliocene and Pleistocene age characterize the active Tofua arc. Between both structures, the western Lau Basin is floored by thinned arc crust and shows a succession of horsts and grabens, where relief extends from less than 1,000 m to more than 3,300 m water depth (Figure 8.21). The eastern part of the basin is floored by oceanic crust, magnetic anomalies indicating maximum age of about 5.5 Ma. To the south, the Eastern Lau Spreading Center (ELSC) consists in an axial trough deeper than 3,000 m that shoals southward. To the north, the Central Lau Spreading Center (CLSC) consists in an axial ridge shallower than 2,300 m, which is flanked by elongate linear basins. Both structures are separated by a shear zone (Figure 8.21).

The Lau Basin formed as a response to the subduction process at the Tonga Trench by splitting and rifting of a Fiji/Lau/Tonga volcanic arc followed by the inception of a spreading axis, its subsequent segmentation and the development of a new volcanic arc. Volcanic activity intensified in the middle Miocene around 13–14 Ma in the Fiji/Lau/Tonga volcanic arc, which had separated into the Lau and Tonga ridges by 6 Ma, because of persistent rifting. As extension increased, arc volcanism progressively ceased in the eastern Lau Ridge which became a remnant arc, but persisted in the central rift where the ELSC formed in the Peggy Ridge area around 5.5 Ma and expanded southward during the Pliocene. By the same time, a new volcanic arc (Tofua arc) developed along the western Tonga Ridge. A jump in spreading occurred around 1.5 Ma and the newly created CLSC started propagating southward from the Peggy Ridge area.

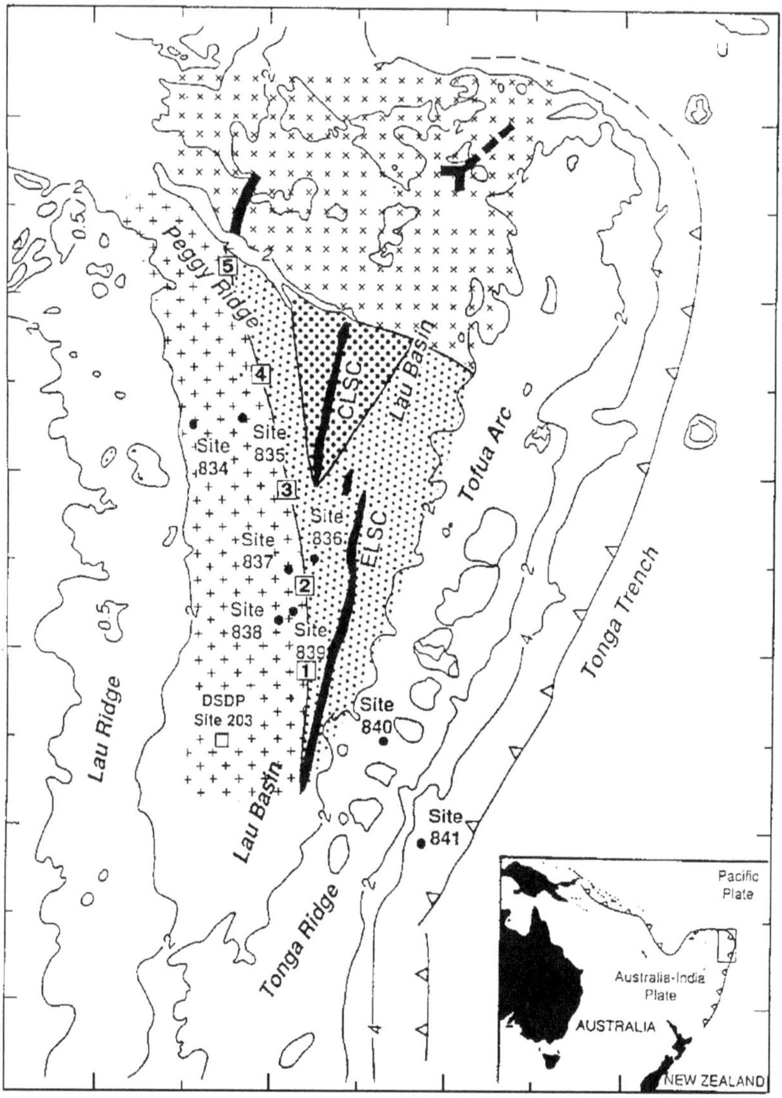

Figure 8.20 Major traits of the Tonga active island arc and Lau Basin backarc systems, and location of DSDP/ODP sites. x, thinned lithosphere of the northern Lau Basin; +, thinned arc lithosphere of the western Lau Basin; stippled patterns, lithosphere generated at backarc spreading centers; ELSC, Eastern Lau Spreading Center; CLSC, Central Lau Spreading Center; numbers in boxes, ages (in Ma) of the lithosphere generated at backarc spreading centers. Note southward propagation of spreading centers. Reprinted from Hawkins, J.W., Parson, L.M., Allan, J.F., Resig, J., Weaver, P. (Editors), 1994. Proceedings of the Ocean Drilling Program, Scientific Results, volume 135. Ocean Drilling Program, College Station, TX.

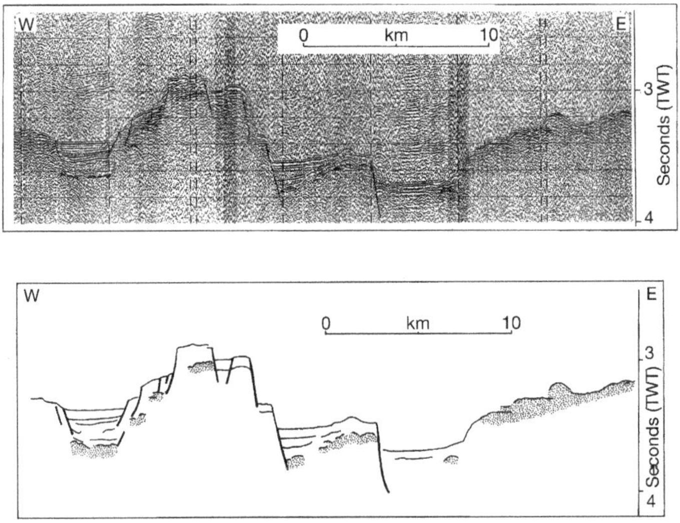

Figure 8.21 Seismic profile across the western Lau Basin at the latitude of ODP Site 839 (top) and its interpretation (bottom). Note the presence of horsts and grabens, and the transition to oceanic crust (right). Modified from Hawkins, J.W., Parson, L.M., Allan, J.F., Resig, J., Weaver, P. (Editors), 1994. Proceedings of the Ocean Drilling Program, Scientific Results, volume 135. Ocean Drilling Program, College Station, TX.

8.2.2. Sediment Deposits of the Tonga Trench and Forearc Area

The oceanic crust being currently subducted in the Tonga Trench is of Late Cretaceous age. Accretion on the inner wall looks limited, and there is some evidence of tectonic erosion and related westward migration of the arc/trench system. The sediments of the Tonga forearc system have been drilled at ODP Sites 840 and 841, to investigate the history of the Tonga Ridge (Figure 8.22).

Drilled at 4,800 m water depth, ODP Site 841 bottomed in a rhyolitic complex at about 780 m below seafloor. There, the products of rhyolitic, subaerial volcanism accumulated as pyroclastic tuffs. They are capped by calcareous (bioclastic) volcanic sandstones which contain large foraminifers and yielded Late Eocene to Early Oligocene ages (Unit V): the volcaniclastics were sporadically emitted from a nearby volcanic center, and accumulated in a shallow marine environment. Above an unconformity spanning approximately 13 Myr and marked by a fault breccia, about 70 m of volcanic sandstones and siltstones accumulated in the form of turbidites principally, early in the Middle Miocene (Unit IV). A major fault separates Unit IV from Late Miocene sediments (Unit III). About 400 m of Late Miocene sediments accumulated at ODP Site 841. They consist of volcanic conglomerates and breccias (Unit III), grading upward to sandstone, siltstone and clayey siltstone (Unit II). It is likely that late Miocene sediments accumulated rapidly at ODP Site 841, based on the frequency of turbiditic (and other gravity flow) facies and biostratigraphy.

Late Miocene sediments are about 250 m thick at ODP Site 840 drilled at shallow water depth (743 m) of the Tonga Ridge. They consist of indurated vitric

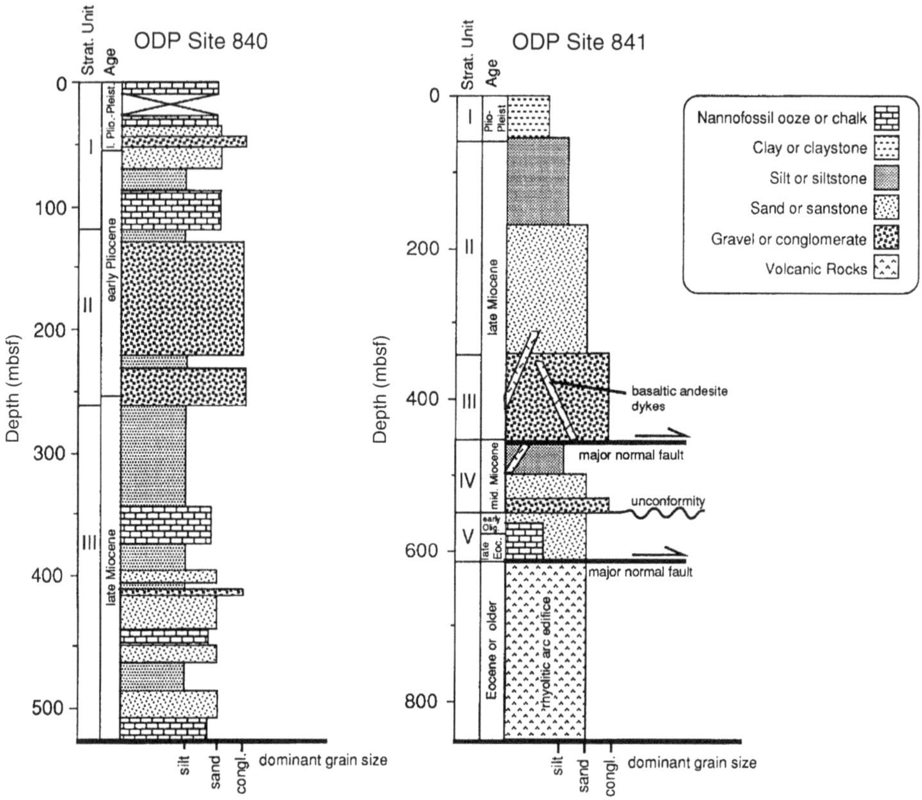

Figure 8.22 Schematic sediment successions of the Tonga forearc. Note the dominance of volcaniclastic facies, and the scarcity of biogenic facies. Modified from Hawkins, J.W., Parson, L.M., Allan, J.F., Resig, J., Weaver, P. (Editors), 1994. Proceedings of the Ocean Drilling Program, Scientific Results, volume 135. Ocean Drilling Program, College Station, TX.

and volcanic siltstone and sandstone turbidites, alternating with highly bioturbated clayey siltstone and nannofossil chalk (Figure 8.23). Individual turbidites decrease in thickness and average grain-size upwards, indicating that emission of large flows from adjacent volcanic arc was followed by a significant decrease in activity.

About 60 m of hemipelagic clays and clays with minor intercalations of volcanic turbidites and thin ash layers accumulated at ODP Site 841 since the Early Pliocene (Unit I). By comparison, Pliocene and Pleistocene sediments at ODP Site 840 are about 250 m thick. They consist of unlithified pumiceous gravel and minor vitric silt/siltstone (Unit II), grading to a diversified assemblage of vitric silt and sand, pumiceous gravel and pyroclastic deposits, alternating with clayey and vitric nannofossil oozes (Unit I). Paleodepth estimates based on benthic microfaunas and trace fossils suggest that major intervals of subsidence occurred around 35 and 5 Ma, and possibly at 16 Ma.

Forearc deposits record a complex interplay of tectonic and subsidence in controlling the sedimentation. Large accumulations of volcaniclastic sediments coincide with intervals of enhanced activity and major steps in the evolution of the

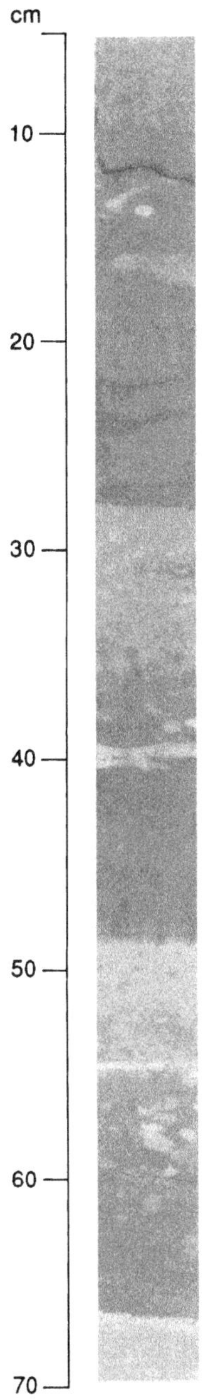

Figure 8.23 Typical Late Miocene (Unit III) sediments of the Tonga forearc (Section 135-840B-45X-2): three volcaniclastic turbidites (dark gray) are overlain by pelagic to hemipelagic sediments (light gray). Note significant bioturbation extending down into volcaniclastic turbidites and the presence of water-escape structures at 20 cm. Reprinted from Hawkins, J.W., Parson, L.M., Allan, J.F., Resig, J., Weaver, P. (Editors), 1994. Proceedings of the Ocean Drilling Program, Scientific Results, volume 135. Ocean Drilling Program, College Station, TX.

active island arc. For example, the Middle Miocene increase of volcanic arc activity coincides with the inception of volcanic turbidite accumulation at ODP Site 841. Subsequent accumulation of thick Late Miocene turbidites occurred during an interval of arc rifting which led to the separation of the Lau and Tonga ridges by 6 Ma. The transition from rifting to spreading in the newly created Lau Basin and related subsidence, and the development of a new active volcanic arc (Tofua arc) further east in the Pliocene, coincide with a drastic decrease in the sedimentation rates and proportion of volcaniclastics at ODP Site 841. By comparison, the shallow ODP Site 840 was relatively sheltered from turbidity currents and other gravity flows in the Late Miocene as suggested by intervals of nannofossil chalk, but was strongly affected by the development of the Tofua arc in the Pliocene as attested by the accumulation of diversified volcaniclastics.

8.2.3. Sediment Deposits of the Western Lau Basin

The sedimentary series of the western Lau Basin have also been investigated during ODP Leg 135. Five sites are located in discrete sub-basins of thinned arc lithosphere and one site is located on the crust generated at the ELSC. The sediments recovered from the backarc Lau basin range from the Late Miocene to the Pleistocene. Overall, the succession of sediment facies is rather identical at all sites (Figure 8.24). However, they vary in age and thickness from site to site. The lower interval principally consists of volcaniclastics which accumulated rapidly as turbidites or debris flows, interbedded with nannofossil clays and/or clayey nannofossil oozes (Units II and III). The volcaniclastics predominantly consist of vitric sand and silt but also gravel, derived from local volcanoes and from basement highs and ridges that separate the sub-basins. The upper interval consists of clayey nannofossil ooze principally, with minor calcareous turbidites (Unit I).

The transition from the lower to the upper interval is rather abrupt and occurs at different ages at the different sites (Figure 8.24). This suggests that the sedimentation in the western Lau Basin was controlled by local volcanic and tectonic processes for most of its history. The sedimentation became more homogeneous and principally controlled by biological processes only recently in the Pleistocene, with average sedimentation rates of about 1.5 cm/kyr. However, most of the upper biogenic sediment at ODP Site 835 is redeposited in the form of turbidites and coherent rafted blocks and mud-clast conglomerates interpreted as muddy debris-flow deposits. The frequency of redeposited sediments underlines four main episodes of instability during the past 3 Myr. The reasons for such intervals of instability are unknown, but the most recent interval (the past 0.4 Myr) coincides with the propagation of the CLSC to the latitude of ODP Site 835 and beyond.

Volcaniclastic material is the main component of backarc Lau Basin sediments, and consists of five main types.

- Fallout tephras have been emitted during subaerial explosive eruptions. These occur mainly as discrete layers, a few centimeters to several tens of centimeters in thickness, interbedded with hemipelagic and biogenic sediments, and sometimes disrupted through bioturbation. The active Tofua arc is the most likely source for

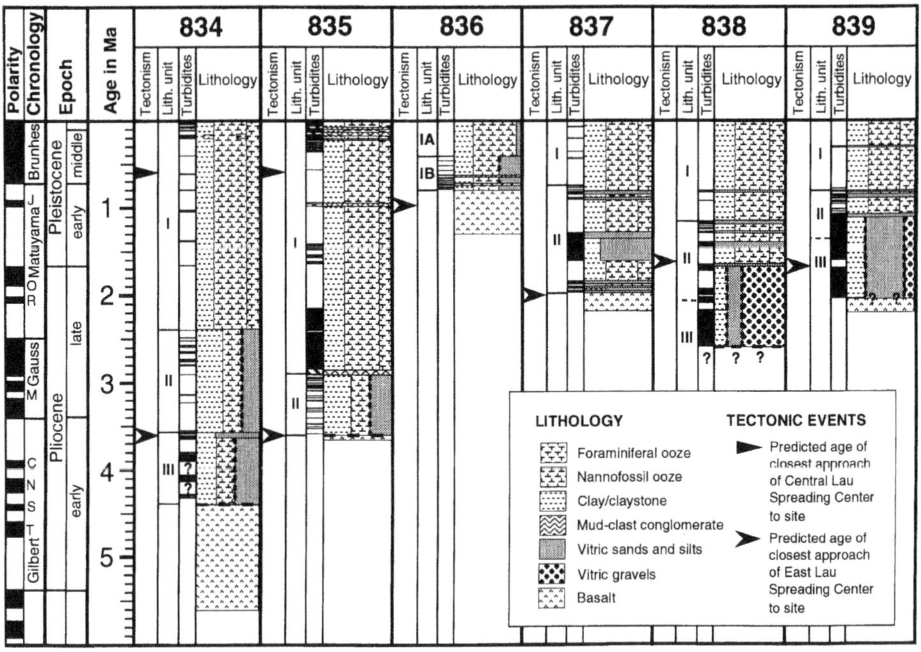

Figure 8.24 Schematic sediment successions of the backarc Lau Basin. Note similarity of main lithologies at all sites but differences in age, and concentration of turbidite facies during specific intervals. Reprinted from Hawkins, J.W., Parson, L.M., Allan, J.F., Resig, J., Weaver, P. (Editors), 1994. Proceedings of the Ocean Drilling Program, Scientific Results, volume 135. Ocean Drilling Program, College Station, TX.

many ash layers. However, some contribution from more distant volcanic areas of the western Pacific Ocean is probable, especially for the thinnest ash layers.

- Epiclastic deposits have been redeposited by turbidity currents and debris flows principally. They usually have eroded bases and sometimes show cross beddings and laminar beddings. High concentrations of volcaniclastic grains in the basal part of the deposits commonly grade to hemipelagic sediments upwards. The volcaniclastic grains (Figure 8.25) produced by local eruptions initially accumulated in shallow areas, and were subsequently reworked because of slope instability.
- Subaqueous fallout and pyroclastic flow deposits have been produced during explosive submarine eruptions. They consist of beds of great thickness (up to several tens of meters) and variable grain size, in the gravel to block size range at the base of the beds. They may include channelized facies and water escape structures.
- Hyaloclastites result from the mechanical fragmentation of the surface of pillows and lava flows during cooling and crystallization.
- Isolated pumice dropstones are especially observed in hemipelagic and biogenic sediments. They are derived either from subaerial or submarine eruptions, and dispersed by oceanic currents.

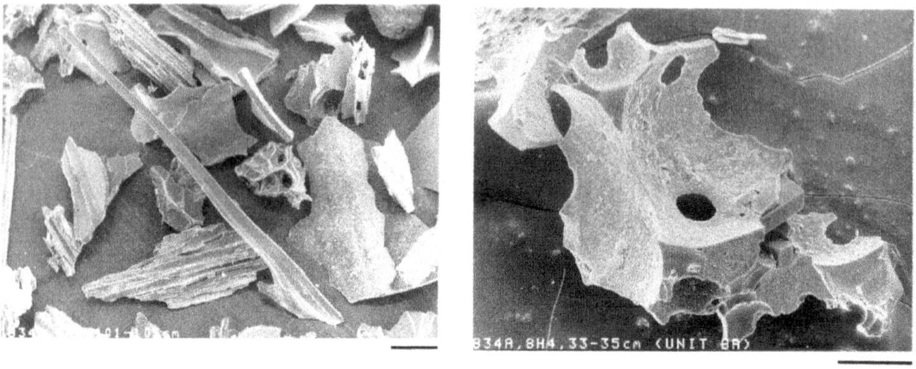

Figure 8.25 Scanning electron microscope photographs of rhyolitic shards (left) from Sample 135-834A-9H-6, 101–103 cm and basaltic shards (right) from Sample 135-834A-8H-3, 33–35 cm. Scale bars represent 1 μm. Reprinted from Hawkins, J.W., Parson, L.M., Allan, J.F., Resig, J., Weaver, P. (Editors), 1994. Proceedings of the Ocean Drilling Program, Scientific Results, volume 135. Ocean Drilling Program, College Station, TX.

The volcanic grains range in composition from basaltic andesite to rhyolite. The whole array of chemical compositions is present in pre-rift and syn-rift deposits, whereas post-rift volcanic grains show a bimodal distribution of chemical elements. It is probable that a change in the chemistry of the volcanism occurred in the Late Pliocene, between 3.6 and 2.1 Ma.

8.3. Example of an Eroded Active Margin: The Middle America Subduction Zone

The oceanic lithosphere of the Cocos plate is produced at the East Pacific Rise, and subducted beneath the Caribbean plate along the Middle America Trench. Convergence rates vary from 70 mm/yr off Guatemala to 90 mm/yr off Costa Rica. Continental areas adjacent to the Middle America Trench principally form a volcanic arc of Cenozoic age which developed over Mesozoic oceanic lithosphere covered by Cretaceous and Early Paleogene sediments. The area was subsequently uplifted from the Late Paleocene to the Early Miocene, especially off Costa Rica where the Cocos Ridge is currently being subducted. As a result of uplift, forearc series outcrop in the Nicoya and Osa peninsulas of Costa Rica.

Seismic reflection profiles across the active margin show a prominent reflection that separates the slope apron from deeper terranes. This Base of Slope Sediment (BOSS) reflector varies in depth from 500 to 2,000 mbsf, is of regional extension, and can be traced from the lowermost forearc slope up to the Nicoya coast off Costa Rica (Figure 8.26). Early investigation of the Middle America subduction area during DSDP Legs 67 and 84 demonstrated that siliciclastic slope sediments were underlain below the BOSS reflector by a disrupted, ophiolitic pre-Eocene basement at several locations off Guatemala. No Neogene accreted sediments were drilled. In addition, only minor compressional structures were evidenced from the cores and

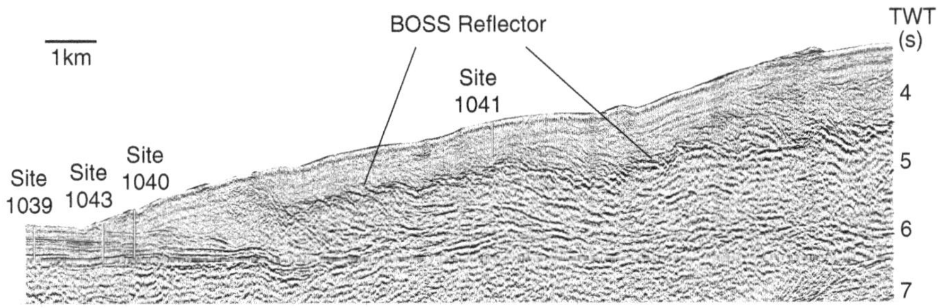

Figure 8.26 Seismic profile across the Middle America subduction zone off Costa Rica. Note the strong and irregular BOSS reflector, which separates the slope apron from its substrate. Modified from Silver, E.A., Kimura, G., Shipley, T.H. (Editors), 2001. Proceedings of the Ocean Drilling Program, Scientific Results, volume 170. Ocean Drilling Program, College Station, TX.

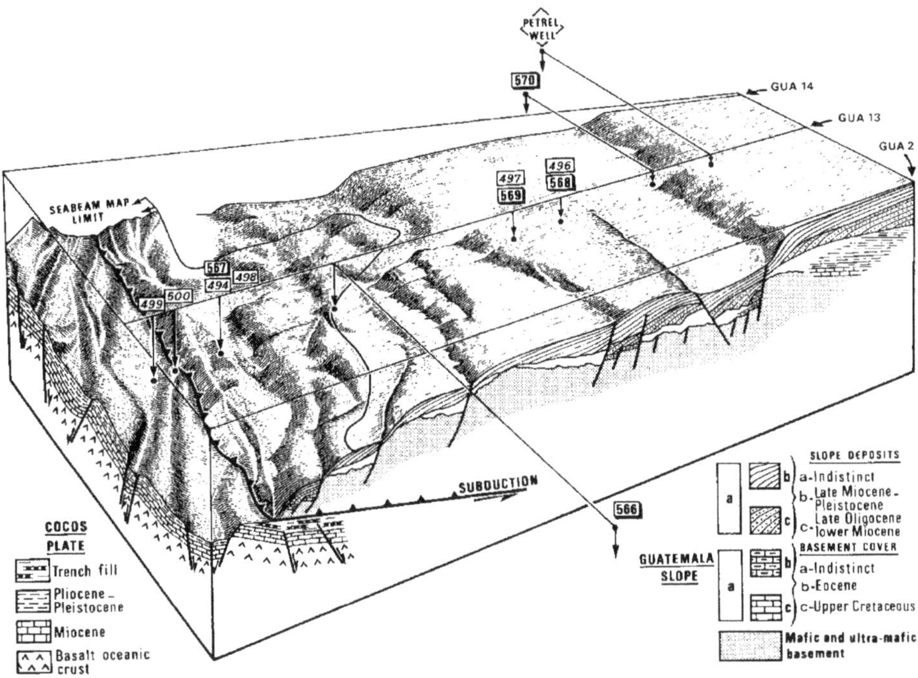

Figure 8.27 Schematic interpretation of the Middle America subduction zone from seismic profiles and DSDP Sites off Guatemala. Note the absence of accretionary wedge, and the presence of extensional structures within the brittle substrate of the overriding plate. Reprinted from Aubouin, J., 1989. Some aspects of the tectonics of subduction zones. Tectonophysics, 160, 1–21.

seismic profiles. It was inferred that accretionary processes and structures could be locally absent from active margin areas (Figure 8.27) and that the questions raised by such conclusions could be answered by coordinated on-land and offshore investigations.

8.3.1. The Nature and Composition of Trench and Forearc Sediments

ODP Legs 170 and 205 were designed to detail the structure and dynamics of the Middle America subduction zone off Costa Rica. The objectives were to sample the incoming sediment series of the Cocos plate, the slope apron, the subducted sediment series and to identify the terranes below the BOSS reflector. The incoming sediment series drilled at ODP Site 1039 in the Middle America Trench off Costa Rica is rather identical to those drilled at DSDP Site 495 off Guatemala (see Section 8.1 and Figures 8.8 and 8.10). The reference sediment series drilled at ODP Site 1039 essentially consists of 378 m of pelagic sediments of Middle Miocene to Holocene age, overlying a gabbro intrusion. Basal calcareous and diatom ooze and breccia grade upwards to siliceous nannofossil ooze, calcareous clay, silty clay and diatom ooze, illustrating the migration of the site from the spreading ridge to the deep basin and to coastal areas of high productivity. The slope apron has been sampled at ODP Site 1041 where it consists of Late Miocene to Pleistocene siliciclastics increasing in grain size and compaction downwards. Biogenic elements are rare. Volcanic ash is concentrated in four thick intervals which represent major, multiple eruptive events of the adjacent Midddle America arc. The upper part of the sequence, which is dominated by claystones, shows variable bedding with dip angles ranging from 5° to 60°, local concentration of microfaults, and a downward increase in fissility. The lower part of the sequence, which is dominated by siltstones and sandstones, has a coherent bedding with 20–45° dip angles but does not show microfaults and structural fabric. Overall, the tectonic deformation of slope apron sediments is of minor importance.

Sediment series rather similar to those of the slope apron have been retrieved from the Middle America Trench at ODP Sites 1040 and 1043, at the toe of the deformed wedge. However, the average grain size of the siliciclastics is somewhat lower than those at ODP Site 1041 on the slope: claystones dominate largely, and sandstones are restricted to some intervals. Besides, the siliciclastic sediments are significantly younger in the trench where they range from Late Pliocene to Pleistocene in age, than on the slope where they span from the Late Miocene to Pleistocene interval. Just like on the slope, deformation is indicated by inclined bedding, fissility and microfaults (Figure 8.28). In addition, fracture networks as well as kink-like bands (sharp deflections of the initial sedimentary fabric) and deformation bands (shear zone where minerals are oriented parallel to the edges) are observed (Figure 8.29).

Deformation is closely related to water release and sediment consolidation: at ODP Site 1040, porosity decreases from 45–50% at 150 mbsf to 40% at 200 mbsf as deformation bands increase in number in a shear zone. It is likely that fluid overpressure induces local dilatation as shear strain increases, and related development of dewatering channels. Clay-sized particles are easily transported to the channels where they accumulate as the channels eventually collapse. Authigenic minerals may ultimately develop within the channels. Permeability also decreases with progressive development of the structures, and areas of dense deformation bands act as barriers to fluid flow. It is noteworthy that the concentration of deformation bands increases strongly around 335–350 mbsf at ODP Site 1040, near the decollement.

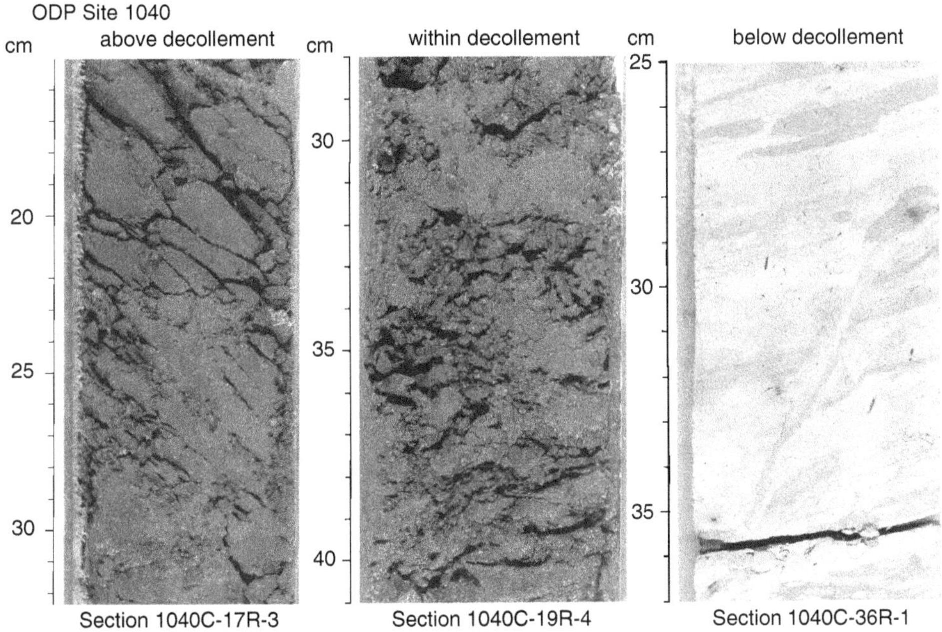

Figure 8.28 Typical lithologies and deformations at ODP Site 1040. Note differences in deformation styles of the silty claystone above the decollement (strong preferred orientation of fissility and fractures) and within the decollement (intense fracture network and lenticular fragments), whereas minor faults and moderately deformed burrows are visible in the nannofossil chalk below the decollement. Modified from Kimura, G., Silver, E.A., Blum, P. et al., 1997. Proceedings of the Ocean Drilling Program, Initial Reports, volume 170, Ocean Drilling Program, College Station, TX.

The decollement occurred at the transition from diatom oozes of high porosity and permeability to siliciclastics of lower porosity and permeability (Figure 8.30). The decollement zone is 9 m thick at ODP Site 1043 at the toe of the wedge and 38 m thick at ODP Site 1040 in the inner trench, and can be subdivided in two domains. Fracture networks break the sediment into lenticular to blocky fragments on the millimeter to centimeter scale in the upper brittle domain where veinlets filled with carbonates are also observed (Figure 8.28), whereas the lower ductile domain consists of soft, plastic silty clay highly deformed during drilling. Silt and clay are relatively separated in the upper brittle domain where fractures and deformation bands separate areas of relatively undeformed and cemented clay, which decrease in size at the transition to the lower ductile domain. Areas of high and low porosity alternate in the decollement zone, low porosities being the consequence of mineral authigenesis and cementation (Figure 8.31). The presence of euhedral calcite and rhodocrosite results from influxes of exogenous fluids in the carbonate-poor sediments of the decollement zone. The style of deformation of the decollement zone seems to result from the association of deformation bands and hydraulic brecciation, the dewatering being mainly channeled. The potential for fluid circulation increases downwards within the decollement zone, to a maximum at the transition from the brittle to the ductile domain. However, model simulations

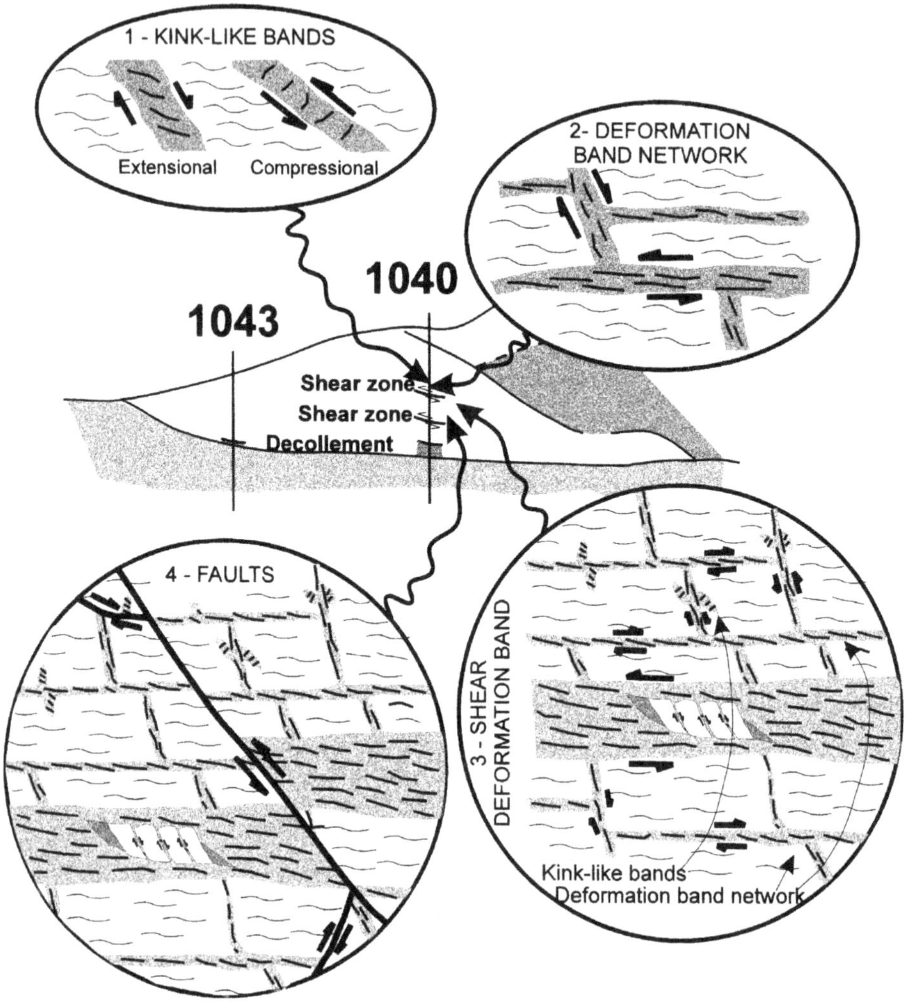

Figure 8.29 Deformation history of the silty claystone above the decollement at ODP Site 1040. Gray areas are zones of reoriented phyllosilicates. Arrows indicate location of main strain. Reprinted from Vannucchi, P., Tobin, H., 2000. Deformation structures and implications for fluid flow at the Costa Rica convergent margin, ODP Sites 1040 and 1043, Leg 170. Journal of Structural Geology, 22, 1087–1103.

suggest that fluid influx to the decollement is controlled by the permeability of the sediments below the decollement zone (biogenic and hemipelagic sediments), rather than the permeability of the decollement zone.

At both ODP Sites 1040 and 1043, the sediment sequence below the decollement is similar to those at the reference site (ODP Site 1039) on the incoming Cocos plate. Late Pliocene siliciclastics are underlain below the decollement zone by Late Pleistocene diatomites, because the lower sequence is underthrusted. The term "diatomite" is being used here in the place of "diatom ooze" because of significant

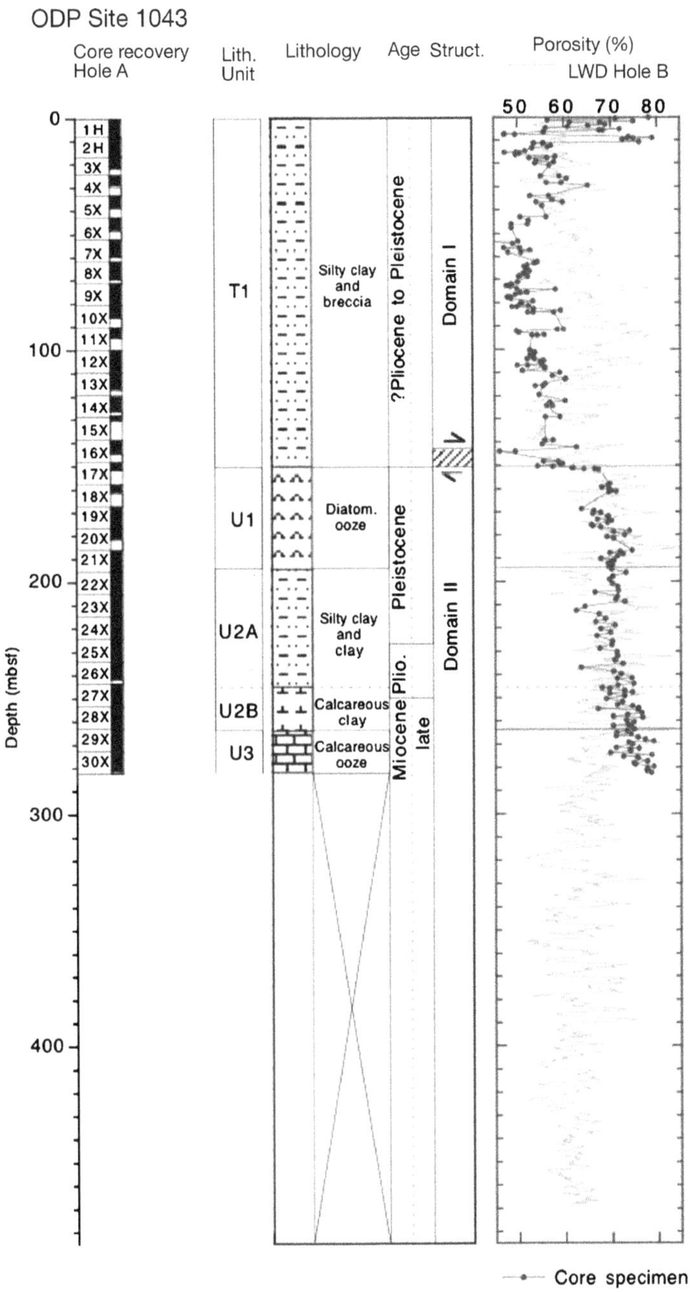

Figure 8.30 Lithostratigraphy and porosity data from logging (LWD) and core samples, ODP Site 1043. Arrows in the Structure column indicate the decollement zone. Note coincidence of low porosity and decollement. Modified from Kimura, G., Silver, E.A., Blum, P. et al., 1997. Proceedings of the Ocean Drilling Program, Initial Reports, volume 170, Ocean Drilling Program, College Station, TX.

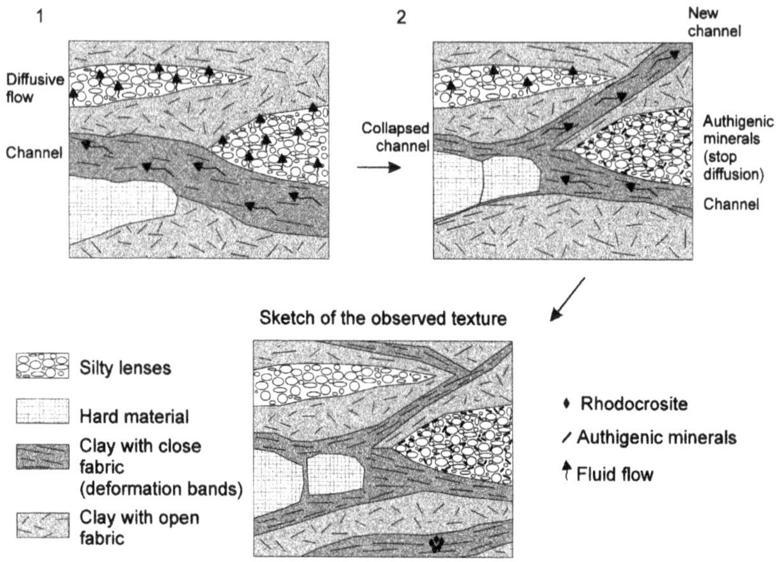

Figure 8.31 Evolution of the decollement deformation and fluid flow through diffusion and successive channel opening. Note minor importance of diffusive flow and successive opening and collapse of dewatering channels. Reprinted from Vannucchi, P., Tobin, H., 2000. Deformation structures and implications for fluid flow at the Costa Rica convergent margin, ODP Sites 1040 and 1043, Leg 170. Journal of Structural Geology, 22, 1087–1103.

lithification of the underthrust sequence. Similarly, underlying lithologies are claystones and nannofossil chalks which are derived from the consolidation of Early Pleistocene to Middle Miocene clays and nannofossil oozes. Detailed correlation between sites shows that the entire sedimentary succession of the incoming Cocos plate is being subducted beneath the Caribbean plate. However, the upper diatomite has lost 33% of its original thickness, whereas the lower nannofossil chalk has lost 20% only. This is a consequence of compaction and related expulsion of pore water which is explained by a loss of about 8 m^3/yr of water per linear meter, including 5.4 m^3/yr from the diatom ooze/diatomite section.

The related decrease in porosity of the upper diatom ooze/diatomite is associated with a strong decrease in permeability, to 0.5% of initial values. Therefore, diffuse fluid expulsion to the decollement slows significantly as the incoming sediments are underthrust. It is probable that fluids are principally collected and channeled to the decollement via major regional fault discontinuities which have been evidenced from the seismic profiles. Besides a reorganization of pore structure associated to dewatering and compaction the underthrust section globally looks undeformed (Figure 8.28), with the exception of minor intervals of deformation bands which could be related to seismic events. Only near the contact with the gabbro, the sedimentary breccia shows normal faulting and fluid escape structures suggesting extension and flattening of the lowermost sediments.

Although slope apron sediments have been sampled at DSDP Site 565 and ODP Sites 1041 and 1042, the forearc basement has not been penetrated yet. Below the slope apron, the BOSS reflector marks an abrupt change in seismic wave velocity which increases significantly in the forearc basement and looks hardly compatible with sediment series. The surface of the reflector is rather smooth and regular below the upper slope, but irregular below the middle and lower slope with landward dipping segments offsets by seaward dipping zones which could correspond to listric faults. The BOSS reflector has been sampled at ODP Site 1042 in the lower slope area, near the termination of the forearc seismic unit. There, the reflector coincides with a succession of two breccias which accumulated in shallow, nearshore conditions. The upper, carbonate cemented breccia is of latest Early Miocene age (16–17 Ma) and principally contains angular fragments of a neritic calcarenite, commonly associated to volcanic fragments. The lower, polymictic breccia includes clasts of doleritic basalt, palagonite, devitrified pumice, red and white chert and schist (Figure 8.32). The regional continuity of the BOSS reflector suggests that equivalent sediments and suitable sources for the clasts might be exposed on the adjacent Nicoya Peninsula (Figure 8.33).

8.3.2. A Synthesis of Onshore and Offshore Data

The lower series exposed on the peninsula (Lower Nicoya Complex) consists of pillow and flow basalt typical of oceanic crust, associated to pelagic sediments (radiolarian chert principally) of Early Jurassic to Late Cretaceous age, which overlie the upper series as a nappe. The upper series (Upper Nicoya Complex) consists of island arc basalt and minor quantities of pelagic sediment of Jurassic and Cretaceous age. The Nicoya Complex is interpreted as an ancient intraoceanic accretionary prism associated to an active island arc system. Younger Late Cretaceous and Paleogene sediments unconformably overlie the Nicoya complex. The overlap series start with a basal breccia made of basaltic clasts principally. The breccia is overlain by pelagic sediments (limestones and radiolarian cherts) of latest Cretaceous age, grading upwards to siliceous and calcareous, distal and proximal turbidites of Late Santonian to Late Eocene age (Garza Supergroup).

Above an unconformity, calcareous and siliciclastic sediments (mainly sandstones) accumulated sporadically in shallow water and continental environments from the Late Eocene to the Pleistocene (Mal Pais Supergroup) following regional uplift. Locally, the lower Mal Pais Supergroup represented by the neritic Late Oligocene Punta Pelada Formation where calcarenite dominates, unconformably overlies the Garza Supergroup via a basal breccia. In other areas, the transgressive shallow-water Montezuma Formation of Miocene/Pleistocene age directly overlies the Nicoya Complex along a highly irregular unconformity, via a conglomerate where basalt and chert clasts dominate. The main intervals of uplift occurred in the Late Eocene, from the Late Oligocene to the Early Miocene, and since the Middle Miocene (and continuing).

A comparison of ODP Site 1042 and field outcrops in the Nicoya peninsula shows that most lower breccia clasts at ODP Site 1042 are derived from the Nicoya Complex. The BOSS horizon looks tied to the Punta Pelada and Montezuma unconformities which are buried by the neritic sediments of the Mal Pais Supergroup. Seismic data and drilling and field data indicate that the Costa Rica

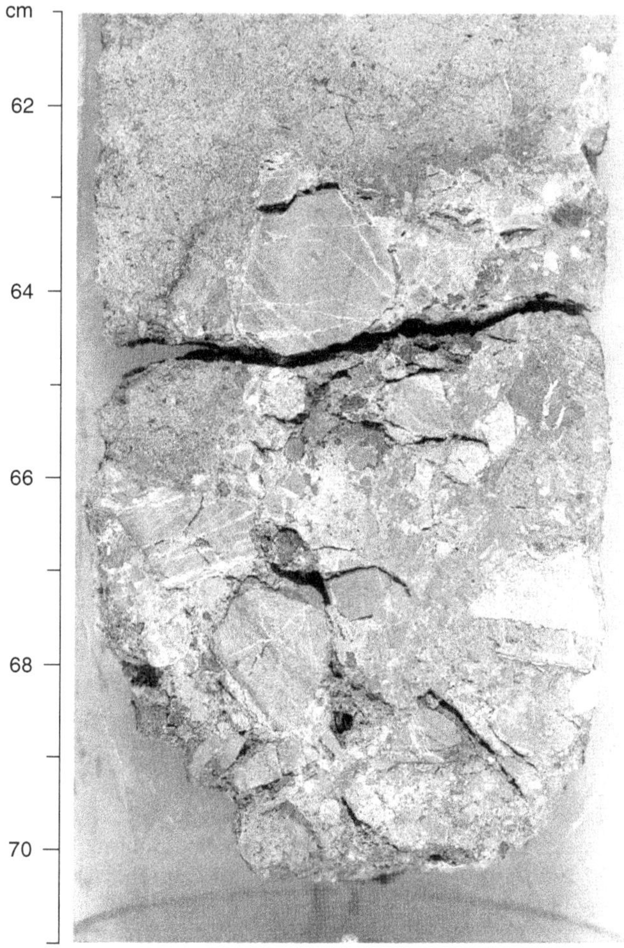

Figure 8.32 Lower polymictic breccia corresponding to the BOSS reflector, ODP Site 1042. Note angular clasts of dolerite, chert and laminated siliciclastics (some with calcite veins). Reprinted from Kimura, G., Silver, E.A., Blum, P. et al., 1997. Proceedings of the Ocean Drilling Program, Initial Reports, volume 170, Ocean Drilling Program, College Station, TX.

active margin lacks a typical accretionary prism. The decollement occurs in the lowermost trench siliciclastics and the entire incoming sedimentary series are being subducted, whereas the trench siliciclastics form a minor deformed wedge which is underthrust beneath the forearc basement. The forearc basement looks very comparable to those of the Nicoya Peninsula and is probably inherited from an active island arc complex of Mesozoic age. The forearc basement does not show traces of uplift, suggesting that the trench siliciclastics are not underplated, and compressive structures are limited to the outermost 10–20 km where slope and trench siliciclastics are underthrust beneath the basement.

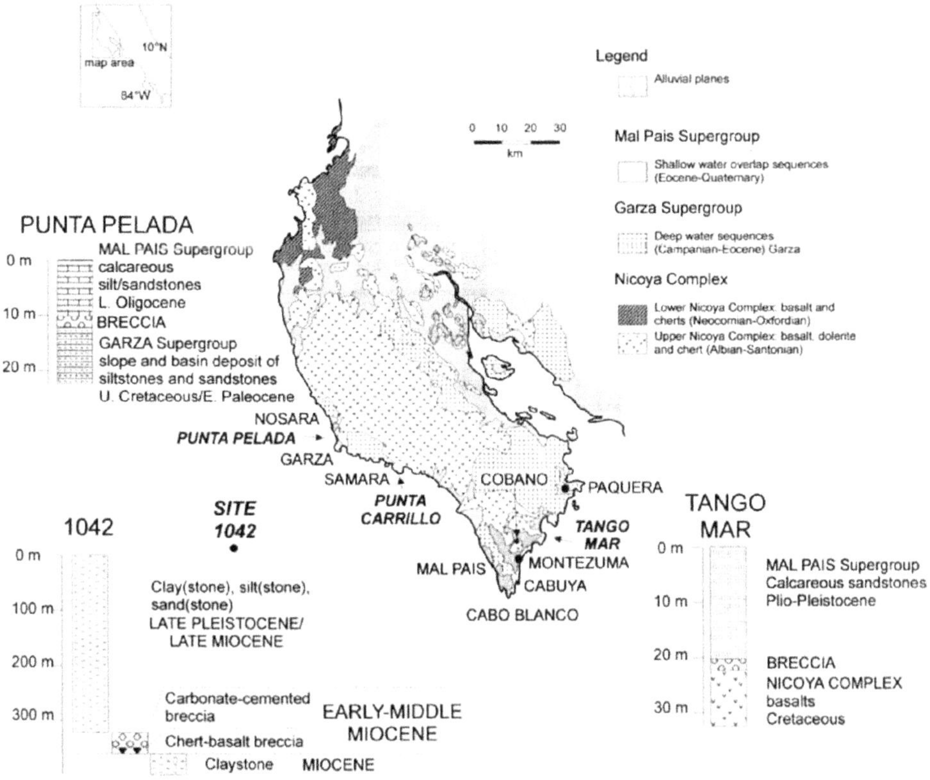

Figure 8.33 Synthetic geological map of the Nicoya Peninsula of Costa Rica. Characteristic sediment successions overlapping the Nicoya complex are compared to those at ODP Site 1042. Reprinted from Vannucchi, P., Scholl, D.W., Meschede, M., McDougall-Reid, K., 2001. Tectonic erosion and consequent collapse of the Pacific margin of Costa Rica: combined implications from ODP Leg 170, seismic offshore data, and regional geology of the Nicoya Peninsula. Tectonics, 20, 649–668.

Conversely, most of the forearc basement shows extensional structures, and benthic microfaunas from ODP Site 1041 on the slope apron indicate that water depth increased from middle bathyal (500–1,500 m) in the Middle Miocene to abyssal (> 4,000 m) in the Pleistocene. The subsidence is inferred to record crustal thinning caused by subduction erosion; i.e., the removal of megalenses (10–15 km long and 1.5–2 km high) from the base of the overriding Caribbean plate which are transferred to the subducted Cocos plate. The average volume loss is about 35 km^3/Myr per linear kilometer. This process leads to a progressive steepening and seaward tilting of the forearc, and a landward migration of the entire active margin (Figure 8.34) which is highlighted by a 40–50 km northeastward shift of the active volcano chain of Costa Rica since the Miocene. Seamount subduction does not seem to be a condition for sustaining subduction erosion. However, the subduction of the Cocos Ridge beginning around 5–3.5 Ma led to regional deformation and uplift. The exhumation of a tectonic mélange (a product of subduction erosion) in

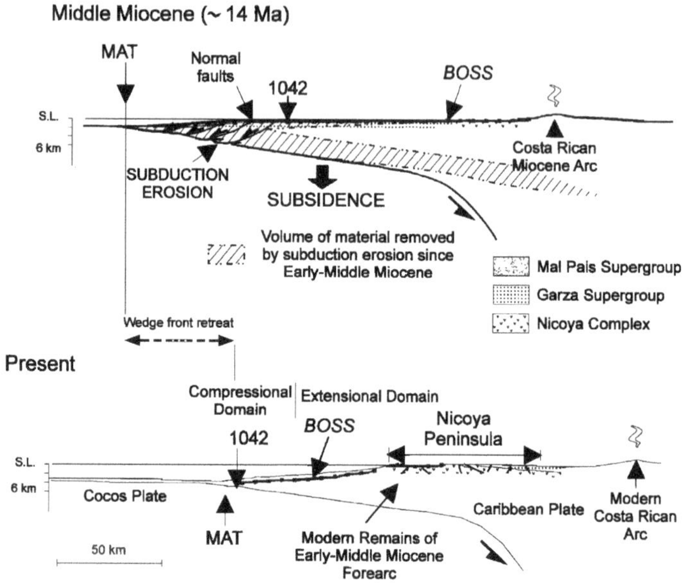

Figure 8.34 Schematic evolution of the Middle America subduction zone (Costa Rica margin) since the Middle Miocene. MAT, Middle America Trench; S.L., Sea level. Note removal of overriding terranes by subduction erosion, related subsidence of the active margin and landward migration of the coastline. Modified from Vannucchi, P., Scholl, D.W., Meschede, M., McDougall-Reid, K., 2001. Tectonic erosion and consequent collapse of the Pacific margin of Costa Rica: combined implications from ODP Leg 170, seismic offshore data, and regional geology of the Nicoya Peninsula. Tectonics, 20, 649–668.

the Osa Peninsula of Costa Rica suggests subsequent erosion of 10–15 km of forearc rocks and sediments. The interaction of the Cocos Ridge and Middle America subduction complex may have played a significant role in the final closure of the Panama seaway.

8.4. Example of an Accreted Active Margin: The Nankai Trough Accretionary Prism

The Philippine plate is being subducted with a dip angle of 3–7° beneath the volcanic arc of Japan (which is part of the Eurasian plate) along the Nankai Trough, at an average rate of about 40 mm/yr. The subduction proceeds largely through large magnitude earthquakes with an average recurrence of about 180 years, the current west-northwest migration of the forearc at a rate of 20–50 mm/yr suggesting a coupling of both lithospheric plates during interseismic periods. However, the prism is far from inactive during interseismic intervals, as indicated by transient pressure anomalies associated to very low frequency earthquakes: they locally relieve stress along the subduction thrust, loading adjacent areas. The Philippine plate entering the Nankai trough carries the undeformed sediments of the Shikoku Basin, but off cape Muroto to the southeast of Shikoku Island, an

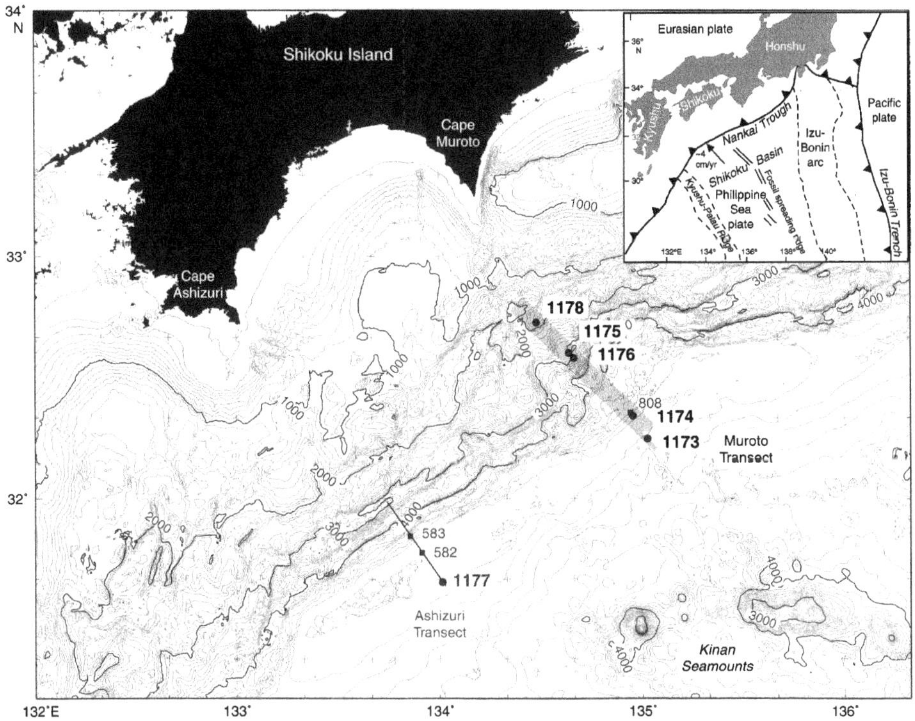

Figure 8.35 Morphology of the Nankai Trough and accretionary wedge, and location of DSDP and ODP sites. Contour interval = 100 m. The inset shows the main traits of the Philippine plate. Modified from Moore, G.F., Taira, A., Klaus, A. et al., 2001. Proceedings of the Ocean Drilling Program, Initial Reports, volume 190. Ocean Drilling Program, College Station, TX.

extinct spreading center marked by the Kinan seamounts is being subducted (Figure 8.35). The subduction of the seamounts produced an indentation of the rather linear inner wall of the Nankai Trough, the Tosa Bae embayment.

8.4.1. Incoming Sediments and the Structure of the Forearc Area

Incoming sediments of the Shikoku Basin have been drilled at ODP Sites 1173 and 1177 where 725 m and 831 m of Middle Miocene to Pleistocene and Early Miocene to Pleistocene sediments have been drilled, respectively (Figure 8.36). At both sites, the igneous basement is covered by a basal sedimentary unit of dominant volcaniclastics and by the lower Shikoku Basin facies of Early Miocene to Early Pliocene age which consist principally of hemipelagic claystones and siltstones. The main difference between both sites resides in the presence of Miocene siliciclastic turbidites with abundant woody organic matter most probably derived from southern Japan at ODP Site 1177. The upper Shikoku Basin facies of Pliocene and Pleistocene age also consist of dominant hemipelagic claystones and siltstones, but contain abundant layers of volcanic ash derived from the volcanic arc of Japan. The distinction between lower and upper Shikoku Basin facies looks based on the diagenetic

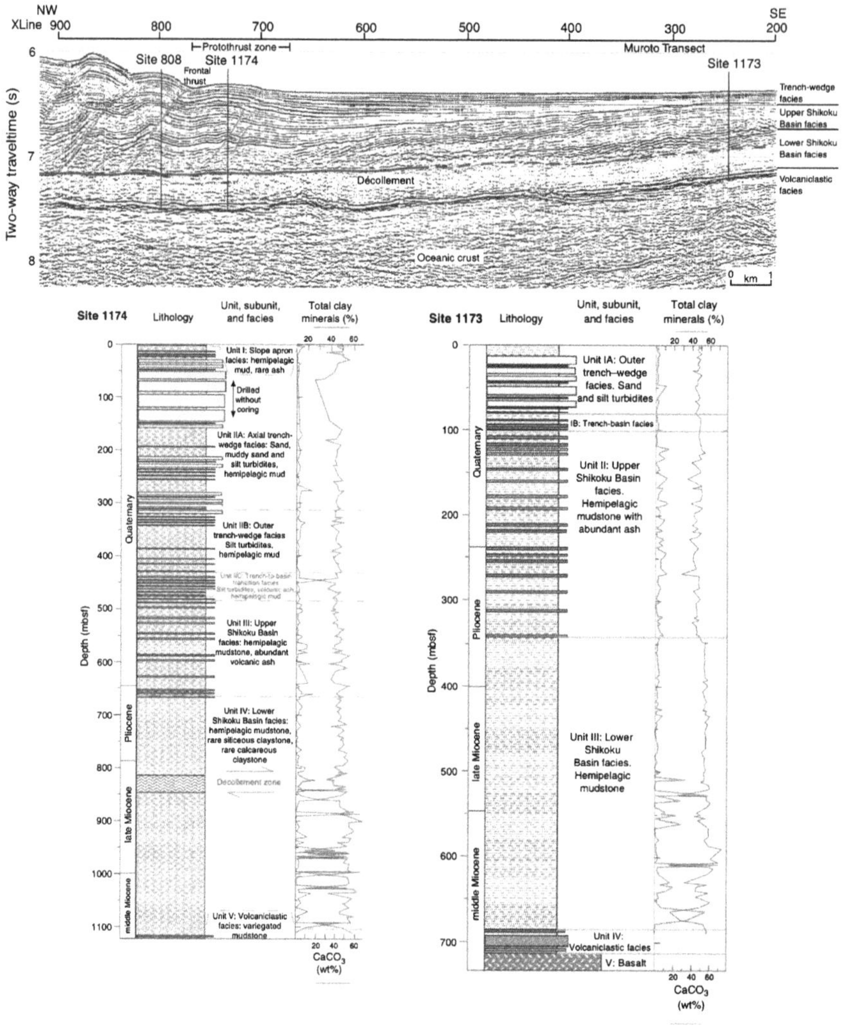

Figure 8.36 Top, seismic reflection profile across the Nankai Trench. The decollement, proto-thrust and frontal thrust zones are highlighted. Bottom, lithologic summaries of ODP Site 1173 in the outer trench and ODP Site 1174 in the proto-thrust area. Note the scarcity of carbonates and the dominance of clay minerals in the bulk sediment. Modified from Moore, G.F., Taira, A., Klaus, A. et al., 2001. Proceedings of the Ocean Drilling Program, Initial Reports, volume 190. Ocean Drilling Program, College Station, TX.

alteration of volcanic ash, discrete ash layers being conspicuously absent from the lower facies where they are replaced by interbeds of siliceous claystone of dominant smectite.

The Pleistocene upper Shikoku facies are capped by about 490 m of Pleistocene Trench deposits at ODP Site 1174, drilled in the proto-thrust zone of the Nankai Trough, past the deformation front (Figure 8.36). Increasing upward frequency of silt turbidites within hemipelagic claystones and siltstones, followed by occurrences of thick sand turbidites characteristic of the axial zone and a thin veneer of slope apron deposits, highlight the migration of the site from the basin to the inner trench during the Pleistocene. The Pleistocene trench turbidites originate principally from the Isu collision zone of Honshu Island. Active sediment accretion is presently taking place along the Nankai Trough. The main characteristic of the inner trench is the development of a decollement surface at a depth of about 150 m within the lower Shikoku facies; i.e., within rather homogeneous siliciclastic sediments (Figure 8.36). The lowermost 250–300 m of the incoming sediment series are being subducted, whereas most of the Shikoku Basin facies as well as the whole trench siliciclastics and volcaniclastics eroded from the volcanic arc of Japan are accreted to the Nankai accretionary wedge.

Landward of the proto-thrust zone, a frontal thrust with vertical displacements of 50–200 m marks the outer limit of the accretionary wedge (Figure 8.37). The outer wedge is characterized by an imbricate thrust zone where a series of thrust faults associated to anticlinal structures delimit distinct sediment packets, and bottom simulating reflectors indicative of methane migration and gas hydrate formation. All major thrust faults are parallel to the frontal thrust and have comparable dips. About 20 km northwest of the deformation front the wedge is cut by a second, younger fault complex which is related to the accretion of thick axial sand turbidites and defines an out-of-sequence thrust zone. The main thrust generates a prominent ridge which retains forearc sediments in a piggy-back basin. The region upslope from the out-of-sequence zone shows landward dipping reflectors which represent thrust boundaries and tilted sedimentary layering, and progressively fade landward. Above the prism, several hundreds of meters of alternating siliciclastic and hemipelagic sediments locally accumulated in piggy-back

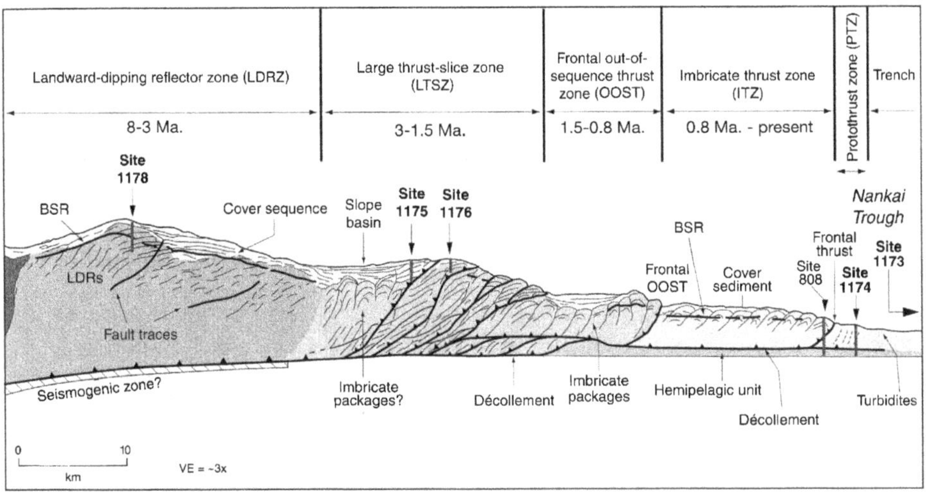

Figure 8.37 Schematic interpretation of a seismic profile across the Nankai accretionary wedge off Cape Muroto. The major thrusts are highlighted. BSR, Bottom simulating reflector; LDR, Landward dipping reflector. Note the extension of the wedge. Reprinted from Moore, G.F., Taira, A., Klaus, A. et al., 2001. Proceedings of the Ocean Drilling Program, Initial Reports, volume 190. Ocean Drilling Program, College Station, TX.

basins where they were subsequently remobilized and deformed. The Nankai prism grades landward to the Shimanto belt terranes which provide lithic fragments to slope and trench siliciclastic sediments.

Accretionary prism sediments identified in the Shimanto belt of Japan and beyond suggest that the Nankai margin has been active since the Jurassic. In the Cretaceous Shimanto belt, repeated sequences of trench turbidites and tectonic mélange including fragments of pillow basalts, pelagic limestone, radiolarian cherts and hemipelagic shales, are locally metamorphosed in the zeolite and greenschist facies. Such a metamorphism taking place at a depth of about 30 km, the Shimanto terranes have been subsequently uplifted and exhumed, probably because of active underplating. In fact, accretion of Shimanto terranes was not continuous but persisted until the Early Miocene.

The current phase of subduction started with the rifting of the Izu-Bonin arc (Figure 8.35) in the Oligocene which progressively isolated the remnant Kyushu-Palau Ridge beyond the Shikoku Basin where seafloor spreading lasted until 15 Ma approximately (Figure 8.38). A resumption of volcanic activity within the forearc of Japan from 17 to 12 Ma marks the initial subduction of Shikoku Basin lithosphere. Extensive volcanism along the forearc and the development of a fold belt in SW Honshu between 8 and 5 Ma are consequences of increased convergence and related subduction of the Shikoku Basin, and collision of the Izu-Bonin arc against Japan. Continuous subduction then progressively shaped the modern margin including the Nankai Trough and accretionary wedge.

Turbidity currents derived from the adjacent margins filled topographic lows and by 4 Ma an accretionary wedge mainly composed of turbidites was formed. By 2 Ma the subduction of a seamount off cape Muroto produced an indentation of the accretionary wedge, whereas reinforced tectonism in the Isu collision zone of Honshu and related erosion led to increased transport of siliciclastics to the trench where axial turbidites and channel deposits became dominant. Since then, rapid accumulation of sediments helps restoring the zone of frontal accretion across the Tosa Bae embayment: accumulation rates of trench turbidites may locally reach 842 m/Myr and the estimated seaward growth of the accretionary wedge is 40 km/Myr on the average.

8.4.2. The Nature and Formation of the Decollement

A remarkable characteristic of the Nankai accretionary wedge is the development of a decollement surface within the rather homogeneous lower Shikoku facies. The decollement zone corresponds to a bright seismic reflector extending seaward below the proto-thrust zone, to the deformation front. The decollement zone was first identified between 945 and 964 mbsf at ODP Site 808 in the Nankai Trench across the frontal thrust, and then between 808 and 841 mbsf at ODP Site 1174 near the deformation front (Figure 8.36). At both sites, the decollement zone is characterized by intense brittle fracturing in the form of finely spaced fractures that break the sediment into small (a few millimeters to centimeters in size) angular fragments (Figure 8.39). The brecciation increases downwards, to a peak above the very sharp base of the decollement zone. In contrast, there is little evidence for a

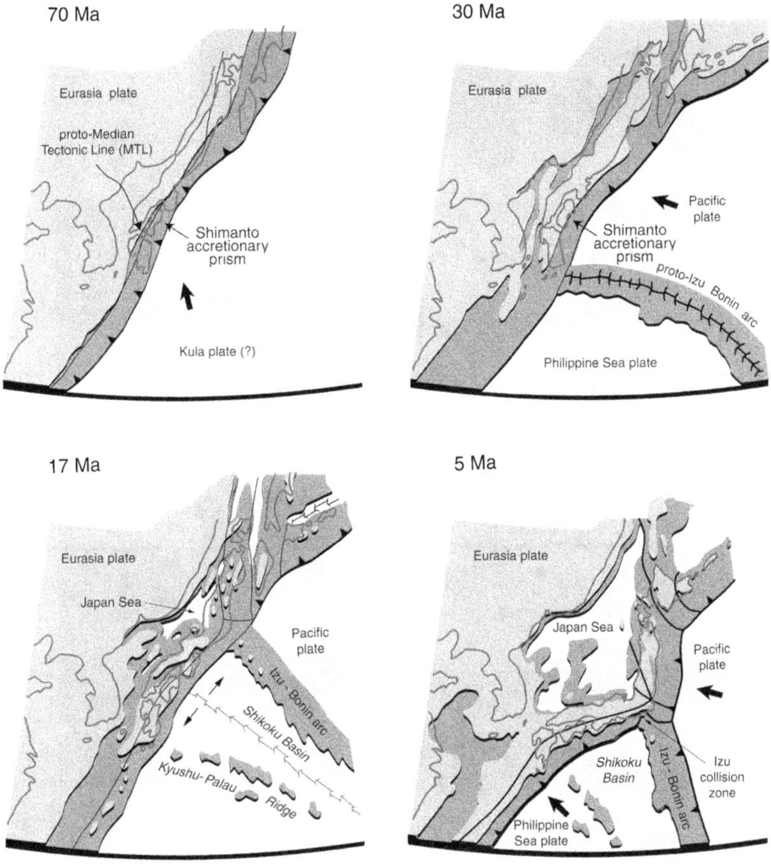

Figure 8.38 Paleogeographic reconstructions of the active margins of Japan. Oceanic basins are in white. Note the creation of the Philippine plate in the Oligocene and similarities between the Pliocene and modern configurations. Reprinted from Moore, G.F., Taira, A., Klaus, A. et al., 2001. Proceedings of the Ocean Drilling Program, Initial Reports, volume 190. Ocean Drilling Program, College Station, TX.

protodecollement zone at ODP Site 1173 which has been drilled 12 km seaward of the deformation front in the outer trench, suggesting that the decollement develops from an homogeneous interval of hemipelagic claystone and siltstone to an horizon of intense brittle deformation which separates the sedimentary column into two intervals of distinct physical and mechanical regimes. Correlations between seismic reflection profiles and drilling sites at regional scale suggest that the decollement develops within the same stratigraphic horizon in the entire area, regardless of significant differences in the thickness of overlying trench turbidites.

On the incoming plate at ODP Sites 1173 and 1177, sediment porosities within the upper Shikoku Basin facies vary from 57% to 69% but do not show significant variation with depth. Porosity values are significantly lower in the lower Shikoku facies where they vary from 36% to 50% and decrease with depth, following a typical compaction

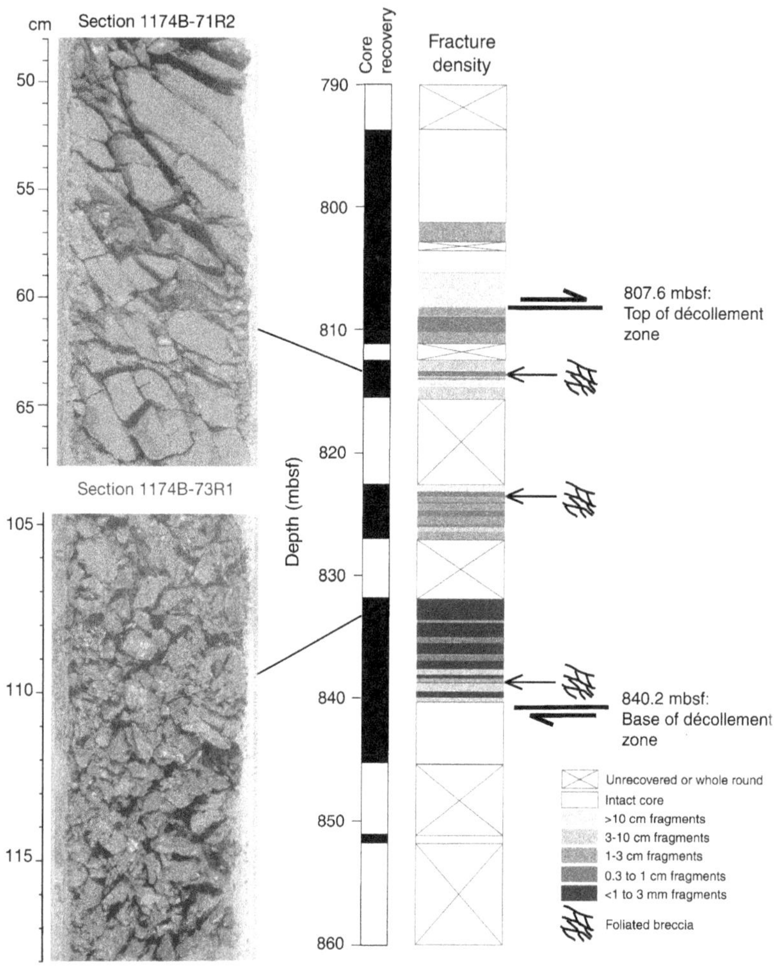

Figure 8.39 Detailed structure of the decollement zone, ODP Site 1174. Note variability and downward increase of fracturing. Modified from Moore, G.F., Taira, A., Klaus, A. et al., 2001. Proceedings of the Ocean Drilling Program, Initial Reports, volume 190. Ocean Drilling Program, College Station, TX.

trend (Figure 8.40). In the proto-thrust and outer accretionary wedge areas where the uppermost trench and wedge facies reach thicknesses of 450 m to 600 m, porosities within the upper Shikoku Basin range from 35% to 45% at ODP Sites 808 and 1174. Slightly lower porosity values are recorded within the lower Shikoku Basin facies, where they range from 30% to 40%, decreasing with depth. A comparison of sediment porosity data in both areas highlights two major points (Figure 8.40).

- A sharp decrease in porosity is clearly visible at the transition from the upper to the lower Shikoku Basin facies on the incoming plate but is of minor importance in the accretionary wedge area.

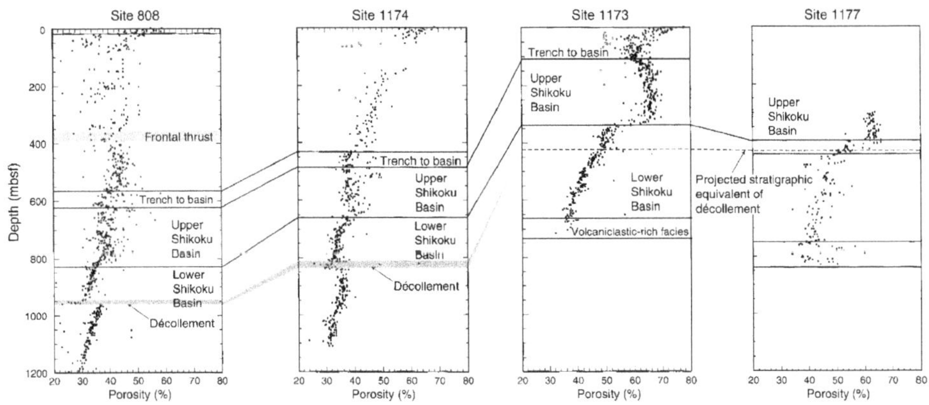

Figure 8.40 Variability of porosities across the Nankai Trench, from the outer trench (right) to the frontal thrust of the accretionary wedge (left). Note major reduction of the porosity contrast between the upper and lower Shikoku Basin facies and concurrent increase of the porosity shift within the decollement zone, from the outer trench toward the frontal thrust area. Modified from Moore, G.F., Taira, A., Klaus, A. et al., 2001. Proceedings of the Ocean Drilling Program, Initial Reports, volume 190. Ocean Drilling Program, College Station, TX.

- A sharp offset to greater porosity across the decollement is clearly visible in the accretionary wedge area, whereas the stratigraphic equivalent of the decollement on the incoming plate does not show any significant variation in porosity.

Sediment porosity data suggest that during the migration of incoming Shikoku Basin sediments from the outer to the inner trench, the upper Shikoku Basin facies undergo considerable consolidation and compacted layers develop between the upper to lower Shikoku Basin facies and the stratigraphic equivalent of the decollement. The upper Shikoku Basin facies initially have high porosity and potentially high permeability and release large volumes of fluids as they experience stress and compaction in the trench. Fluid release may induce hydrofracturing (vertical faults) in the proto-thrust area and a reduction in intergranular porosity, and permeability. Rapid dewatering of the uppermost lower Shikoku Basin facies underneath may generate compacted layers which in turn diminish the rate of dewatering and delay the consolidation of underlying sediments, beneath the horizon that becomes the decollement.

It is probable that the accumulation of trench siliciclastics at high sedimentation rates on top of the sedimentary column and heavy mass loading of the accretionary wedge provide the pressure required for such a rapid compaction. Maximum stress is parallel to the direction of convergence, and is released in specific zones such as the frontal thrust and the decollement where tensile fractures tend to form. However, a localized decrease in P-wave velocity and minor fluctuations in porosity at ODP Site 1173 suggest that a subtle mechanical discontinuity could contribute to the localization of the decollement in this interval. By comparison, sediments below the decollement experience only minor strain (Figure 8.41).

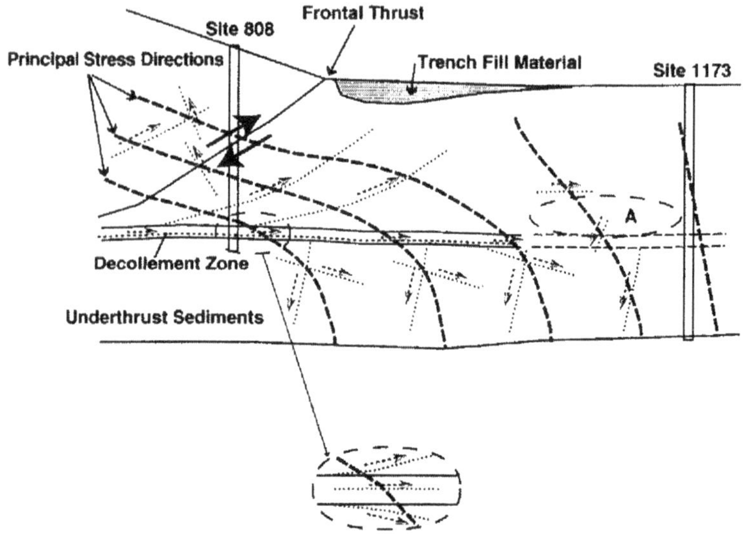

Figure 8.41 Model of stress field in the Nankai trench. Because of heavy mass loading of the accretionary wedge, the trend of compaction deviates toward the wedge. (A) possible zone of tensile field. (Inset) the model suggests that tensile fractures may develop within the decollement zone. The principal directions of stress explain oblique deformation in the accreted series. Reprinted from Ienaga, M., McNeill, L.C., Mikada, H., Saito, S., Goldberg, D., Casey Moore, J., 2006. Borehole image analysis of the Nankai accretionary wedge, ODP Leg 196: Structural and stress studies. Tectonophysics, 426, 207–220.

Physical properties and sediment microstructure at ODP Site 1173 on the incoming plate suggest that the upper Shikoku Basin facies is dominated by an open texture with random distribution of coarse particle aggregates and clay particles distributed throughout the matrix. However, those sediments also look partially supported by a cement, which may result from opal diagenesis and/or alteration of volcanic glass to authigenic smectite and delays the compaction. The degree of phyllosilicate preferred orientation increases locally in areas adjacent to deformation structures, which consist of compactive shear bands principally.

Overall, the upper Shikoku Basin facies in the area of ODP Site 1173 looks underconsolidated with a relatively high permeability, which probably helps with the dewatering and consolidation of the underlying lower Shikoku Basin facies. By comparison, the lower Shikoku Basin facies at ODP Site 1173 exhibits more homogeneous textures and distributions of clay-sized particles, which commonly have a high degree of preferred orientation, especially below the stratigraphic equivalent of the decollement. Clay-sized particles principally consist of smectite and maximum smectite contents of 60% are recorded in the uppermost lower Shikoku Basin facies, gradually decreasing below the stratigraphic equivalent of the decollement. In fact a phase transition does occur at this level; i.e., an alteration of smectite to illite/smectite mixed-layered clays which remains however moderate, with no ordering. Liberation of water from interlayers shrinks the mineral volume, whereas the growth of well-aligned illite particles contributes to the cementation of the lower Shikoku Basin facies. Because of the absence of cement, the uppermost lower Shikoku Basin facies may compact to a greater degree than the cemented intervals above and below as the incoming sediment migrates to areas of greater stress of the inner trench. Phyllosilicate preferred orientations are more intense at ODP Site 1174 in the proto-thrust zone of the trench, and the sediment fabric begins to homogenize at a higher stratigraphic level than at ODP Site 1173. Also, the alteration of smectite to illite/smectite mixed-layered clays is more important (although there is no evidence of ordering) and starts at a higher stratigraphic level, above the decollement.

Seismic impedance (velocity × density) derived from data inversion of a seismic reflection profile across the trench suggests that the uppermost lower Shikoku Basin facies consolidates rapidly past the location of ODP Site 1173, whereas the upper Shikoku Basin facies experiences only minor changes in density (and porosity). Then, the upper Shikoku Basin facies consolidates substantially near the location of ODP Site 1174 (Figure 8.42). Related decreases in permeability strongly impede the expulsion of fluids, creating overpressured conditions beneath the compacted sediments.

In detail, the alteration of smectite at ODP Site 1173 begins at 390 mbsf, whereas the stratigraphic equivalent of the decollement extends from 390 mbsf to 420 mbsf. This interval shows a random orientation of sediment particles where clay aggregates bind particles together, acting as a cement which maintains pore spaces. The fabric is locally highly porous, as suggested by anomalous individual values of porosity. A random orientation of sediment particles is also observed below the stratigraphic equivalent of the decollement, but intergranular bonding looks limited in those horizons of decreased porosity. Within the decollement zone at ODP Site 1174, the microstructures of the brecciated fragments and unbroken intervals generally show a random orientation of particles where clay aggregates are commonly broken, but locally maintain open pore spaces (Figure 8.43). Discontinuous, bifurcated or anastomosed zones of highly developed preferred orientation of phyllosilicates are observed locally: pore spaces have collapsed within the bands of reoriented phyllosilicates, which define the slip surfaces of the brecciated fragments. It is probable that shear strain was accommodated by phyllosilicate rotation and porosity collapse along those slip surfaces producing compactive deformation and local overconsolidation. These observations account for the wide range of porosities within the decollement zone.

To summarize, the decollement seems to propagate within a porous interval characterized by a high degree of cementation associated with intergranular bonding. A collapse of the clay cement is caused by shear stress conditions related to increased fluid pressure as compaction of overlying horizons progresses, reducing the rate of fluid escape. Besides, resistivity images collected at ODP Site 808 show the presence of conductive, fluid-filled and dilatant fractures within the decollement. Dilatant fractures may form as overconsolidated horizons reach conditions for brittle failure because of tectonic and fluid overpressure. Increased pressure would initially produce compactive shear bands as stress conditions reach the domain of ductile strain, reducing the permeability of the sediment. Because of additional increase in pressure and related stress, the sediment would reach the

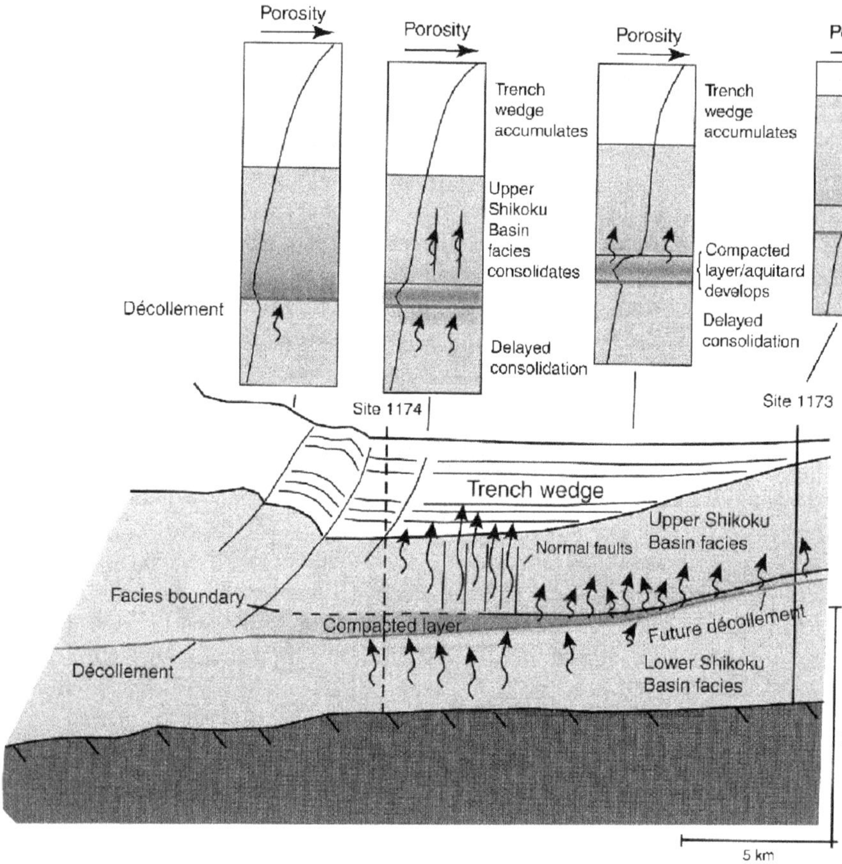

Figure 8.42 Schematic model of compaction of the upper and lower Shikoku Basin facies across the Nankai Trench. Arrows represent fluid flow. Note the formation of a compacted layer as fluids are still able to be expelled upwards, followed by compaction of overlying sediments. This helps an overpressured decollement zone to develop below the compacted layer. Reprinted from Taira, A., Moore, G.F., Becker, K., Mikada, H., Casey Moore, J., Klaus, A. (Editors), 2005. Proceedings of the Ocean Drilling Program, Scientific Results, volume 190/196. Ocean Drilling Program, College Station, TX.

domain of brittle failure, increasing fracture porosity and permeability. Related increase in fluid transfer may cause a reduction of effective stress and an increase in fracture dilatancy. The association of dilatant fractures and fluid overpressure may lubricate the decollement zone, in turn generating the mechanical decoupling of the accretionary wedge from the subducted sediments in accordance with the stress field model.

The investigations conducted on the Nankai and on the Middle America subduction zones highlight the role of the composition and diagenesis of sediment series in the location and initiation of the decollement zone, and therefore in the relative importance of the accretionary wedge and morphology of the active margin.

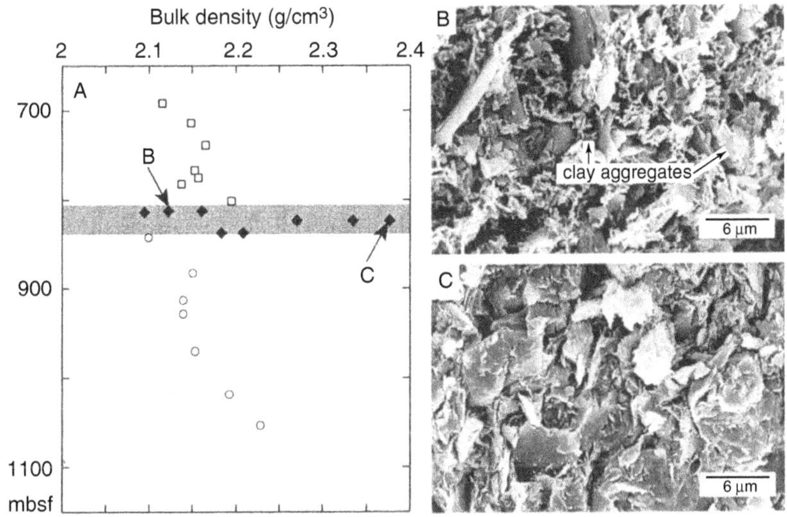

Figure 8.43 Relationships between sediment microstructure and density/porosity, ODP Site 1174. (A) Variability of the bulk density within the decollement (gray band). Arrows indicate location of SEM micrographs. (B) Microstructure of low-density (high porosity) sediment. (C) Microstructure of high-density (low porosity) sediment. Note the presence of clay aggregates and porous (dark) intervals in (B). Modified from Ujiie, K., Hisamitsu, T., Taira, A., 2003. Deformation and fluid pressure variation during initiation and evolution of the plate boundary decollement zone in the Nankai accretionary prism. Journal of Geophysical Research, 108, doi 10.129/2002JB002314.

FURTHER READING

Boillot, G., Huchon, P., Lagabrielle, Y., 2003. Introduction à la géologie: la dynamique de la lithosphere. Dunod, Paris.

Cadet, J.P., Uyeda, S. (Editors), 1989. Subduction zones: The Kaiko project. Tectonophysics, 160: 1–337.

Debelmas, J., Mascle, G., 2000. Les grandes structures géologiques. Dunod, Paris.

Einsele, G., 1992. Sedimentary basins, evolution, facies, and sediment budget. Springer, Berlin.

Kennett, J.P., 1982. Marine geology. Prentice-Hall, Englewood Cliffs, NJ.

Le Pichon, X. et al. (Editors), 1987. Project Kaiko. Earth and Planetary Science Letters, 83: 181–375.

Other references used in this chapter

Bourlange, S., Henry, P., Casey Moore, J., Mikada, H., Klaus, A., 2003. Fracture porosity in the decollement zone of Nankai accretionary wedge using Logging While Drilling resistivity data. Earth and Planetary Science Letters, 209: 103–112.

Davis, E.E., Becker, K., Wang, K., Obara, K., Ito, Y., Kinoshita, M., 2006. A discrete episode of seismic and aseismic deformation of the Nankai Trough subduction zone accretionary prism and incoming Philippine Sea plate. Earth and Planetary Science Letters, 242: 73–84.

Fryer, P., Coleman, P., Pearce, J.A., Stokking, L.B. (Editors), 1992. Proceedings of the Ocean Drilling Program, Scientific Results, volume 125. Ocean Drilling Program, College Station, TX.

Gulick, S.P.S., Bangs, N.L.B., Shipley, T.H., Nakamura, Y., Moore, G., Kuramoto, S., 2004. Three-dimensional architecture of the Nankai accretionary prism's imbricate thrust zone off cape Muroto,

Japan: prism reconstruction via en echelon thrust propagation. Journal of Geophysical Research, 109:, doi 10.129/2003JB002654.

Hawkins, J.W., Parson, L.M., Allan, J.F., Resig, J., Weaver, P. (Editors), 1994. Proceedings of the Ocean Drilling Program, Scientific Results, volume 135. Ocean Drilling Program, College Station, TX.

Huchon, P., Tokuyama, H. (Editors), 2002. Japan–France Kaiko-Tokai project: tectonics of subduction in the Nankai Trough. Marine Geology, 187: 1–220.

Ienaga, M., McNeill, L.C., Mikada, H., Saito, S., Goldberg, D., Casey Moore, J., 2006. Borehole image analysis of the Nankai accretionary wedge, ODP Leg 196: structural and stress studies. Tectonophysics, 426: 207–220.

Kastner, M., Le Pichon, X., (Editors), 1992. Fluids in convergent margins. Earth and Planetary Science Letters, 109: 275–506.

Lancelot, Y., Larson, R., Fisher, A. et al., 1990. Proceedings of the Ocean Drilling Program, Initial Reports, volume 129. Ocean Drilling Program, College Station, TX.

Mascle, A., Moore, J.C., Taylor, E., Underwood, M.B. (Editors), 1990. Proceedings of the Ocean Drilling Program, Scientific Results, volume 110. Ocean Drilling Program, College Station, TX.

Meschede, M., Zweigel, P., Kiefer, E., 1999. Subsidence and extension at a convergent plate margin: evidence for subduction erosion off Costa Rica. Terra Nova, 11: 112–117.

Meschede, M., Zweigel, P., Frisch, W., Völker, D., 1999. Mélange formation by subduction erosion: the case of the Osa mélange in southern Costa Rica. Terra Nova, 11: 141–148.

Moore, G.F., Taira, A., Klaus, A. et al., 2001. Proceedings of the Ocean Drilling Program, Initial Reports, volume 190. Ocean Drilling Program, College Station, TX.

Screaton, E.J., Saffer, D.M., 2005. Fluid expulsion and overpressure development during initial subduction at the Costa Rica convergent margin. Earth and Planetary Science Letters, 233: 361–374.

Silver, E.A., Kimura, G., Shipley, T.H. (Editors), 2001. Proceedings of the Ocean Drilling Program, Scientific Results, volume 170. Ocean Drilling Program, College Station, TX.

Taira, A., Moore, G.F., Becker, K., Mikada, H., Casey Moore, J., Klaus, A., 2005. Proceedings of the Ocean Drilling Program, Scientific Results, volume 190/196. Ocean Drilling Program, College Station, TX.

Tamaki, K., Suyehiro, K., Allan, J., Mc Williams, M. (Editors), 1992. Proceedings of the Ocean Drilling Program, Scientific Results, volume 127/128. Ocean Drilling Program, College Station, TX.

Ujiie, K., Hisamitsu, T., Taira, A., 2003. Deformation and fluid pressure variation during initiation and evolution of the plate boundary decollement zone in the Nankai accretionary prism. Journal of Geophysical Research, 108:, doi 10.129/2002JB002314.

Vannucchi, P., Tobin, H., 2000. Deformation structures and implications for fluid flow at the Costa Rica convergent margin, ODP Sites 1040 and 1043, Leg 170. Journal of Structural Geology, 22: 1087–1103.

Vannucchi, P., Scholl, D.W., Meschede, M., McDougall-Reid, K., 2001. Tectonic erosion and consequent collapse of the Pacific margin of Costa Rica: combined implications from ODP Leg 170, seismic offshore data, and regional geology of the Nicoya Peninsula. Tectonics, 20: 649–668.

Von Huene, R., Aubouin, J. et al., 1985. Initial Reports of the Deep Sea Drilling Project, volume 84, U.S. Gov. Print. Office, Washington.

Westbrook, G.K., Carson, B., Musgrave, R.J. et al., 1994. Proceedings of the Ocean Drilling Program, Initial Reports, volume 146. Ocean Drilling Program, College Station, TX.

CHAPTER NINE

Basins in a Context of Plate Collision

9.1. Structure, Tectonics and Sedimentation of Collision Areas

9.1.1. The Evolving Structure and Morphology of Collision Areas

Lithospheric plates may collide when convergence is not anymore compensated by subduction. Subduction is principally driven by a contrast of density between incoming and overriding lithospheres and the underlying asthenosphere, but ceases when lithospheres of similar densities are in contact. Collision events may therefore involve continental as well as oceanic lithospheres, and active island arcs as well as active continental margins. In most cases however, lithospheric plates include areas of oceanic and continental lithospheres, and plate convergence is accomodated through the subduction of oceanic lithosphere beneath continental lithosphere in active margin areas. When the oceanic lithosphere is completely subducted, the continental lithosphere engages within the subduction plane but resists entrainment, because of its low density and important thickness. Internal stress being still transmitted to plate boundaries, convergence is accomodated via thrusting, reverse faulting and folding, progressively building a collision prism (Figure 9.1). For example, northward migration of India as the Indian Ocean opened, was initially associated to rapid subduction of Eurasian oceanic lithosphere and reduction of the Tethys Ocean (see Section 2.1.6). Subduction ceased as the continental lithosphere of Eurasia resisted entrainment beneath India. Persistent convergence was then accomodated via major thrusts during a series of collision stages, leading to the formation of the Himalayas. However, collision is not uniform and areas of thrust belt may alternate with areas of subduction along active continental margins of convergent oceans. For example, collisional events started during the Cretaceous in

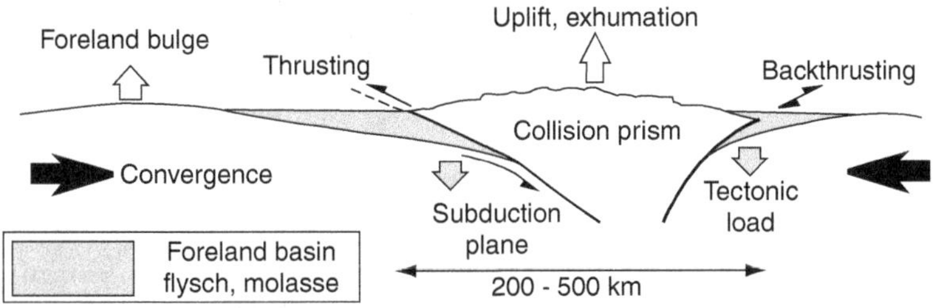

Figure 9.1 Schematic structure of a collision prism. Convergence is no longer accomodated via subduction, but via thrusting, backthrusting and uplift principally. Note overthrusting of the collision prism above the foreland basins. Modified from Lemoine, M., De Graciansky, P.-C., Tricart, P., 2003. De l'océan à la chaîne de montagnes. Gordon and Breach, Paris.

the Central and Eastern Alps, while subduction of European lithosphere beneath Apulia persisted to the West in the Paleogene.

During collision events the converging lithospheres are shortened and thickened. The deformation principally affects the underlying (previously subducted) lithospheric plate. Deep seismic investigations in the Western Alps (ECORS-CROP Project) suggest that deformation involves the upper lithospheric mantle in the inner parts of the collision prism (lithospheric prism) toward the overriding plate, the continental crust in the central part of the collision prism (crustal prism) and the sediment series only in the outer collision prism (sedimentary prism) toward the incoming plate (Figure 9.2). Only the brittle terranes of the lithosphere are found in the collision prism in most cases, the dense and ductile terranes of the lower crust being thinned and/or detached from their substrate and entrained at depth. Such a structure is the probable consequence of an increasing compression and a succession of reverse and thrust faults and decollements that affect deeper terranes from the outer to the inner parts of the incoming plate. Reverse faulting occurs within the brittle terranes of the upper crust and sediment series, for example in limestone and sandstone units. Thrusting and decollement principally affect the ductile terranes of the lower crust and sediment series, for example marlstones, claystones and evaporites. Therefore, the structure and morphology of collision prisms depend for a part on the composition of the sediment series of the incoming plate.

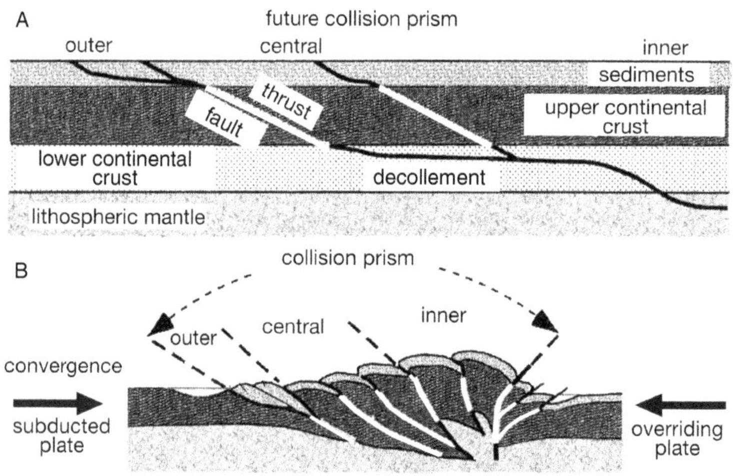

Figure 9.2 Schematic model of a collision prism. (A) Initially, a succession of major thrusts and decollements occur within the brittle and ductile parts of the lithosphere, respectively. (B) The distribution of major thrusts and decollements defines the main parts of the collision prism. The inner prism includes sections of lithospheric mantle, upper continental crust and sediments; the central prism includes sections of upper continental crust and sediments; the outer prism includes deformed sediments only. Note the absence of ductile lower continental crust from the collision prism, because of (almost) total subduction. Modified from Lemoine, M., De Graciansky, P.-C., Tricart, P., 2003. De l'océan à la chaîne de montagnes. Gordon and Breach, Paris.

The organization of collision prisms suggest that blocks of lithosphere, crust and/or sediments are detached along major fault systems, and underthrust beneath preceding structural blocks as collision progresses (Figure 9.3). As a consequence, the inner blocks are progressively uplifted as new blocks are detached and underthrust through time, and the continental crust significantly increases in thickness in the entire collision area. Hence, the collision prism grows from areas adjacent to the plates boundary to outer areas of the incoming plate. Maximum thicknesses are observed in the inner parts of the collision prism adjacent to the plates boundary,

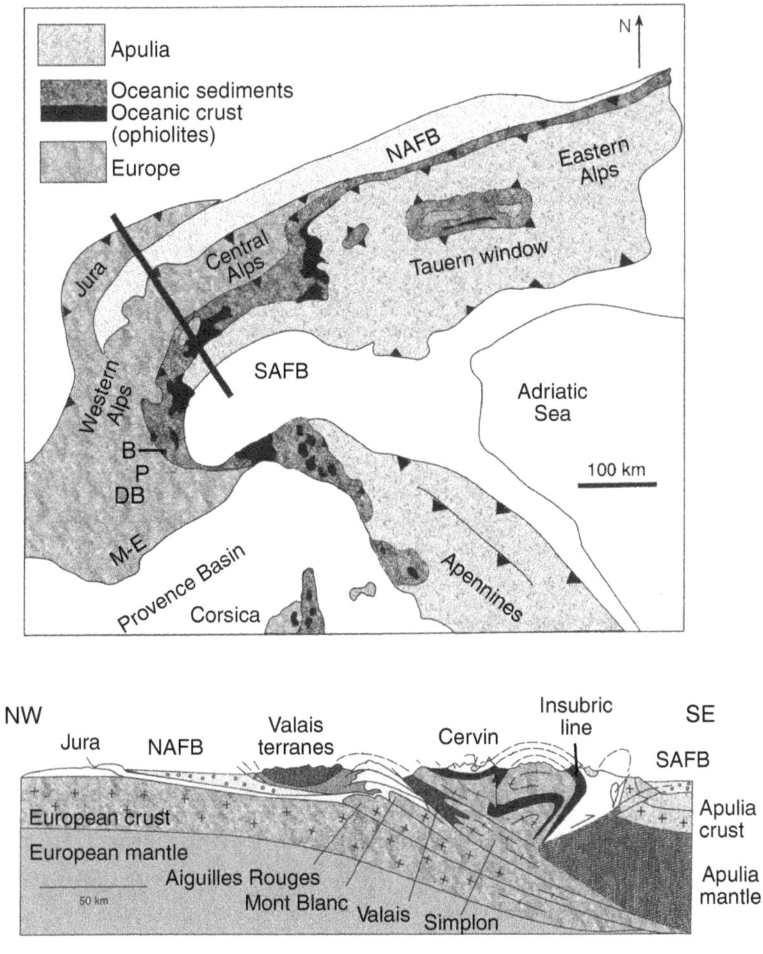

Figure 9.3 Schematic structure of the collision prism in the Western and Central Alps. (top) Distribution of incoming (European) and overriding (Apulian) terranes in the collision area. Note minor outcrop of deep oceanic sediments and oceanic crust. (bottom) Simplified cross section of the Central Alps. B, Briançonnais; DB, Digne Basin; M-E, Maures-Esterel; NAFB, North Alpine Foreland Basin; P, Pelvoux; SAFB, South Alpine Foreland Basin. Note the succession of underthrust slabs, the total thickness of continental crust in the central and inner parts of the prism, and backthrusting of European terranes over Apulia. Modified from Debelmas, J., Mascle, G., 2000. Les grandes structures géologiques. Dunod, Paris.

which also coincide with maximum relief. For isostatic reasons, however, the subaerial relief represents only a minor part of the increases in thickness and is associated to a deep root of continental crust, and related subsidence of the Moho to depths of 50–70 km. As collision progresses, crustal terranes and to some extent sedimentary rocks are transported to deeper areas of higher temperature and pressure where they are metamorphosed. Because of the diversity of initial lithologies and conditions of pressure and temperature, the rocks are altered into a variety of facies ranging from zeolite to greenschist, blueschist, amphibolite and eclogite. In the last stages of collision, backthrusting principally affects the inner parts of the collision belt (Figures 9.1 and 9.3). Although very simplistic, this conceptual model gives an idea of the main characteristics of a collision prism and the succession of major events that control its evolution. In most cases, the terranes of the underlying plate principally are involved in the construction of the collision prism, and are locally separated from the terranes of the overriding plate by ophiolite slices of oceanic crust. In some cases however, overthrust nappes of oceanic crust (ophiolites) or terranes derived from the overriding plate may overlap the collision prism (Figure 9.3).

9.1.2. The Accumulation of Sediments in Collisional Areas

Because of additional load exerted by the thickened continental crust during the construction of the collision prism, an increased subsidence affects a larger portion of lithosphere through time, creating narrow and elongated basins on each side of the collision belt: the foreland basins. To compensate, more distal areas of the undeformed foreland slightly uplift (Figures 9.1 and 9.3). Such basins are dissymetric, being deeper in areas contiguous to the collision prism and grading to shallower depths toward the undeformed foreland. Sedimentation in foreland basins may take place in marine or continental environments, as determined by the degree of evolution of the collision prism and related connection to the open sea. Because of active erosion of a growing collision belt, sediments in foreland basins are essentially siliciclastic in nature.

During the transition from subduction to collision and the early stages of collision, foreland sediment series are very similar to trench fills and mostly consist of turbidites and other gravity flows (flysch). The flysch generally consist of fine-grained series where shales dominate. Also, the nature of siliciclastic particles changes through time, as intense erosion of the collision prism causes the progressive exhumation of older and deeper crustal terranes. As a consequence of exhumation, the collision prism is further uplifted to re-establish isostasic equilibrium. As collision progresses, flysch units of the deep foreland basin are progressively thrusted and incorporated into the collision prism, and piggy-back basins may develop on top of the deformed flysch units. The advancing thrust fronts locally generate tectonic and sedimentary melanges, which include coarse-grained and conglomeratic elements (wildflysch).

During the advanced stages of collision, the flysch grade to shallow marine to continental sedimentary units including delta fan and alluvial fan facies principally, which fill the foreland basins and prograde over the undeformed foreland through time: the molasse. Early molasse maybe deformed as the collision prism progresses,

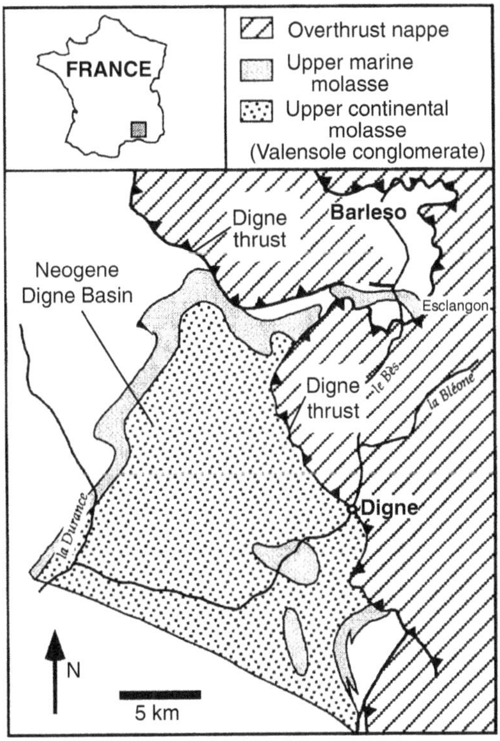

Figure 9.4 Simplified geological map of the Alpine Foreland Basin of SE France. Note that the nappe overlaps the foreland basin. Modified from Couëffé, R., Tessier, B., Gigot, P., Beaudoin, B., 2001. Le temps préservé sous forme de sédiments: résultats semi-quantitatifs obtenus dans la *molasse marine miocène* du bassin de Digne (Alpes-de-Haute-Provence, Sud-Est de la France). Comptes-Rendus de l'Académie des Sciences de Paris. Earth and Planetary Sciences, 332, 5–11.

or capped by overthrust nappes. For example, in the Neogene Digne Basin of southeastern France (Figure 9.4), Early to Middle Miocene marine molasse consists of shallow marine, tidal and estuarine deposits, capped by the continental and fluviatile proximal facies of the Late Miocene and Pliocene Valensole Conglomerates. The facies of the Digne Basin may reach up to 3,000 m in thickness, and are in part covered by Alpine overthrust nappes. In the late stages of collision, molasse basins form alluvial plains adjacent to the collision prism.

9.2. Closure of the Tethys and Collision in the Alps: A Brief Summary

9.2.1. Extension between Laurasia and Gondwana, and the Formation of the Western Tethys

The Alps are the result of a succession of collision events between two continental margins (Europe and Apulia) originally separated by an ocean, the Ligurian Tethys.

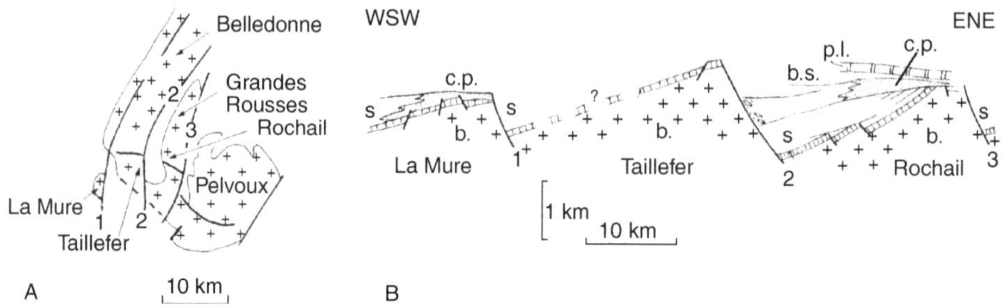

Figure 9.5 Reconstruction of the structure of the Ligurian Tethys passive margin in SE France. (A) Simplified structural map of the Pelvoux-Belledonne massif, indicating the current position of Early Jurassic structural highs. (B) Simplified cross section of the Tethys passive margin. b, basement; b.s., black-shale facies (Middle Jurassic); c.p., carbonate platforms (Early Jurassic); p.l., pelagic limestones (Late Jurassic); s., siliciclastics (Triassic to Early Jurassic). Note shortening of the margin. Modified from Boillot, G. et al., 1984. Les marges continentales actuelles et fossiles autour de la France. Masson, Paris.

Rifting started in the Triassic in the Ligurian Tethys and in the Central Atlantic, then parts of a single structure. The association of extension and strike-slip activities led to a succession of semi-isolated basins in the Ligurian Tethys (Figure 9.5). Syn-rift deposits include a variety of siliciclastics and breccias interbedded with volcanic products, evaporites (anhydrite, gypsum and mainly halite), dolomites and limestones, currently preserved between crustal blocks of the central collision prism (e.g., in the Briançonnais area of southeastern France). Rifting and related erosion accelerated near the Triassic/Jurassic boundary and, together with increased regional subsidence and expansion of marine environments in the Early Jurassic, led to thick accumulations (up to 2,000 m) of siliciclastics in the grabens (Figure 9.6). However, large portions of structural highs were emergent, as indicated by the local presence of karsts, bauxites and coal units, for example in the Briançonnais area. Between subaerial and marine environments, carbonate platforms progressed in nearshore environments on top of crustal blocks, where they persisted till the Middle Jurassic. By that time, shallow marine environments and carbonate platforms extended from the margins of the Tethys to the London and Paris Basin (see Section 4.3.3).

The crustal fissure stage probably started in the Middle Jurassic in the Ligurian Tethys (as in the Central Atlantic, see Section 6.3.1), based on the age of the oldest oceanic crust found in the Western Alps. The crustal fissure stage is principally characterized by the accumulation of thick Jurassic black-shale units (now included in the Terres Noires and Schistes Lustrés facies) in the basins. The transition to the oceanic stage occurred in the Late Jurassic, with the widespread transgression of pelagic facies (nannofossil limestones, radiolarites) above a major unconformity (Figure 9.5). The ocean reached maximum extension in the earliest Cretaceous (Figure 9.7) and was characterized by the presence of extensive carbonate platforms on continental shelves, and the dominance of hemipelagic sediments in the basins.

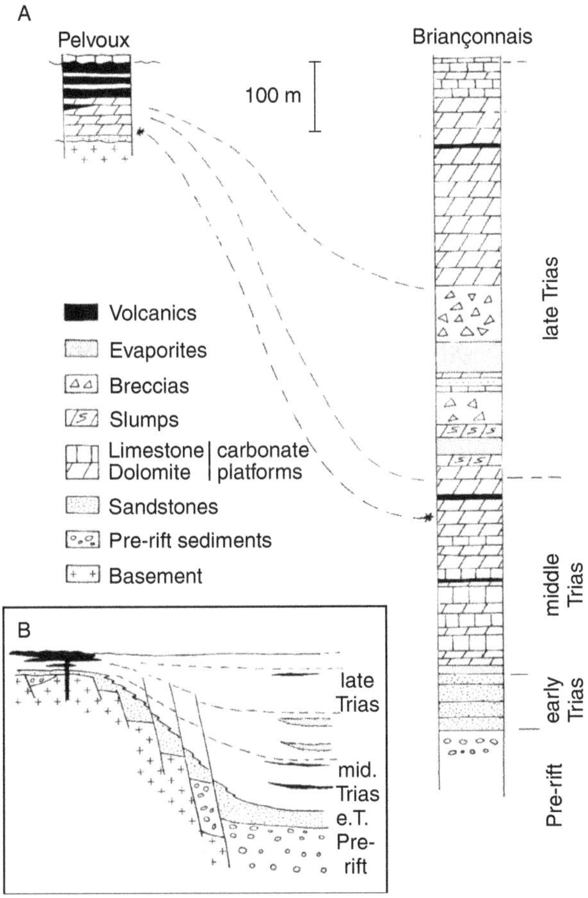

Figure 9.6 Syn-rift sediments of the Ligurian Tethys. (A) Comparison of syn-rift deposits in a graben basin (Briançonnais) and on a structural high (Pelvoux). (B) Reconstructed cross section highlighting the extensional regional context. Note similarities with modern rifts. Modified from Boillot, G. et al., 1984. Les marges continentales actuelles et fossiles autour de la France. Masson, Paris.

Beginning in the Early Cretaceous, extension was not solely accomodated via the production of oceanic crust, and a stage of rifting led to the formation of pull-apart basins (the future Valais Ocean) north of the Ligurian Tethys. This in turn caused a cessation of seafloor spreading followed by early compression within the Ligurian Tethys. The Valais Ocean extended from the northern part of the Tethys, or Meliata-Hallstatt Ocean (now the Carpathians) to the Bay of Biscay. Its formation is coeval with the separation of Iberia from the Grand Bank of Newfoundland and extension of the North Atlantic northward, through the Labrador Sea. The Valais Ocean isolated a region of continental crust extending from the Briançonnais to Corsica, Sardinia and Iberia beyond oceanic environments (Figure 9.7). However, oceanic crust is known from the wider, eastern part of the

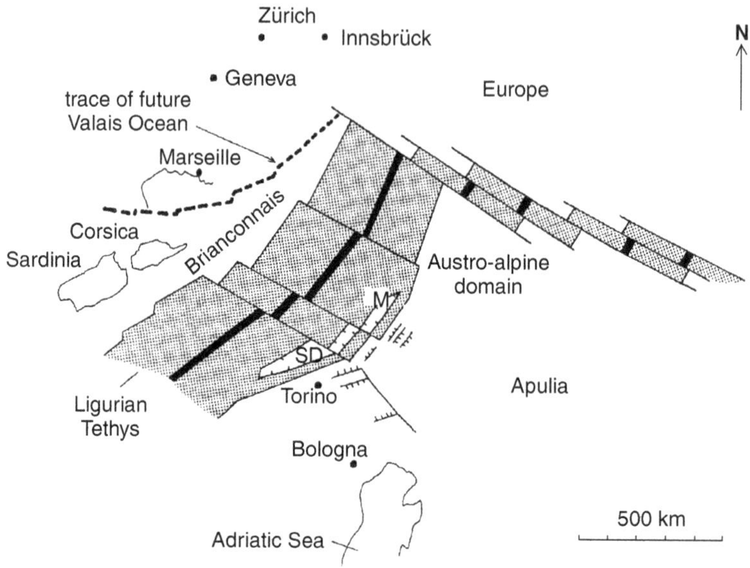

Figure 9.7 Paleogeographic reconstruction of the Ligurian Tethys Ocean at the time of maximum extension, near the Jurassic–Cretaceous boundary. M and SD, Margna and Sesia-Dent Blanche extensional areas. Note the succession of passive and transform margin areas. Modified from Schmid, S.M., Pfiffner, O.A., Froitzheim, N., Schönborn, G., Kissling, E., 1996. Geophysical-geological transect and tectonic evolution of the Swiss-Italian Alps. Tectonics, 15, 1036–1064.

Valais Ocean and from the Bay of Biscay only, the rest of the structure consisting of thinned continental lithosphere.

9.2.2. Convergence in the Western Tethys, and Related Collisional Events of Cretaceous and Paleogene Age

Subduction of the Ligurian Tethys beneath Apulia was already active in the Middle Cretaceous when convergence between Europe and Africa began around 100 Ma, progressively leading to complex tectonics in an area which now extends from central Europe to North Africa. Early convergence was NE–SW oriented and was associated to the development of significant accretionary wedges in the Eastern and Central Alps, which later gave flysch nappes (e.g., the rheno-danubian flysch). The convergence also produced a bulge of the subducted plate, leading to the emersion of the shallowest marine areas, for example in Provence (SE France). The early stage of convergence ended with a collision event of Cenomanian–Turonian age, which principally affected the Eastern Alps and Carpathians where it corresponds to the closure of the Meliata-Hallstatt Ocean. Collisional deformation was associated to the earliest Alpine nappes.

Convergence resumed in a N–S direction after a reorganization in the rotation of the plates, and compression principally concerned the Central and Western Alps for the remaining of the Late Cretaceous whereas some extension occurred in the

Eastern Alps where grabens locally formed and filled with siliciclastics (e.g, the Gosau basins of Austria). Flysch series accumulated along the active margins of the Ligurian Tethys, and in the Valais Ocean, which closed during the Late Cretaceous. Collision deformation started in the Western and Southern Alps around 80 Ma in the Campanian, and persistent convergence led to major collisional events in the Paleogene. The closure of the Ligurian Tethys was complete in the Paleocene, when Briançonnais terranes entered the subduction zone (Figure 9.8). Briançonnais continental terranes were totally underthrust in the Early Eocene, as well as parts of the Valais Ocean. By 50 Ma, the penetration of the southern tip of continental Europe in the subduction zone marks the beginning of a major collision. Deeply underthrust Briançonnais terranes reached their peak depth and pressure at that time. This succession of major collisional and related events lasted till 35 Ma in the Late Eocene (Pyrenean tectonic phase, or Meso-Alpine stage), but post-collisional tectonics persisted till the Early Miocene. Early phases of exhumation, most likely related to forced extrusion of underthrust slabs and subsequent erosion, were associated to decreased pressure and the onset of temperature-dominated metamorphism at 40–35 Ma. Late Eocene collisional events are observed from the Alps to the Pyrenees and the Betics in Europe, and to the Atlas in North Africa.

Early foreland basins developed in the Late Cretaceous, as an isostatic response to an increasing load applied by a thickened continental crust to the lithosphere. The onset of foreland basin sedimentation in the North Alpine Foreland Basin is marked by a change from hemipelagic deposits to siliciclastic flysch deposits on incoming European terranes. The oldest, onlapping foreland deposits overlay unconformities and decrease in age northward in western Switzerland, from the Santonian in the inner areas to the Middle Eocene in the outer areas of the incoming plate: as collision progressed, flexural subsidence migrated toward the European foreland ahead of the orogenic load. However, deep-water environments prevailed till the Eocene (underfilled stage) in the inner basin, whereas shallow carbonate platform environments characterized the outer basin. The oldest molasse deposits (lower marine molasse) consist of shallowing and coarsening upwards siliciclastics, which accumulated in shallow water and delta fan environments (overfilled stage) during the Late Eocene and the Early Oligocene (Figure 9.9). In the Piedmont foreland basin of Italy, south of the Central Alps, the oldest siliciclastics are of Late Eocene age. They consist of continental to shallow marine conglomerates and sandstones to the west, and of deep marine, turbiditic sandstones in the eastern sector of the basin. Beginning during the Late Meso-Alpine stage in the Late Eocene, foreland sedimentation clearly reflects flexural basin conditions related to the orogenic load of the collision prism on both sides of the Central Alps.

In southeast France, then separated from the North Alpine Foreland Basin by an Apulian indentor, most areas have been emergent since the latest Cretaceous. There, the oldest foreland basin sediments are of Middle Eocene age. The base of the sediment series is marked by a unconformity, locally capped by alluvial conglomerates. The Middle Eocene to earliest Oligocene succession starts with the Nummulitic Limestones, which range in age from Lutetian in the inner areas of the basin to Priabonian in the outer foreland and accumulated in transgressive shelf

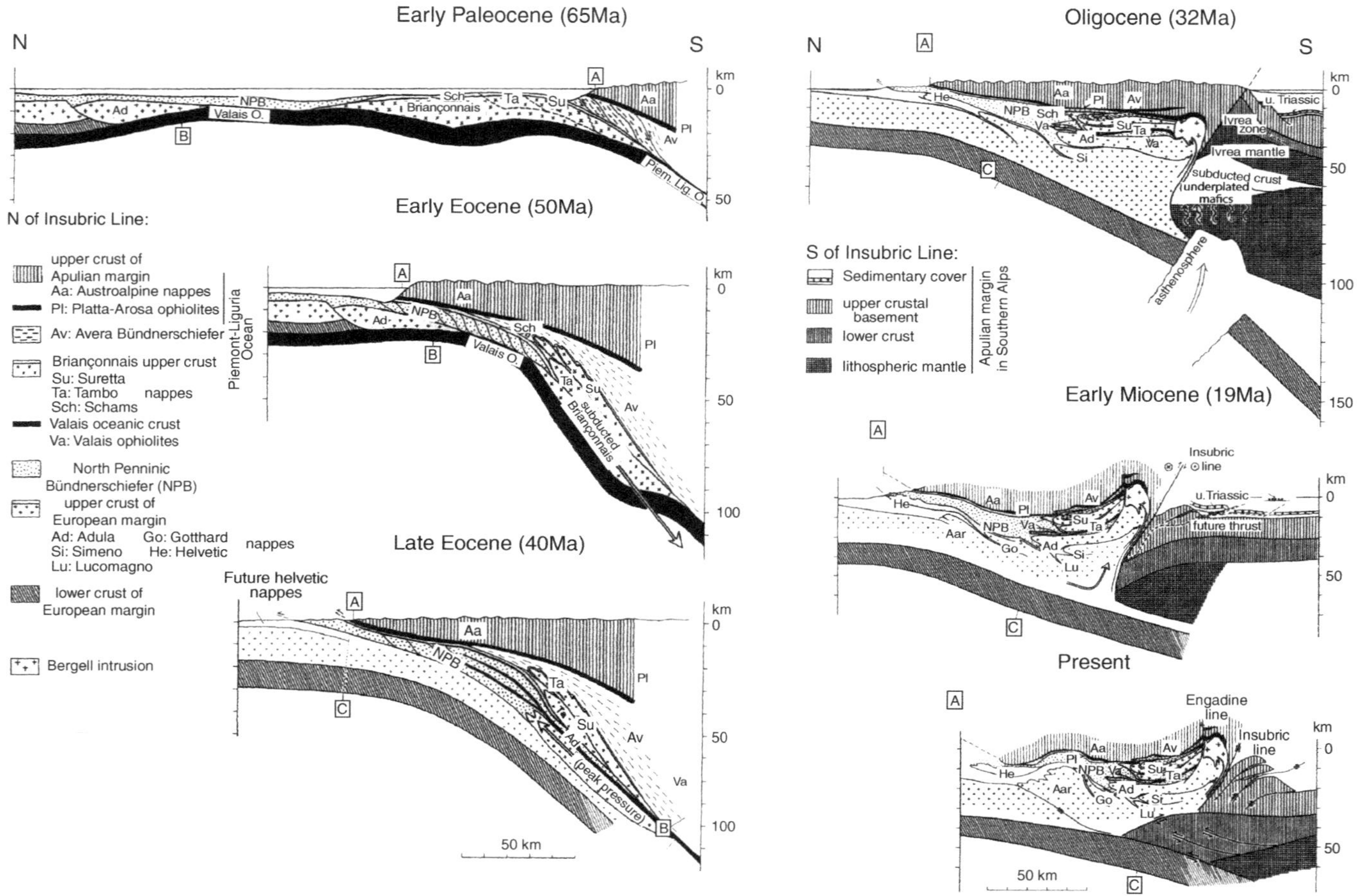

Figure 9.8 Kinematic evolution of the Central Alps, from Early Paleocene subduction to Late Eocene collision and subsequent post-collisional shortening. Letters in boxes are guide marks for estimating the degree of convergence and shortening during the Cenozoic. Modified from Schmid, S.M., Pfiffner, O.A., Froitzheim, N., Schönborn, G., Kissling, E., 1996. Geophysical-geological transect and tectonic evolution of the Swiss-Italian Alps. Tectonics, 15, 1036–1064.

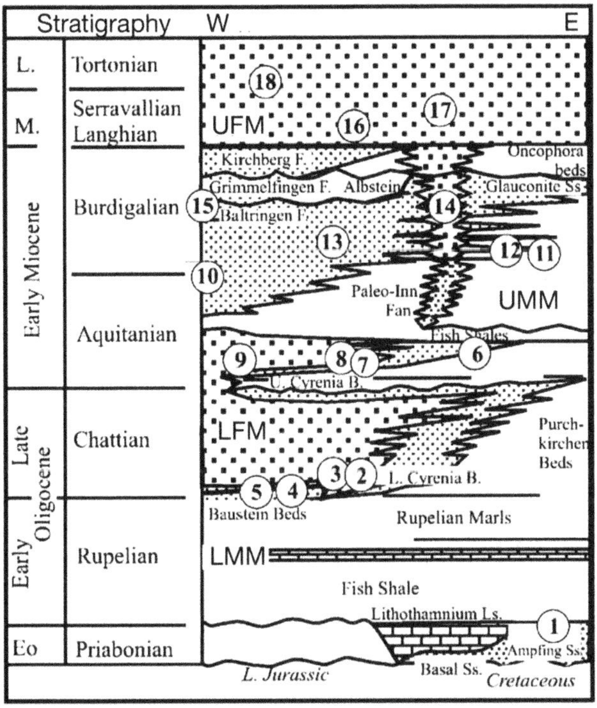

Figure 9.9 Typical succession of facies in the North Alpine Foreland basin. LMM, lower marine molasse; LFM, lower freshwater molasse; UMM, upper marine molasse; UFM, upper freshwater molasse. Numbers represent the position of the samples used for age determination. Note the presence of limestones (bricks) in the LMM facies, the abundance of siliciclastic facies (dots) to the west and overall upward increase in siliciclastic grain-size in each marine/freshwater cycle from marl/claystone (white) to fine sand (small dots) and coarse sands/conglomerates (big dots). Modified from Kuhlemann, J., Dunkl, I., Brügel, A., Spiegel, C., Frisch, W., 2006. From source terrains of the Eastern Alps to the Molasse Basin: detrital record of non-steady-state exhumation. Tectonophysics, 413, 301–316.

environments. The Nummulitic Limestones are overlain by the hemipelagic Globigerina Marls, which reflect neritic to middle bathyal environments of 100–900 m, increasing in depth through time. The Globigerina Marls are overlain by thick (up to 1,200 m), turbidite-dominated marine siliciclastics, the Annot and Champsaur Sandstones. The age of the transition is diachronous, being Priabonian in the inner areas of the basin and Rupelian in the outer foreland. The siliciclastics are derived from the Maures-Esterel massif to the south and from the Pelvoux massif to the north, which were uplifted and exhumed during the Late Meso-Alpine tectonic phase. The Middle Eocene to earliest Oligocene interval was principally characterized by marine environments in most foreland basin areas north and west of the Alps.

The Meso-Alpine collision stage was associated to a reorganization of the plates relative motion, which involved a counterclockwise rotation of Apulia and a change in convergence direction from N–S to NW–SE. As a consequence, the Oligocene was a time of post-collisional shortening in the Central Alps where northward progression of the deformation was associated to backthrusting along the Insubric Line around 32 Ma (Figure 9.8). This backthrusting led to rapid exhumation and erosion of the Penninic nappes of metamorphosed upper continental crust. Because of increased erosion, prograding siliciclastic facies accumulated in alluvial fan environments principally (lower freshwater molasse), onto shallow marine sediments of the North Alpine Foreland Basin (Figure 9.9). In the Piedmont Basin of Italy (or South Alpine Foreland Basin) a thick clastic wedge including coarse and fine siliciclastics accumulated in a subsident marine environment, as a mechanical response to backthrusting and emplacement of the Southern Alpine nappes.

In the Western Alps of southeast France, thrust sheets made of Jurassic and Cretaceous carbonate lithologies (initially parts of the Tethys passive margin) and flysch were translated toward the southwest above Triassic evaporites during the entire Late Alpine convergence. The Embrunais-Ubaye nappes, derived from inner areas of the prism, progressed in submarine conditions onto the foreland. However, the Alpine orogenic load no longer caused a flexure of the European plate in this area and the foreland was characterized by WNW–ESE extension, like other areas of the European plate. Extensional stress within the European plate generated the Cenozoic European Rift System (see Section 3.2) and controlled the opening of the Ligurian Sea, i.e. the Provence Basin of the Western Mediterranean. As a consequence, sedimentation in most of the foreland was limited to isolated, small and structurally controlled rift basins and synclines where Oligocene deposits never exceed 600 m in thickness, and ceased in the Early Miocene. Siliciclastic facies of predominantly continental affinities are locally associated to volcanic materials of calc-alkaline affinity. In the outer foreland, however, graben-type basins like the Digne Basin accomodated up to 2,000 m of Oligocene lacustrine and evaporitic sediments.

Further south, the Late Eocene collision of Apulia and Iberia (now attached to Europe) along the Betic Front was associated to an interval of deformation (folding, reactivation of ancient faults) in the Atlas and followed by a major change in plate motion: eastward subduction of Europe beneath Apulia ceased and was replaced by westward subduction of Apulia's oceanic lithosphere beneath the continental lithosphere of Alboran, Kabylia, Peloritan and Calabria (AlKaPeCa terranes) which remained attached to Europe (Figures 9.10 and 9.11). This subduction zone (Apenninic subduction zone) started migrating eastward in the Late Oligocene and the related accretionary wedges allowed the accumulation of the Algibe and Numidian flysch, now uplifted from the Betics to Sicily.

9.2.3. Neogene Collisional Events and the Formation of the Western Mediterranean Sea

A third stage of post-collisional shortening took place during and after the Early Miocene in the Central Alps and thrusting also involved the Molasse Basin. The thickness of the collision prism increased significantly, as more parts of the European margin entered the subduction zone: elements of upper continental crust and overlying sediments were detached and incorporated into the collision prism,

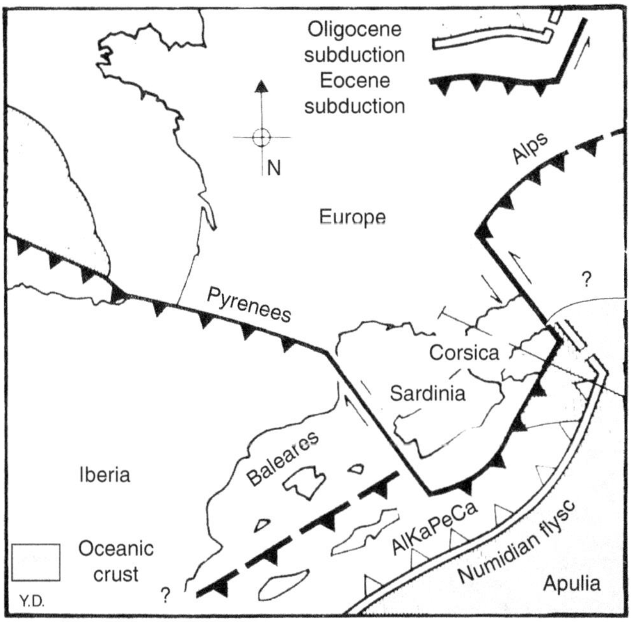

Figure 9.10 Schematic paleogeographic reconstruction of the Southern Alpine and Mediterranean domains in the Oligocene. Note integration of Iberia and AlKaPeCa (Alboran, Kabylia, Peloritan and Calabria) terranes to Europe in the Late Eocene, and related changes in subduction. Modified from Rehault, J.-P., Boillot, G., Mauffret, A., 1984. The Western Mediterranean basin geological evolution. Marine Geology, 55, 447–477.

whereas elements of ductile lower continental crust were still subducted (Figure 9.8). As a consequence, erosion and tectonic exhumation of deep terranes accelerated (especially from the Aar to the Mont-Blanc and Belledonne massifs), to a maximum of Burdigalian to Serravallian age (Figure 9.12). Terrigenous elements derived from those areas shed into the North Alpine Foreland Basin around 20 Ma in the Burdigalian, rapidly filling the marine Molasse Basin (Figure 9.9). The upper marine molasse accumulated till 17–18 Ma when shallow water environments were replaced by alluvial fan deposits of the upper freshwater molasse. In the Piedmont foreland basin (South Alpine Foreland Basin), marine environments persisted for the entire Early and Middle Miocene. However, a succession of submarine lobes made of massive sandstones coarsening upward are overlain by about 1,000 m of alternating conglomerates, sandstones and mudstones typical of proximal delta fan environments, and capped by a submarine channel-levee complex made of mixed siliciclastics of Middle Miocene age.

Further south, extensional stress persisted within the European foreland: Oligocene grabens west of the migrating Apenninic subduction zone (Provence Basin, Valencia Trough, Alboran Basin) evolved into initial back-arc basins during the Early Miocene (Figure 9.11), whereas a series of grabens developed in a fore-arc position (e.g., the Vavilov Basin of the Tyrrhenian Sea) and filled with syn-rift series

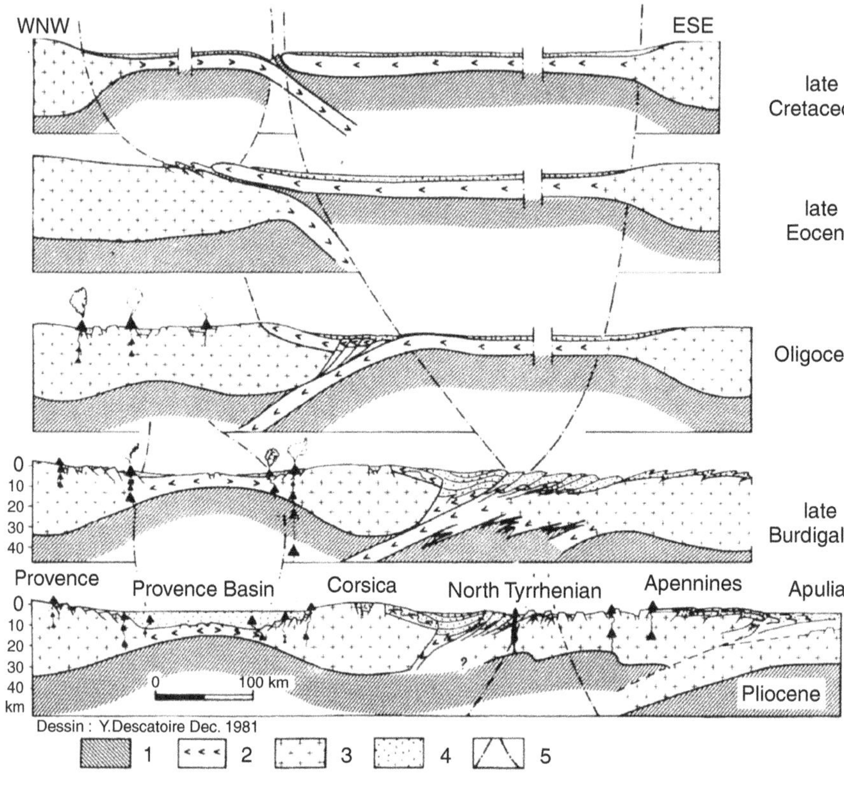

Figure 9.11 Geodynamic evolution along a WNW–ESE transect extending from Provence to Corsica and the Apennines (see Figure 9.10). 1, lithospheric mantle; 2, oceanic crust; 3, continental crust; 4, sediments; 5, limits of basins. Note the inversion in the direction of subduction after the Late Eocene Alpine collision and Corsican obduction and successive openings of the Provence Basin and Tyrrhenian Basin in back-arc position as the subduction area migrated eastward. Modified from Rehault, J.-P., Boillot, G., Mauffret, A., 1984. The Western Mediterranean basin geological evolution. Marine Geology, 55, 447–477.

of Burdigalian age. For example, oceanic crust formed in the Provence Basin from about 19 Ma to 16–17 Ma in the Burdigalian, while the Corsica-Sardinia block rotated by about 60° counterclockwise. At DSDP Site 372 near Menorca, the lowermost sediment consists of siltstones and mudstones, which graded to marlstones and to nannofossil marls during the Burdigalian as the Provence Basin expanded and subsided (Figure 9.13). Thermal subsidence at regional scale also facilitated a marine transgression that migrated northwards from the Gulf of Lion to the North Alpine Foreland Basin in the Burdigalian. About 1,000 m of marine siliciclastics, locally interbedded with fluvial or lacustrine facies, accumulated from the Burdigalian to the Serravallian in the Digne Basin of SE France (Figure 9.4), where those upper marine series unconformably overlay Albian sedimentary units. Spreading ceased in the Provence Basin at 16–17 Ma when the Corsica-Sardinia and AlKaPeCa blocks collided with the continental lithosphere of Apulia, creating the Apennines.

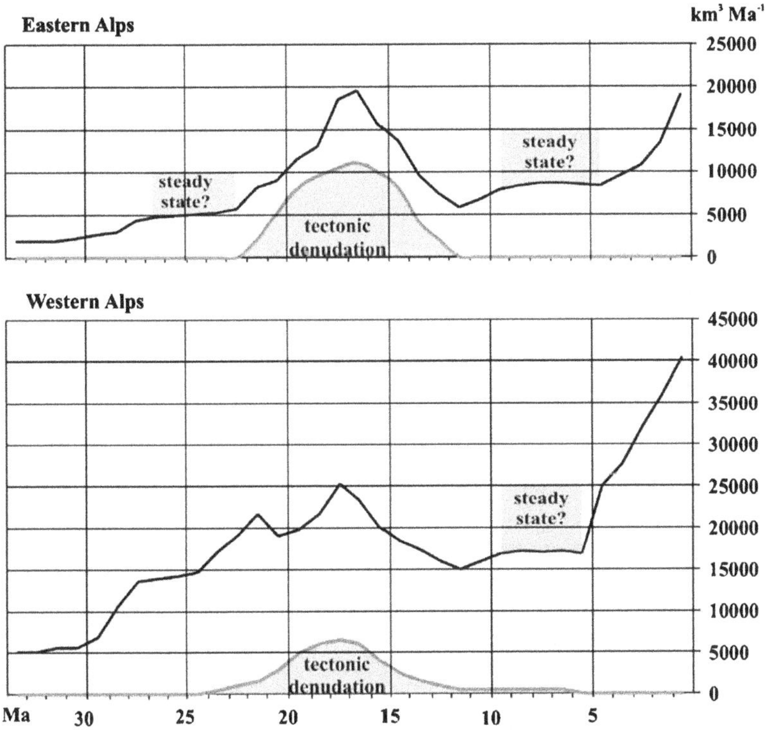

Figure 9.12 Bulk denudation rates of the Eastern and Western Alps, separated for tectonic denudation (exhumation) and erosion. Note Early to Middle Miocene peak of tectonic denudation and erosion, and strong increase in erosion since the latest Miocene. Reprinted from Kuhlemann, J., Dunkl, I., Brügel, A., Spiegel, C., Frisch, W., 2006. From source terrains of the Eastern Alps to the Molasse Basin: detrital record of non-steady-state exhumation. Tectonophysics, 413, 301–316.

After the collision of Kabylia and Africa in the Burdigalian, a further extensional episode around 15–17 Ma preceded the opening of the Algeria Basin of the Western Mediterranean Sea. The Algeria Basin developed south of the Provence Basin by westward and eastward rollback of the subduction zone. Westward rollback generated the Gibraltar arc and related migration of the Rif and Alboran Basin to their current positions. Strike-slip motion occurred along the south Balearic and the Maghrebian margins. Eastward rollback generated the Calabria arc and oceanic crust expanded west and east from the Hannibal Ridge spreading center. More than 400 km of oceanic crust were already created in the Western basin and only 160 km in the Eastern basin when westward motion ceased in the Late Tortonian around 8 Ma (Figure 9.14). The Tortonian was a time of major reorganization in the Western Mediterranean: because of a collision event between the Kabylia blocks and Africa, the convergence of Africa and Europe was no longer absorbed within the Numidian Trough. A major compressive phase resulted in large-scale thrusting and faulting in the Maghreb, whereas the northern Kabylia margin became transpressive. By the same time, eastward rollback of the Calabria arc toward the last

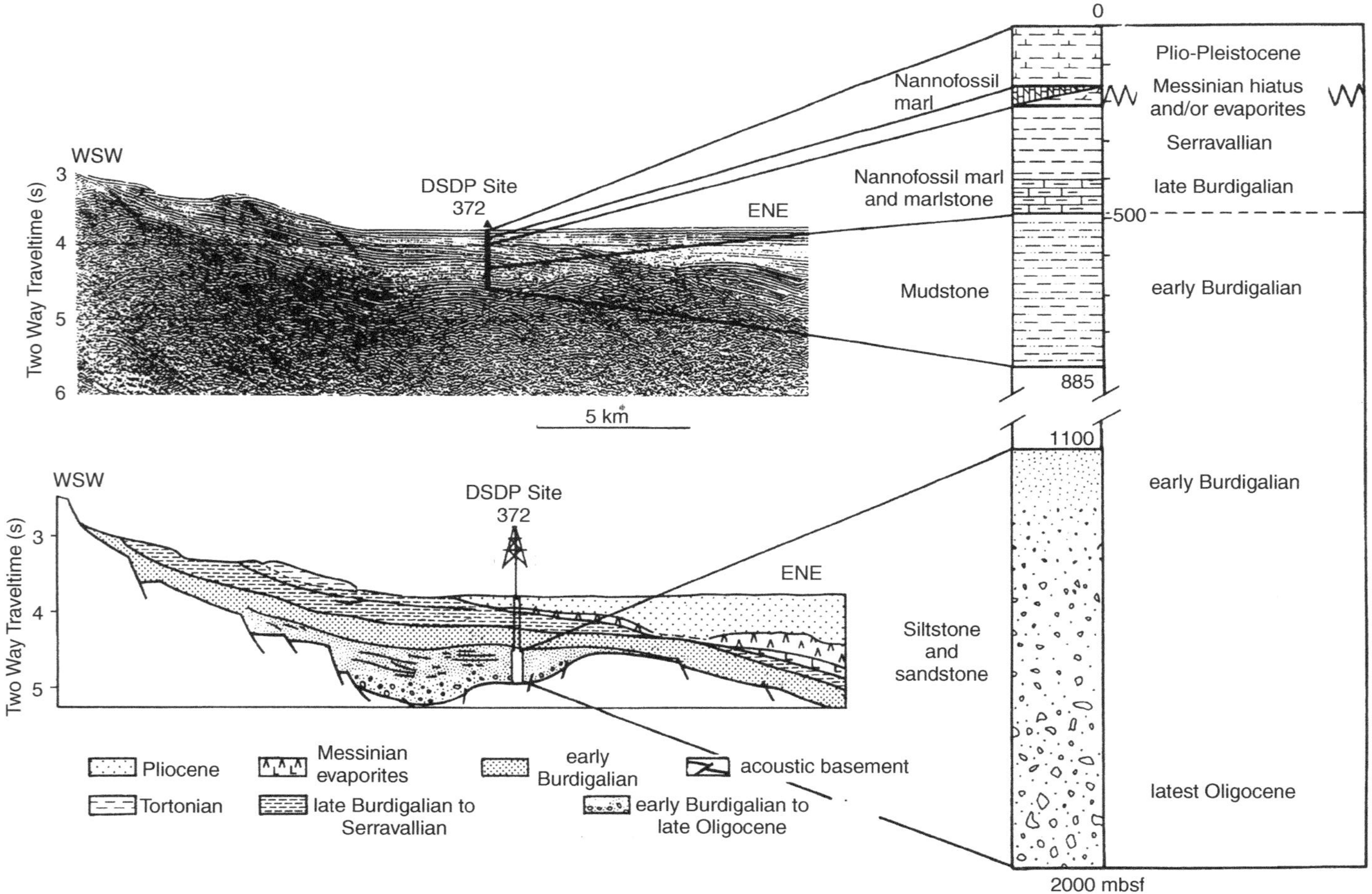

Figure 9.13 Structure of the eastern Menorca margin of the Provence Basin and major sedimentary facies at DSDP Site 372. Note Late Oligocene filling of grabens, Early Miocene subsidence and transgression, and Messinian evaporites in continuity with an erosional surface. Reprinted from Rehault, J.-P., Boillot, G., Mauffret, A., 1984. The Western Mediterranean basin geological evolution. Marine Geology, 55, 447–477.

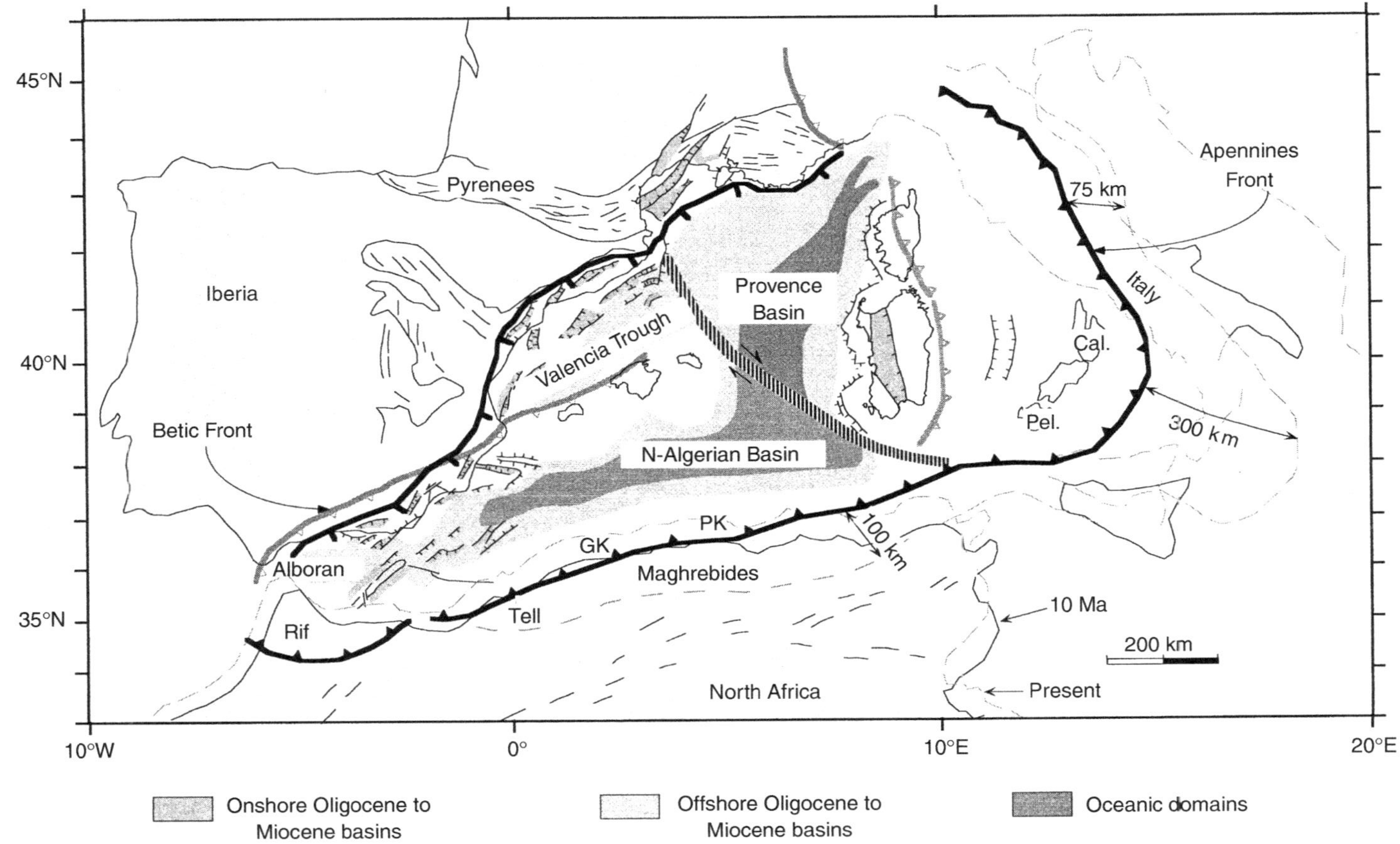

Figure 9.14 Late Tortonian paleogeography of the Western Mediterranean. GK and PK, Kabylia; Pel, Peloritan; Cal, Calabria. Note full opening of the Provence and Algeria basins, collision of Kabylia and Africa, and distance of the eastern Calabria arc of the Apennine subduction zone from its current position. Reprinted from Gueguen, E. Doglioni, C., Fernandez, M., 1998. On the post-25 Ma geodynamic evolution of the western Mediterranean. Tectonophysics, 298, 259–269.

remaining area of Mesozoic oceanic crust in the Ionian Sea accelerated. Related back-arc extension further separated the Corsica-Sardinia block from the Peloritan-Calabria blocks and created the Tyrrhenian Sea (Figure 9.11), where oceanic crust was generated during the Late Miocene to Early Pliocene in the northwestern Vavilov Basin, and during the Late Pliocene in the southeastern Marsili Basin.

The Western Mediterranean succession of terrigenous and hemipelagic sediments was interrupted in the Late Messinian when thick evaporitic series (gypsum, anhydrite, dolomitic marls) accumulated in most basins, which were more than 3,000 m deep at that time. On the margin of the Provence Basin near Menorca, evaporitic facies at DSDP Site 372 (Figure 9.13) contain lagoonal benthic faunas indicating paleodepths of 300–500 m. The association of geodynamical and paleoenvironmental data indicates that the Messinian evaporitic series accumulated in relatively shallow conditions of an already deep (crustal fissure to mature oceanic stage) oceanic basin. Low sea levels caused the development of the Messinian erosional surface, which truncated the continental margins, leading to the formation of deep canyons. Traces of Messinian erosion are found in the entire drainage basins, up to the Alpine foreland (Figure 9.15). In the shallower Tyrrhenian basins still in their rift to crustal fissure stage, the salinity crisis started at ODP Site 654 with the deposition of laminated, dolomitic, organic-rich and predominantly siliciclastic sediments grading upwards to alternances of evaporites and fine siliciclastics (Figure 9.16). Hemipelagic facies were restored in the earliest Pliocene, as Mediterranean sea level rose rapidly. It is probable that geodynamic conditions, i.e. the Late Tortonian cessation of the western rollback of the Gibraltar arc and compressive events in the Maghreb and the Betics, created a morphology favorable to an isolation of the Western Mediterranean. However, the trigger of the Mediterranean salinity crisis was a significant drop of sea level, which started around 5.9–5.8 Ma in the Late Messinian. Since low sea levels persisted after 5.3 Ma, it is likely that the restoration of open marine conditions and hemipelagic facies in the Western Mediterranean at 5.33–5.32 Ma was rather of tectonic origin. Marine environments progressed into deeply incised drainage basins, far in the continental hinterland.

During the Miocene, tectonic activity of the Alpine system progressively concentrated on Mediterranean regions that accomodated most of the plate convergence. However, exhumation persisted in the Central and Western Alps where erosion significantly increased in the Pliocene, further feeding the continental molasse basins (Figures 9.9 and 9.12). In SE France, the reactivation of the Digne thrust system in the Late Tortonian, synchronous with the exhumation of the Argentera and Pelvoux massifs, caused a new influx of coarse siliciclastics into the Digne Basin (Figure 9.4). The basin was rapidly filled, and the accumulation of conglomerates and sands persisted in alluvial fan environments (upper continental series, or Valensole Conglomerate) during the Pliocene.

The closure of the Tethys was a long and complex geodynamical process, which started in the Cretaceous and is not completed yet. The N–S convergence of Africa and Europe initiated the process and exerted a major control on subduction and collision till the Eocene. However, the importance of NW–SE and E-W convergence exceeded by five to eight times that of N–S convergence since

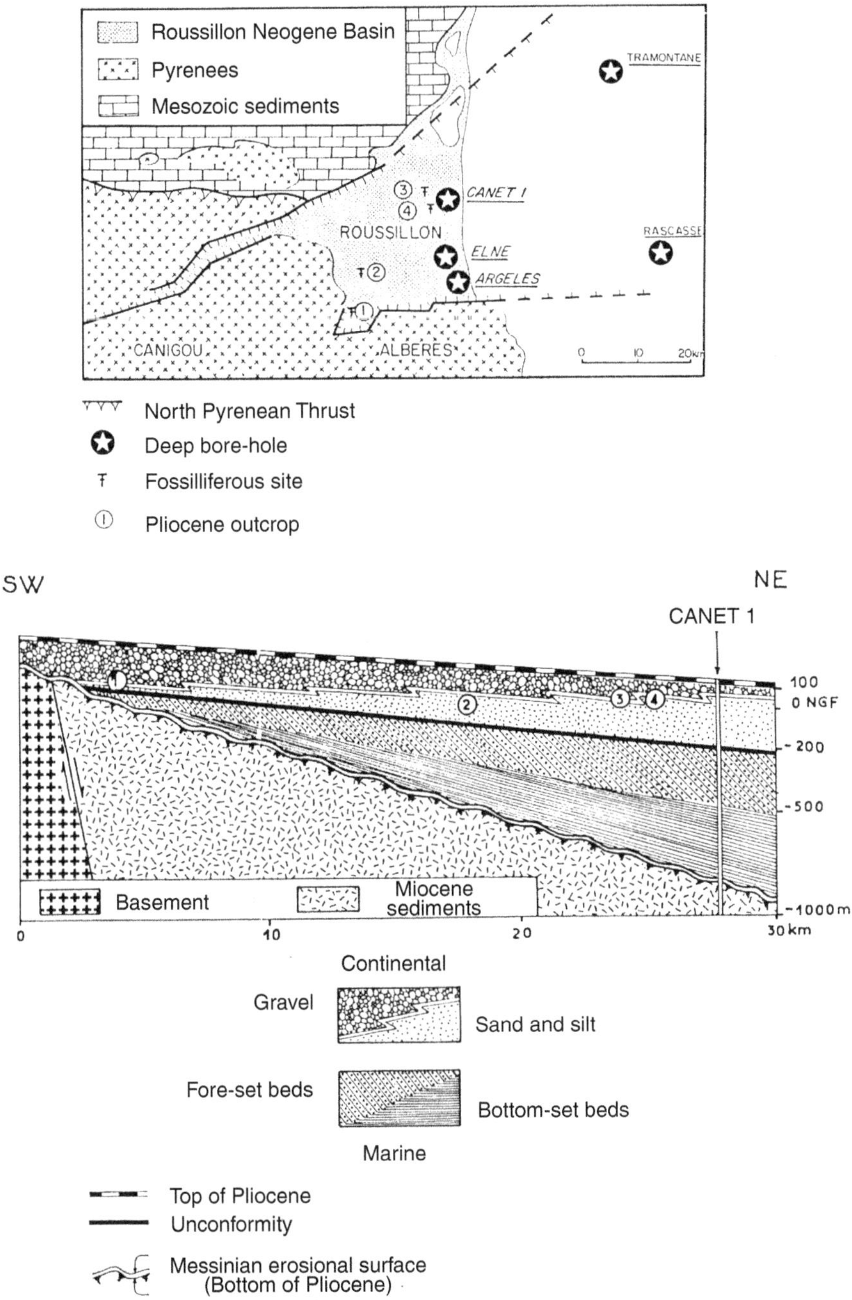

Figure 9.15 The Messinian erosional surface in the Roussillon Basin (Catalonia, SE France). (Top) Map of the Roussillon Neogene Basin and site location. (Bottom) Cross section of the Roussillon Basin highlighting the Messinian Erosional Surface and the Pliocene transgressive facies. The Roussillon Basin is a graben-like basin of Oligocene age filled with continental and Early/Middle Miocene marine siliciclastics eroded from the Pyrenees. Messinian low sea levels favored a deep erosion of the basin, the erosional surface being capped by Pliocene transgressive marine facies. Siliciclastics prograded in marine and continental environments later in the Pliocene. Modified from Clauzon, G., Aguilar, J.-P., Michaux, J., 1987. Le bassin pliocène du Roussillon (Pyrénées-Orientales, France): exemple d'évolution géodynamique d'une ria méditerranéenne consécutive à la crise de salinité messinienne. Comptes-Rendus de l'Académie des Sciences de Paris. Earth and Planetary Science, 304, 585–590.

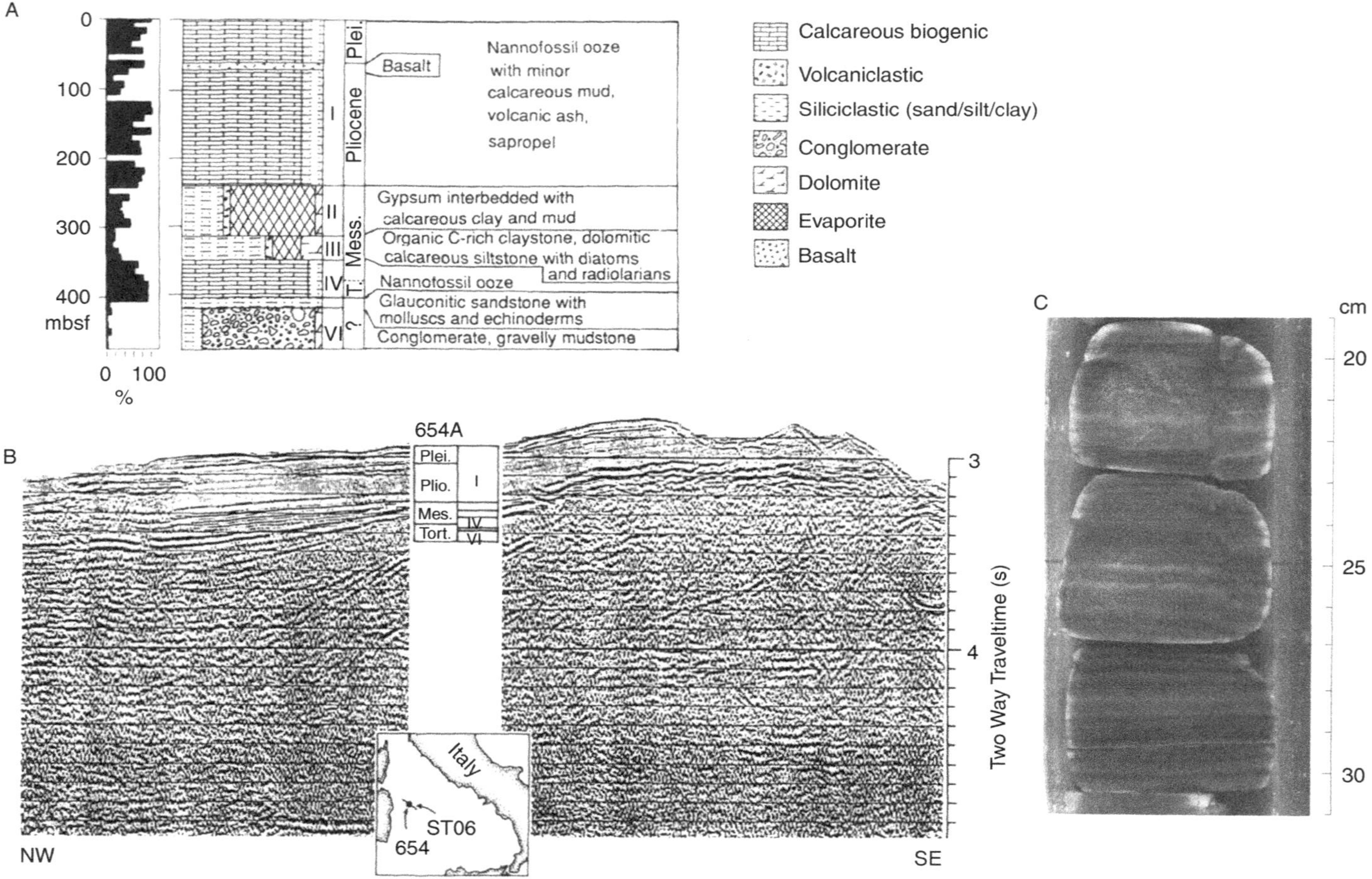

Figure 9.16 Sediments of the Tyrrhenian Sea. (A) Lithostratigraphy of sediments drilled at ODP Site 654 on the Sardinia margin (left, core recovery). (B) Reflection seismic profile across the Sardinia margin. Lithostratigraphy and seismic profile suggest that the transition from syn-rift to post-rift series occurs at 300–320 mbsf in the Late Messinian. (C) Typical finely laminated balatino-type gypsum, Section 107-654A-28R1, 260 mbsf, Late Messinian. Modified from Kastens, K.A., Mascle, J., Auroux C. et al., 1990. Proceedings of the Ocean Drilling Program, Scientific Results, volume 107, Ocean Drilling Program, College Station, TX.

the Oligocene, and shaped the current morphology of Southern Alpine and Mediterranean regions.

9.3. Example of a Paleo-Margin: The Mesozoic African-Tethyan Margin of Tunisia

9.3.1. A Brief History of the Tunisian Margin of the Tethys

Modern Tunisia is located at the intersection of the margin of the Mediterranean Ionian Basin, which has been passive since the Cretaceous at least, and the Atlas thrust belt which principally result from the Neogene collision of AlKaPeCa terranes with Africa (Figure 9.17). The collision belt extends from northern Tunisia to Algeria (Tell) and Morocco (Rif). To the south, the Tunisian Atlas is mainly the deformed foreland of the collision prism. In the Tunisian Atlas, folded sediment series made of carbonates principally have been detached from their substrate along Cretaceous and Triassic decollement zones. This domain is bounded to the south by the South Atlas Fault, and to the East by the North–South Axis (Figure 9.17). The South Atlas Fault marks the limit of the Saharan Platform, a monocline slightly dipping southwestward. The northern Saharan Platform (Dahar Plateau) was part of the southern Tethyan margin during the Mesozoic and most of the Cenozoic. East of the Tunisian Atlas, the North–South Axis forms the deformation front at the

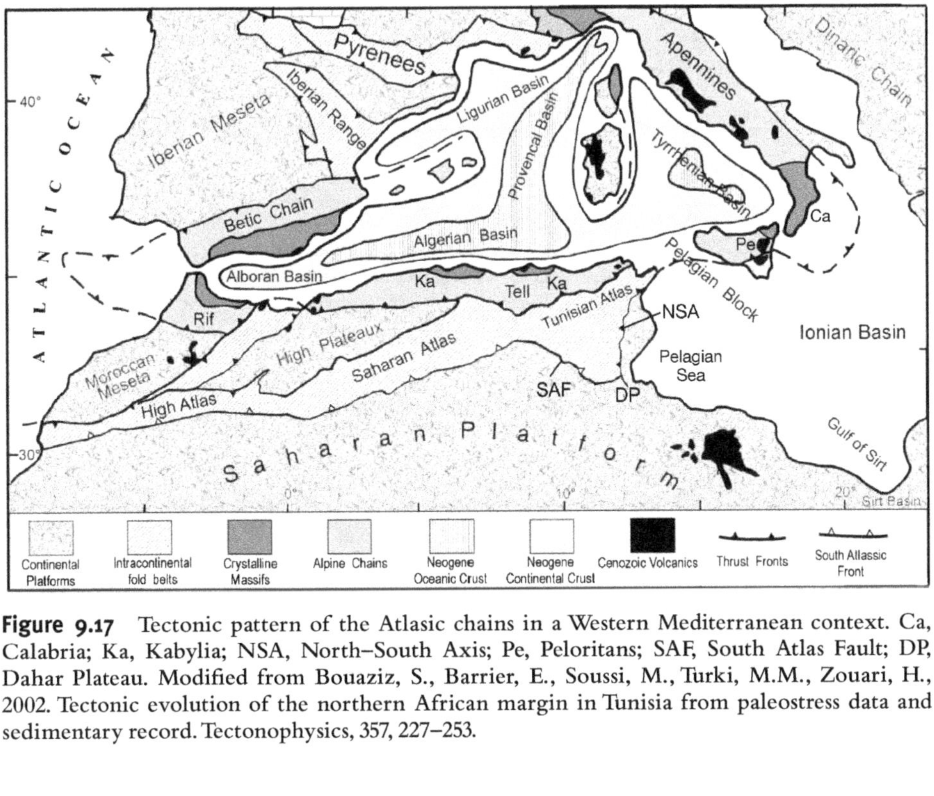

Figure 9.17 Tectonic pattern of the Atlasic chains in a Western Mediterranean context. Ca, Calabria; Ka, Kabylia; NSA, North–South Axis; Pe, Peloritans; SAF, South Atlas Fault; DP, Dahar Plateau. Modified from Bouaziz, S., Barrier, E., Soussi, M., Turki, M.M., Zouari, H., 2002. Tectonic evolution of the northern African margin in Tunisia from paleostress data and sedimentary record. Tectonophysics, 357, 227–253.

margin of the Pelagian Platform, a region of broad and low-amplitude folding, which extends eastward beneath the Pelagian Sea.

Extension started north of Africa in the Triassic and rifting accelerated during Liassic times, like in other areas of the Ligurian and Western Tethys Ocean (Figure 9.18). Most of Tunisia was a subsiding shallow-water domain where evaporites, siliciclastics and sabkha carbonates accumulated. Subsidence increased later in Liassic to Middle Jurassic times, in probable relation to a transition of rift to crustal fissure stage (see Section 9.2). Deep marine carbonates predominantly

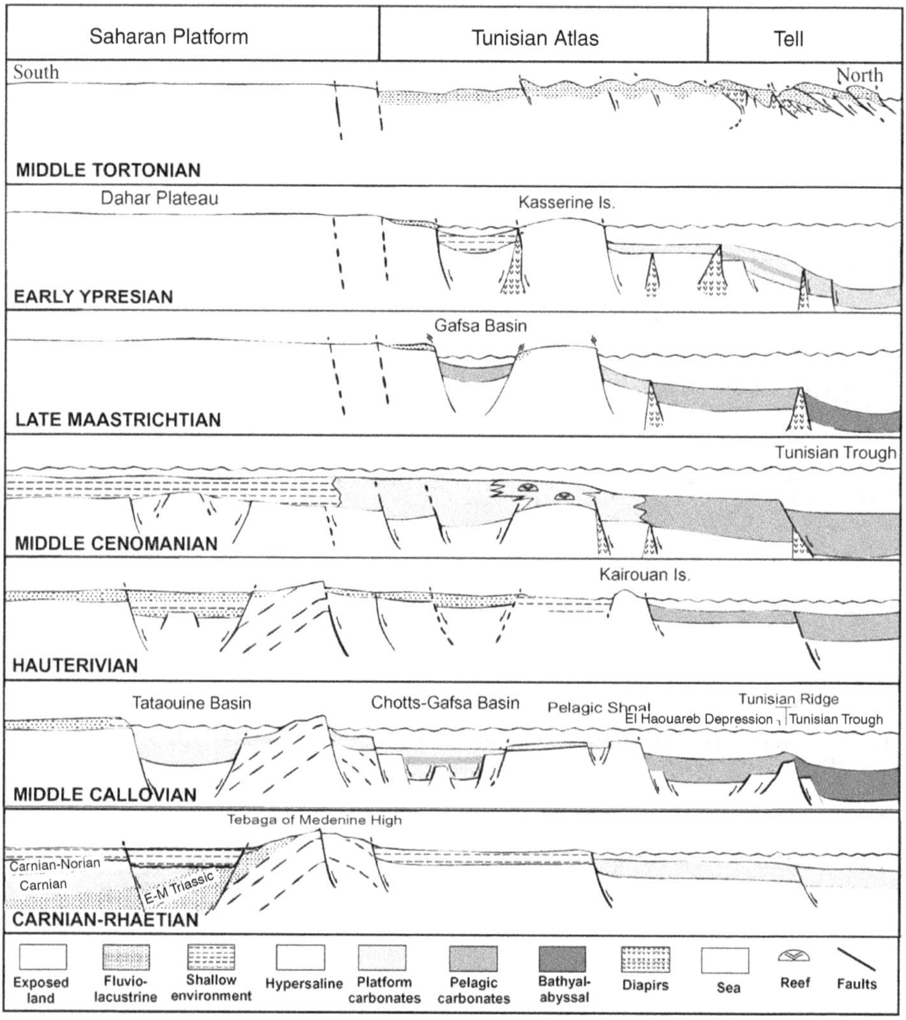

Figure 9.18 Interpretative N–S cross sections of the north African paleo-margin of the Tethys Ocean. Note Cenomanian maximum of transgression and salt tectonics, latest Cretaceous compression, and Late Miocene collisional deformation. Modified from Bouaziz, S., Barrier, E., Soussi, M., Turki, M.M., Zouari, H., 2002. Tectonic evolution of the northern African margin in Tunisia from paleostress data and sedimentary record. Tectonophysics, 357, 227–253.

accumulated in most areas during the Middle and Late Jurassic, grading to deep marine radiolaritic facies in the northern Tunisian domain. Accumulations of siliciclastics and local development of carbonate platforms characterized the African margin to the south. Extension and subsidence persisted in the Early Cretaceous, but clastic sedimentation sporadically occurred during regressive periods. A transpressional event of Early Albian age marked the beginning of a transition period characterized by a strong variability of extension and subsidence, local compressional events and salt tectonics, during the early stages of convergence between Europe and Africa. However, an interval of accelerated regional subsidence occurred in the Cenomanian and Turonian, coeval with increased extension (Figure 9.18). By this time, a marine transgression extended over many African margins, and seawater probably connected the Tethys to the South Atlantic via parts of the West and Central African Rift System, and the Niger and Chad basins (see Section 7.2.1). During the Late Cretaceous, evaporitic environments and carbonate platforms developed in many shallow coastal areas (northern Saharan Platform) and prominent structural blocks of the starved African margin (Central Tunisian High and Kasserine Island of the Tunisian Atlas, Kerkennah High of the Pelagian Platform). To the North, pelagic carbonates and marls accumulated on the deeper margin and in the basins of the Tethys Ocean. It is remarkable that during most of the Late Cretaceous, the African plate essentially responded to extensional stress and related opening processes in the Atlantic Ocean rather than to N–S convergence, which was principally accomodated between Europe and Apulia.

The oldest compressional event caused by the convergence of Europe and Africa occurred during the Late Maestrichtian and Early Paleocene in Tunisia where it is marked by local uplift. Major tectonic phases occurred in the Late Eocene (Atlasic phase), in the Middle to Late Miocene and in the Pleistocene (Figure 9.18). Tectonic phases were separated by intervals of extensional deformation and basin subsidence: the Late Paleocene to Early Eocene interval led to further development of platform and pelagic carbonates, whereas younger intervals led to the accumulation of siliciclastics in shallow water and continental environments.

9.3.2. Cretaceous Carbonate Platforms of Tunisia

Shallow water carbonate platforms were widespread on the northern, starved passive margin of Africa for most of the Late Cretaceous. However, some basement faults as well as E–W transcurrent fault systems were still active, suggesting that this area of the Tethys Ocean accomodated part of the stress generated by opening processes in the Atlantic. In Tunisia, intervals of carbonate platform development occurred from the Late Albian to the Late Cenomanian, from the Early Turonian to the Middle Coniacian, and from the Early Campanian to the Early Maestrichtian (Figure 9.19). Intervals of carbonate platform development were separated by intervals of pelagic carbonates in the Turonian and pelagic shaly limestones and marls in the Santonian. Accumulations of pelagic sediments do not coincide with high stands of sea level, and were rather caused by an accelerated regional subsidence and extensional stress.

A resumption of normal faulting along tilted structural blocks probably facilitated the development of an evaporitic trough in nearshore areas of the Saharan

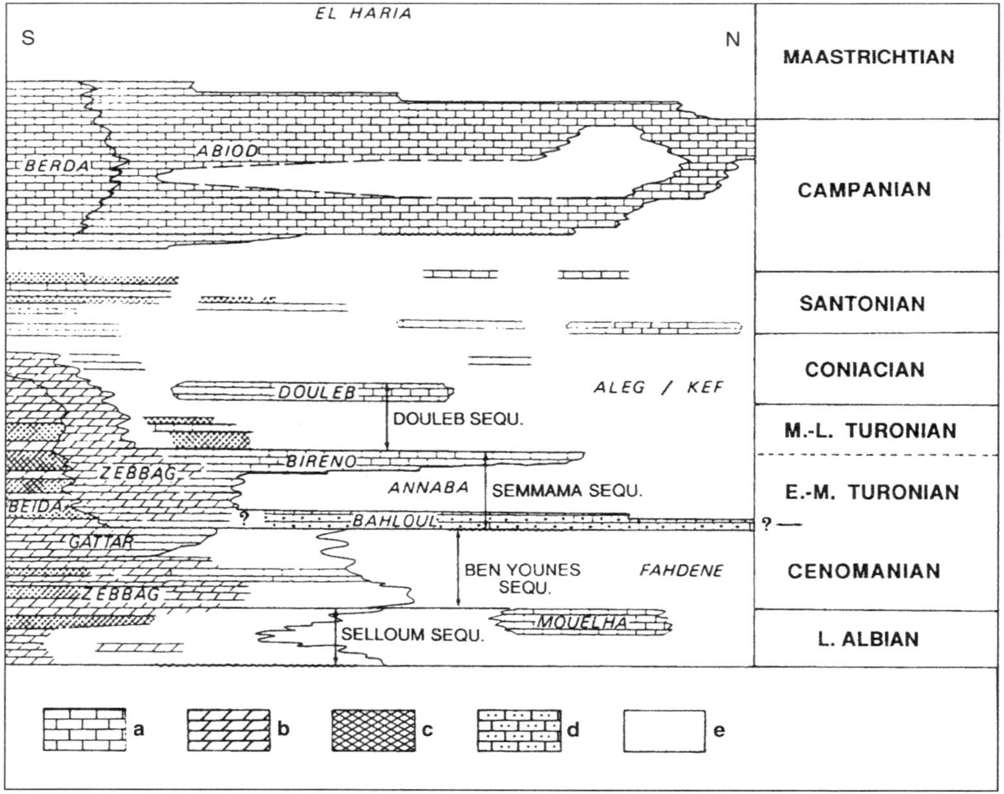

Figure 9.19 Late Cretaceous intervals of carbonate platform development on the Tunisian margin of the Tethys Ocean. a, Platform carbonates (limestones); b, Platform carbonates (dolomites); c, evaporites; d, pelagic limestones; e, pelagic shaly limestones and marls. Arrows represent major sequences of platform development. Modified from Camoin, G., 1991. Sedimentologic and paleotectonic evolution of carbonate platforms on a segmented continental margin: example of the African Tethyan margin during Turonian and early Senonian times. Palaeogeography, Palaeoclimatology, Palaeoecology, 87, 29–52.

margin, early in the Turonian (Figures 9.19 and 9.20). The trough principally filled with alternating carbonates (Stromatolites, etc.) and evaporites, locally replaced by breccias. Then, rudist frameworks developed in relatively shallow, gradually decreasing water depths of the margin. Conditions favorable to the development of carbonate platforms were met on top of structural highs and rising diapirs of Triassic evaporites (Figure 9.21). By the Middle Turonian carbonate platforms were widespread on the African margin of Tunisia, extending to nearshore areas where they replaced evaporitic environments. Turonian carbonate platform sediments decrease in thickness from the outer margin where they commonly reach 200 m, to the inner parts of the margin where they are about 100 m thick on the average.

Rudist bioconstructions are associated to bioclastic and oolitic deposits, and to mud mounds, which principally developed in outer shelf to upper slope areas. Mud mounds generally consist of subcircular lenses up to 20 m wide, which may coalesce to form extensive carbonate banks up to 100 m wide. All mud mounds show a roughly similar vertical distribution of facies, which reflects successive stages of development. The basal bioclastic and oncoidal facies also include cyanobacterial bushes. The core of the mud mounds consists of massive micritic, locally dolomitic and/or ferruginous micrite, which includes calcispheres, foraminifers, fragments of rudists and corals, calcareous sponges, echinoids and gastropods. Stromatolites are also fairly constant constituents. The mud mounds are capped by crestal bioclastic limestones where corals and/or rudists are associated to cyanobacterial bushes and fragments of red and green algae, and gastropods. The succession of facies suggests depositional environments up to 100 m water depth. The carbonate platforms receded in the Late Turonian, when they were replaced by fine siliciclastics in many areas of the outer margin. Because platform carbonates persisted in nearshore areas and no significant deepening was recorded at that time, the decline of carbonate platforms on the outer margin was tentatively attributed to cooler surface waters.

Carbonate platforms further developed in the Early Coniacian (Figures 9.19 and 9.20), especially in Central Tunisia and in the southern Pelagian Sea where their average thickness is about 100 m. The Coniacian platforms are characterized by a relatively uniform and gentle slope where nearshore high-energy facies grade to deeper, low-energy sediments, and an apparent lack of reworking. In the Coniacian like in the Turonian, the development of rudist frameworks was facilitated in areas of submarine relief such as diapirs, and mud mounds accounted for a significant part of the carbonate platforms. From the Turonian to the Coniacian, the carbonate platforms of Tunisia evolved from the "rimmed-shelf" to the "carbonate-ramp" type (Figure 9.20) and this evolution was likely driven by a change in seawater conditions. An increase in subsidence later in the Coniacian led to the demise of the carbonate platforms in the entire region. Intense carbonate productivity occurred during the Cretaceous in many starved margin areas of the Tethys, where the development of carbonate platform (and evaporitic) environments probably created conditions locally favorable to the formation of dense, warm waters, which circulated at deep water depths of the ocean (see Section 2.2.5).

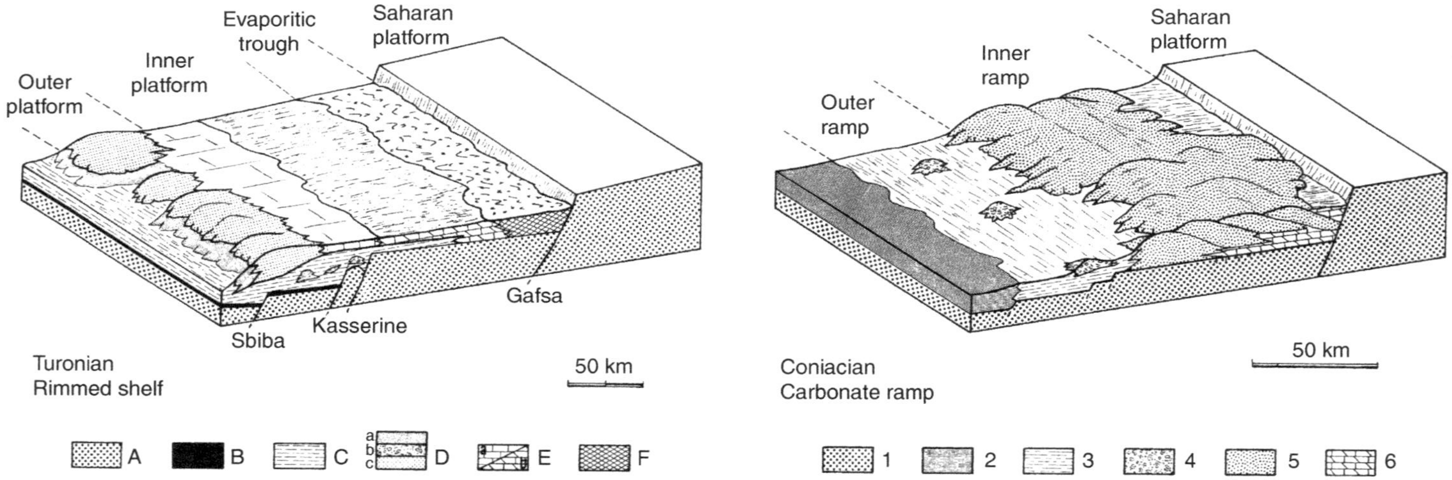

Figure 9.20 Reconstitution of Late Cretaceous carbonate platforms of the Tunisian margin of the Tethys Ocean. (Left) Turonian platform of rimmed-shelf type. A, Basement; B, Organic pelagic limestones; C, Shaly limestones and marls; D, Platform carbonates (a, slope facies; b, mud mounds; c, oolitic/bioclastic deposits and rudist frameworks); E, Inner platform facies (a, limestones; b, limestones and dolomites); F, Evaporites. (Right): Coniacian platform of carbonate-ramp type. 1, Basement; 2, Pelagic carbonates and marls; 3, Shaly limestones; 4, Mud mounds; 5, Oolitic/bioclastic facies; 6, Restricted platform facies. Modified from Camoin, G., 1991. Sedimentologic and paleotectonic evolution of carbonate platforms on a segmented continental margin: example of the African Tethyan margin during Turonian and early Senonian times. Palaeogeography, Palaeoclimatology, Palaeoecology, 87, 29–52.

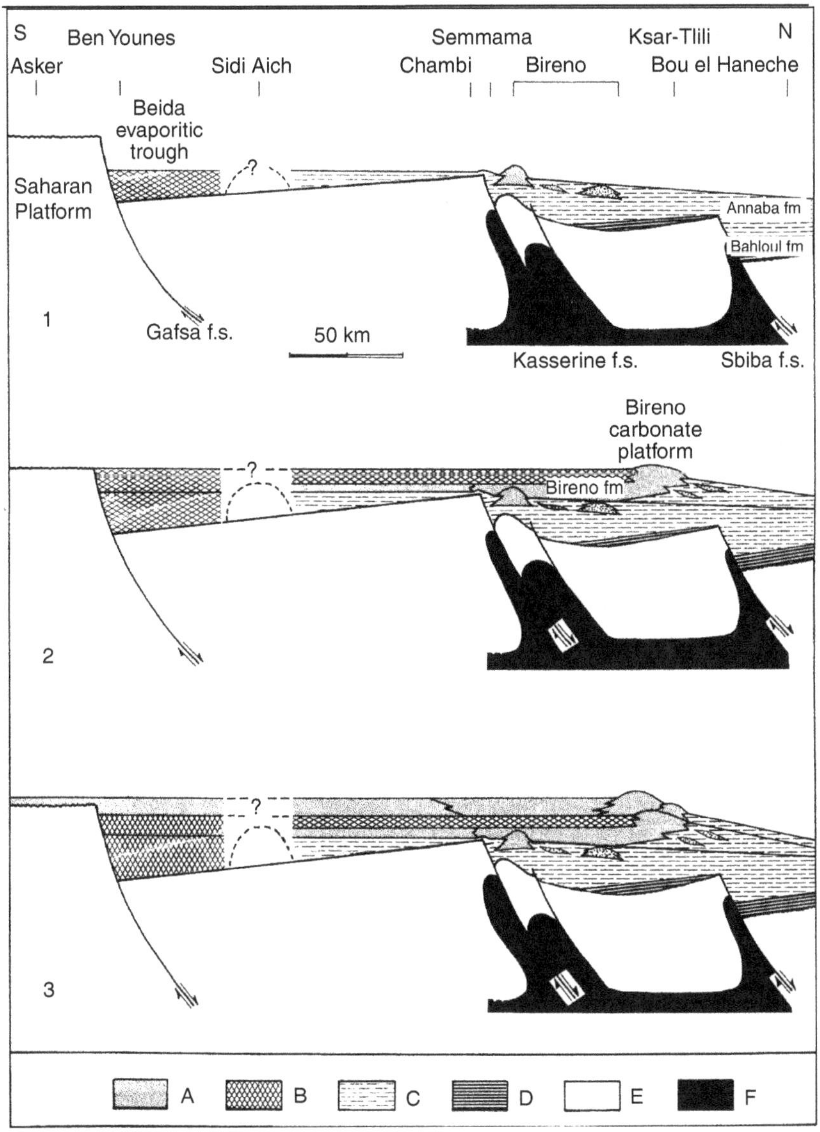

Figure 9.21 Three stages of carbonate platform development on the Tunisian margin of the Tethys Ocean during the Early and Middle Turonian. Fm, Formation; A, Triassic evaporites; B, Pre-Turonian terranes; C, Pelagic limestones (Bahloul fm); D, Shaly limestones and marls (Annaba fm); E, Evaporites (Beida fm); F, Platform carbonates (Bireno fm). Note early development of an evaporitic trough and isolated carbonate mounds, and further extensions of carbonate platforms and evaporites. Camoin, G., 1991. Sedimentologic and paleotectonic evolution of carbonate platforms on a segmented continental margin: example of the African Tethyan margin during Turonian and early Senonian times. Palaeogeography, Palaeoclimatology, Palaeoecology, 87, 29–52.

FURTHER READING

Debelmas, J., Mascle, G., 2000. Les grandes structures géologiques. Dunod, Paris.

Einsele, G., 1992. Sedimentary basins, evolution, facies, and sediment budget. Springer, Berlin.

Kennett, J.P., 1982. Marine geology. Prentice-Hall, Englewood Cliffs, N.J.

Lemoine, M., De Graciansky, P.-C., Tricart, P., 2003. De l'océan à la chaîne de montagnes. Gordon and Breach, Paris.

Other references used in this chapter

Bouaziz, S., Barrier, E., Soussi, M., Turki, M.M., Zouari, H., 2002. Tectonic evolution of the northern African margin in Tunisia from paleostress data and sedimentary record. Tectonophysics, 357: 227–253.

Camoin, G., 1991. Sedimentologic and paleotectonic evolution of carbonate platforms on a segmented continental margin: example of the African Tethyan margin during turonian and early senonian times. Palaeogeography, Palaeoclimatology, Palaeoecology, 87: 29–52.

Camoin, G., 1995. Nature and origin of Late Cretaceous mud mounds, North Africa. Special Publications of the International Association of Sedimentologists, 23: 385–400.

Ford, M., Lickorish, W.H., Kusznir, N.J., 1999. Tertiary foreland sedimentation in the southern subalpine chains, SE France: a geodynamic appraisal. Basin Research, 11: 315–336.

Golonka, J., 2004. Plate tectonic evolution of the southern margin of Eurasia in the Mesozoic and Cenozoic. Tectonophysics, 381: 235–273.

Kuhlemann, J., Dunkl, I., Brügel, A., Spiegel, C., Frisch, W., 2006. From source terrains of the Eastern Alps to the Molasse Basin: detrital record of non-steady-state exhumation. Tectonophysics, 413: 301–316.

Lemoine, M., Bas, T., Arnaud-Vanneau, A., Arnaud, H., Dumont, T., Gidon, M., Bourbon, M., De Graciansky, P.-C., Rudkiewicz, J.-L., Megard-Galli, J., Tricart, P., 1986. The continental margin of the Mesozoic Tethys in the Western Alps. Marine and Petroleum Geology, 3: 179–199.

Mauffret, A., Frizon de Lamotte, D., Lallemant, S., Gorini, C., Maillard, A., 2004. E-W opening of the Algerian Basin (Western Mediterranean). Terra Nova, 16: 257–266.

Patriat, M., Ellouz, N., Dey, Z., Gaulier, J.M., Ben Kilani, H., 2003. The Hammamet, Gabes and Chotts basins (Tunisia): a review of the subsidence history. Sedimentary Geology, 156: 241–262.

Pfiffner, O.A., Schlunegger, F., Buiter, S.J.H., 2002. The Swiss Alps and their peripheral foreland basin: stratigraphic response to deep crustal processes. Tectonics, 21: doi: 10.1029/2000TC900039.

Rehault, J.-P., Boillot, G., Mauffret, A., 1984. The Western Mediterranean basin geological evolution. Marine Geology, 55: 447–477.

Schmid, S.M., Pfiffner, O.A., Froitzheim, N., Schönborn, G., Kissling, E., 1996. Geophysical-geological transect and tectonic evolution of the Swiss-Italian Alps. Tectonics, 15: 1036–1064.

Part 3: Formation and Transformation of Oceanic Sediments

Ocean basins and margins are filled with sediments. Exceptions include submarine cliffs along basement blocks and trench walls, areas of newly formed oceanic crust and regions of the South Pacific isolated from current and wind transport, and below the carbonate compensation depth. Soft uppermost sediments, which consist of a mixture of sediment particles and seawater, grade to stiff and firm sediments with depth. During the process, they loose most of their pore-waters, decrease in volume and porosity, and increase in density. Sediment particles are derived from continental processes and marine biological activity principally, but may also include volcanic and/or authigenic elements. In areas of low sedimentation rates and/or isolated from terrigenous and biogenic influences, hydrogenous sediments may develop. Sediment particles are formed in specific environmental conditions, and are commonly used to reconstruct some aspects of past environments. However, the efficiency of sediment particles as proxies of past environments decreases as diagenetic alteration progresses with depth. Chapters 10–13 deal with the genesis, transport and accumulation of sediment particles, as well as with the formation and diagenetic alteration of the major types of oceanic sediments: terrigenous, biogenic, organic and hydrogenous.

CHAPTER TEN

Terrigenous Sediments

Terrigenous sediments principally consist of particles derived from the alteration of the continents surface through physical and/or chemical weathering processes. They are eroded by running waters, winds and glaciers and transported to the ocean where they principally accumulate in continental margin areas. The finest terrigenous particles are partly reworked and further transported by oceanic currents to distal areas where they are mixed with other types of sediment particles (see Section 2.3).

10.1. Physical and Chemical Weathering of the Earth's Surface

10.1.1. Processes of Physical and Chemical Weathering

Weathering processes are interrelated but the relative importance and distribution of physical and chemical weathering are strongly affected by surface temperature, precipitation and biological activity acting upon parent rocks of variable composition, and the morphology of the continent surface. Continental weathering progresses using lines of weakness of the continental surface like fractures and expansion joints. Those weaknesses are produced during exhumation of the parent rock to lower pressures because of continuous erosion and/or tectonic activity. Stress release and expansion of the parent rock cause joints and fractures to form, following directions roughly parallel to the surface of the outcrop. Fractures and expansion joints greatly increase the specific surface of the uppermost parent rock, i.e. the surface area exposed to weathering processes. For example, a square surface 10 cm × 10 cm exposes 100 cm^2 to weathering whereas a cube of similar square surfaces, containing fractures and expansion joints, may expose about 600 cm^2 to weathering.

10.1.1.1. Physical weathering

Physical weathering breaks the surface of the parent rock into small fragments with no changes in chemical composition, and is caused by several processes

- Large temperature variations such as those observed between day and night in tropical deserts (up to 50°C) cause expansion and contraction of minerals of different volumetric coefficients of thermal expansion. The resulting stresses applied to the minerals interface may overcome the yield strength of the parent rock, resulting in fragmentation.
- Water from rain or melting snow circulates in the fractures and joints, and penetrates within the pores of the uppermost parent rock. Subsequent freezing of water causes a 9% increase in volume and ice growth exerts considerable pressure

on the walls of the cavities, facilitating the fragmentation of the parent rock. This process is more intense in saturated rocks, in high-latitude and mountain areas. Precipitation and expansion of evaporite crystals in arid and semi-arid areas and salt in nearshore environments also facilitate the fragmentation.
- Biological activity, principally through growing roots, may also facilitate the fragmentation of the uppermost parent rock.

Physical weathering processes are especially efficient in regions of extreme climatic conditions such as arid to semi-arid tropical deserts, and cold high-latitude and mountain areas.

10.1.1.2. Chemical weathering

Chemical weathering is an alteration of the uppermost parent rock by chemical reactions. The minerals react with rainwater, which has a pH of about 5.7 and pO_2 and pCO_2 in equilibrium with those of the atmosphere. In fact, oxygen is a Group 6 element (like S, Se, Te and Po) of small radius and high electro-negativity. When combined to hydrogen to form water many hydrogen bondings develop between the molecules, which have an open structure. This results in atypical properties, like a very high boiling point and a solid phase of lower density than the liquid phase. The molecules of water have a strong polarity (Figure 10.1) and a capacity to develop hydrogen bondings with anions, and water is therefore a unique solvent of ionic compounds. Moreover, water can give or accept protons and has the characteristics of both an acid and a base. At the Earth's surface, hydrodynamism, microenvironments, biomass may influence the behavior of water and three major trends of weathering are defined according to Eh–pH conditions (Figure 10.2). The acid trend occurs where climate and drainage facilitate a net leaching and evacuation of solutions whereas the alkaline trend occurs in environments of high evapotranspiration and net water deficit, the reduced trend being found in water-saturated conditions. The acid trend is currently the dominant one in weathering: the molecules of water behave as dipoles which attract the elements at the surface of the minerals (hydrolysis reaction), and the intensity of the attraction force defines

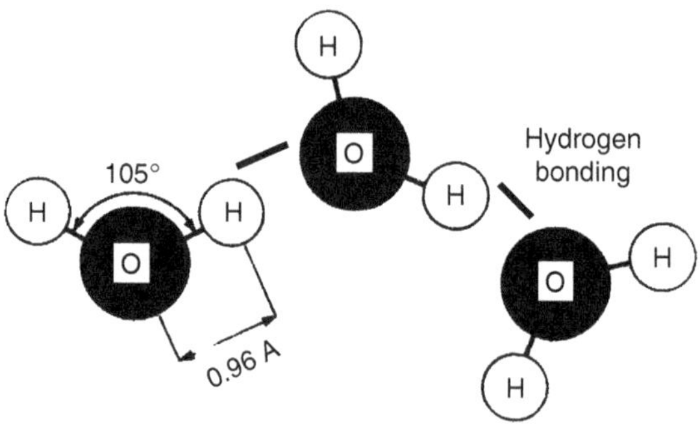

Figure 10.1 Structure and arrangement of water molecules in a liquid phase. Note hydrogen bondings between molecules.

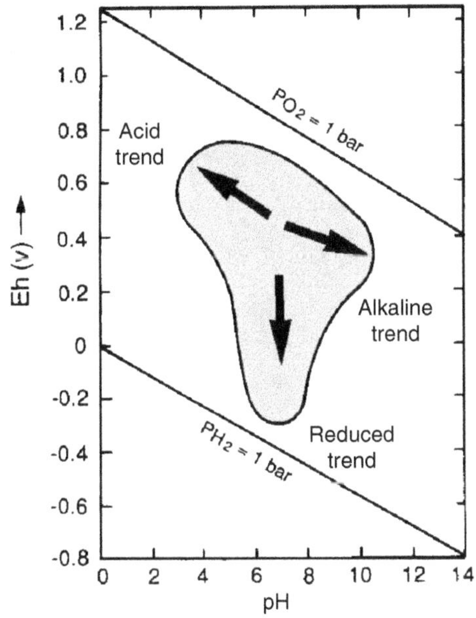

Figure 10.2 Major trends of chemical weathering, as defined by Eh–pH conditions. Modified from Martini, I.P., Chesworth, W. (Editors), 1992. Weathering, soils and paleosols. Elsevier, Amsterdam.

the solubility of the elements. The charge (*e*) and the radius (*r*) of the ions are used to estimate their degree of solubility as expressed by the Goldschmidt potential (e/r):

- $e/r < 3$ (Soluble cations). Because of their weak attraction force they combine to molecules of water in the form $(SC^nH_2O)^n$ where SC is any element of the domain $e/r < 3$ (Figure 10.3). The soluble cations are removed together with running or percolating waters. However, the bondings of the soluble cations of very low potential ($e/r < 1$) are fragile and those elements are often retained within the weathering profile. This is for example the case of K, an important component of clay minerals derived from continental weathering such as illite, and its chemical equivalent Cs.
- $3 < e/r < 10$ (Hydroxiles). The attraction force between those elements and water being higher, one of the bondings of the molecules H_2O collapses. Hydroxile ions of XOH type are created, where X is any element of the domain $3 < e/r < 10$ (Figure 10.3), and H^+ is released. The resulting products are insoluble and accumulate within residual weathering profiles. For example, Al and Fe may combine to H_2O to form gibbsite and goethite, which are major hydroxide components of the bauxites.
- $e/r > 10$ (Soluble anions). Because of the strong attraction force between those elements and water, both bondings of the molecules H_2O collapse. Compounds of SAO^- type are created, where SA is any element of the domain $e/r > 10$ (Figure 10.3), H^+ being released. The resulting products are removed by running or percolating waters. They include CO_3, SiO_2 and SO_4 among others, which

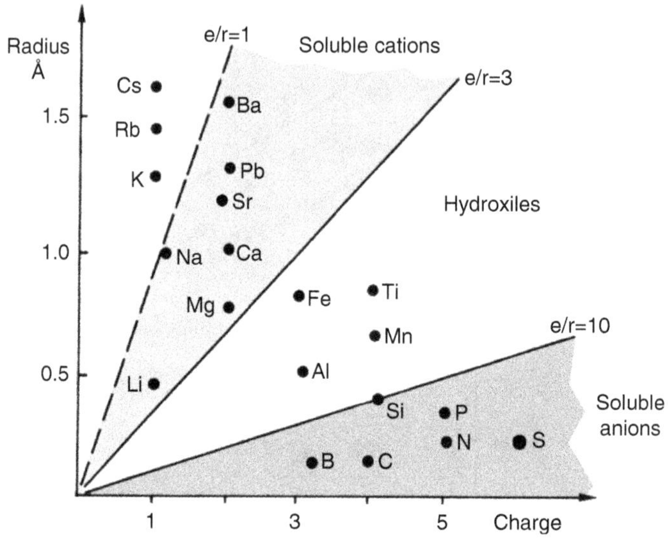

Figure 10.3 Solubility of chemical elements frequently found at the Earth's surface, as expressed by their Goldschmidt potential (e/r). Gray areas represent the domains of soluble elements, whereas white areas represent the domains of insoluble to poorly soluble elements which accumulate in weathering profiles. Modified from Pomerol, C., Lagabrielle, Y., Renard, M., 2005. Eléments de Géologie. Dunod, Paris.

are of major importance for the functioning of the ocean system and widespread components of all types of rocks and sediments.

To summarize, the Goldschmidt potential explains the removal of many elements from the Earth's surface by rainwater and the accumulation of others like Al and Fe, in an acid context. It also explains the solubility of many carbonates, which combine cations and anions of high solubility.

Silicate minerals are largely dominant in the rocks and sediments which outcrop at the Earth's surface. The substances derived from the chemical weathering of silicates consist of clay minerals principally, and hydroxides. Clay minerals are sheet silicates made of alternating tetrahedral and octahedral layers (Figure 10.4). The tetrahedral layers associate Si and O, Si being in a central position and O at the apexes of the tetrahedra. The octahedral layers associate Al in a central position to O and/or OH at the apexes of the octahedra. Because of occasional substitution of Si by Al in the tetrahedra and Al by Mg or Fe in the octahedra, free valences are filled by exchangeable cations in an interlayer position. The number of layers and their chemical composition define three major types of clay minerals:

- Clay minerals of 1/1 type are 7Å thick, the sheet units being made of one tetrahedral and one octahedral layers, with no exchangeable cations. The most common clay mineral of this type is kaolinite.
- Clay minerals of 2/1 type are of variable thickness (10–15Å), the sheet units being made of two tetrahedral and one octahedral layers. Free valences in tetrahedral layers are filled by K and to some extent Na in illite, a close analog of mica,

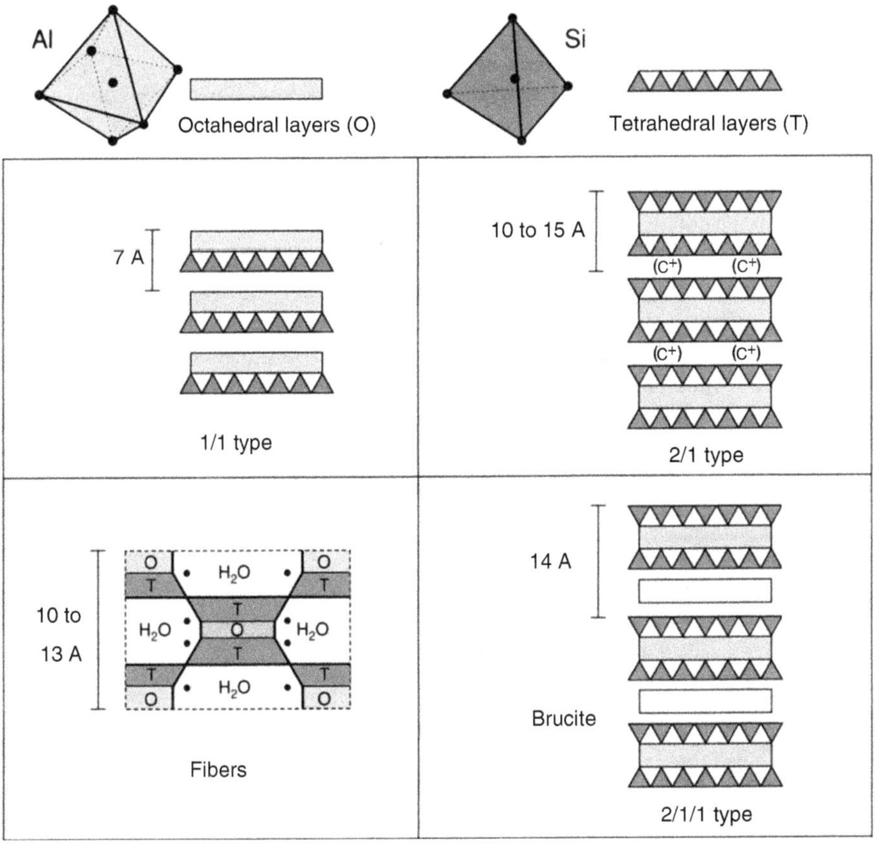

Figure 10.4 Major types of clay minerals. C^+: soluble cations. Note the potential of clay minerals for incorporating chemical elements and water in their interlayers. Modified from Pomerol, C., Lagabrielle, Y., Renard, M., 2005. Eléments de Géologie. Dunod, Paris.

whereas they are filled by Ca and Mg in vermiculite. Free valences in both tetrahedral and octahedral layers are filled by a variety of cations in smectites.

- Clay minerals of 2/1/1 type are about 14Å thick. The sheet units include two tetrahedral and one octahedral layers, and an additional layer of $Mg(OH)_6$ (brucite) in an interlayer position. The most common clay mineral of this type is chlorite.
- Other clay minerals are made of two tetrahedral and one octahedral layers organized in regular sequences (fibers) which delimit spaces filled with molecules of water. The most common clay minerals of this type are palygorskite and sepiolite.

Hydroxides are non-silicate minerals which most of the time consist of double layers of octahedra where Al and/or Fe are located in a central position and in octahedral interstices, and O at the apexes of the octahedra. Examples include boehmite and goethite. Other hydroxides, such as gibbsite, consist of layers of Al ions sandwiched between two layers of closely packed hydroxil ions.

10.1.2. The Variability of Continental Weathering

The vulnerability of silicate minerals to chemical weathering is highly variable. For example, quartz has a very low solubility. Its structure consists of very densely packed tetrahedra where Si is in a central position and O at the apexes. Because of the absence of substitution by other elements than Si and O, the related absence of free valences, the hardness and lack of cleavage of the crystals, quartz is very resistant to chemical weathering. Conversely, silicate minerals such as olivine, amphiboles and pyroxenes contain chemical elements of relatively close radius and physical properties (e.g. Si and Al or Na and Ca) which make possible a relatively high degree of substitution within the mineral's structure. Differences in the ionic charge of these elements and related presence of free valences (especially at the surface of the minerals) facilitate their bonding to the dipoles of rainwater and make the minerals highly sensible to chemical weathering (Figure 10.5).

The reaction of rainwater with surface minerals being gradual, chemical weathering progresses through time as more soluble elements are removed.

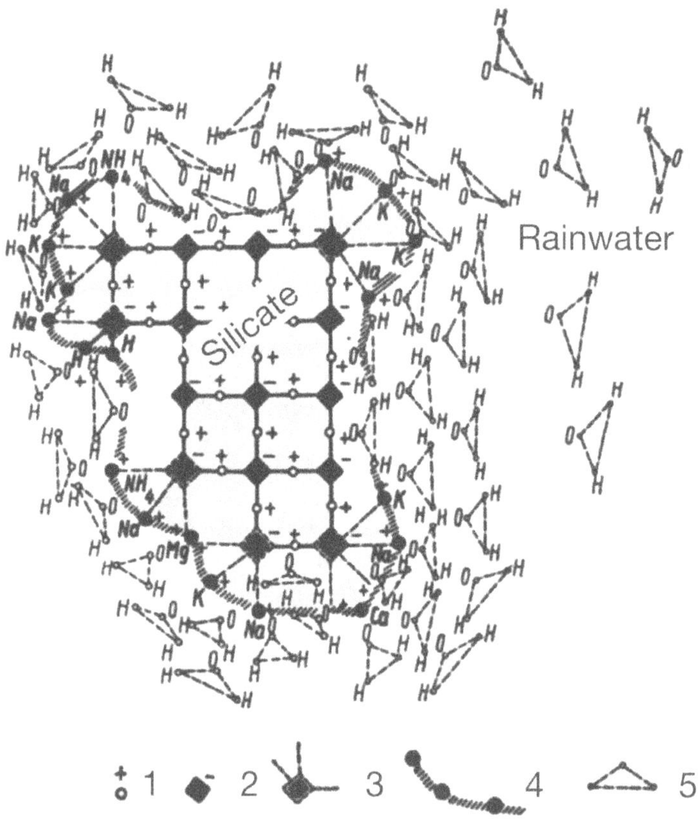

Figure 10.5 Weathering of silicate minerals by rainwater. 1, Cations from the mineral; 2, Anions from the mineral; 3, Peripheral anions with free valences; 4, Adsorbed cations, bonded to water and 5, Dipoles of water. Modified from Pomerol, C., Lagabrielle, Y., Renard, M., 2005. Eléments de Géologie. Dunod, Paris.

For example, orthose is a feldspar where the substitution of Si by Al within tetrahedra is compensated by an association with K. The formula of orthose is Si_3AlO_8K. The mineral contains soluble cations (Si and K) and a hydroxile element (Al) and its ratio Si/Al = 3. During the early stages of chemical weathering, clay minerals of the illite group are formed. The formula of illite is $KAl_4(Si_8O_{20}).OH_4$ and its ratio Si/Al = 2 (bisiallitization). Depending on climate and drainage conditions clay minerals of the smectite group, of similar structure but enriched in exchangeable elements such as Fe, Na and Mg, may form. As chemical weathering progresses, clay minerals of the kaolinite group are formed. The formula of kaolinite is $Si_2O_5Al_2(OH)_4$ and its ratio Si/Al = 1 (monosiallitization). In the ultimate stages of chemical weathering, hydroxides such as gibbsite are formed. The formula of gibbsite is $Al(OH)_3$ and its Si/Al ratio = 0 (lateritization). As a result, weathering profiles are characterized by occurrences of early chemical weathering products near the transition to the parent rock, whereas advanced chemical weathering products are found in the upper part of the profiles (Figure 10.6).

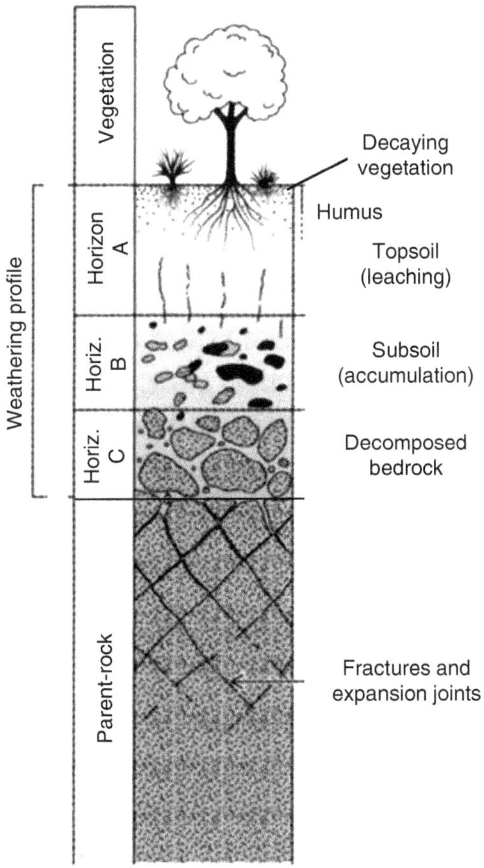

Figure 10.6 Weathering profile for a temperate climate. Note the transition from fresh parent rock to topsoil through a sequence of horizons capped by decomposed organic material. Modified from Pomerol, C., Lagabrielle, Y., Renard, M., 2005. Eléments de Géologie. Dunod, Paris.

As for any chemical reaction, chemical weathering processes accelerate with increased temperatures and availability of water. This is the reason why chemical weathering is more intense in low latitude areas of high precipitation where kaolinite and laterites commonly develop. At low latitudes, average temperatures around 26°C are recorded at 1.5 m depth of the weathering profiles, but may locally reach 60°C at the surface during the day. In those conditions, water contains six times more free H^+ ions than in temperate regions and its action as an agent of chemical weathering is strongly reinforced. For example, traces of kaolinite were found in volcanic ashes of the Krakatoa volcano of Indonesia, 50 years after their emission in 1883. Conversely, early chemical weathering products only are found in arid and high-latitude areas where the lack of precipitation and/or low temperatures limit the chemical reactions (see Sections 2.2 and 2.3). For example, silicate bedrocks exposed at the surface during the retreat of the Scandinavian ice cap barely weathered in 10,000 years.

Dissolved gases may reinforce the action of water as an agent of chemical weathering. Oxygen, dissolved in rainwater or from the atmosphere, may accelerate the alteration of sulfurs and the formation of oxides and hydroxides. Also, dissolved CO_2 may intervene in the hydrolysis reaction. For example, calcium carbonates (calcite, aragonite) include a soluble cation (Ca) and a soluble anion (CO_3) which in theory should be easily separated and removed (Figure 10.3). However, a very minor proportion of cation and anion sites in calcite are vacant and the solubility of calcite in pure water at 20°C is very low (120–140 ppm), which is very comparable to the solubility of quartz. In fact, carbonic acid derived from the combination of rainwater with carbon dioxide of atmospheric origin or produced by soil activity reacts with calcite to form calcium and bicarbonate ions which are removed in solution:

$$CaCO_3 + H_2CO_3 \rightarrow Ca^{2+} + 2HCO_3^-$$

As for all gases, the solubility of CO_2 in water increases as temperature decreases, leading to higher concentrations of carbonic acid. As a consequence, the solubility of calcite strongly increases at low temperatures (to about 2,000 ppm at 0°C) and the chemical weathering of carbonate rocks is more intense in temperate to sub-polar regions.

Si and Al are major components of silicate minerals which dominate in the lithosphere, and Si is soluble whereas Al is insoluble in the neutral to slightly acidic conditions which currently prevail at the Earth's surface. However, the solubility of most chemical elements changes with pH conditions: Al is much more soluble than Si in strongly acidic conditions, whereas the solubility of both Si and Al strongly increases in alkaline conditions (Figure 10.7). The alkaline trend of chemical weathering may occur in arid to semi-arid regions, where downward leaching of soluble elements during the rainy season is succeeded by capillary rise and evaporation during the dry season (Figure 10.8). In those conditions, basic cations such as Na, Ca, Mg and oxyanions such as CO_3 and SO_4 may accumulate in the weathering profiles. The alkaline trend of chemical weathering may ultimately lead to the formation of clay minerals enriched in basic cations (smectites, palygorskite, sepiolite), alkaline carbonates, zeolites and evaporitic minerals. The reduced trend

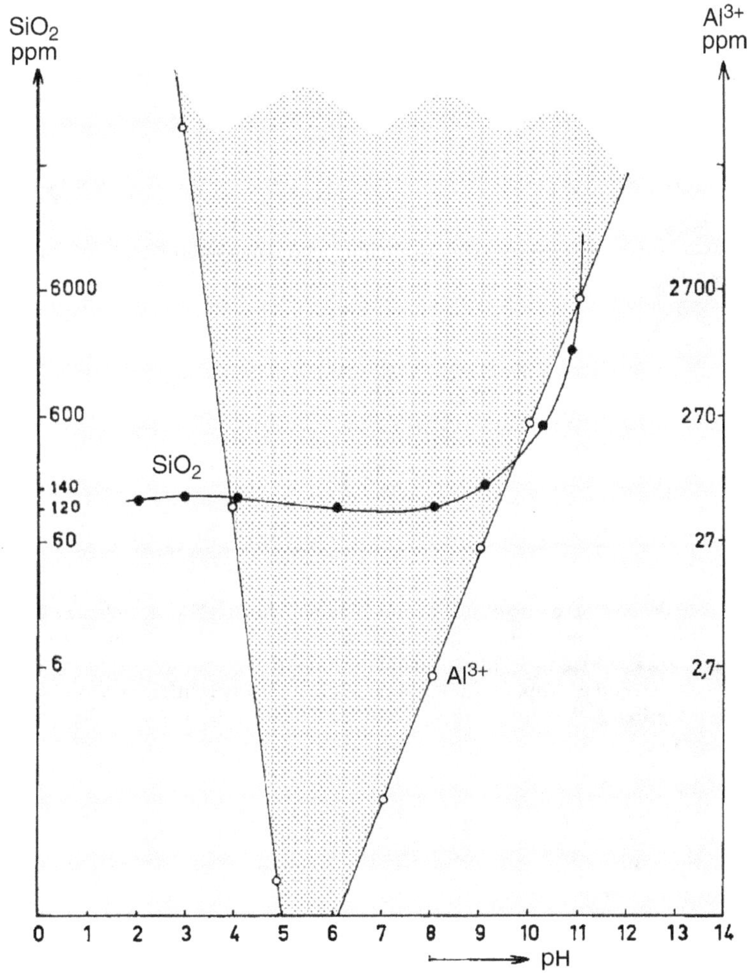

Figure 10.7 Compared solubilities of Si and Al as pH increases. Note that Al is insoluble for pH values between 4.8 and 6.2 approximately. Modified from Millot, G., 1964. Géologie des argiles. Masson, Paris.

of chemical weathering occurs in areas where an excess of water results in episodic or total flooding of the weathering profiles (Figure 10.8). In those conditions, elements such as Fe and Mn are present in a reduced state and easily mobilized. The reduced trend of chemical weathering may lead to the formation of Fe-rich clay minerals (smectites) and carbonates (siderite), pyrite and authigenic iron silicates. Fluctuations in the level of the water table and related alternations of reduced and oxidizing conditions may result in the formation of hydroxides in the form of concretions and indurated layers.

The products of chemical and physical weathering form the regolith, which principally consists of soils in areas of active chemical weathering. The regolith is easily separated in fragments and individual particles, which are removed by run-off,

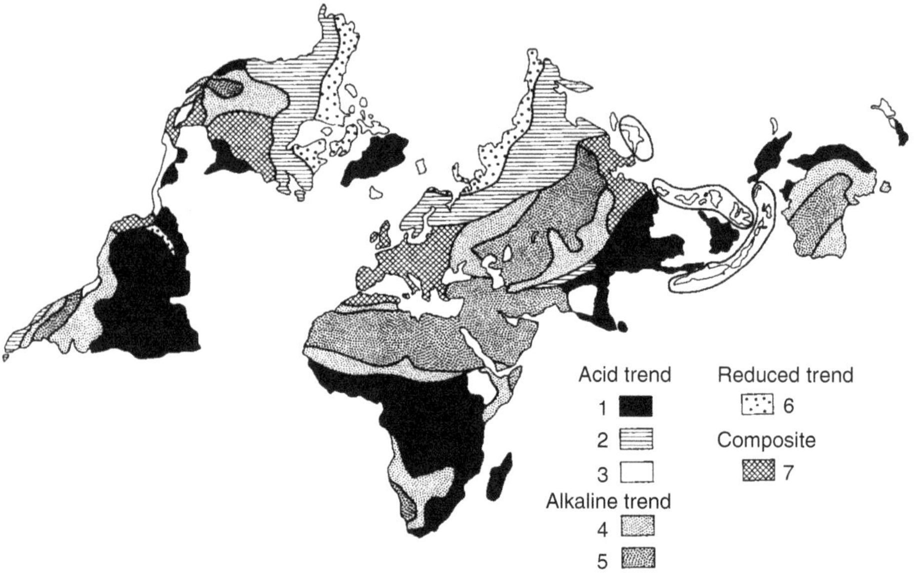

Figure 10.8 Distribution of major trends of chemical weathering. Composite indicates regions where a clearly dominant trend is not defined. Numbers indicate the dominant soil type. Note the dominance of the acid trend in humid tropical and cool temperate regions and the dominance of the alkaline trend in tropical and temperate areas of low rainfall and high evaporation. Modified from Martini, I.P., Chesworth, W. (Editors), 1992. Weathering, soils and paleosols. Elsevier, Amsterdam.

wind and ice. The regolith overlays pristine parent rock and varies in thickness according to the rates of continental weathering and erosion. For example, thin regoliths commonly occur in cold areas of high relief and thick regoliths in tropical areas of low relief. Also, intervals of stable climate characterized by the development of thick soils and a well established vegetation (state of biostasis) are sometimes succeeded by intervals of poor vegetation and intense erosion (state of rhexistasis): such a succession may occur during intervals of abrupt climate change, but may also result from human activities involving the destruction of the vegetation.

10.2. The Removal and Transport of Terrigenous Elements

10.2.1. Erosion and Transport by Run-Off

A driving force of continental weathering, rainwater is also a major agent of erosion and evacuation of terrigenous particles to the ocean. The efficiency of rainwater as an agent of erosion is directly related to the intensity of the precipitation and the velocity of raindrops, which increases with the diameter of the drops and the velocity of the wind. The kinetic energy of raindrops (see Section 2.2) is released in the form of heat and shear stress as they reach the ground, the proportion of shear stress being controlled by the availability of loose particles at the surface of the regolith. Where present the vegetation fragments the raindrops, lessening their kinetic energy to the ground and protecting the regolith from erosion. Also, the vegetation retains part of the precipitation for later evaporation. In the absence of vegetation, loose particles are easily detached by raindrop impact and removed by run-off.

Rainwater first percolates through the regolith and the parent rock. Water starts accumulating at the surface when the infiltration capacity of the ground is exceeded, eventually leading to run-off which evacuates the particles detached by raindrop impact. Very low run-off velocities (below 1 cm/s) are sufficient to remove particles below 10 μm in size (see Section 2.3.7 and Figure 2.14) which dominate in the soils, illustrating the importance of the vegetation as a protection against erosion. Surface run-off flows downslope and focuses within discontinuous shallow channels which converge to form deeper gullies, the velocity of run-off increasing with the slope. Gullies develop preferentially where the slope is disrupted, and expand by headward retreat and scouring principally. The energy (e) generated by the flow being in the form

$$e = 1/2^{*}mv^2$$

where m is the mass of water and v its velocity, the erosion of the regolith by running waters starts as soon as they produce enough shear stress at the surface, the efficiency of run-off as an agent of erosion increasing downslope. As the velocity of the flow increases it may transport sand- to gravel-sized particles as bedload, which are used as abrasives for further eroding the substrates. It is clear that the nature of the regolith and underlying parent rock, the topography and the vegetation, as well as the intensity and frequency of rainfall control the erosion of the continents surface by raindrop impact and run-off. Therefore, this is no surprise that maximum suspended loads are recorded for tropical rivers of extensive drainage basins and/or draining mountain areas of high relief (Figure 10.9), such as the Amazon in South America, the Ganges in India and the Fly River of New Guinea.

Erosion characterizes the upper section of the drainage basins principally. Erosion decreases with slope gradient and flow velocity. Flow velocity is minimum on the sides of the river channel where particles may lay down and build a succession of channel levees through time. Slope gradients are very low, and the related stream velocity reduced, in the lower section of the drainage basins. Water may flow over channel levees during high floods and expand in the floodplains where a fraction of the terrigenous load is released. However, the vast majority of the suspended load is released at the river mouth and river systems are currently the most efficient agent of erosion and transport of terrigenous particles to the ocean.

River mouths are transitional environments influenced by both the ocean and the river system. The modern river mouths developed during the last deglaciation, as the rising marine waters penetrated within the lower section of the glacial river systems. The modern estuaries and deltas form a continuous series where the degree of evolution is determined by the importance of the rivers terrigenous load principally. Well-developed estuaries are characteristic of river systems of low terrigenous load (Figure 10.10). As the terrigenous load increases, estuaries are

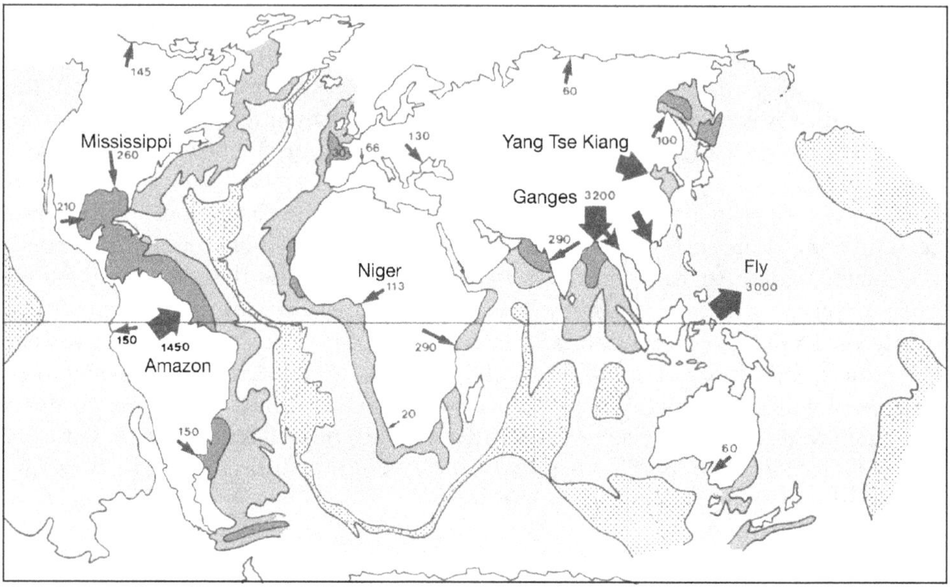

Figure 10.9 Terrigenous load of some major river systems (in million tons per year). Darker gray patterns indicate greater thicknesses of sediments. Note local coincidence between high terrigenous loads and high thicknesses of sediments. Modified from Dercourt, J., Paquet, J., 1995. Géologie: objets et méthodes, Dunod, Paris.

progressively filled of sediments. Deltas are characteristic of river systems of very high terrigenous load, and their morphology may vary with the importance of the tides and the action of the waves. The accumulation of sediment at or near river mouths is principally the consequence of interaction (mixing) between fluvial and marine waters:

- fluid motion is driven by the decreasing gradient of density between fluvial and marine waters;
- the entrainment of fluvial waters into marine waters causes a deceleration of the flow;
- the mixing of both water masses is associated to physical and geochemical changes facilitating solution, precipitation and flocculation.

The coarsest particles settle by gravity when the velocity of the river flow decreases at the contact with seawater. In theory, fine particles of low settling velocities should be evacuated to the ocean and further transported by oceanic currents. In reality, most fine particles flocculate and accumulate in or near river mouth areas (see Section 2.3.7). For electrostatic reasons these negatively charged particles are maintained at such a distance that molecular attraction by the forces of van der Waals remains ineffective in freshwater. Only a reduction of their negative charge could help maintaining the terrigenous particles at a distance sufficient for the forces of van der Waals to operate. Such a neutralization is made possible in the presence of free, positively charged cations which increase in concentration with

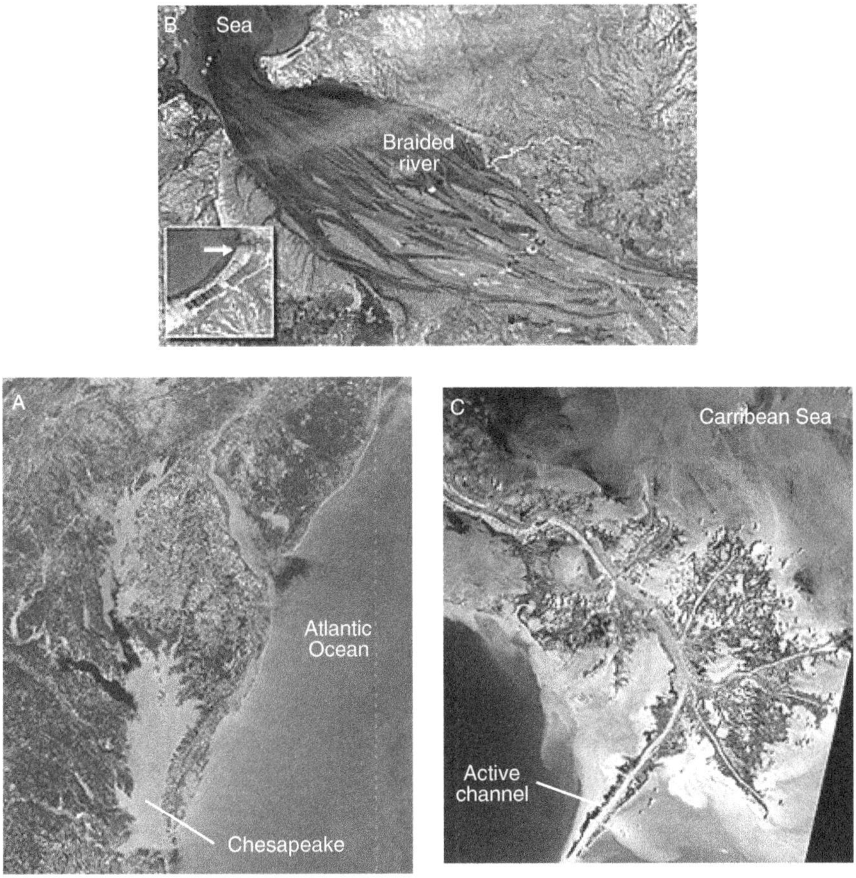

Figure 10.10 Degrees of evolution of river mouths. (A) Well-developed estuary of the Chesapeake estuary of North America. Note penetration of ocean water in the lower valley. (B) Filled estuary of the Betsiboka river of East Madagascar. Arrow in insert indicates the location of the river mouth. (C) Active area of the delta of the Mississippi River of North America. Courtesy of NASA, http://visibleearth.nasa.gov

salinity. Flocculation may start as soon as salinity increases in river mouth areas, where particles come in close proximity and flocs develop in turbulences principally. Flocs generally increase in size as long as they are maintained in areas of high particle concentration and within turbulent flows, but preferentially settle as they reach areas of sluggish circulation.

The nature of the suspended material, as well as local specificities of the environment may significantly affect sedimentation processes at or near river mouths. In the Pearl River estuary of China, the vertical gradient of salinity and related water stratification seem to exert a major control on the development and residence time of flocs near the interface between freshwater and saltwater, especially during the wet season. In the Scheldt estuary of northern Europe, floc formation is principally controlled by tidal current velocity and concentration in suspended matter. In addition, seasonal variations in floc composition and density affect their degree of cohesion and settling velocities, as well as the capacity of tidal currents to erode freshly deposited sediment layers. In the Rhône delta area of southern Europe, a poor reactivity of suspended material regarding flocculation is associated to significant differential settling and dilution of most particles within the river plume. The minor importance of flocculation there could be partly related to the nature of the terrigenous suspended load, where authigenic calcite grains, quartz and illite dominate. Conditions of sedimentation in and near river mouths are very complex, but can be reviewed following a simplified scheme.

For river systems of low discharge and/or terrigenous load, sedimentation processes are principally controlled by interaction between river outflow and tide magnitude (Figure 10.11):

- Stratified estuaries develop in areas of limited tidal range (microtidal). Freshwater of low density flows above seawater which forms an almost stationary salt wedge decreasing in thickness upstream. Shear stress forces develop at the interface between both water masses, in turn generating turbulences which facilitate the partial flocculation of the riverine suspended load. The remaining suspensions are entrained to the ocean where they are either dispersed or aggregated as fluvial and seawater mix together. Examples include small river systems of the Mediterranean and the Carribeans.
- Partially mixed estuaries develop in areas of moderate tidal range (mesotidal, 2–4 m) principally. In this case the salt wedge moves upstream and downstream with the tides, carrying sediment particles eroded from the substrate through the tidal channels. The velocity of tidal currents decreases upstream, to near zero where fluvial and seawaters meet at the extremity of the salt wedge (null point). As a consequence, maximum concentration of suspended particles occurs near the null point (turbidity maximum). Shear stress forces generate significant turbulence at the interface between freshwater and seawater and on the estuary floor, facilitating the mixing of waters and the flocculation of sediment particles. Flocculation is especially active in the area of maximum turbidity, which moves with the tides and is sometimes expelled out at sea during high tides and/or flood events. Examples include the Seine and Thames estuaries of western Europe (Figure 10.12).
- Well-mixed estuaries are large structures which develop in areas of high tidal range (macrotidal, over 4 m). Water circulation is dominated by the flow path of tidal currents and fluvial waters, influenced by the force of Coriolis (see Section 2.2.5). The vertical gradient of salinity is replaced by a lateral gradient with low salinity waters dominating on one side of the estuary. Turbulences are widespread and the turbidity maximum reaches utmost extension. Examples includes the Rio de la Plata estuary of South America and the Gironde estuary of western Europe where the turbidity maximum concentrates about 2 million tons of sediment particles.

Studies conducted on the Seine estuary of western Europe have evidenced a seasonality of sedimentary processes. Accumulation prevails during intervals of high river discharge (winter) when the terrigenous load is maximum and part of the sediment is stored in the mudflats which are continuously flooded. Erosion prevails

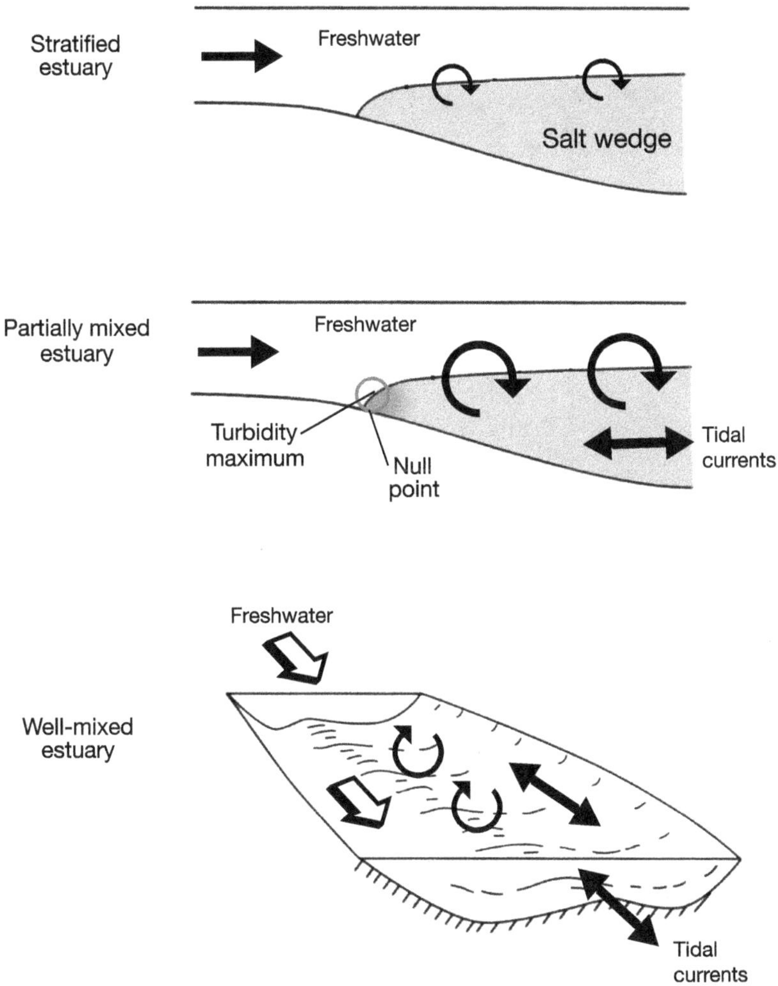

Figure 10.11 Classification of estuaries according to their degree of mixing. Note that the null point and turbidity maximum move with tidal currents in partially mixed estuaries and that freshwater and tidal currents are deflected by the force of Coriolis in well-mixed estuaries. Modified from Allen, P.A., 1997. Earth surface processes. Blackwell, London.

during intervals of low river discharge (summer) when the terrigenous load is minimum and the mudflats are exposed to the action of tidal currents principally.

For river systems of high discharge and terrigenous load characterized by a delta, sedimentation processes are principally controlled by the dynamics of the fluvial outflow. To simplify the outflow is compared to a turbulent axisymmetric jet entering a large reservoir, where it expands into a conical plume while mixing with seawater. Therefore, sedimentary processes including the decantation of coarse particles and the flocculation of fine particles occur in the ocean principally. Whereas axisymmetric jets expand at constant angles in theory, fluvial outflows may not be able to expand in all directions, depending on the morphology of coastal

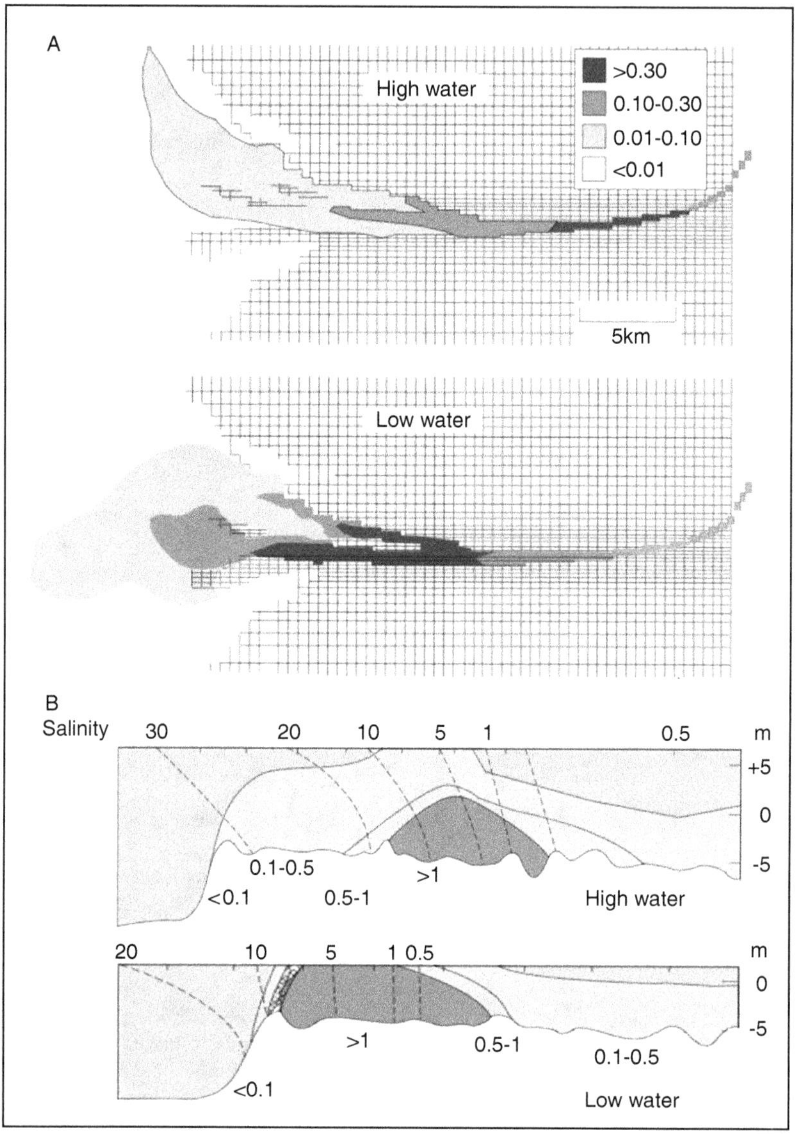

Figure 10.12 Concentration of suspended matter (kg/m^3) in the Seine estuary of western Europe: (A) from modeling experiments and (B) from measurements. Salinity in g/l. Note the shift of the turbidity maximum between high tide and low tide. Modified from Brenon, I., Le Hir, P., 1999. Modelling the turbidity maximum in the Seine estuary (France): identification of formation processes. Estuarine, Coastal and Shelf Science, 49, 525–544.

areas and continental shelves. In areas where the water depth does not constrain the expansion of the plume, the angle of expansion remains narrow and velocities decrease slowly with distance from the outlet (Figure 10.13). In this case, fluvial waters and their terrigenous load are transported relatively far into the ocean. For example, the plume of the Amazon River extends over the outer shelf and upper

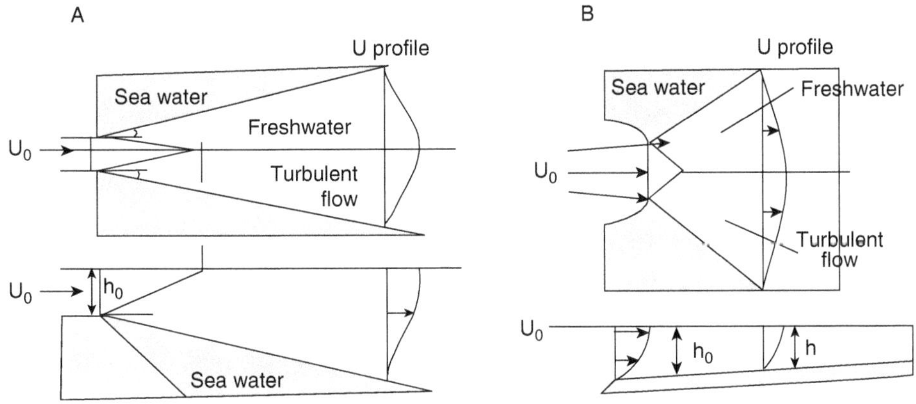

Figure 10.13 Plan view (top) and longitudinal cross-section (bottom) of turbulent axisymmetric river mouth jets. (A) Fully turbulent jet, not constrained by the water depth. (B) Turbulent jet with bed shear, constrained by the water depth. Modified from Wright, L.D., 1977. Sediment transport and deposition at river mouths: a synthesis. Geological Society of America Bulletin, 88, 857–868, 1977.

continental slope of South America, where most of its terrigenous load accumulates. In areas where the expansion of the plume is constrained by shallow water depths, important shear stress forces develop on the seafloor. As a consequence, the velocity of the flow decreases rapidly and the plume expands on the inner shelf, where most of the terrigenous load accumulates. For example, the plume of the Mississippi River is constrained by shallow water depths of the Gulf's shelf. Resulting nearshore accumulation of sediments produces radial bars which further limit the expansion of the flow and facilitate repeated channel avulsion.

Other factors controlling the sedimentation processes are the magnitude of the tides and the action of the waves. Their relative importance shapes the morphology of the deltas (Figure 10.14):

- River-dominated deltas develop in areas of limited tidal range (microtidal) and wave action. The river flow continues into the sea and sedimentary processes progressively build sub-aqueous levees, in continuity with the channel, which increase in size through time. Channel avulsion during high flows creates new distributaries, progressively giving a characteristic "bird-foot" aspect to the delta. However, the morphology of the abandoned channels and related levees is progressively modified by marine processes. A typical example is the Mississippi delta of North America.
- Tide-dominated deltas develop in areas of significant tidal range (macrotidal). Tidal currents interfer with the river flow and the areas of sediment accumulation vary with the tides, part of the sediment accumulating in mudflats. Tidal currents control the erosion of freshly deposited sediments, progressively shaping a delta where elongated mudflats and banks are separated by a number of tidal channels. A typical example is the Ganges delta of India.
- Wave-dominated deltas develop in areas of low tidal range where the energy of the waves is important. As the waves and river flow converge, the wavelength and

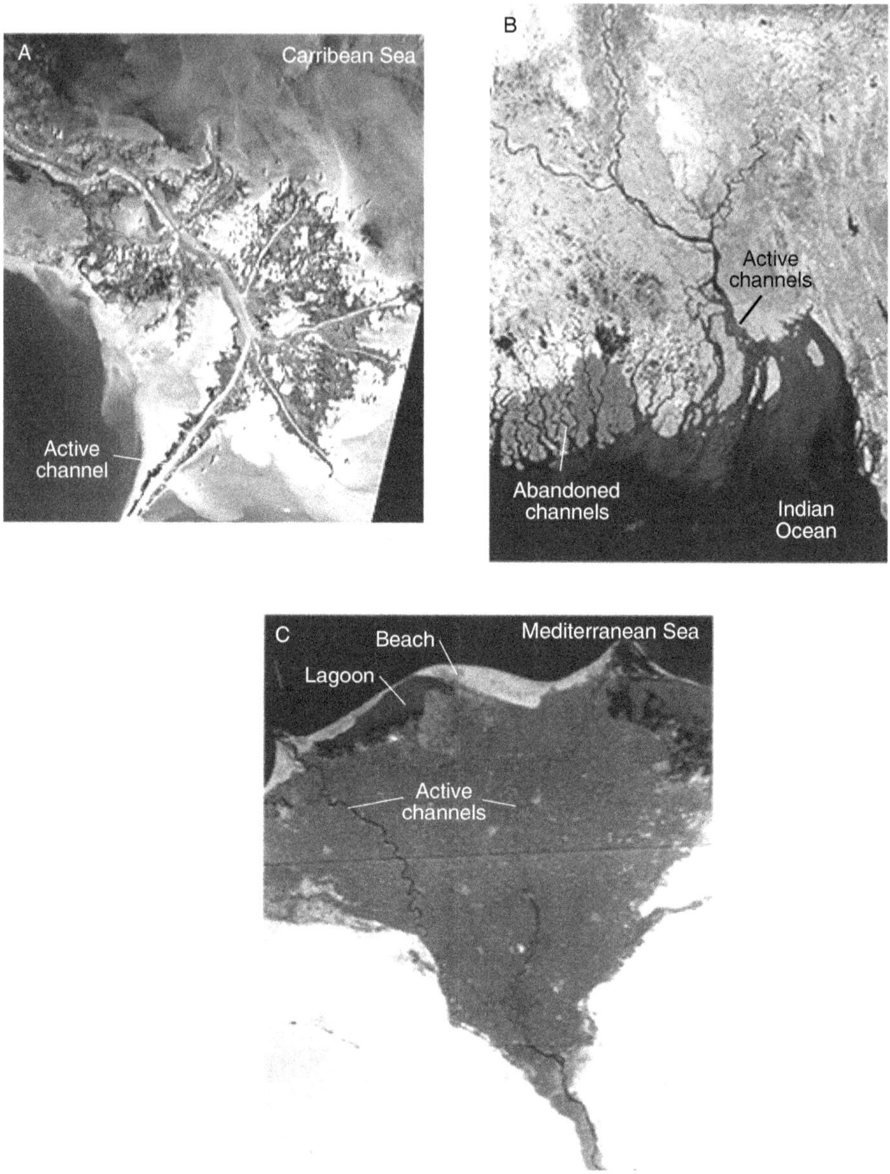

Figure 10.14 Morphology of major types of deltas. (A) River-dominated delta of the Mississippi River of North America. Note the "bird-foot" aspect of the active part of the delta. (B) Tide-dominated delta of the Ganges river of India. Note the elongated islands in the active part of the delta. (C) Wave-dominated delta of the Nile river of North Africa. Note the large beaches isolating coastal lagoons. Courtesy of NASA, http://visibleearth.nasa.gov

velocity of the waves decrease while they increase in height. Also, the refraction of the waves concentrates their energy around the plume of fluvial water. As a consequence, the waves may break to some distance from the shore, facilitating the mixing of fluvial and marine waters and related sedimentary processes. A proportion of the sediment accumulates as mouth bars which are progressively reworked alongshore with changing wave regimes. Reworked sediments form a succession of beach spits which in some cases isolate coastal lagoons from the open sea. However, mouth bars may hamper the discharge of fluvial waters and facilitate channel avulsion, especially during flood events. Examples include the Rhône delta of southern Europe and the Nile delta of North Africa.

As a delta progresses it builds a delta plain where active and abandoned channels coexist and where vegetation develops (Figure 10.15). Sediments accumulate at the delta front principally which progrades onto the continental shelf, whereas the most buoyant particles are transported further away to the prodelta which grades to typical oceanic sediments. Fresh sediments of high water content are instable, especially when they accumulate at high sedimentation rates and/or rest on a slope (see Section 2.3.8). Therefore, sediments which accumulate in the delta front area are easily removed as gravity flows, and especially as turbidity currents. Many modern river systems are in continuity with submarine canyons, where turbidity currents flow downslope to deep-sea fans where they release their terrigenous load in the form of turbidites (see Section 6.1.2). It is remarkable that major deltas and deep-sea fans are generally characteristic of the subsiding passive margins of mature

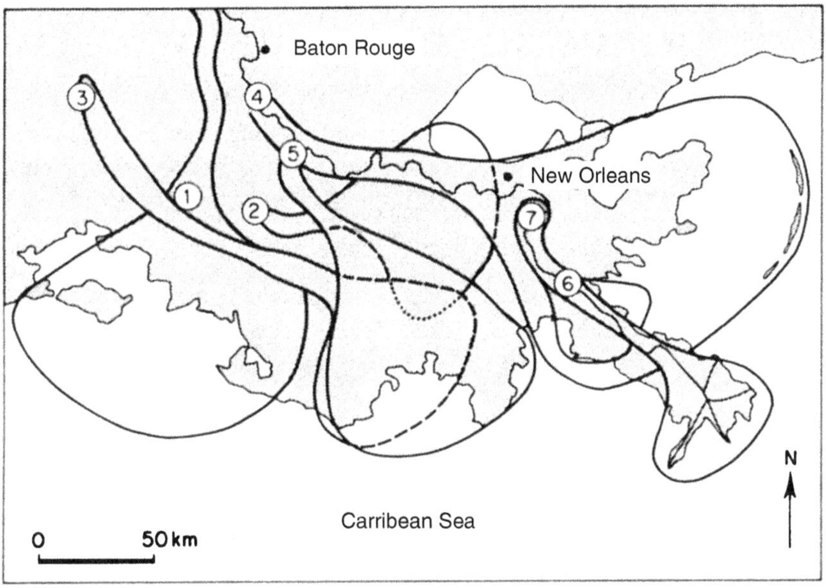

Figure 10.15 Successive (1–7) Quaternary lobes of the Mississippi delta. The currently active lobe is 7. Compare the "bird-foot" aspect of the active lobe to that of the abandoned lobes which have been reshaped by nearshore processes. Modified from Selley, R., 2000. Applied sedimentology. Academic Press, London.

divergent oceans of low latitudes, where broad drainage basins reaching far into a hinterland of high relief are associated to high precipitation.

10.2.2. Erosion and Transport by Ice

In cold areas where precipitation occurs in the form of snow, alternation of sublimation and recrystallization induced by daily variations in temperature progressively transform snow crystals into ice crystals. As snow and ice accumulate, the pressure they exert facilitates the expulsion of air from the ice which decreases in porosity until a mass of compact ice is formed. For instance, the porosity decreases from 90% to about 10%.

As ice forms, the hydrogen bondings which already connect the molecules of water in the liquid phase increase in number until each molecule is connected to four adjacent molecules (Figure 10.16). Because of the configuration of individual molecules, their connection via hydrogen bondings creates hexagonal structures isolating empty spaces. One consequence is that ice is of lower density (and higher volume) than liquid water and floats on its own liquid phase. Another consequence is that an increased pressure may weaken the hydrogen bondings, in turn slightly decreasing the temperature of the melting point of ice (Figure 10.17). In other words an increased pressure facilitates the transition from a solid to a liquid state, providing that the temperature of ice is sufficiently close to that of its melting point. When ice melts, the hydrogen bondings decrease in number within a transitional layer where fragments of ice and liquid water coexist. This transitional layer acts as a lubricant, facilitating the motion of objects along the melting surface.

Glaciers are ice streams which erode the continental surface and transport terrigenous elements which accumulate where the ice melts. Glaciers flow under the pressure of their own weight as ice accumulates, and may coalesce to form ice fields. If high precipitation is associated to limited melting, the ice fields may grow as to form an ice cap. Ice under pressure behaves as an extremely viscous fluid, and at high viscosities the flow is laminar. To simplify, glaciers are considered as viscous,

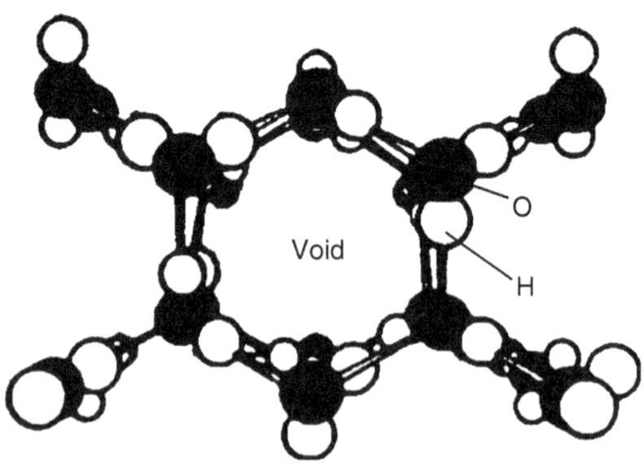

Figure 10.16 Arrangement of water molecules in a solid state. Note the hexagonal structure isolating empty spaces. Modified from Van Vliet-Lanoë, B., 2005. La planète des glaces: histoire et environnements de notre ère glaciaire. Vuibert, Paris.

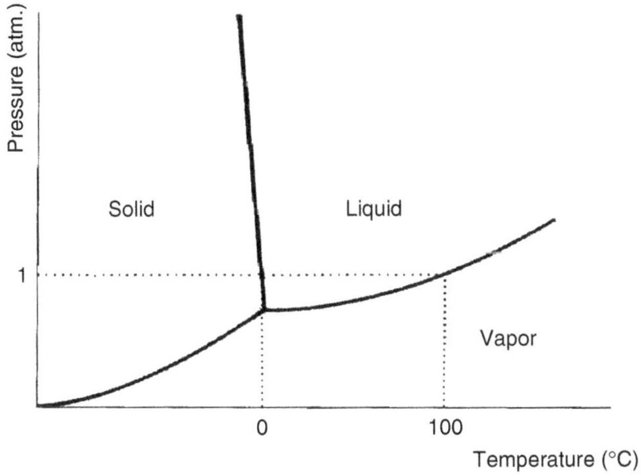

Figure 10.17 Phase diagram of water. Note that an increased pressure facilitates the transition from solid to liquid. Modified from Atkins, P., 1992. General chemistry. Scientific American Books, New York.

laminar flows of low velocity. Velocities of a few hundred meters a year are common, but may increase by a factor 100 during glacial surges. Many surges occur during intervals of higher precipitation (which increase the quantity of ice in upstream areas and the pressure exerted on the glacier) and higher temperature (which increase the plasticity of the ice and the volume of melting waters). Velocities also vary vertically within the ice flow and horizontally with the continental morphology, extending and compressing the ice where shear planes and fractures develop. Where the ice flow is obstructed by a hill for example, the pressure exerted on the obstacle facilitates ice melting and the production of meltwater which flows at the base of the glacier and increases basal slip and downstream velocity. As the ice melts, energy is required from the environment to overcome molecular attraction (80 cal/g). This loss of energy decreases the temperature of the remaining ice, facilitating the persistence of glaciers at temperatures close to that of the melting point in temperate environments. A water table may develop in areas where melting occurs not only at the base of the glacier but also over its entire surface, the water being released at the end of the glacier. In high-latitude areas, many glaciers reach sea level and break into icebergs which drift away from the continent before melting. This is especially the case in Antarctica where glaciers coalesce and float over the ocean's surface (Figure 10.18), feeding the ice shelves which ultimately break into tabular icebergs, some of them of vast dimensions (Figure 10.19).

Glaciers are powerful agents of erosion which operate in two different ways. Glacial plucking is the removal of loosened fragments of bedrock which range in size from sand and gravel to large boulders. The fragments are incorporated into the ice and are especially abundant at the base of the glacier (sub-glacial load) where they actively participate in the abrasion of the substrate. Glacial abrasion is an effective agent of erosion and may remove large quantities of bedrock as sand-sized

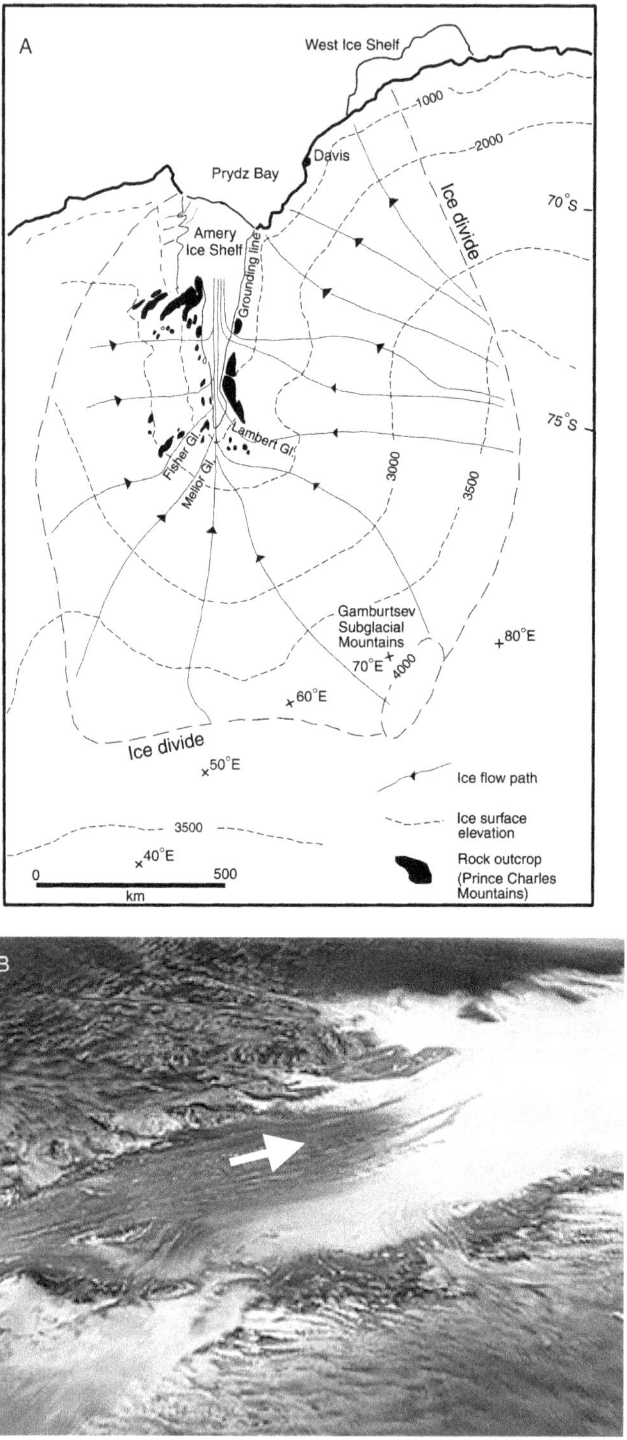

Figure 10.18 The Lambert Glacier and Amery Ice Shelf of Antarctica. (A) Drainage basin of the Lambert Glacier and adjacent glaciers. Note the convergence of flow paths and integration of glaciers to the ice shelf. Reprinted from O'Brien, P.E., Cooper, A.K., Richter, C. et al., 2001. Proceedings of the Ocean Drilling Program, Initial Reports, volume 188, Ocean Drilling Program, College Station, TX. (B) Satellite photo of the downstream part of the Lambert Glacier. Courtesy of NASA, http://visibleearth.nasa.gov

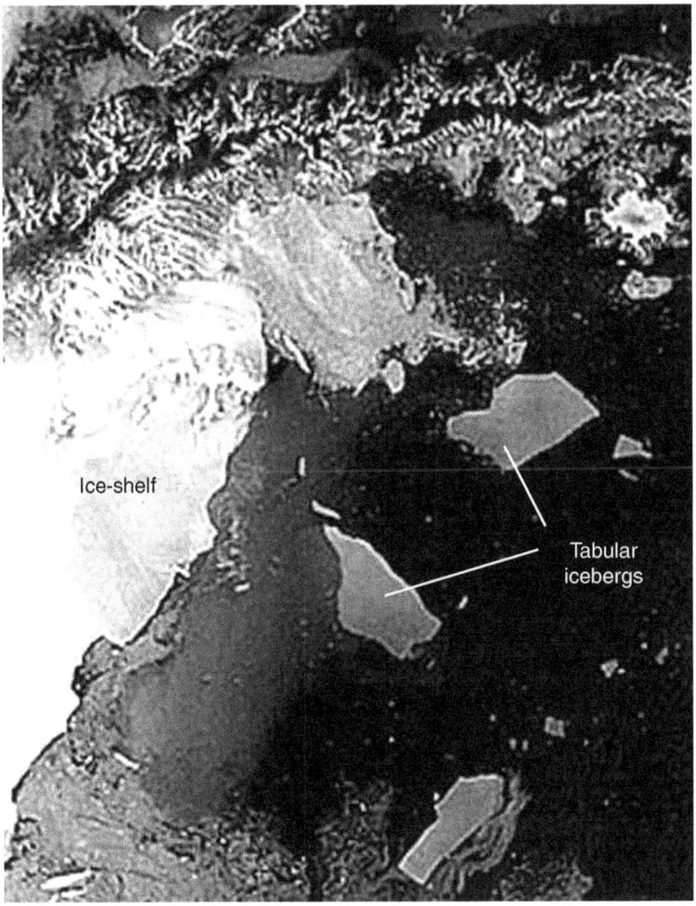

Figure 10.19 Tabular icebergs detached from the Antarctic Peninsula ice shelf, and drifting into the Weddell Sea. Courtesy of NASA, http://visibleearth.nasa.gov

and especially silt- to clay-sized particles (rock flour). The efficiency of glacial abrasion increases with the concentration of coarse fragments of bedrock at the base of the glacier, the contrasts of hardness between these abrasives and the substrate as the glacier moves to areas of different lithologies, the velocity of the glacier, and the quantity of meltwater at the base of the glacier. Meltwater increases the velocity of the glacier but also transports part of the rock flour downstream, improving the effectiveness of abrasion. Besides, glaciers may also transport at their surface materials eroded from adjacent slopes (supra-glacial load). Sub-glacial as well as supra-glacial debris may move along shear planes and fractures, to the interior of the ice mass (englacial load). Chemical weathering being limited in cold areas and impossible below ice cover, glaciers principally erode and transport fragments of bedrock, and the clay minerals found in glacial deposits are largely derived from outcrops of ancient sediment series.

Because of the extremely high viscosity of ice flows, glaciers transport a wide range of sizes of terrigenous elements, and glacial deposits are unsorted. Glacial sediments accumulate as ice melts, i.e. at the end and on the sides of glaciers in temperate areas (moraines). Glacial deposits accumulate as massive diamictites in continental shelf to slope environments in cold, high-latitude regions where glaciers reach sea level or build ice shelves (Figure 10.20). However, a significant proportion of rock flour remains in suspension and is further transported by oceanic currents, flocculation being generally limited by the scarcity of negatively charged clay minerals

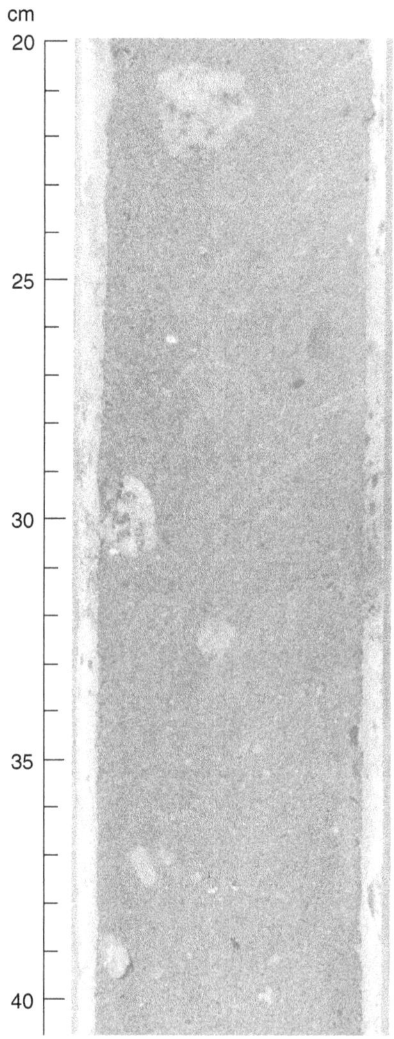

Figure 10.20 Typical massive diamictite with assorted gneissic clasts sampled at ODP Site 739, Prydz Bay (Section 119-739C-4R-3). Reprinted from Barron, J.A., Larsen, B., Baldauf, J., et al., 1989. Proceedings of the Ocean Drilling Program, Initial Reports, Volume 119, Ocean Drilling Program, College Station, TX.

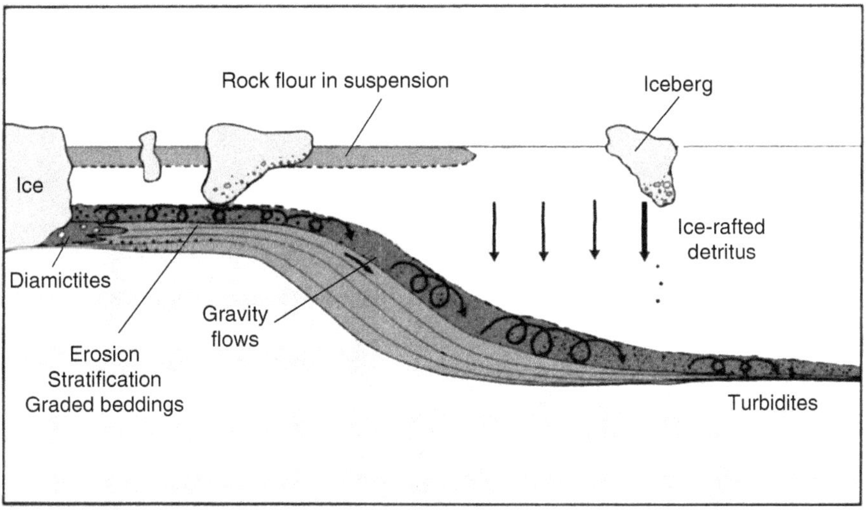

Figure 10.21 Summary of major glacial and glacio-marine processes of sedimentation. Note increasing influence of marine processes with distance from ice shelf. Modified from Van Vliet-Lanoë, B., 2005. La planète des glaces: histoire et environnements de notre ère glaciaire. Vuibert, Paris.

in glacial loads. Fresh glacial sediments of high water content are unstable and likely to be reworked as gravity flows and by storm waves and marine currents. The influence of marine processes increases with the degree of reworking. The resulting glacio-marine sediments show evidence of erosion, graded beddings, cross-stratification and bioturbation. They range from stratified diamictites to beach deposits in coastal areas, and to channel fills and turbidites (Figure 10.21) grading to hemipelagic and pelagic sediments in distant areas of the outer margin (see Section 2.3). Calving glaciers continuously produce icebergs which drift away under the combined influence of surface currents and winds, and melt as marine and atmospheric temperatures increase. The unsorted terrigenous load of icebergs is released in the deep ocean as ice melts. These ice-rafted detritus fall under the influence of gravity and are incorporated into the sediment underneath. Ice-rafted detritus are generally present as layers or isolated dropstones into finer-grained terrigenous, hemipelagic, or biogenic sediments (Figure 10.22). They increase in abundance in areas of intense ice melting, such as the polar convergence where cold air and surface water masses meet warmer air and surface water masses of sub-tropical origin.

10.2.3. Erosion and Transport by Winds

Air masses are fluids of very low density and viscosity which most of the time behave as turbulent flows in the lower part of the atmosphere, i.e. the troposphere which extends from the Earth's surface to an altitude of 10–16 km. The circulation of air masses is caused by pressure and temperature gradients and affected by the same parameters as the circulation of water masses (see Section 2.2):

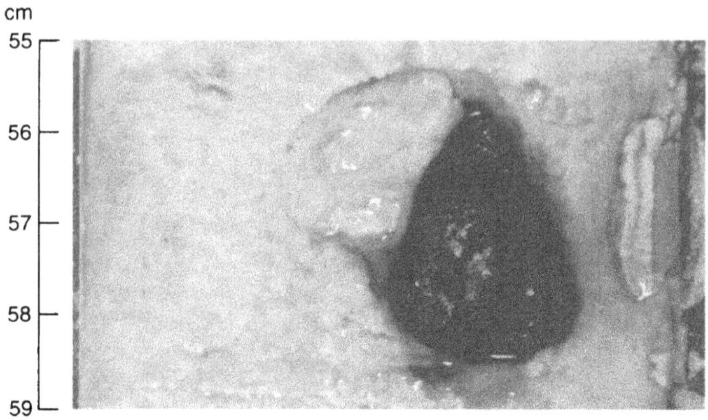

Figure 10.22 Isolated dropstone in a Pliocene diatom ooze, sampled at ODP Site 689 on Maud Rise in the Weddell Sea (Section 113-689D-1H-2). Reprinted from Barker, P.F., Kennett, J.P., O'Connell, S., et al., 1988. Proceedings of the Ocean Drilling Program, Initial Reports, volume 113, Ocean Drilling Program, College Station, TX.

- Circulating air masses are deflected by the Coriolis force, resulting in geostrophic winds which flow parallel to the isobars.
- The conservation of angular momentum increases wind velocities at high latitudes.
- The friction between air layers modifies the wind speed and direction, in the same way as friction between water layers generates Ekman's spirals.

Because air masses have much lower densities and viscosities than water masses they circulate at much higher velocities and the friction they exert on the ground generates a much thicker boundary layer (500–1,000 m compared to 1–10 m). The position of the adsorbed layer (where velocities decrease to near zero) varies with the nature of the substrate, from near surface in the case of still waters and bare bedrocks to an altitude of several meters (and up to 20–30 m) in the case of dense forests which shelter the ground from the action of winds. In accordance with the physical properties of air masses and the primary climatic control of their pattern of circulation, higher surface wind velocities are recorded over oceanic surfaces of low frictional resistance, especially at high latitudes, whereas wind energy is principally spent in coastal areas.

As air masses circulate at high velocities, loose particles of regoliths experience high lift forces and are transported with high kinetic energies (see Section 2.3). Because of the low viscosity of air masses, their capacity to transport particles is limited to the smallest grain-sizes (fine sand to clay) and the terrigenous load of air masses is well-sorted. Winds are able to remove silt- and clay-sized particles as soon as they reach velocities of 0.1–0.3 m/s, but wind velocities of 20–30 m/s are required to move particles around 2 mm in size.

Particles of density close to that of quartz (i.e. d ‡ 2.65 g/cm^3) and more than 0.5 mm in size (coarse and very coarse sand) generally do not lose contact with the surface where they roll and slide, except for short trajectory paths during storms of

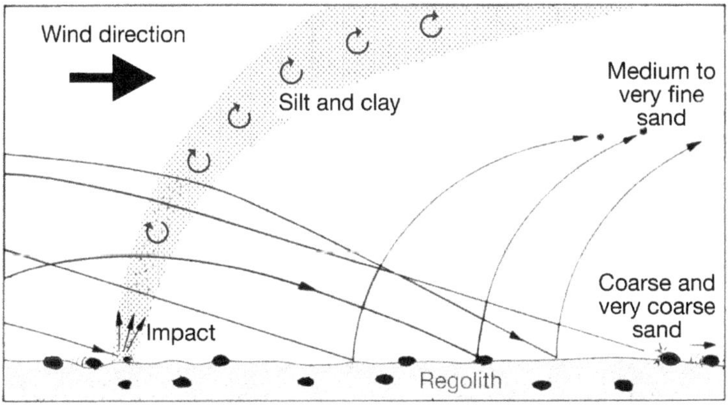

Figure 10.23 Summarized trajectory paths of windblown particles. Note the sorting of grain-sizes and the role of saltating sand mobilizing loose particles of the regolith. Modified from Hamblin, W.K., Christiansen, E.H., 2004. Earth's dynamic systems. Prentice-Hall, Upper Saddle River, NJ.

anomalously high wind intensities (Figure 10.23). Medium to very fine sand, which range in size from 0.5 to 0.063 mm are principally transported by saltation and modified saltation (i.e. including intervals of suspension depending on the wind strength) at elevations ranging from a few centimeters to about 1 m above the ground and over short distances. The high kinetic energy of the particles is spent as they reach the ground, further facilitating the removal of particles from the regolith. Saltating particles may progressively build well-sorted sand dune fields (ergs) in desert areas. Particles below 0.063 mm in size are principally transported in suspension. Coarse and medium silt (0.063–0.016 mm in size) are transported as short-term suspensions at elevations up to several hundreds meters and over distances ranging from several tens to several hundreds of kilometers. They generally settle as the wind speed decreases, in areas adjacent to the deserts where they may form thick loess deposits. For example, Chinese loess were windblown from the Asian deserts during glacial stages of the Pleistocene. Only particles below 0.016 mm in size (fine silt, clay and colloid) are transported at elevations of several kilometers (and up to the tropopause) and over distances of several thousands of kilometers. These particles may remain in suspension for weeks and form the bulk of the eolian dust which circulates over the oceans. The finest particles can hardly settle, even for very low wind velocities. They generally act as nuclei of condensation, and are washed to the ground together with the precipitation (see Section 2.2). For example, African dust is at the origin of "red rain" and "red snow" events which commonly occur in Mediterranean regions and beyond. Because most winds are unable to move particles over 2 mm in size, they may progressively accumulate at the surface where they form desert pavements (regs).

Desert pavement protects the regolith against further wind erosion (Figure 10.24), limiting the potential of tropical arid areas as durable sources of eolian dust. Besides, surface moisture improves the cohesion of the regolith and drastically increases the critical velocity allowing particle removal. However, some

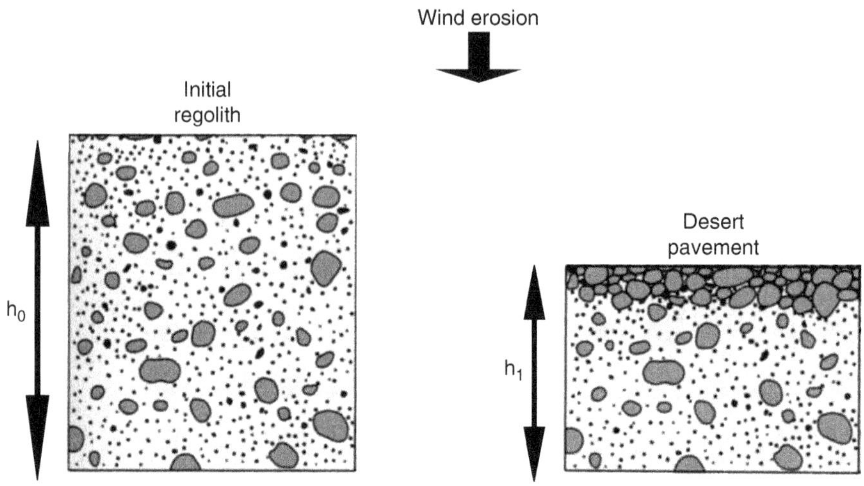

Figure 10.24 Removal of sand- to clay-sized particles from a regolith, and formation of desert pavement. The desert pavement results from the concentration of the coarsest particles at the surface and protects the regolith from further wind erosion. Modified from Hamblin, W.K., Christiansen, E.H., 2004. Earth's dynamic systems. Prentice-Hall, Upper Saddle River, NJ.

arid Saharan regions are currently major sources of eolian dust (Figure 10.25). Among them, the Bodélé depression is located in a very dry area where streams from adjacent mountains transported large quantities of terrigenous silt which accumulated in an expanded Holocene lake Chad. Other Saharan sources include alluvial plains which formed during humid Holocene intervals. Potential sources of eolian dust are principally found in tropical semi-arid areas, where intervals of precipitation ensure some degree of chemical weathering and the formation of small-sized clay minerals in the regolith. Subsequent aridity and a limited vegetation allow the removal of particles by wind activity. The succession of humid and arid intervals may either have a seasonal character or be caused by periodic changes in atmospheric circulation, like El-Niño Southern Oscillation (ENSO) events. Therefore, major and durable sources of eolian dust include tropical regions where semi-arid climatic conditions coincide with significant wind activity. Typical examples are found in the western Saharan and Namibian regions of Africa, and in the Sonoran region of North America.

Wind erosion principally occurs during dust storms (Figure 10.26). Most dust storms are caused by the horizontal atmospheric circulation (see Section 2.2.4) and more particularly by low-pressure corridors where the saturation vapor pressure of the air increases as they flow over warm continental areas. The winds associated to these dry depressions lift and transport the loose particles of the regolith. Dust storms are also caused by monsoonal winds in some East African and Asian regions, and by katabatic winds in mountain areas such as northern Chile and California. Local conditions may increase the efficiency of winds as agents of erosion. For example, the removal of Holocene silt from the Bodélé depression of southeast Sahara is made easier because surface winds increase in velocity as they are channeled between the mountains adjacent to the depression. Eolian dust falls and

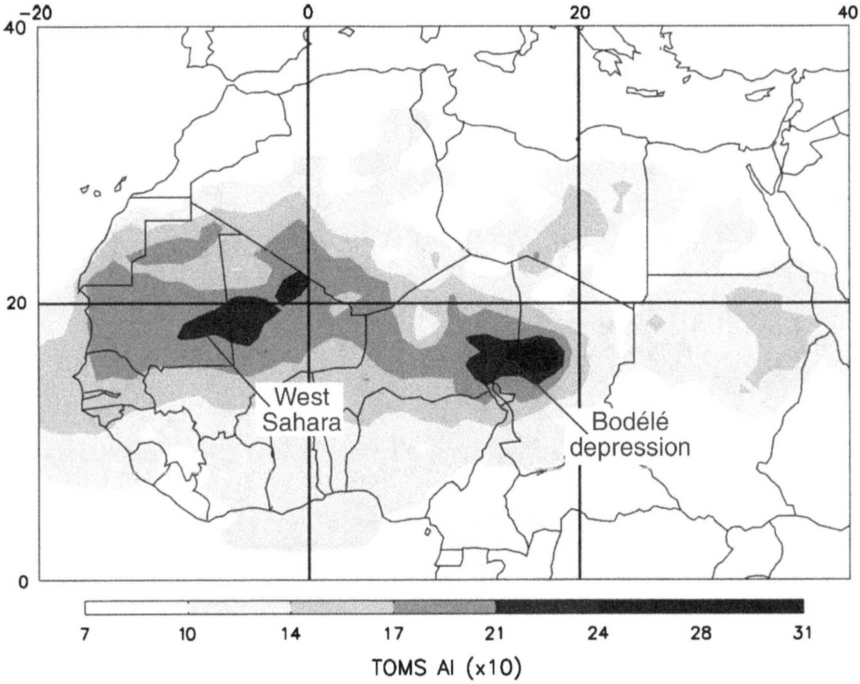

Figure 10.25 Major sources of Saharan dust, from the distribution of average values (for the interval 1980–1992) of the Total Ozone Mapping Spectrometer absorbing Aerosol Index (TOMS AI) over West Africa. Highest concentrations define major source areas. The Bodélé depression is currently the most important source of dust emissions in the world. Reprinted from Engelstaedter, S., Tegen, I., Washington, R., 2006. North African dust emissions and transport. Earth Science Reviews, 79, 73–100.

accumulates in significant quantities in ocean basins close to major source areas, decreasing with distance from the source areas (Figure 10.27). For example, annual mass budgets show that most eolian dust of Saharan origin settles in the Atlantic Ocean within 2,000 km from the African coast, but that about 20% of the dust still remains in suspension beyond 5,000 km. As a consequence, Saharan dust accumulates at rates up to 20 cm/kyr within 2,000 km from the African coast, decreasing to 0.2 cm/kyr across the ocean.

10.3. The Fine Terrigenous Fraction in the Ocean

Most terrigenous elements brought to the ocean by river (and glacier) systems accumulate by decantation and flocculation near river mouths (and glacier calving areas), and are further reworked by gravity flows. However, part of the terrigenous load is maintained in suspension and further transported by oceanic currents. As water masses circulate in the ocean, they may also collect windblown dust falling from the atmosphere. This composite suspended load consists of silt- to clay-sized particles

Figure 10.26 Dust storm over West Africa and the Atlantic Ocean. Note the extension of the plume. Courtesy of NASA, http://visibleearth.nasa.gov

which include a significant proportion of clay minerals derived from continental weathering. Clay minerals are sheet silicates and their platy morphology gives them an excellent buoyancy. In theory, most clay minerals settle for flow velocities below 1 cm/s and therefore they can be transported over great distances by oceanic currents. Most of the time, flocculation facilitates their accumulation on the seafloor where they are mixed with other terrigenous elements and especially with biogenic components in hemipelagic and pelagic sediments. The proportion of terrigenous elements varies from 70% to 30% in hemipelagic sediments, but may be as low as 2–3% in pelagic sediments. However, high proportions of fine terrigenous elements are locally recorded in pelagic sediments below the carbonate compensation depth, because of the dissolution of calcareous biogenic elements (see Section 11.1.3). The distribution of clay minerals in the ocean reflects the zonation of continental weathering, which is primarily controlled by climatic conditions. For example, kaolinite (an end-product of chemical weathering) dominates in many oceanic sediments from low latitudes, whereas chlorite (which forms during early stages of chemical weathering) increases in abundance in many mid- to high-latitude areas. This is especially visible from the distribution of the kaolinite/chlorite ratio in Atlantic sediments (Figure 10.28). Yet, this primary distribution of clay minerals is altered in different ways:

- Rivers of extended watersheds may carry to the ocean terrigenous elements eroded from regions of contrasted continental weathering conditions. For example, the Amazon brings kaolinite from well-drained tropical areas to the Atlantic Ocean, in accordance with regional conditions of chemical weathering and climate. However, the Amazon also removes illite and chlorite (together with

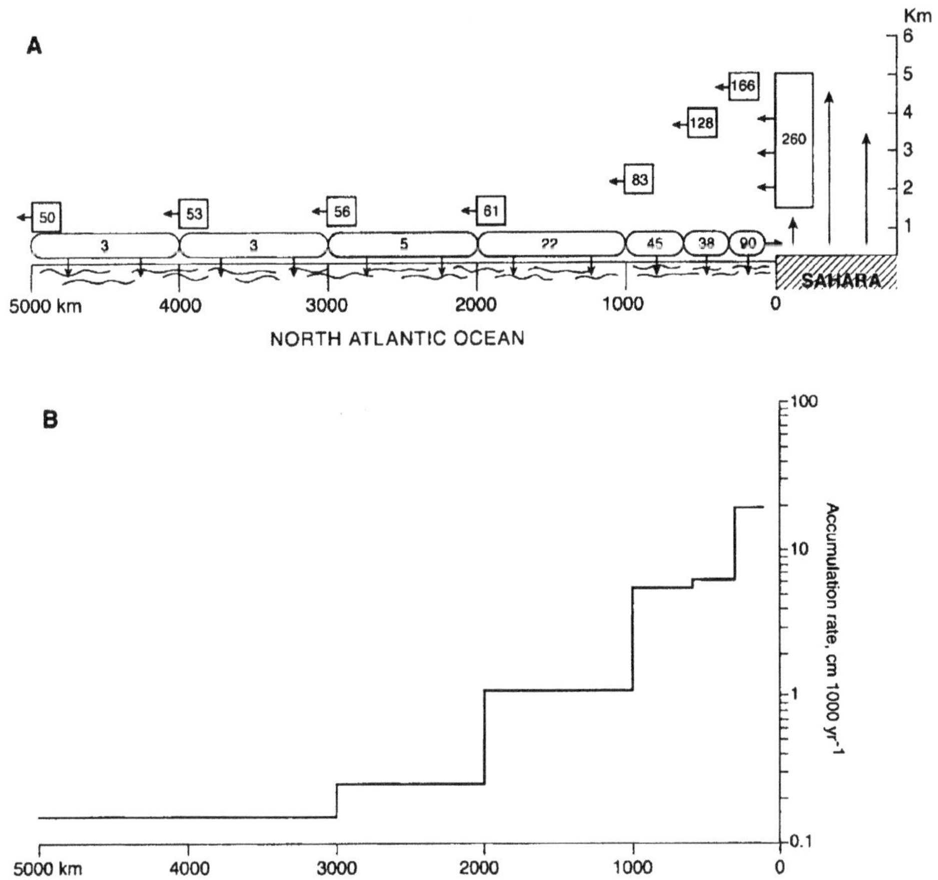

Figure 10.27 Dust transport and accumulation offshore from the Sahara. (A) Mass budget of dust transport over the Atlantic at 15–24°N (10^6 tons/year). (B) Accumulation rates of Saharan dust in marine sediments. Reprinted from Goudie, A.S., Middleton, N.J., 2001. Saharan dust storms: nature and consequences. Earth Science Reviews, 56, 179–204.

other terrigenous elements) from high-altitude areas of the Andes where chemical weathering is limited and physical weathering is dominant, as well as smectite which locally develops in floodplain areas.

- Major oceanic currents may transport significant quantities of terrigenous particles over great distances. For example, kaolinite of tropical origin is transported southward along the continental margin of South America by the Brazil current. Illite and chlorite eroded from Greenland and Canada are transported southward along the continental margin of North America by the Labrador current and the North Atlantic Deep Water mass. Also, illite and chlorite from southern high-latitudes are transported by surface and intermediate waters as far as shallow depths of the Walvis Ridge in the tropical South Atlantic.
- Rivers, glaciers and oceanic currents may carry terrigenous elements eroded from ancient sediment series. For example, Alaska rivers bring to the Arctic Ocean a

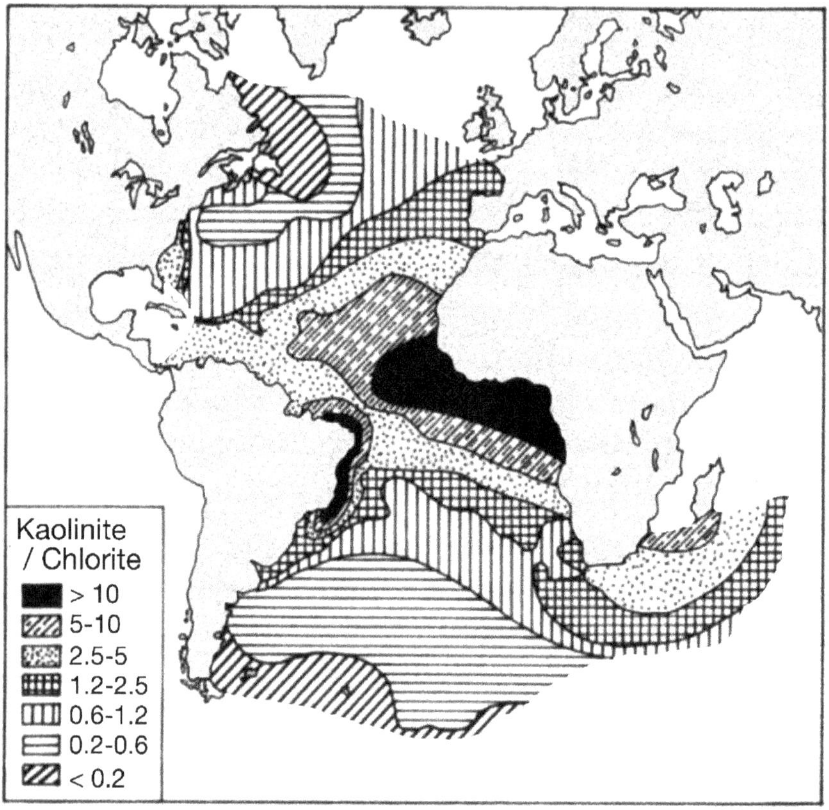

Figure 10.28 Distribution of the kaolinite/chlorite ratio in the Atlantic Ocean. Note high concentrations of kaolinite at low latitudes and on the path of western boundary currents. Higher concentrations of chlorite are visible at high latitudes as well as on the path of cold surface and intermediate currents. Modified from Chamley, H., 1999. Clay sedimentology. Springer, Berlin.

proportion of kaolinite eroded from Eocene sediments. Also, Antarctic Bottom Water currents bring to the deep basins of the South Atlantic significant proportions of smectite eroded from Cretaceous series of the Antarctic margin principally.

- Locally, submarine alteration of basaltic materials to smectite may alter the climatic zonation of clay minerals.

10.4. Diagenesis of Terrigenous Sediments

Coarse siliciclastics have relatively low porosities (frequently below 60%), and therefore resist compaction. They also have high permeabilities, which facilitate the circulation of pore fluids. As a consequence, chemical diagenesis plays a key role in the diagenetic alteration of coarse siliciclastic sediments such as sands (see Section 2.3.10). During early diagenetic stages, micro-crystals of Fe sulfides may develop within the pore space of the sulfate reduction zone. Where carbonates are available, siderite may also develop. As pressure increases, dissolution of siliciclastic grains and resulting pressure solution may favor the formation of silica crystals such as quartz within the pore space. However, circulating pore fluids provide the largest part of the chemical elements which precipitate and progressively cement coarse siliciclastic deposits, which decrease in porosity and permeability. Quartz, carbonates, sulfates and Al-silicates are among the most common diagenetic cements of coarse siliciclastic sediments. The nature of the cement may change through time with the origin and circulation path of pore fluids, for example in active margin areas. The changing characteristics of pore fluids may also enhance the dissolution of primary siliciclastic grains and/or diagenetic cement, increasing in turn the porosity of the sediment (secondary porosity).

Conversely, fine siliciclastic sediments where clay minerals dominate have relatively high porosities (frequently around 80%) and very low permeabilities. Clay minerals are sheet silicates which show long and thin, platy morphologies. Platy particles present edge-to-edge and edge to face contacts, isolating relatively important, but barely interconnected, pore intervals (Figure 10.29). A preferred orientation of sheet silicates progressively develops as pressure increases, perpendicular to the dominant stress direction. As a consequence, the porosity of the sediment strongly decreases. Fine siliciclastic sediments are therefore very responsive to compaction. Because of very low permeabilities, clay-rich, fine

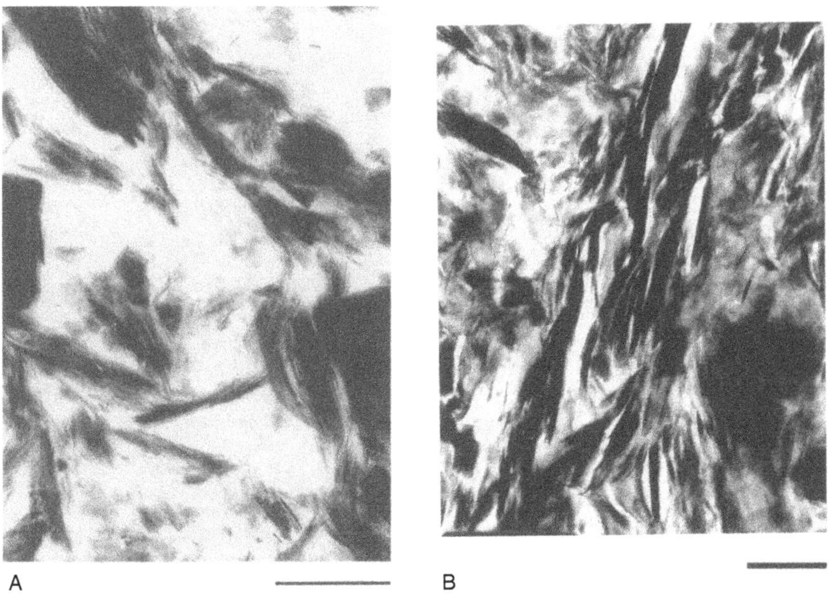

Figure 10.29 Transmission electron micrographs of ultrathin sections, ODP Site 697, Weddell Sea. (A) Highly random orientation of clay particles in an early Pliocene mud. Note the relatively high porosity and edge-to-face contacts. Scale bar: 0.5 μm. (B) Preferred particle orientation in a Pleistocene mud. Note the relatively low porosity. Scale bar: 2 μm. Reprinted from Barker, P.F., Kennett, J.P., O'Connell, S., Pisias, N.G. (Editors), 1990. Proceedings of the Ocean Drilling Program, Scientific Results, volume 113, Ocean Drilling Program, College Station, TX.

siliciclastic sediments are barely sensitive to chemical diagenesis. For example, rare earth elements compositions of clay minerals generally reflect strong continental influences, even in significantly compacted mudstones. In some cases however, clay minerals characterized by free valences in both tetrahedral and octahedral layers, such as smectites, may recrystallize in the form of fine laths (Figure 10.30). The recrystallization of smectite particles is not associated to significant changes in their chemical composition, and is attributed to minor intake in silica from the pore fluid and related chemical reorganization of the particles.

Chemical diagenesic alteration of clay minerals occurs in high-pressure areas, for example in accretionary prisms and at important burial depths (beyond 2 km) of passive margin and basin areas. There, the concentration in chemical elements such as K, Na and Si may increase in the pore environment because of the presence of pressure solutions or circulating fluids. Those chemical elements may come from the dissolution of other minerals, for example terrigenous feldspars. They tend to fill the free valences in clay mineral layers. Clay minerals of 2/1 type with free valences in both tetrahedral and octahedral layers, such as smectites, are the most vulnerable to chemical diagenetic alteration. In some cases, the incorporation of K,

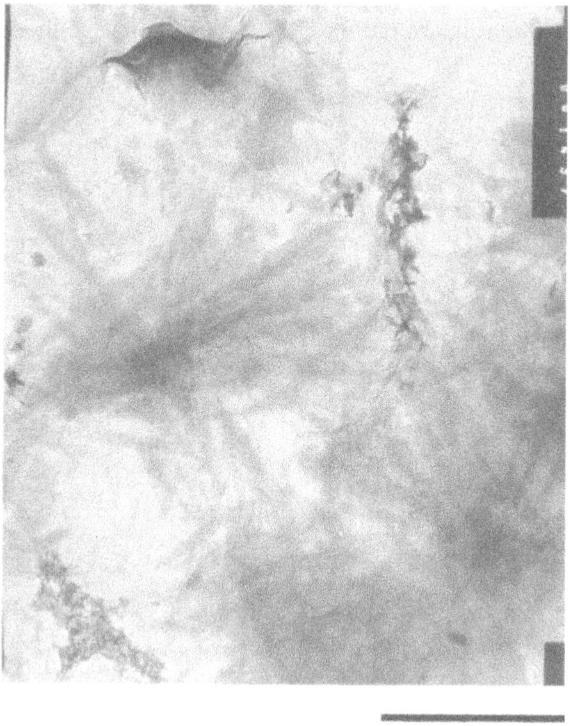

Figure 10.30 Very thin laths of smectite, sometimes associated in figures with 60° angles, in a middle Eocene mudstone from the Weddell Sea (Section ODP 113-696B-62R-4). Scale bar : 1 μm. Reprinted from Barker, P.F., Kennett, J.P., O'Connell, S., Pisias, N.G. (Editors), 1990. Proceedings of the Ocean Drilling Program, Scientific Results, volume 113, Ocean Drilling Program, College Station, TX.

Si and Na, and related reorganization of the clay mineral layers which decrease in thickness, progressively transform smectite layers into illite layers. In other cases, alternating illite and smectite layers, or illite layers only, may grow directly from solution onto a substrate. Illite development is facilitated by increased temperatures, but may start at temperatures as low as 50–100°C. Circulating pore fluids, and therefore the permeability of the sediment series, principally control the diagenetic development of illite.

FURTHER READING

Allen, P.A., 1997. Earth surface processes. Blackwell, London.
Chamley, H., 1999. Clay sedimentology. Springer, Berlin.
Erhart, H., 1967. La genèse des sols en tant que phénomène géologique. Masson, Paris.
Hamblin, W.K., Christiansen, E.H., 2004. Earth's dynamic systems. Prentice-Hall, Upper Saddle River, NJ.
Kennett, J.P., 1982. Marine geology. Prentice-Hall, Englewood Cliffs, NJ.
Martini, I.P., Chesworth, W. (Editors), 1992. Weathering, soils and paleosols. Elsevier, Amsterdam.
Pomerol, C., Lagabrielle, Y., Renard, M., 2005. Eléments de géologie. Dunod, Paris.
Van Vliet-Lanoë, B., 2005. La planète des glaces: Histoire et environnements de notre ère glaciaire. Vuibert, Paris.
Weaver, C.E., 1989. Clays, muds, and shales. Developments in sedimentology. Elsevier, Amsterdam.

Other references used in this chapter

Atkins, P., 1992. General chemistry. Scientific American Books, New York.
Brenon, I., Le Hir, P., 1999. Modelling the turbidity maximum in the Seine estuary (France): identification of formation processes. Estuarine, Coastal and Shelf Science, 49: 525–544.
Chen, M.S., Wartel, S., Temmerman, S., 2005. Seasonal variation of floc characteristics on tidal flats, the Scheldt estuary. Hydrobiologia, 540: 181–195.
Chen, M.S., Wartel, S., Van Eck, B., Van Maldegem, D., 2005. Suspended matter in the Scheldt estuary. Hydrobiologia, 540: 79–104.
Delofre, J., Lafite, R., Lesueur, P., Lesourd, S., Verney, R., Guézennec, L., 2005. Sedimentary processes on an intertidal mudflat in the upper macrotidal Seine estuary, France. Estuarine, Coastal and Shelf Science, 64: 710–720.
Engelstaedter, S., Tegen, I., Washington, R., 2006. North African dust emissions and transport. Earth Science Reviews, 79: 73–100.
Garnaud, S., Lesueur, P., Clet, M., Lesourd, S., Garlan, T., Lafite, R., Brun-Cottan, J.-C., 2003. Holocene to modern fine-grained sedimentation on a macrotidal shoreface-to-inner-shelf setting (eastern bay of the Seine, France). Marine Geology, 202: 33–54.
Goudie, A.S., Middleton, N.J., 2001. Saharan dust storms: nature and consequences. Earth Science Reviews, 56: 179–204.
Selley, R., 2000. Applied sedimentology. Academic Press, London.
Thill, A., Moustier, S., Garnier, J.-M., Estournel, C., Naudin, J.-J., Bottero, J.-Y., 2001. Evolution of particle size and concentration in the Rhône river mixing zone: influence of salt flocculation. Continental Shelf Research, 21: 2127–2140.
Xia, X.M., Li, Y., Yang, H., Wu, C.Y., Sing, T.H., Pong, H.K., 2004. Observations on the size and settling velocity distributions of suspended sediment in the Pearl River estuary, China. Continental Shelf Research, 24: 1809–1826.

CHAPTER ELEVEN

Biogenic Sediments

Biogenic sediments contain a large majority of mineral fragments which are secreted by living organisms and preserved after their death. Biological activity is principally concentrated in the photic zone, i.e. the upper 100–200 m of the water column (see Section 2.3). Benthic organisms dominate in shallow, coastal areas, whereas pelagic organisms prevail in the deep ocean. Biological activity is especially important in areas where nutrients (Si, P, K, SO_4, NO_3, Fe, Mo, etc.) are made available from the deep ocean by upwellings (coastal upwellings, oceanic divergences), and from the continents by runoff. At ocean scale, the large majority of biogenic particles found in the sediment is derived from marine microorganisms. Drifting, planktonic species dominate over bottom-living, benthic species. Among a variety of marine microorganisms, a few groups produce mineral shells which are preserved as microfossils in marine sediments. Most shells are calcareous (calcite, Mg-calcite, aragonite) or siliceous (amorphous opal), but some organisms agglutinate their shells using terrigenous and organic fragments. Calcareous microfossils principally consist of foraminifers (zooplankton) and coccoliths (phytoplankton) but also include species from other groups such as ostracods. Siliceous microfossils principally consist of radiolarians (zooplankton) and diatoms (phytoplankton), but also include species from other groups such as silicoflagellates. In dysoxic to anoxic environments, organic-walled microorganisms, such as dinoflagellate cysts, also contribute to the sediment.

11.1. Calcareous Microfossils: Formation, Preservation, and Transformation

11.1.1. Foraminifers

Marine foraminifers are a very diversified group of protozoans. They belong to the Sarcodina, a class which includes the simplest protozoans. The foraminifer cell consists in a protoplasm which contains a nucleus (or several nuclei), Golgi bodies, mitochondria, ribosomes and vacuoles, and emits pseudopods which play a role in nutrition and excretion. Although the foraminifers comprise a few planktonic families and a variety of benthic families, the planktonic forms are by far the most abundant in the ocean. Foraminifers are present in all marine environments at all latitudes. They range approximately from 50 to 400 μm in size and secrete a calcareous test made of calcite, with the exception of some deep benthic species (Textularinae) which agglutinate their test using terrigenous and organic fragments. Proteins at the surface of the protoplasm probably control the construction of the test. The test probably plays a role in the stabilization of the organisms, as well as in their protection against predators and adverse environmental conditions.

The oldest known foraminifers are simple, agglutinated, shallow benthic forms of Cambrian age. Radiations led to increased diversification and complexity later in the Paleozoic, including the development of multi-chambered and calcareous tests. In the Permian, the fusilinid group probably consisted of more than 5,000 species, with individuals up to 10 cm long. Following mass extinction at the end of the Paleozoic, diversification in the Early Mesozoic was slow until a major step in the evolution of the foraminifers occurred in the Jurassic: some groups migrated to deeper water depths of the ocean, whereas other groups adapted to a planktonic mode of life.

11.1.1.1. Planktonic foraminifers

Planktonic foraminifers first appeared in the Late Jurassic and developed considerably during the Cretaceous. Since then, more than 40 genera and 400 species have evolved. Because of this pattern of evolution, the planktonic foraminifers are widely used in stratigraphy. About 30 species, grouped into two families, are present in the modern ocean. Planktonic foraminifers have a life span of a few weeks, are non-motile, and drift. Their small size and surface rugosity, as well as the presence of spines, increase their specific surface area and therefore their frictional drag and buoyancy (Figure 11.1). As seawater decreases in viscosity with increasing temperatures, planktonic foraminifer species from low latitudes are characterized by a fine and porous shell with large apertures and well-developed spines for a better buoyancy. In contrast, species from cooler areas are characterized by a thicker and less porous shell, with small apertures and less developed spines. Because of the correlation between temperature, salinity and viscosity of seawater, planktonic foraminifers are reliable proxies for water mass conditions.

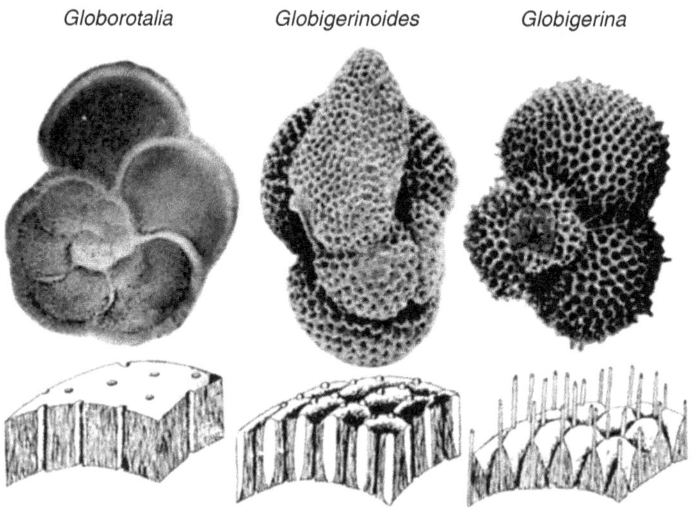

Figure 11.1 Morphology of the three main groups of planktonic foraminifers, and their wall texture. Note differences in test thickness and specific surface between the spinose *Globigerina* and *Globigerinoides*, and the non-spinose *Globorotalia*. Modified from Haq, B.U., Boersma, A. (Editors), 1978. Introduction to marine micropaleontology. Elsevier, Amsterdam.

Many planktonic foraminifer species occupy specific depth habitats or migrate vertically during their life cycle, and must therefore control their buoyancy. To do so, they adjust the size and thickness of their test, and the importance of their spines principally. For example, the spinose *Globigerinoides* live at shallower water depth (above 100 m), whereas the non-spinose *Globorotalia* live at greater water depth (below 100 m). Observations of the tropical species *Globigerinoides sacculifer* have shown that individuals reproduce around full moon, probably within the chlorophyll maximum, at approximately 80 m water depth. Although most juveniles die rapidly, some of them grow in size as they ascend to the surface. They reach the surface when they are around 100 μm in size, in about 18 days. They live near surface approximately 2–3 days until they are about 200 μm in size, and then steadily descend within 9–10 days to the reproduction depth where they undergo gametogenesis.

Laboratory cultures suggest that many planktonic foraminifer species increase in abundance and build a larger test in nutrient-rich waters where phytoplankton is plentiful. However, high productivity often limits light penetration by increasing the rate of diffusion and absorption of wavelengths. As a result, the production of oxygen by endosymbiont species decreases and reduces the respiration rate of symbiont-bearing foraminifers which decrease in size, balancing the consequences of high fertility. When high fertility is not combined with low oxygen contents, all species build larger than normal foraminiferal tests. Besides water fertility, planktonic foraminifers are highly sensible to changes in water temperature whereas they appear more tolerant to changes in salinity. The temperature ranges of planktonic foraminifers in culture are very similar to those in the ocean (Figure 11.2), but their salinity ranges in culture are much wider than those observed in the ocean. This observation highlights the dominant control of temperature on the geographic distribution of foraminifer species. Highest reproduction rates are reached for optimal temperature conditions, but beyond thresholds the species are unable to reproduce or die. Also, maximum porosities of foraminiferal tests are reached for high-temperature and low-salinity conditions (i.e. lowest viscosities of seawater) via an increase of the pore area and concomitant decrease in pore density.

In the ocean, the distribution of planktonic foraminifer species is mainly controlled by seawater temperatures. This has been known since the Challenger expedition (1872–1876), which established the latitudinal distribution of foraminifer assemblages (Figure 11.3). The diversity decreases from the low latitudes where the tropical assemblage includes 25 species, to the high latitudes where polar assemblages include one species only (*Neogloboquadrina pachyderma*). However, the latitudinal distribution of foraminifer assemblages is biased by the pattern of surface circulation, which carries warm waters to higher latitudes west of the oceans, and cool waters to lower latitudes east of the oceans. In fact, surface water masses are clearly distinguished by their planktonic foraminifer assemblages. For example, in the Northwest Pacific off Japan, the warm waters of the Kuroshio Current contain a diversified planktonic foraminifer assemblage (Figure 11.4) characterized by *Globigerinoides ruber*, *Globigerinoides sacculifer* and *Neogloboquadrina dutertrei*. In contrast, the cool waters of the Oyashio Current contain a poorly diversified

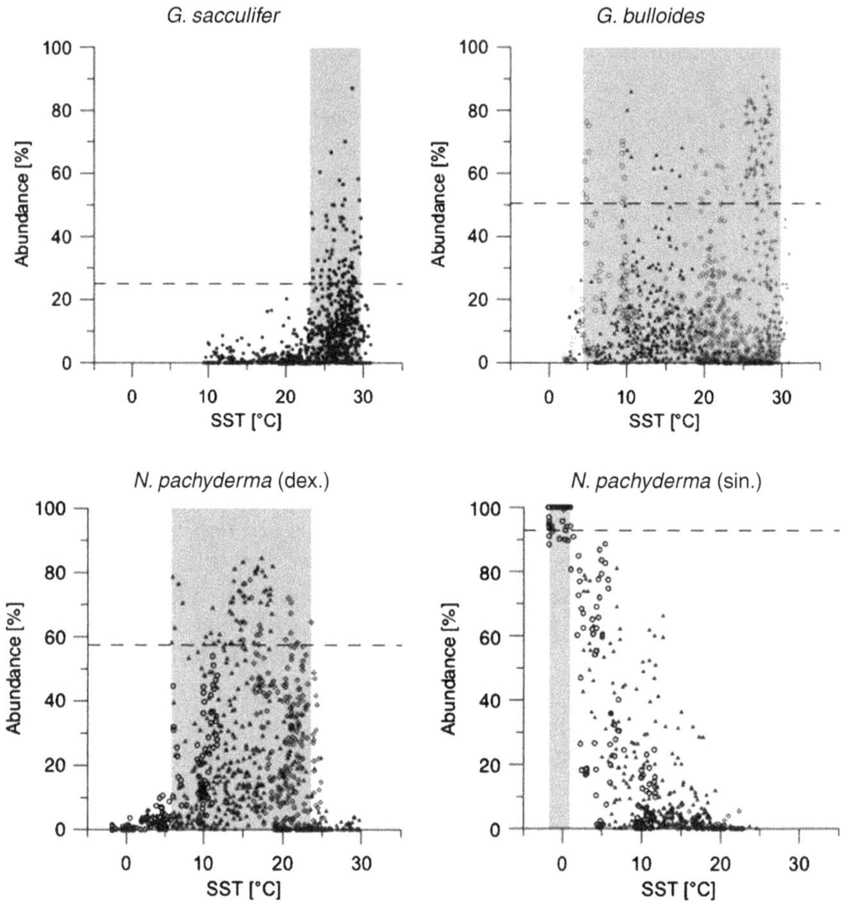

Figure 11.2 Temperature ranges of common species of planktonic foraminifers in the ocean. Optimal temperature conditions in gray. Note the broad range of *Globigerina bulloides*, which is more dependent on food availability and dominates in upwelling areas. Modified from Zaric, S., Donner, B., Fischer, G., Mulitza, S., Wefer, G., 2005. Sensitivity of planktic foraminifera to sea surface temperature and export production as derived from sediment trap data. Marine Micropaleontology, 55, 75–105.

assemblage, dominated by *Neogloboquadrina pachyderma* and *Neogloboquadrina incompta*.

In detail, foraminifer species may show daily to seasonal and interannual variations in abundance. For example, monthly sampling of living plankton during spring and summer in the North Atlantic has shown significant seasonal changes in the content of *Globigerinita glutinata,* which is most abundant during spring (Figure 11.5). Maximum abundances of *Globigerinita glutinata* follow the development of phytoplankton (especially diatoms which constitute its main food source) and coincide with intervals of deep mixed-layer and relatively low surface temperatures. During summer when sea surface temperatures increase, as well as during years of abnormally high sea surface temperatures, abundances of

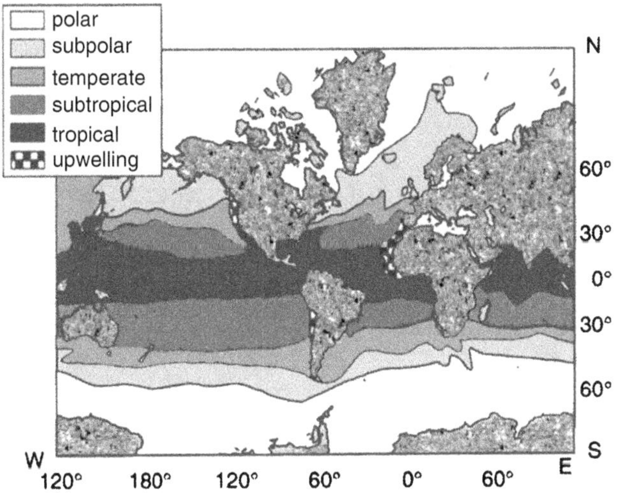

Figure 11.3 Latitudinal distribution of planktonic foraminifer assemblages. Note the bias induced by warm western boundary currents (Gulf Stream in the Atlantic, Kuroshio in the Pacific) and cool eastern boundary currents (California Current in the Pacific). Modified from Schmidt, D.N., Lazarus, D., Young, J.R., Kucera, M., 2006. Biogeography and evolution of body size in marine plankton. Earth-Science Reviews, 78, 239–266.

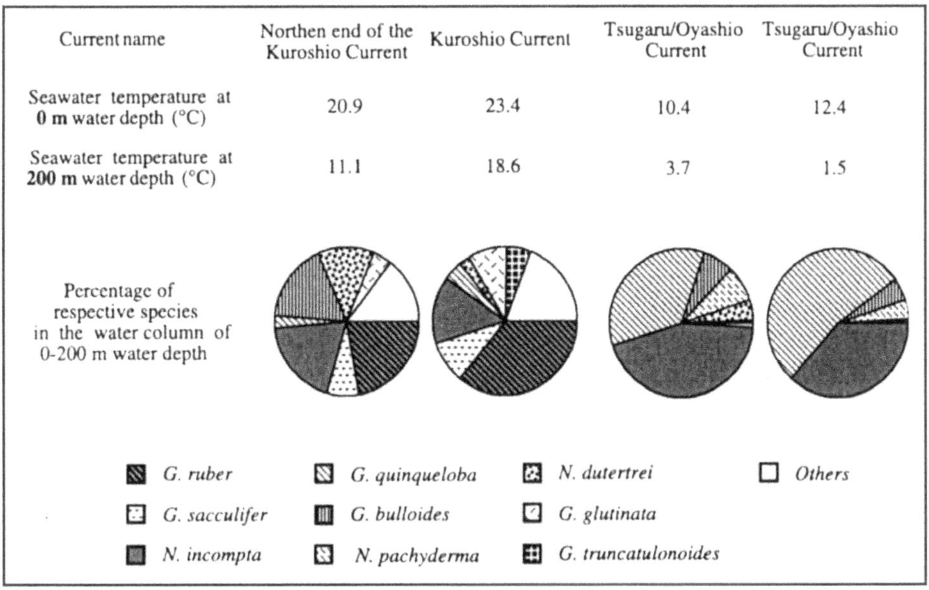

Figure 11.4 Foraminifer assemblages and seawater temperatures of the upper 200 m of the water column, at four stations of the warm Kuroshio and cool Oyashio currents off Japan. Note the clear distinction between water masses, from their microfaunal content. Modified from Kuroyanagi, A., Kawahata, H., 2004. Vertical distribution of living planktonic foraminifera in the seas around Japan. Marine Micropaleontology, 53, 173–196.

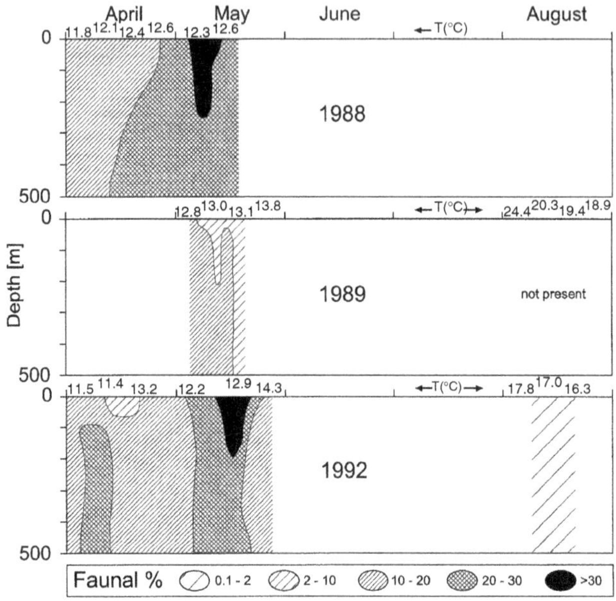

Figure 11.5 Distribution of *Globigerinita glutinata* in the upper 500 m of the water column of the North Atlantic, in spring and summer. Note high abundances in spring and low abundances in summer and intervals of abnormally high water temperatures. Reprinted from Schiebel, R., Hemleben, C., 2000. Interannual variability of planktic foraminiferal populations and test flux in the eastern North Atlantic Ocean (JGOFS). Deep-Sea Research II, 47, 1809–1852.

Globigerinita glutinata decrease significantly. However, the dominant control of sea water temperature may be modulated by other parameters, as suggested by laboratory cultures. For example, *Globigerinoides ruber* includes two morphotypes (*Globigerinoides ruber* s.s. and *Globigerinoides ruber* s.l.) which show different vertical distributions: *Globigerinoides ruber* s.s. predominates in uppermost, warmer surface waters (0–20 m) because it is more dependent upon symbionts, whereas *Globigerinoides ruber* s.l. occupies deeper habitats (down to 120 m) because it is more dependent on food availability.

Planktonic foraminifers generally decrease in size and diversity in oceanographic front areas, where the contact between surface waters of different characteristics represents unfavorable conditions. However, opportunistic species such as *Globorotalia inflata* increase in importance. Size and diversity also decrease in upwelling areas, together with the importance of symbiont-bearing species which are sensible to turbidity. In contrast, planktonic foraminifer species which tolerate a broader range of temperatures and are principally controlled by the availability of food, such as *Globigerina bulloides* (Figure 11.2), dominate.

The organic matter of dead foraminifers generally dissolves rapidly, and the spines are separated from the test which sinks at velocities of a few hundreds of meters a day. The tests reach the seafloor within a few weeks, and are considered as representative of upper water mass conditions in their area of sedimentation. However, the relative importance of the species may locally vary according to the vulnerability of the tests to dissolution (see Section 11.1.3). Planktonic foraminifer assemblages on the seafloor consist in a number of species living at different depths of the surface layer, controlled by a variety of parameters where seawater temperature and trophic conditions dominate.

11.1.1.2. Benthic foraminifers

Benthic foraminifers are found at almost all water depths of the ocean, from nearshore to deep basin areas. They represent a few percent of all foraminifers on average, but are especially concentrated in shallow environments where the seafloor is located within the photic zone. Most benthic foraminifers live in proximity to the sediment–water interface, where food is the most abundant (Figure 11.6). Many species are infaunal, and able to move in the sediment. About 75% of those infaunal

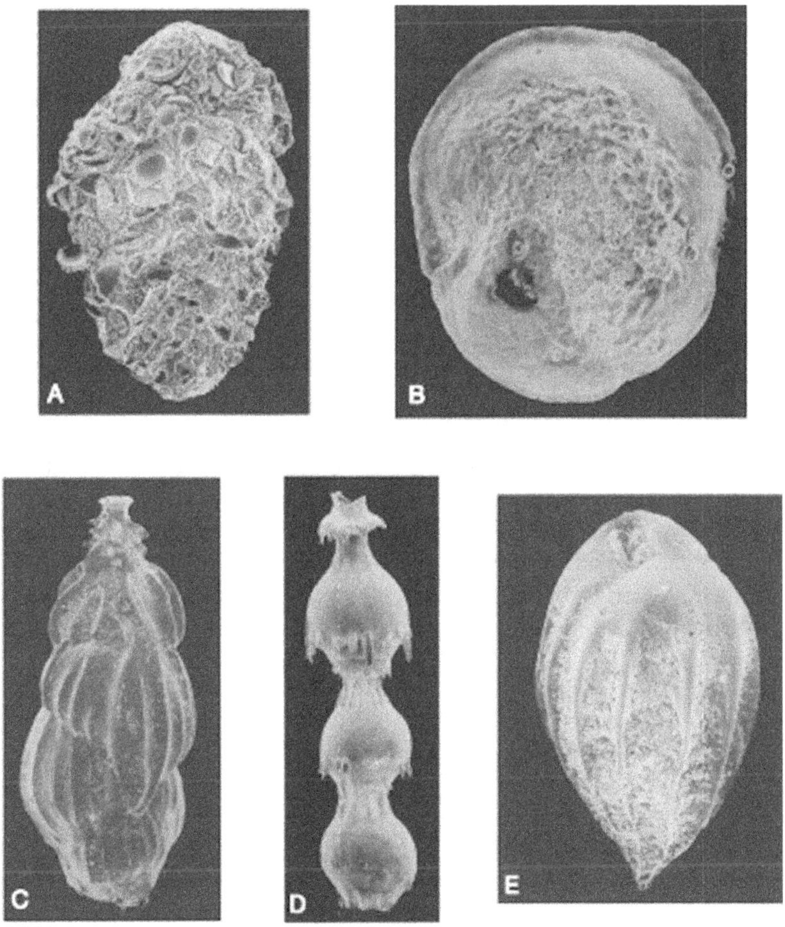

Figure 11.6 Examples of benthic foraminifers. A: *Textularia*, B: *Cibicidoides*, C: *Uvigerina*, D: *Stilostomella* and E: *Bulimina*. Modified from Kennett, J.P., von der Borch, C.C. et al., 1986. Initial Reports of the Deep Sea Drilling Project, volume 90, U.S. Gov. Print. Office, Washington D.C.

species are found within the first centimeter of sediment. This is probably due to a decrease of resources and oxygen content deeper in the sediment, where a few species nevertheless survive. Deep infaunal species are found down to the sulfate reduction front but not beyond, because of lethal production of hydrogen sulfide. Other species are epifaunal, being attached by their pseudopods or by cementation onto a substrate. Some species known as epifaunal are also found in the topmost centimeter of sediment, such as *Cibicidoides pachydermus*. It is however possible that at least a fraction of those species live at the sediment–water interface for a part of the year, and then migrate into the sediment. Most infaunal and epifaunal benthic foraminifers found at shallower water depths feed on organic elements which accumulate on the seafloor, and include a number of burrowing species. Benthic foraminifers found at deeper water depths principally filter the water for food, and epifaunal species predominate. The most frequent groups of benthic foraminifers in the modern ocean are the Textularids, the Miliolids and the Rotalids. The Textularids are agglutinated forms which extend to the deep ocean beneath the lysocline. The Miliolids are characterized by an opaque, porcellaneous test and the absence of pores. The Rotalids are characterized by a hyaline test made of oriented crystals of calcite and the presence of pores.

The distribution of benthic foraminifers is influenced by a variety of parameters including temperature, salinity, type of substrate, turbidity and light, availability of food and oxygen, and hydrostatic pressure. At ocean scale, benthic foraminifers look roughly distributed in bathymetric zones. However, these zones and the related assemblages of benthic foraminifers are defined by the properties of the seawater and sediment, which may vary in depth regionally. In fact, some species are more sensitive to one parameter than to another. As for planktonic foraminifers, temperature and salinity exert a major control on the distribution of some benthic foraminifer species. This is especially the case in the deep ocean and other areas where environmental conditions such as food supply and oxygen content are rather stable. Moreover, stable environmental conditions favor the development of complex and diversified microfaunas. As a consequence, distinct assemblages of benthic foraminifers are associated with different deep-water masses. For example, the benthic foraminifer assemblage associated with the Antarctic Bottom Water (AABW) mass (where agglutinated forms predominate) is found in the deepest parts of most oceanic basins, where its distribution varies with those of the Antarctic flow.

Although most benthic foraminifers are limited to a few hundreds of micrometers in size, larger benthic foraminifers (LBFs) may reach up to 10 mm. LBFs are adapted to low latitude, shallow carbonate (recifal) environments, where terrigenous and seasonal influences are negligible. A majority of species hosts symbiotic algae within their tests. LBFs have evolved many times from ordinary-sized ancestors in the past, principally during intervals of global warming, sea-level rise and expansion of tropical environments. They are strongly controlled by environmental factors, and driven to extinction by severe environmental changes.

Most benthic foraminifer species are tolerant to environmental changes, surviving in small numbers in microhabitats when conditions are adverse and rapidly increasing in importance and expanding their habitats when conditions become optimal. In this case, oxygen content and food supply are dominant parameters for controlling the diversity of the assemblages and the abundance of the species: oxygen content is a limiting factor which controls the presence of the species, whereas food supply controls the importance of the biomass. In oligotrophic and oxic environments (Figure 11.7), the abundance and diversity of benthic foraminifers are rather low and epifaunal species dominate. Diversity reaches a maximum in mesotrophic environments, where assemblages include a variety of epifaunal and infaunal species. However, benthic foraminifer abundances remain moderately high. Diversity is low in eutrophic and dysoxic environments where deep infaunal species dominate and may produce a large number of individuals.

Export production of marine organic matter (derived from photosynthesis and secondary production beneath the photic zone) to the seafloor (see Section 12.1.2) exerts a major control on benthic productivity beneath neritic water depths, and there is generally a good correlation between contents of organic matter in the water column and total abundances of benthic foraminifers on the seafloor. However, benthic foraminifer species are not distributed according to the availability of food. One reason is that the most opportunistic species are also the most vulnerable to a decreased food supply. Also, oxygen demand increases with fluxes of export production and abundances of benthic foraminifers. Beyond a threshold, oxygen deficiency begins to be limiting. Epifaunal species are affected first, deep infaunal species being the most tolerant. Nevertheless, a number of infaunal species can survive in anaerobic conditions.

Assemblages of benthic foraminifers may vary significantly over short time intervals and small geographical areas. This is highlighted by a comparison of winter and summer assemblages at two bathyal stations of the continental slope of the Gulf of Lions in the northwestern Mediterranean Sea (Figure 11.8). One station is located at 920 m water depth in the axis of the Lacaze-Duthier canyon which acts as a trap for sediment and organic matter advected from the shelf, especially during the spring. There, benthic foraminifers are abundant, the assemblages are highly diversified and include a number of deep infaunal species. Oxygen consumption rates are high and the uppermost, oxidized sediment layer is very thin (less than 5 cm). Differences in the abundance and diversity of species between winter and summer are rather low, suggesting a stability of the environment. Benthic foraminifer assemblages are considered as controlled by oxygen contents principally, and then by food. The other station is located at 800 m water depth of the continental slope, and organic fluxes are much lower than in the canyon. There, benthic foraminifers are less abundant than in the canyon, the assemblages are poorly diversified and deep infaunal species are limited. Significant differences in the dominant species are clearly visible between winter (*Saccorhiza ramosa*) and summer (*Uvigerina mediterranea*). The uppermost, oxidized layer of sediment is much thicker than in the adjacent canyon. Benthic foraminifer assemblages are considered as controlled by the availability of food essentially (Figure 11.7).

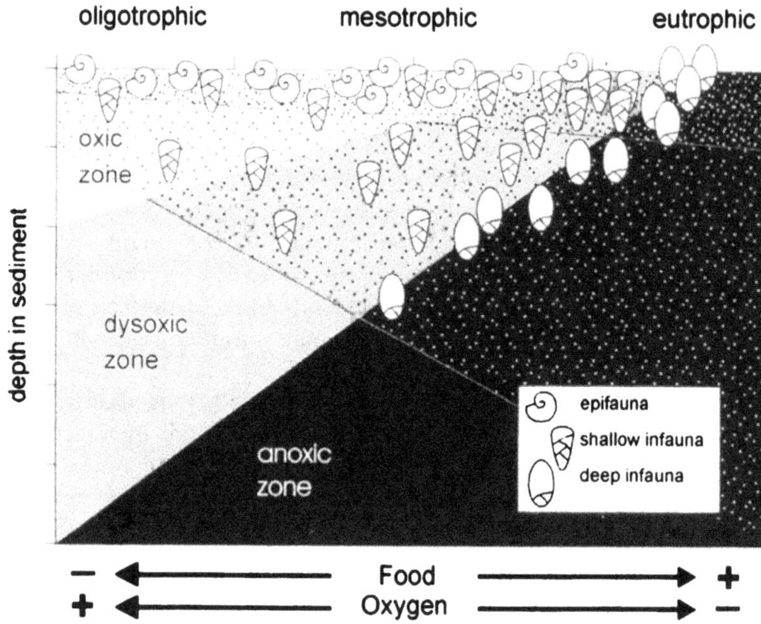

Figure 11.7 Conceptual model explaining the distribution of benthic foraminifer microfaunas according to the availability of food and oxygen. Reprinted from Van der Zwaan, G.J., Duijnstee, I.A.P., den Dulk, M., Ernst, S.R., Jannink, N.T., Kouwenhoven, T.J., 1999. Benthic foraminifers: proxies or problems? A review of paleocological concepts. Earth-Science Reviews, 46, 213–236.

11.1.2. Coccoliths

Coccoliths are tiny circular plates of calcite and magnesian calcite. They are a few micrometers in diameter (1–15 μm) and form an exoskeleton (coccosphere) which surround the living cell of a coccolithophore (Figure 11.9). Coccolithophores are unicellular biflagellate algae which belong to the class Haptophyceae. Being photosynthetic (photoautotrophic) organisms, they make their organic substance from seawater, carbon dioxide and nutrients, using energy from the sun for carbon fixation. However, some species such as *Coccolithus pelagicus* also have the ability to ingest tiny organic molecules and bacteria. The coccolithophore cell consists in a protoplasm bound by double membranes, which includes a prominent nucleus, a reticular body adjacent to the nucleus, two chloroplasts, a Golgi body, mitochondria and vacuoles (Figure 11.10). The chloroplasts change in shape and position according to light intensity.

The formation of coccoliths occurs within the protoplasm, between the reticular body and the nucleus in some species, near the Golgi body in other species. Then, the coccoliths are driven to the surface of the cell. The number of coccoliths at the periphery of individual coccospheres varies between 10 and 150. It increases as individual cells grow, but remains about the same for any given species. For example, the number of coccoliths in *Gephyrocapsa oceanica* varies

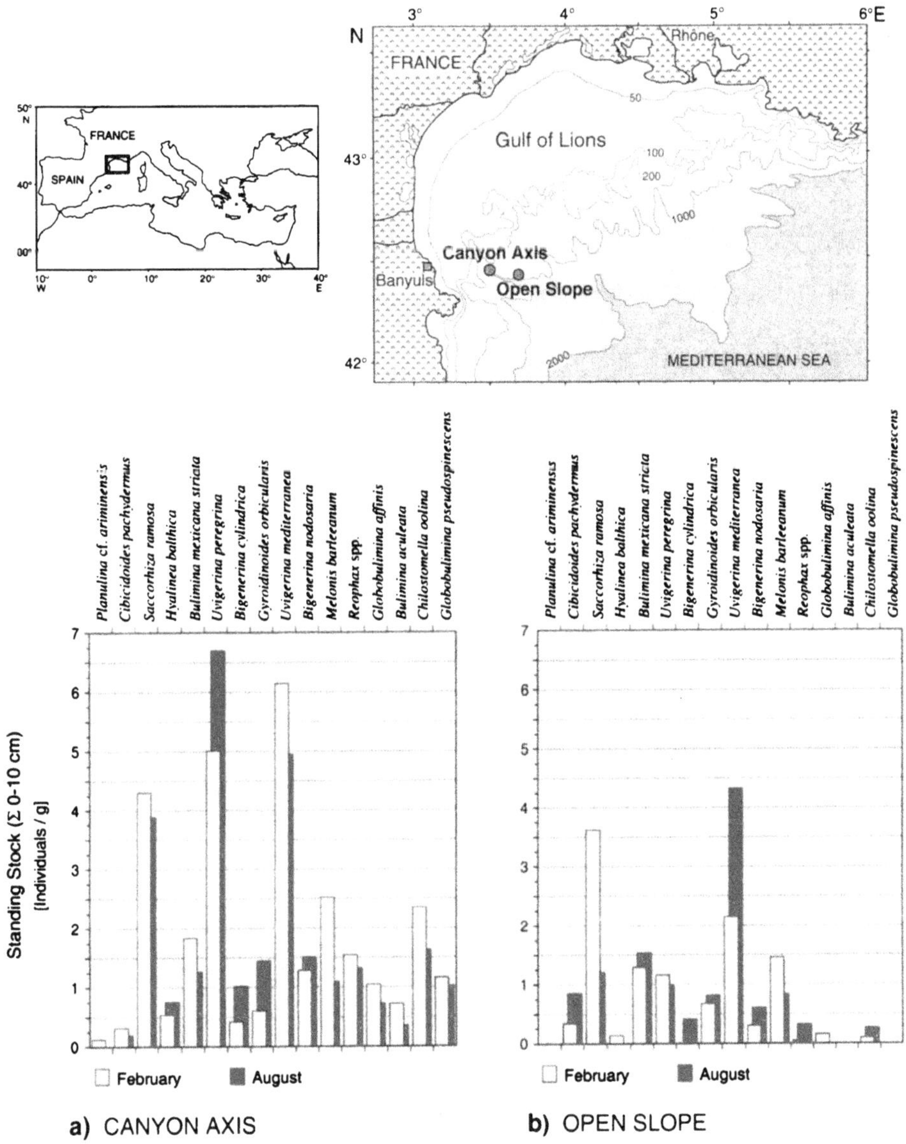

Figure 11.8 Abundances of selected benthic foraminifers in the uppermost 10 cm of sediment in the Lacaze-Duthier canyon area of the Gulf of Lions (Western Mediterranean). Top: Location map. Bottom: Winter and summer abundances of benthic foraminifer species. Note differences in diversity and abundances between the two stations. Modified from Schmiedl, G., de Bovée, F., Buscail, R., Charrière, B., Hemleben, C., Medernach, L., Picon, P., 2000. Trophic control of benthic foraminiferal abundance and microhabitat in the bathyal Gulf of Lions, western Mediterranean Sea. Marine Micropaleontology, 40, 167–188.

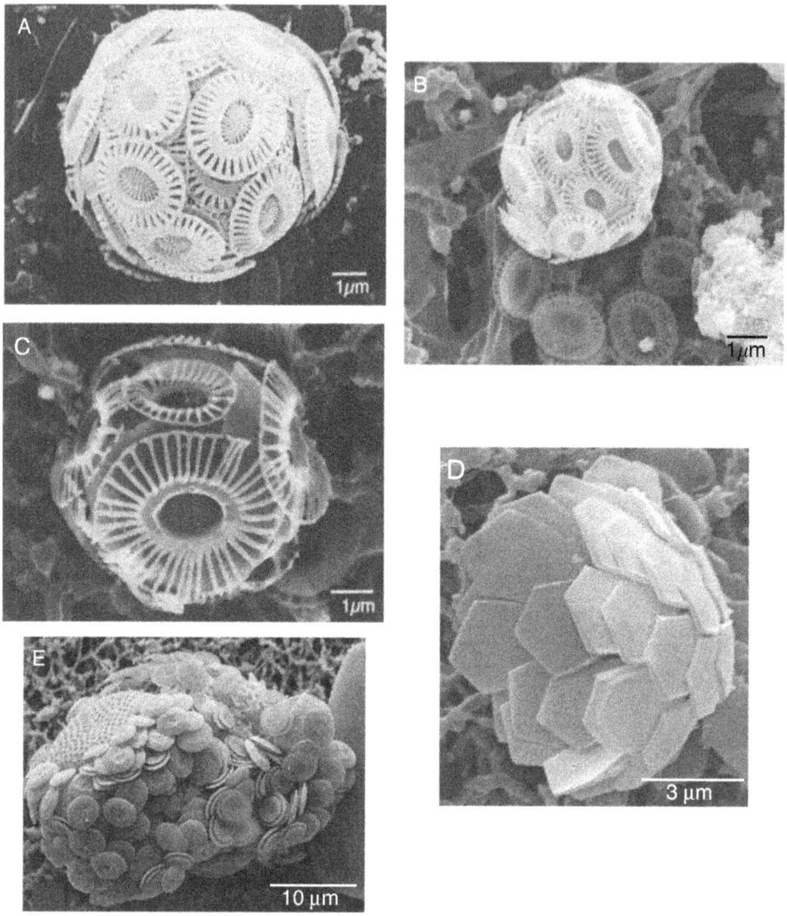

Figure 11.9 Scanning electron micrographs of coccolithophores. A: mature *Emiliana huxleyi* Type A (warm); B: young *E. huxleyi* Type A (warm), and individual coccoliths; C: young *E. huxleyi* Type B (cold); D: *Florisphaera profunda* and E: *Reticulofenestra sessilis*. Modified from Hagino, K., Okada, H., Matsuoka, H., 2005. Coccolithophore assemblages and morphotypes of *Emiliana huxleyi* in the boundary zone between the cold Oyashio and warm Kuroshio currents off the coast of Japan. Marine Micropaleontology, 55: 19–47.; and Andruleit, H., Stäger, S., Rogalla, U., Cepek, P., 2003. Living coccolithophores in the northern Arabian Sea: ecological tolerances and environmental control. Marine Micropaleontology, 49, 157–181.

between 10 and 30. There are two forms of coccoliths: holococcoliths result from the arrangement of similar, tiny euhedral crystals of calcite, whereas heterococcoliths consist of complex, radial arrangements of larger crystals of calcite. The life cycle of coccolithophores includes a haploid phase (motile phase) characterized either by naked cells or the formation of holococcoliths, and a diploid phase (nonmotile phase) characterized by the production of heterococcoliths. Therefore, single coccolithophore species may produce coccoliths of different morphologies during their life span of a few weeks. The fusion of two motile cells is involved in the

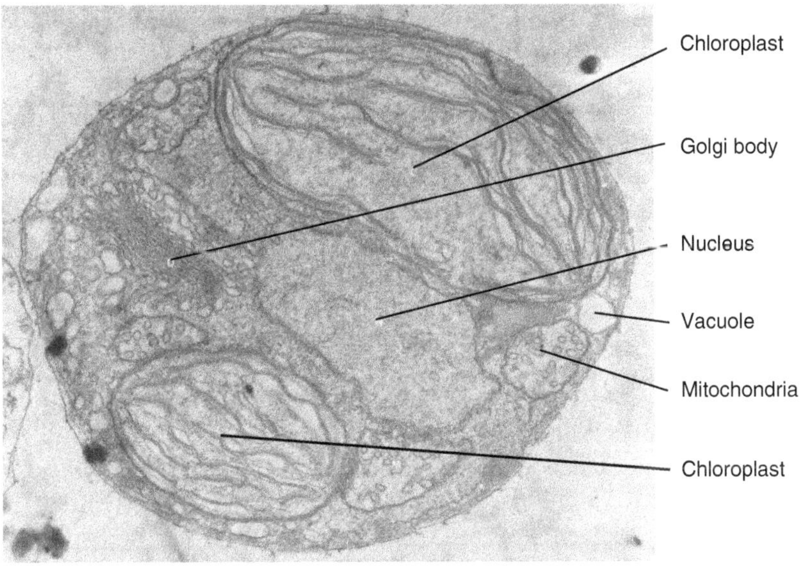

Figure 11.10 Cell components of a living coccolithophore in its haploid phase (naked cell), *E. huxleyi*. Modified from Haq, B.U., Boersma, A. (Editors), 1978. Introduction to marine micropaleontology. Elsevier, Amsterdam.

formation of non-motile cells, which divide at rates of one to five divisions per day, depending on environmental conditions.

Coccolithophores first appeared in the Early Jurassic, and significant evolutive radiation occurred during the Early/Middle Jurassic transition. The diversity decreased, but abundances remained high during the Middle and Late Jurassic. Further diversification occurred in the Cretaceous and very high abundances were recorded in the Late Cretaceous, but only a few species survived the catastrophic events at the Cretaceous/Tertiary boundary. New taxa evolved in the Paleocene, leading to a maximum of diversity in the Early Eocene. Among them, the star-shaped *Discoasters* survived till the end of the Pliocene. Cenozoic evolutive radiations in the Early Paleogene and Middle Miocene were both followed by intervals of decreased diversity. About 150 species are present in the modern ocean. Among them 16 species are abundant enough, and produce sufficient quantities of well-preserved coccoliths, to significantly contribute to the formation of oceanic sediments.

Ecological adaptation is a major characteristic of coccolithophores, which are found in a variety of environments ranging from open sea to nearshore and lagoonal. Laboratory cultures suggest that coccolithophores can survive down to the temperature limit of 0°C, and can tolerate broad variations in salinity. In the ocean, some species have an ample temperature range. *Emiliana huxleyi*, which tolerates water temperatures between 0°C and 30°C, dominates in many cold water environments of high latitudes but is also present in assemblages from low latitude areas. However, minimum temperatures of 2–3°C are required for the coccolithophores to reproduce and develop coccoliths. Other species have a narrower temperature range, which may vary regionally. *Gephyrocapsa muellerae* has a temperature tolerance between 5°C and 14°C in the Greenland Sea, but between 10.3°C and 15.7°C in the Australian sector of the Southern Ocean. At geological timescale, coccolithophores may also change their habitat through time. For example, *Coccolithus pelagicus* first occurred at low latitudes of the Pacific Ocean in the Eocene, extended to higher latitudes where it increased in abundance in the Neogene, and disappeared from low latitudes in Late Pliocene to Early Pleistocene times.

Coccolithophores being photosynthetic organisms, sunlight is essential for their development. Although they are sometimes found at several hundreds of meters below sea level, their development is restricted to the upper 150–200 m of the photic zone and they are most abundant in the uppermost 50 m. It has been postulated that the shape of the coccoliths could focus the light on the chloroplasts of the living cell and facilitate the survival of the organisms in areas where sunlight does not penetrate easily. The availability of sunlight exerts a major control on the development of coccolithophores, and low insolation may account for low abundances or absence. This is the case in high-latitude areas of summer sea ice extent, as well as in turbid areas where high productivity and/or terrigenous particles reduce the penetration of sunlight. As a consequence, rather low coccolithophore densities are found in upwelling areas, where diatoms dominate. In contrast, higher proportions of coccolithophores are found in nutrient-depleted waters, such as those of the stratified, central oceanic gyres. Besides, local oceanographic conditions also control the vertical distribution of coccolithophore species. In the ocean, some species such as *Emiliana huxleyi* are concentrated within a relatively narrow depth range near the thermocline, whereas other species such as *Florisphaera profunda* occur below the thermocline (Figure 11.11). This distribution closely follows the position of the thermocline, rising near surface in areas of shallow thermocline. In fine, the availability of sunlight, the concentration of nutrients and other plankton and the thickness of the mixed-layer principally influence the distribution of coccolithophores in the ocean.

Because of the diversity of parameters influencing the development of coccolithophores, their abundance in the water column may vary greatly with the seasons. Strong seasonal changes in hydrography characterize the eastern margin of the North Pacific Ocean, in coincidence with a variable activity of the California Current. In the San Pedro Basin off southern California, isothermal (well-mixed) conditions in the photic zone during the winter months are generally succeeded by slight water stratification in early spring, and upwelling conditions peaking in late spring to early summer. The highest fluxes of coccolithophores are recorded in winter, when insolation and nutrient concentrations are low, and diatom and foraminifer fluxes are minimum (Figure 11.12). They account for up to 90% of the total carbonate flux. The lowest fluxes of coccolithophores are recorded during upwelling events, when high nutrients levels (especially nitrate and silicate) and increased solar radiation coincide with maximum productivity of diatoms, dinoflagellates and grazing groups such as foraminifers. Nevertheless, carbonate fluxes are maximum during upwelling events, when planktonic foraminifers drastically increase in abundance. Therefore, coccolithophore production is favored by relatively stable hydrographic conditions and low concentrations of nutrients, but decreases when primary productivity overall increases. In detail, *Emiliana huxleyi*

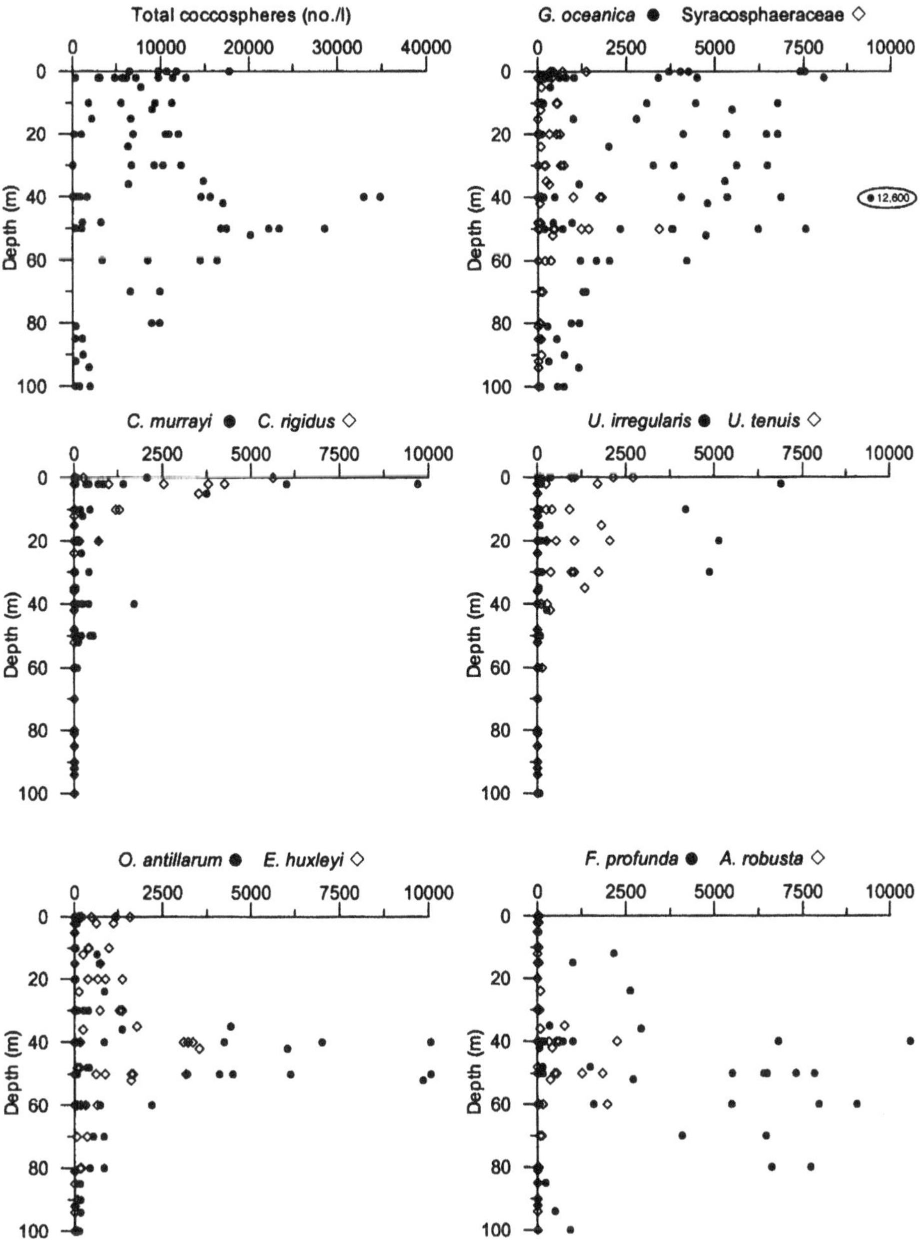

Figure 11.11 Depth zonation and concentration of coccolithophore species in the northern Arabian Sea. Note maximum abundances of *Emiliana huxleyi* (and total coccospheres) at the thermocline which is here at about 40 m depth, and of *Florisphaera profunda* below the thermocline. Reprinted from Andruleit, H., Stäger, S., Rogalla, U., Cepek, P., 2003. Living coccolithophores in the northern Arabian Sea: ecological tolerances and environmental control. Marine Micropaleontology, 49, 157–181.

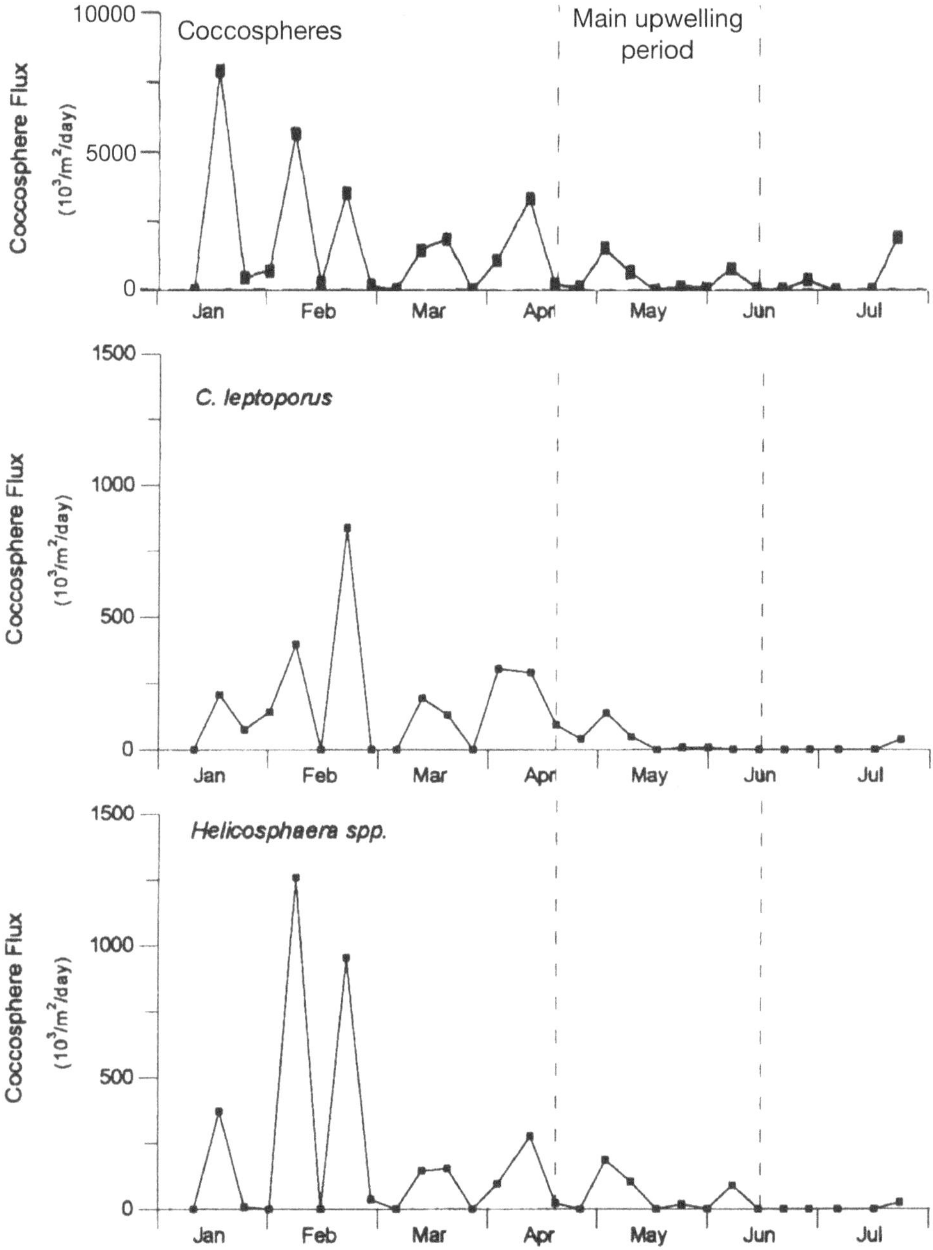

Figure 11.12 Flux records of coccospheres in the San Pedro Basin off Southern California. Note high concentrations of coccospheres in winter with development of low-nutrient species, and low concentrations during the peak of the upwelling season. Modified from Ziveri, P., Thunell, R.C., Rio, D., 1995. Export production of coccolithophores in an upwelling region: results from San Pedro Basin, Southern California Borderlands. Marine Micropaleontology, 24, 335–358.

and *Florisphaera profunda* are the two dominant species in the San Pedro Basin area. Fluxes of *Emiliana huxleyi* do not show important seasonal variability, whereas fluxes of *Florisphaera profunda* increase significantly in winter. Besides, species typical of low to intermediate nutrient concentrations such as *Helicosphaera carteri* and *Calcidiscus leptoporus*, and species which are most abundant in temperate to tropical waters such as *Oolithus fragilis* and *Umbilicosphaera sibogae* develop in winter (Figure 11.12), in coincidence with maximum activity of the warm Davidson Countercurrent.

The majority of coccolithophores are grazed by zooplankton. They are transported to the seafloor through fecal pellets and aggregates principally, and most of the time the fluxes of living coccolithophores and coccoliths are closely related. In waters enriched in carbon dioxide however, selective dissolution of coccoliths on the ocean floor may induce some differences between living and fossil assemblages where the proportion of solution-resistant species slightly increases. The coccoliths are released when the organic elements of the coccosphere are oxidized. If the disintegration of the coccosphere occurs in the water column, individual coccoliths may be transported over great distances by marine currents because of their tiny size and platy shape. For instance, coccoliths of sub-Antarctic origin have been found in Oligocene sediments of the Walvis Ridge in the tropical southeast Atlantic Ocean, at water depths bathed by the Antarctic Intermediate Water (AAIW) mass. Despite an incomplete transport of coccoliths to the sediment, the average relative taxonomic composition of coccolith assemblages in the photic zone, the water column and the sediment are very similar in many areas. This allows the recording of upper water mass conditions (including seasonal variability) in many oceanic sediments.

Many coccolithophore groups and species show some degree of variability in the morphology and size of their coccoliths. Culture experiments have demonstrated that such morphological variations are under genotypic control. A variety of parameters may affect the morphology of the coccoliths, but fertility and temperature of surface waters are probably important factors. For example, simple measurements such as length and bridge angle (Figure 11.13) have been used to distinguish six morphological associations within the genus *Gephyrocapsa*. Among them, small coccoliths with low bridge angle only occur where the average temperature of surface waters is below 14°C (*Gephyrocapsa* cold), whereas large coccoliths with high bridge angle are more abundant in warmer surface waters (*Gephyrocapsa* equatorial). Also, different morphotypes of *Emiliana huxleyi* are recognized, based on the morphology and size of the coccoliths. Initially, two morphotypes were described (Figure 11.9): a more calcified, warm water type, characterized by a solid plate in the central area and a less calcified, cold water type, characterized by a grid structure in the central area. Recently, detailed morphometric measurements including the respective lengths of the proximal and distal shields of *Emiliana huxleyi* coccoliths (Figure 11.14), led to the distinction of four morphotypes. Together with the composition of the coccolith assemblages, these morphotypes allow a precise description of the surface water masses of the North Pacific Ocean off Japan. The Warm North Pacific Central Water is characterized by abundant *Emiliana huxleyi* type A, which is associated to abundant *Gephyrocapsa* in the warm Kuroshio Current regime. The waters of the Tsugaru Warm Current contain abundant *Emiliana huxleyi* type B/C-2 together with

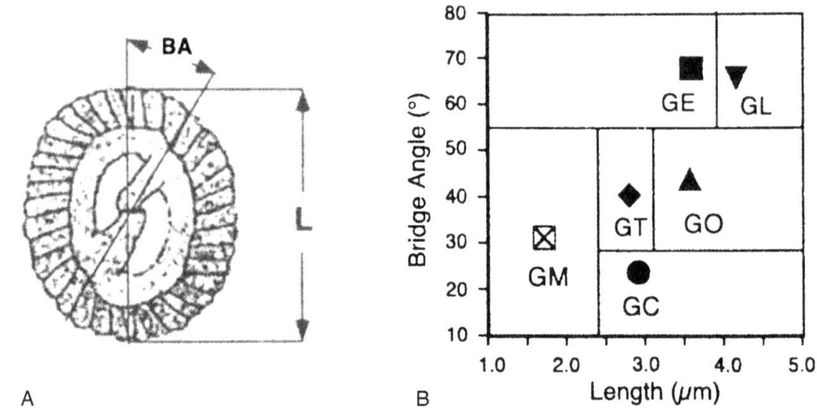

Figure 11.13 Morphological variability of *Gephyrocapsa*. A: measurements of bridge angle (BA) and length (L) made on scanning electron microscope images of single coccoliths; B: morphological associations of *Gephyrocapsa* in Holocene assemblages. GC: cold; GE: equatorial; GL: large; GM: minute; GO: oligotroph; GT: transitional. Modified from Bollmann, J., Brabes, B., 2002. Global calibration of *Gephyrocapsa* coccolith abundance in Holocene sediments for paleotemperature assessment. Paleoceanography, 17, doi 10.1029/2001PA000742.

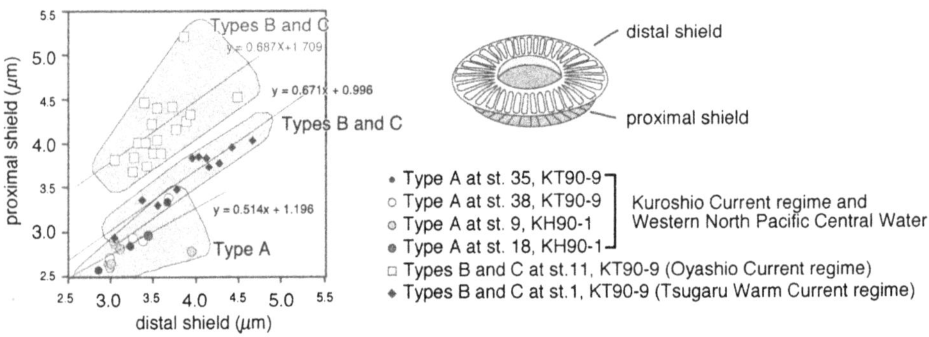

Figure 11.14 Morphological variability of *Emiliana huxleyi* as deduced from distal and proximal shield lengths, measured on images of individual coccoliths. Note distinct correlations among similar types from different water masses. Reprinted from Hagino, K., Okada, H., Matsuoka, H., 2005. Coccolithophore assemblages and morphotypes of *Emiliana huxleyi* in the boundary zone between the cold Oyashio and warm Kuroshio currents off the coast of Japan. Marine Micropaleontology, 55, 19–47.

Gephyrocapsa oceanica and *Braarudosphaera bigelowii*. By contrast, the cold Oyashio Current waters contain an assemblage dominated by *Emiliana huxleyi* type B, associated to *Coccolithus pelagicus* principally (Figure 11.15).

11.1.3. The Dissolution and Preservation of Biogenic Carbonates

Biological activity extracts significant quantities of calcium carbonate from the ocean. Calcium is abundant in the entire water column of the modern ocean where

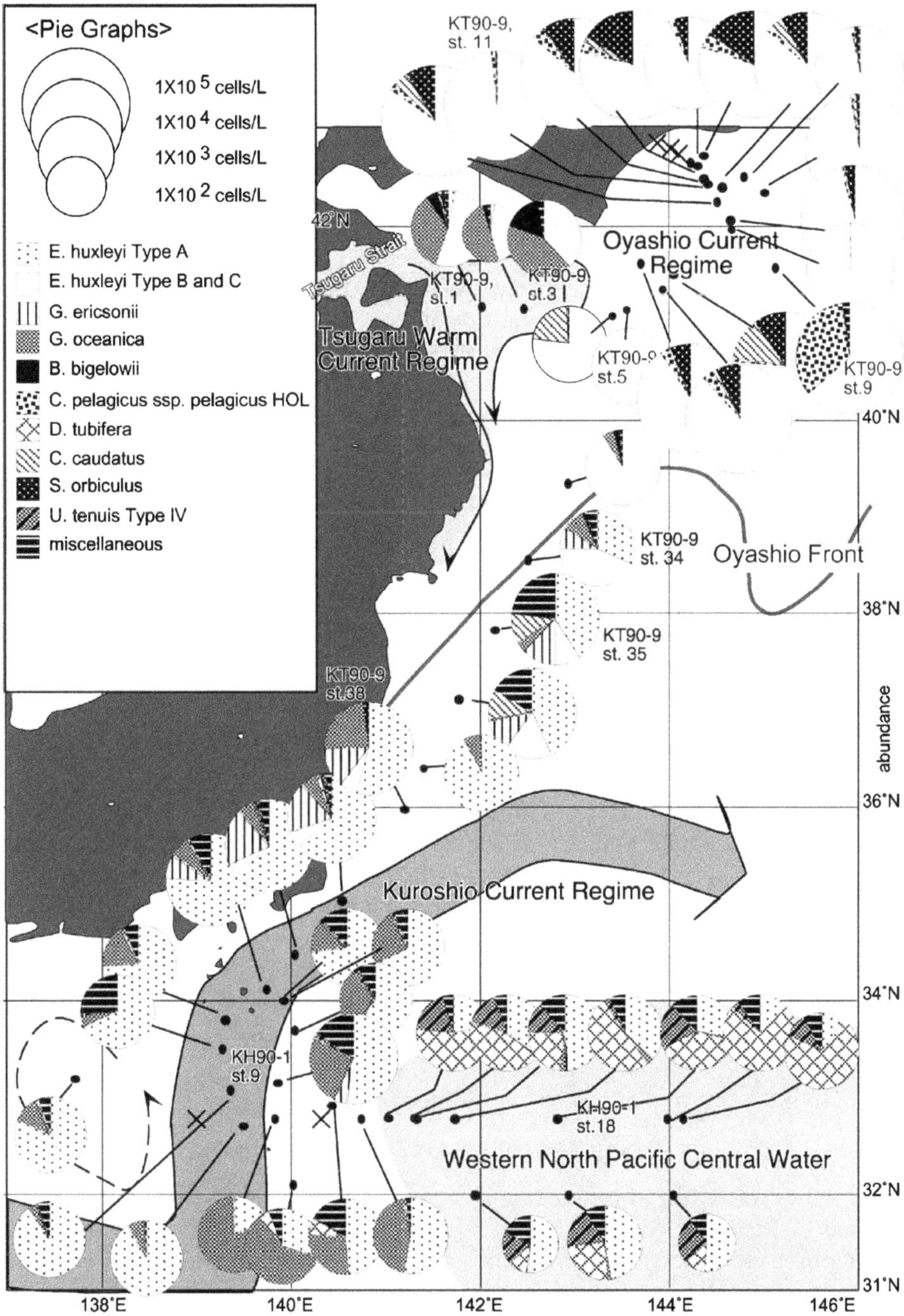

Figure 11.15 Geographic distribution of living coccolithophore assemblages in the Pacific Ocean off Japan. Note the variety of assemblages, and their coincidence with major water masses, i.e. the Western Pacific Central Water, the warm Kuroshio and Tsugaru currents and the cold Oyashio Current. Reprinted from Hagino, K., Okada, H., Matsuoka, H., 2005. Coccolithophore assemblages and morphotypes of *Emiliana huxleyi* in the boundary zone between the cold Oyashio and warm Kuroshio currents off the coast of Japan. Marine Micropaleontology, 55, 19–47.

its average concentration is about 412 mg/l, and its behavior is close to those of conservative chemical elements. Calcium is principally derived from volcanic processes and chemical weathering of continental surfaces. By comparison, the concentration in carbonates is much lower, as suggested by concentrations in dissolved carbon of approximately 28 mg/l. Carbonates are present in seawater as carbonate (CO_3^{2-}) and bicarbonate (HCO_3^-) ions principally. Part of the carbonate and bicarbonate ions is derived from continental weathering (see Section 10.1), i.e. the dissolution of outcropping carbonates by rainwater and subsequent transport of dissolved phases to the ocean by runoff. However, most carbonate and bicarbonate ions are derived from oceanic processes such as the chemical reaction of carbon dioxide and seawater, and the alteration of organic molecules in anoxic environments.

The concentration of carbon dioxide in the ocean is partly controlled by biological activity. Carbon dioxide is produced by the respiration of planktonic and benthic heterotrophic organisms and the oxidation of organic compounds, and consumed during photosynthesis by primary producers. In the modern ocean however, carbon dioxide largely comes from the atmosphere. The dissolution of gases in seawater is an exothermic reaction which releases heat into the environment and is associated to a decrease in volume. Fluxes at the interface between the ocean and the atmosphere are expressed by the formula

$$F = K\Delta C$$

where K is the velocity of the reaction which is controlled by a variety of parameters such as temperature, salinity, pressure and wind velocity, and ΔC the difference in concentration between ocean and atmosphere which also controls the direction of the flux. As a rule, the velocity of the reaction increases in high pressure, low temperature and low salinity conditions. Although the solubility of carbon dioxide is very high, its concentration in seawater remains generally low because of interaction with the chemical reactions of the carbonate cycle. Carbon dioxide reacts with seawater to produce carbonic acid which releases hydrogen ions, subsequently generating bicarbonate and carbonate ions:

$$CO_2 + H_2O \rightleftarrows H_2CO_3 \text{ (carbonic acid)}$$

$$H_2CO_3 \rightleftarrows H^+ + HCO_3^- \text{ (bicarbonate ion)}$$

$$HCO_3^- \rightleftarrows H^+ + CO_3^{2-} \text{ (carbonate ion)}$$

When the quantity of carbon dioxide in seawater increases for some reason, this is followed by increased concentrations in bicarbonate and carbonate, and hydrogen ions. Conversely, a decreased availability of carbon dioxide in seawater may draw the reactions in the opposite direction, leading to decreased concentrations in bicarbonate, carbonate and hydrogen ions. In the modern ocean where carbonate concentration is low these reactions may participate in the regulation of the greenhouse effect, for example by transferring quantities of carbon dioxide from the atmosphere to the ocean, especially in high-latitude regions of low temperature. An additional source of hydrogen ions exists in dysoxic to anoxic environments, where for instance sulfate-reducing bacteria use sulfate from seawater and surface sediment pore waters as an energy source for disintegrating simple organic molecules into hydrogen sulfide and carbonic acid.

Higher quantities of carbon dioxide in the ocean and/or organic matter degradation ultimately lead to an increased concentration of hydrogen ions, which principally controls the solubility of the carbonates:

$$CaCO_3 + H^+ \rightarrow Ca^{2+} + HCO_3^-.$$

Other factors also influence the solubility of the carbonates, which for example augments as pressure increases. Comparison of foraminiferal assemblages from sediment traps and surface sediments indicates that calcareous tests are generally well preserved in the water column and that carbonate dissolution essentially occurs on the seafloor, as well as within a few centimeters of the uppermost sediment. This is a probable consequence of the rapid sinking of many planktonic remains through aggregates or within fecal pellets. Among carbonate shells, those made of calcite (most foraminifers and coccoliths) are more resistant to dissolution than those made of less stable aragonite (pteropods). The solubility of calcareous microfossils also varies with the structure of the shell. For example, thick shells are more resistant to dissolution, whereas the presence of pores increases the specific surface of the shell and facilitates its dissolution. As a consequence, some species (e.g. tropical planktonic foraminifers) are more vulnerable than others, and microfossil assemblages in surface sediments can be quite different from those in the water column. The preservation of foraminifer assemblages in marine sediments, deduced from the proportion of planktonic foraminifer fragments, provides a good estimate of the selective dissolution and the reliability of sedimentary assemblages as proxies for surface water conditions.

The concentration in dissolved carbon is generally lowest in the uppermost water column where carbon dioxide is consumed by photosynthetic organisms while calcareous microorganisms build their shells using carbonate and bicarbonate ions. This is especially the case in highly productive areas such as coastal upwellings and oceanic fronts. As a consequence, carbonate dissolution is limited as long as concentrations in carbonate and bicarbonate ions remain above saturation. Concentrations rapidly increase below the mixed-layer, as most of the carbon dioxide returns to the oceanic reservoir via the oxidation of organic compounds principally (Figure 11.16). The subsequent concentration in hydrogen ions within the oxygen minimum layer (see Chapter 12) may locally increase as to initiate the dissolution of the most vulnerable carbonate shells at the water/sediment interface. A comparison of the microfossil contents of sediment trap (1,265 m water depth) and surface sediment (1,567 m water depth) at a station of the coastal upwelling off Somalia demonstrates the complete dissolution of aragonitic pteropod shells of high solubility and the significant increase in planktonic foraminifer fragments in surface sediments (Figure 11.17). There are intervals during geological times which are characterized by lower contents of biogenic carbonates at most water depths of the deep ocean although planktonic productivity was important as suggested by the development of a mixed-layer and the preservation of organic matter of oceanic origin in the sediment. This has been for example the case in wide areas of the Tethys and Atlantic oceans during Middle Cretaceous Oceanic Anoxic Events

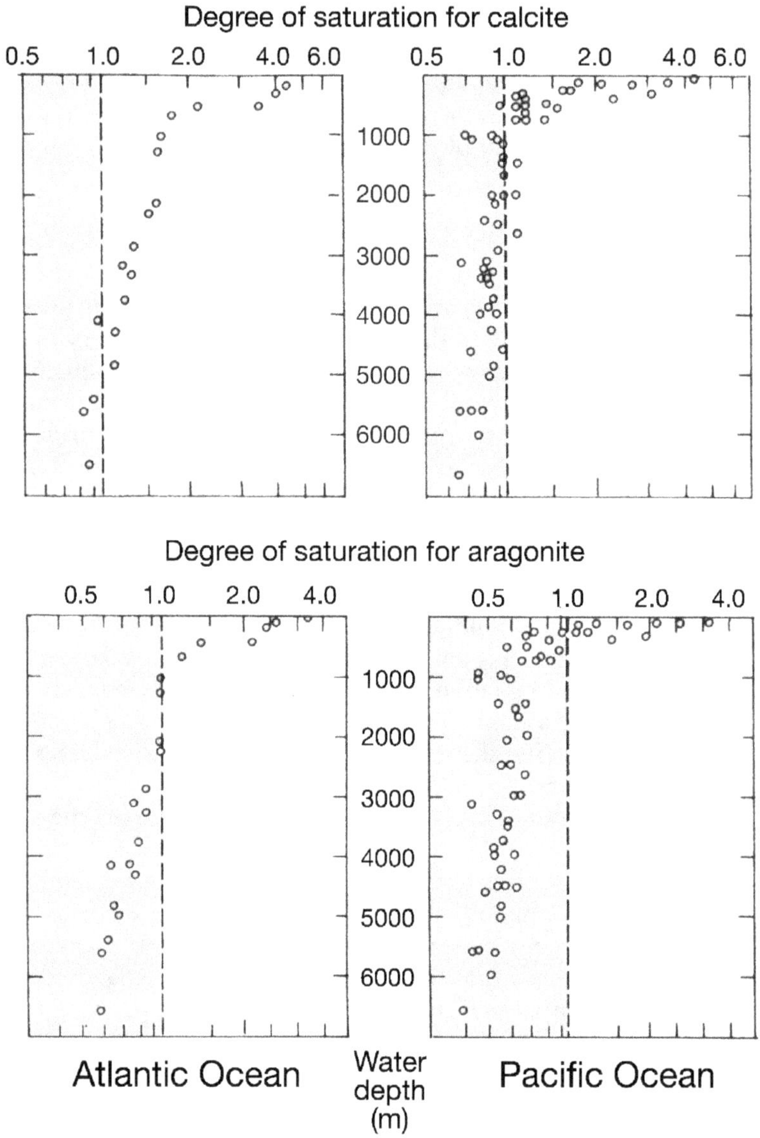

Figure 11.16 Degree of saturation of oceanic waters regarding carbonate minerals. Values below one indicate undersaturation. Note higher degrees of saturation in the Atlantic Ocean than in the Pacific Ocean and in uppermost waters where concentration in dissolved carbon is minimum, lower degree of saturation with respect to aragonite and widespread undersaturation in the deep ocean. Modified from Cojan, I., Renard, V., 1997. Sédimentologie. Masson, Paris.

(OAEs) of black-shale accumulation. The development of an oxygen minimum layer at ocean scale, in a context of water stratification in the deep ocean, was probably responsible for poor preservation of calcareous microfossil assemblages and enhanced carbonate dissolution.

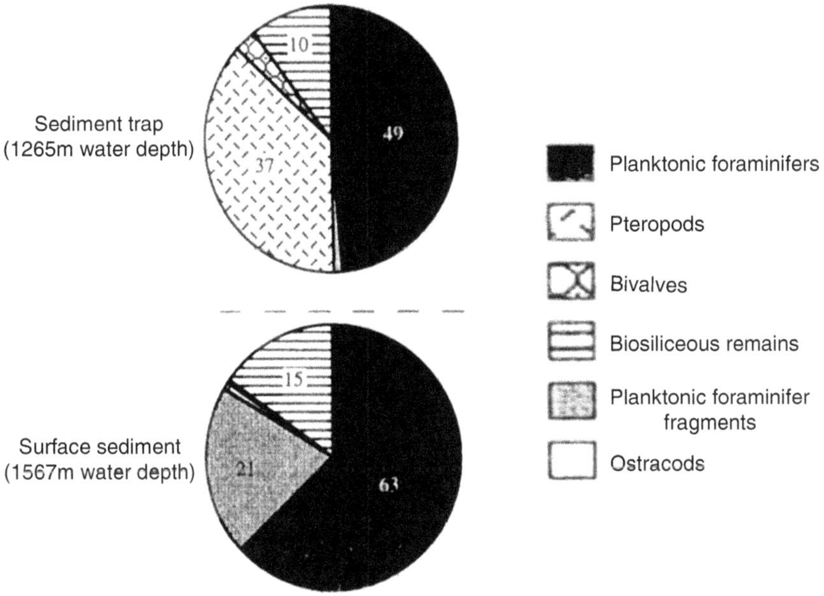

Figure 11.17 Comparison of microfossil assemblages in the water column and surface sediment below a coastal upwelling off Somalia. Note the complete dissolution of aragonitic Pteropods near or at the water/sediment interface, the abundance of planktonic foraminifer fragments and the increase in relative abundance of planktonic foraminifers in surface sediments. Modified from Conan, S.M.-H, Ivanova, E.M., Brummer, G.-J.A., 2002. Quantifying carbonate dissolution and calibration of foraminiferal dissolution indices in the Somali Basin. Marine Geology, 182, 325–349.

Atmosphere to ocean fluxes of carbon dioxide are maximum in high-latitude areas, where sea surface temperature is the lowest. The concentration in hydrogen ions is important in the entire water column, and carbonate dissolution is active at all water depths. As a consequence calcareous biogenic oozes are conspicuously absent from the modern Southern Ocean, south of the Antarctic convergence. Another consequence is that the AABW masses, made of cold waters which dive along the Antarctic margin or subside from the Deep Circumpolar Current (see Section 2.2.5), are highly concentrated in hydrogen ions. As AABW masses circulate in the deep ocean basins they are supplemented in hydrogen ions of metabolic origin, and become very aggressive regarding the carbonates. Other components of the thermohaline circulation which form in high-latitude areas, such as AAIWs and North Atlantic Deep Waters (NADWs), are also enriched in hydrogen ions but to a lesser extent.

In many regions of the modern ocean, with the exception of high latitude and highly productive areas, sediments sampled along depth transects show a good preservation of biogenic carbonates at shallow and intermediate water depth. There is no significant difference between calcareous planktonic assemblages in the water column and on the seafloor, and microfossil assemblages are characteristic of surface water environments. Significant dissolution of biogenic carbonates, as deduced from a poor preservation of microfossil assemblages and decreased percentages of

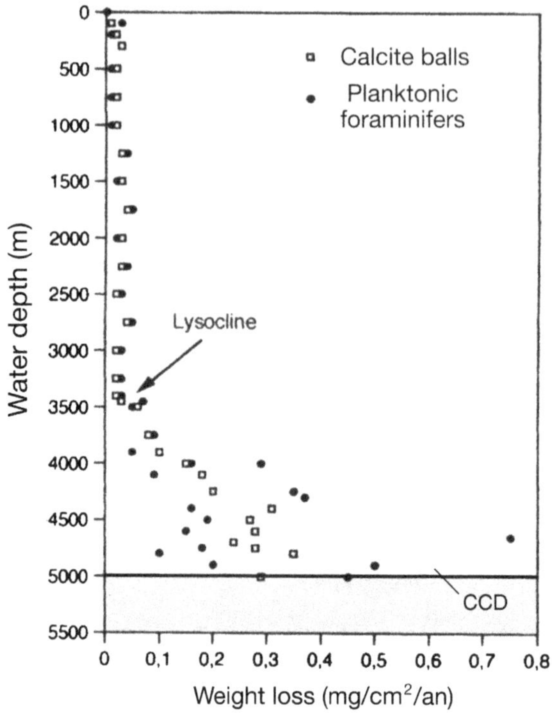

Figure 11.18 Schematic results of an experimental study of calcite dissolution (using known quantities of calcite balls and planktonic foraminifers) in the tropical Pacific Ocean. Note significant increase in dissolution at 3500 m and complete dissolution at 5000 m which characterize the lysocline and carbonate compensation depth (CCD), respectively. Modified from Cojan, I., Renard, V., 1997. Sédimentologie. Masson, Paris.

carbonates in surface sediments, currently starts at approximately 3,700–4,500 m water depth depending on the nature and origin of the water mass in contact with the seafloor. This level corresponds to the lysocline (Figure 11.18). In the western Brazil Basin of the South Atlantic Ocean, the depth of the lysocline corresponds to the boundary between the NADW and the corrosive AABW. Below, dissolution intensifies rapidly with depth. The dissolution of the more vulnerable foraminifers significantly modifies the assemblages (Figure 11.19). The coccoliths generally increase in relative proportion on the seafloor, a probable consequence of their rapid transport as aggregates or within fecal pellets where they are relatively protected against dissolution. The level below which all carbonates are dissolved corresponds to the carbonate compensation depth (CCD). The CCD varies in depth from 4,200–4,500 m in the Pacific Ocean to 5,000–5,500 m in the Atlantic Ocean, following the path and importance of the AABW flow principally. On the average, carbonate dissolution is more important and the lysocline and CCD are shallower in the Pacific Ocean than in the Atlantic Ocean. This has been attributed to the importance of cold deep waters of Antarctic origin in the deep Pacific basins, but also to the release of additional carbon dioxide in volcanic and active margin areas. These conditions have prevailed for most of the Neogene.

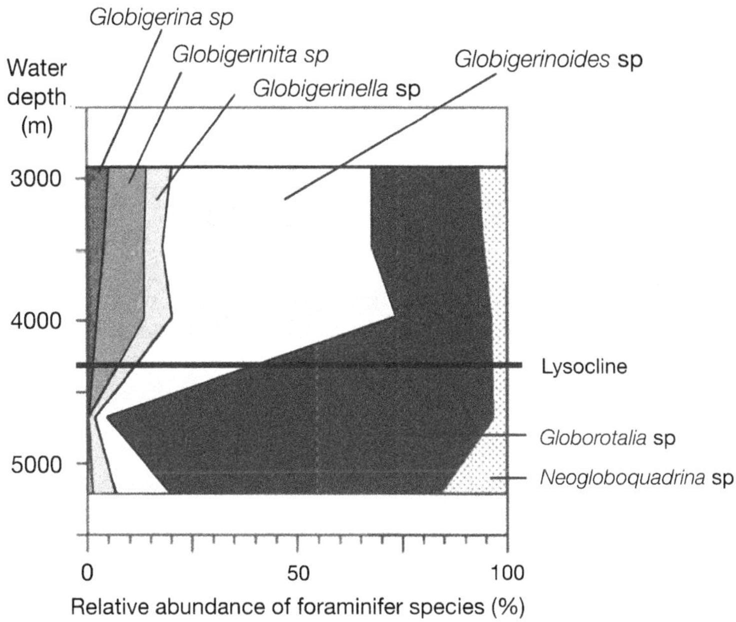

Figure 11.19 Variation of planktonic foraminiferal assemblages in surface sediments along a depth transect of the mid-oceanic ridge in the Equatorial Atlantic Ocean. Note increased relative abundance of highly calcified (thick test) species in the deeper part of the transect and estimated location of the lysocline. Modified from Dittert, N. et al., in Fischer, G., Wefer, G. (Editors), 1999. Use of proxies in paleoceanography. Springer, Berlin, pp. 255–284.

11.1.4. The Diagenesis of Biogenic Carbonates

Carbonate diagenesis is especially active in coastal areas of low latitudes where reef systems and other carbonate environments are abundant. There, dissolution and recrystallization processes result from a complex interplay involving meteoric waters, pore waters and back-and-forth movement of tidal flows principally. They generally lead to the formation of calcite cements. Where pore waters are anomalously concentrated in Mg, dissolution of calcite and aragonite ($CaCO_3$) may be followed by the crystallization of dolomite ($CaMg(CO_3)_2$. This is the case for example in hypersaline sebkhas and lagoons where the precipitation of evaporites such as gypsum ($CaSO_4.2H_2O$) or anhydrite ($CaSO_4$) increases the Mg/Ca ratio of the waters, or during mixing of meteoric and marine waters. In such environments, the diagenetic alteration of carbonates may create additional porosity that can function as a trap for hydrocarbons. This is especially the case in areas of extensive dolomitization of calcite, which is associated to a decrease in volume.

By comparison, the diagenesis of biogenic carbonate oozes in the deep ocean basins is rather straightforward. The porosity of carbonate oozes near the sediment/ocean interface is relatively low (70–75%) and decreases rapidly with sub-bottom depth as pore waters are released. Porosity values below 50% are common at burial depths of 200–300 m. The solubility of calcareous microfossils increases with pressure, especially at contact points between particles. As a result the proportion of

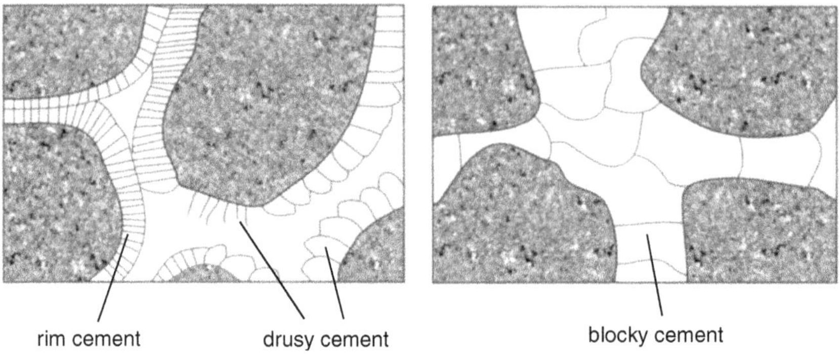

Figure 11.20 Major types of calcite cements. Modified from Biju-Duval, B., 1999. Géologie sédimentaire. Technip, Paris.

foraminifer fragments increases. Calcite reprecipitates in areas of lower pressure and higher concentration in dissolved carbonates, as overgrowths at the surface of the central coccolith shields and foraminiferal chambers. A grain-supporting framework progressively develops, indicating the transition from firm biogenic ooze to friable chalk.

Chemical diagenesis progressively prevails in carbonate chalks, as increased pressure solution and circulating pore waters facilitate the precipitation of calcite onto the surface of particles and within the pore space. Coccoliths, foraminifers and fragments are increasingly filled with secondary calcite. The particles are more and more bonded together by calcite bridges, and the development of cement marks the transition from chalk to limestone (Figure 11.20). In some cases calcite overgrowths around framework particles progressively merge (rim cement) whereas in other cases cementation develops in several steps. Initial formation of needle-like crystals of calcite radiating from surfaces toward the center of the pores (drusy cement) is followed by the crystallization of larger and irregular sparry calcite (blocky cement). In limestones, diagenetic processes may have totally erased the initial structure of the sediment (fabric, shape and size of the particles) as well as its micropaleontological archives. Pressure dissolution, which involved only contact points between particles in firm oozes and chalks, now occurs across broad horizontal surfaces which form pressure solution seams (Figure 11.21). Fine residual siliciclastics may concentrate along those surfaces, which sometimes appear as irregular fine lines (stylolites) and mimic bedding planes.

The formation of dolomite via the diagenetic alteration of deep biogenic carbonate oozes is extremely rare. It is generally considered that the presence of sulfates in seawater and pore waters could inhibit the reaction. This hypothesis is indirectly supported by occurrences of dolomite in sediments containing significant amounts of organic matter, which accumulated in dysoxic to anoxic environments. There, the early diagenetic alteration of calcite to dolomite coincides with active bacterial sulfate reduction. Anaerobic sulfate reduction may also provide alkaline pH conditions favorable to the precipitation of carbonates. In such environments, iron carbonates (siderite) may also form in the presence of high concentrations in Fe,

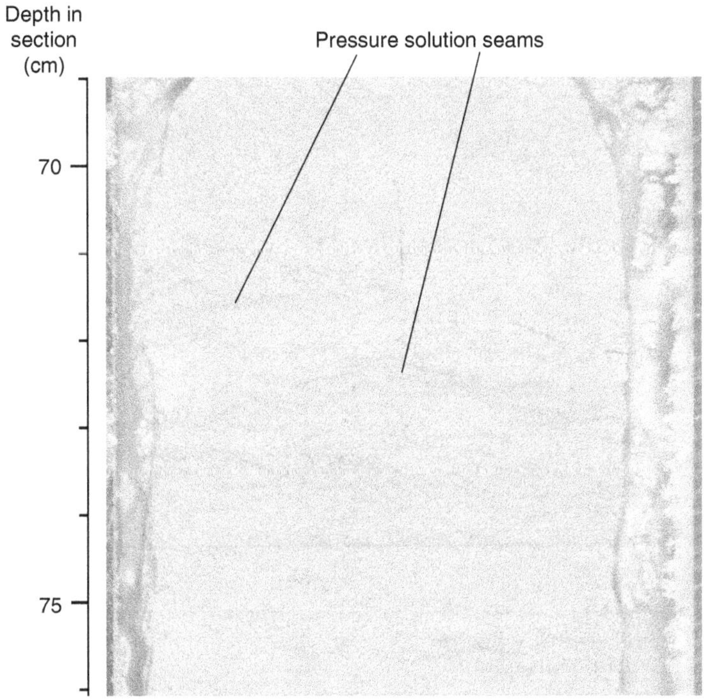

Figure 11.21 Pressure solution seams in Middle Miocene lithified carbonates of the East Tasman Plateau, Southern Ocean (Section 189-1172A-34X-5). Modified from Exon, N., Kennett, J.P., Malone, M. et al., 2001. Proceedings of the Ocean Drilling Program. Initial Reports, volume 189. Ocean Drilling Program, College Station, TX.

especially at deeper sub-bottom depths where sulfate reduction decreases in importance.

11.2. Siliceous Microfossils: Formation, Preservation, and Transformation

11.2.1. Radiolarians

The radiolarians are planktonic protozoans which belong to the class Sarcodina. They are characterized by an elaborate skeleton, which in most cases is made of amorphous silica (opal A). The radiolarians include three major groups: the acantharians (which build a skeleton made of strontium sulfate) and the trypileans are barely preserved after death, whereas the polycystines are commonly found in oceanic sediments. The polycystines include the radially symmetrical spumellarians, and the ring or cone-shaped nasselarians (Figure 11.22). Radiolarians range in size from 40 to 400 μm, but some forms may reach up to 2 mm. The radiolarian cell is characterized by a "chitinous" central capsule which separates the protoplasm into an inner part (endoplasm) and an outer part (ectoplasm). The intracapsular

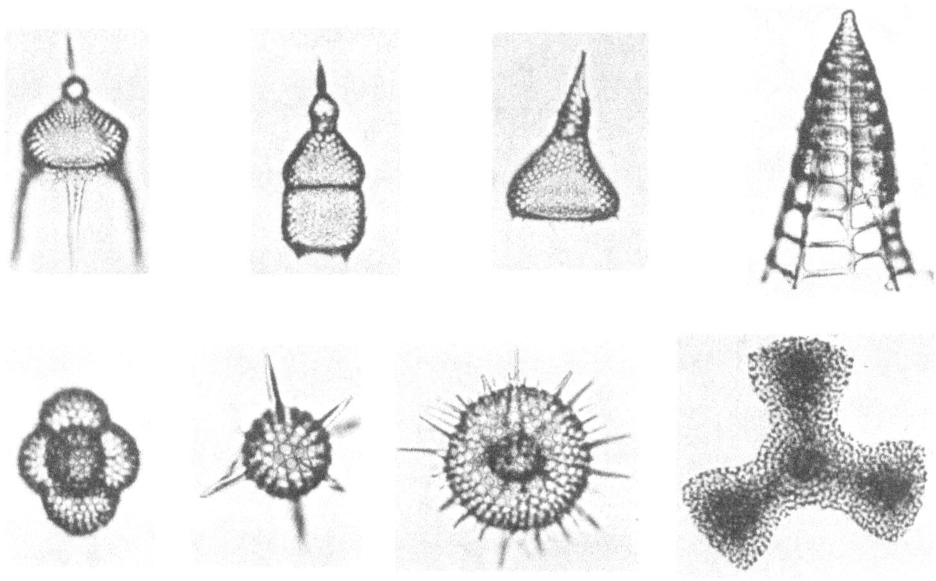

Figure 11.22 Examples of nasselarian (top) and spumellarian (bottom) polycystine radiolarians. Modified from Haq, B.U., Boersma, A. (Editors), 1978. Introduction to marine micropaleontology. Elsevier, Amsterdam.

endoplasm contains the nucleus (or nuclei), the Golgi body, mitochondria, vacuoles, lipid droplets and the axoplast. The axoplast produces the axopods, stiff tubular filaments which play a role in the construction of the skeleton (Figure 11.23). Tiny holes in the central capsule allow exchanges between the endoplasm and ectoplasm, and serve as exits for the axopods. The extracapsular ectoplasm principally contains a number of ellipsoidal alveoli saturated with carbon dioxide (the calymma) which probably are hydrostatic regulators, and emits thin pseudopods (filopods). The ectoplasm of symbiont-bearing species also contains symbiotic algae, principally zooxanthellae.

The skeleton of polycystine radiolarians consists of one or several latticed walls, i.e. networks of bars isolating closely spaced pores, and spines. The skeleton is entirely enclosed within the ectoplasm, including the spines which serve as axones for the pseudopods. Therefore, there is no direct contact between seawater and the siliceous skeleton of living organisms, which is preserved from dissolution in undersaturated surface waters. The skeleton develops step by step, during specific intervals of the radiolarians life cycle and under given temperature conditions (Figure 11.24). The initial spines grow and produce a series of connecting bars during these intervals of silica deposition. Simple cell division is a common reproductive process of radiolarians, but multiple cell divisions sometimes occur. In both cases however, one daughter cell keeps the initial skeleton, whereas the other cells must fabricate a new skeleton. Reproduction also occurs via isosporogenesis, i.e. division of the nucleus and formation of tiny cells with flagella which are released when the central capsule breaks.

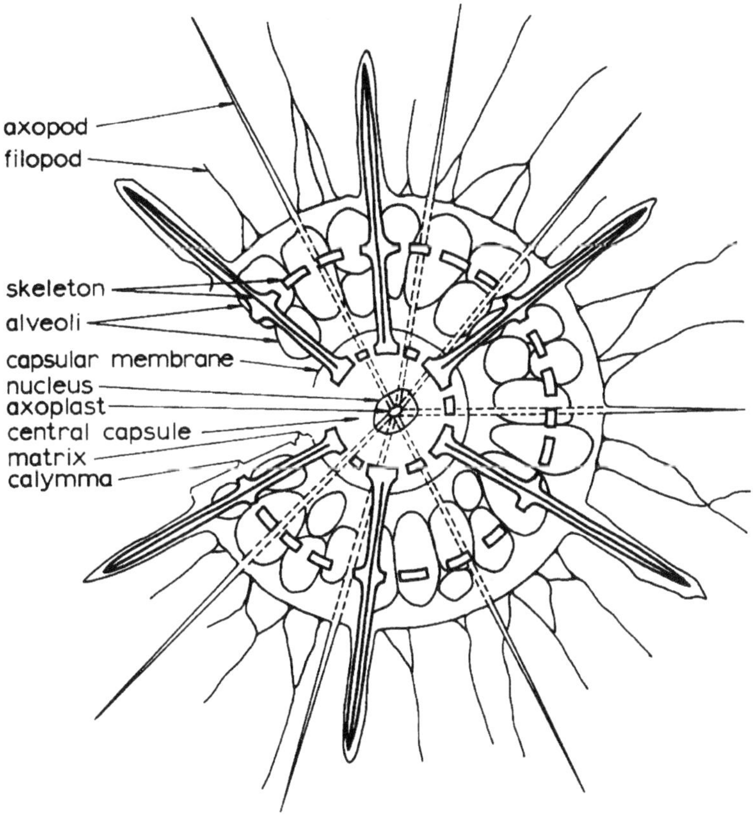

Figure 11.23 Cell and skeleton components of a living polycystine radiolarian. Reprinted from Haq, B.U., Boersma, A. (Editors), 1978. Introduction to marine micropaleontology. Elsevier, Amsterdam.

The life cycle of radiolarians lasts for one to three months approximately. Radiolarians are predominantly found in the photic zone, but some species may live at water depths of several thousands meters. They float, waiting for contact with potential preys which are captured by the filopods. Besides nutrition, the pseudopods are used for temporary fixation on floating objects. Vertical movements up to 350 m in amplitude are commonly observed, the radiolarians using the carbon dioxide content of their alveoli for controlling their density, and therefore their buoyancy.

Radiolarians are currently found in surface waters of all oceanic areas, but are especially abundant in the productive waters of oceanic divergences and coastal upwellings, where they commonly form colonial groups up to a few meters long. However, radiolarians are also found in areas of hydrothermal activity of mid-oceanic ridges (together with other microorganisms) where the chemistry of seawater is significantly modified by the injection of hydrothermal fluids, which for example may contain high concentrations of silica. The ecology of radiolarians is probably dependent on silica, as well as on other environmental parameters such as

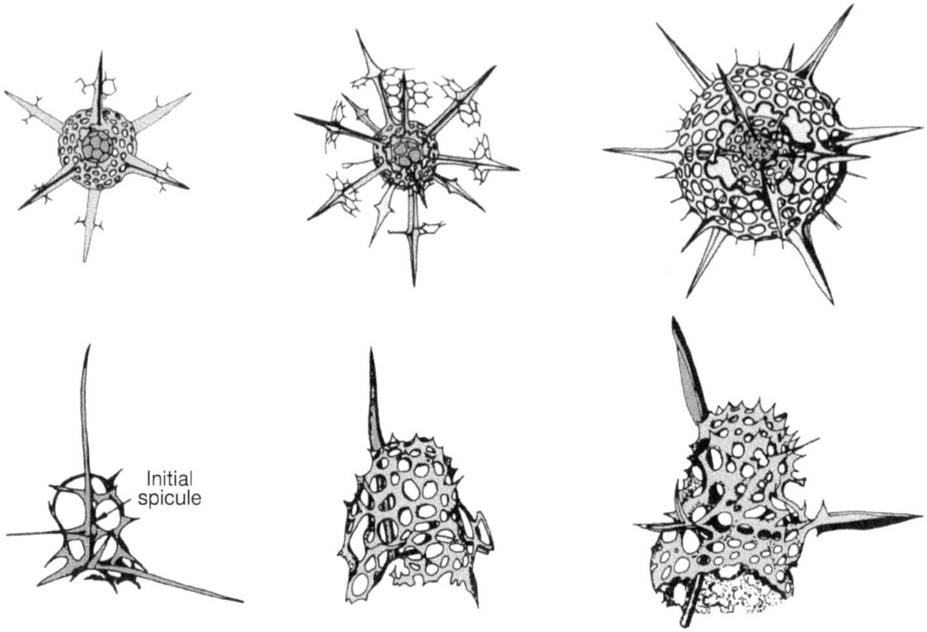

Figure 11.24 Different stages in the growth of spumellarian (top) and nasselarian (bottom) polycystines. Note that the growth of the central capsule and spines precedes those of connecting bars and skeleton. Modified from De Wever, P., Azéma, J., Fourcade, E., 1994. Radiolaires et radiolarites: production primaire, diagenèse et paléogéographie. Bulletin des Centres de Recherches Exploration-Production Elf Aquitaine, 18, 315–379.

nutrients (especially P, N and Fe) and gradients of temperature, which may explain their modern distribution. This is therefore no surprise if the radiolarians were the dominant plankton in the silica-rich oceans of the Paleozoic and Early Mesozoic, where their variability was probably also controlled by the availability of N and Fe principally.

The radiolarians have been known since the Cambrian. Since then, about 7,000 species have evolved, and about 300 species are found in the modern ocean. The oldest species were benthic and epiplanktic, but pelagic species soon appeared in the Cambrian. They may have lived in highly productive coastal waters, in surface oceanic waters of the photic zone with symbiotic algae, or in deeper waters where they fed on detritus. Among the polycystines, the nasselarians developed early in the Paleozoic whereas the spumellarians radiated in the latest Permian. Regardless of severe losses in diversity such as those recorded at the Permian–Trias boundary (when 66% of radiolarian species became extinct), radiolarians remained dominant in the plankton until the Cretaceous when they had to share their habitats with planktonic foraminifers. No mass extinction was recorded at the Cretaceous–Tertiary boundary: only a few families became extinct, and a probable bloom in siliceous plankton was followed among the radiolarians by a dominance of the spumellarians over the nasselarians.

Since they first appeared in the Cambrian, radiolarians have shown intervals of significant increases in diversity and abundance. Major peaks in diversity occurred for example during the Triassic and near the Jurassic–Cretaceous boundary (Figure 11.25). Major peaks in abundance occurred for example during the Albian–Cenomanian and the Middle Eocene. Albian–Cenomanian events occurred at a time of high primary productivity and related dysoxia, together with a number of extinctions and radiations. In detail, extinctions coincide with the preservation

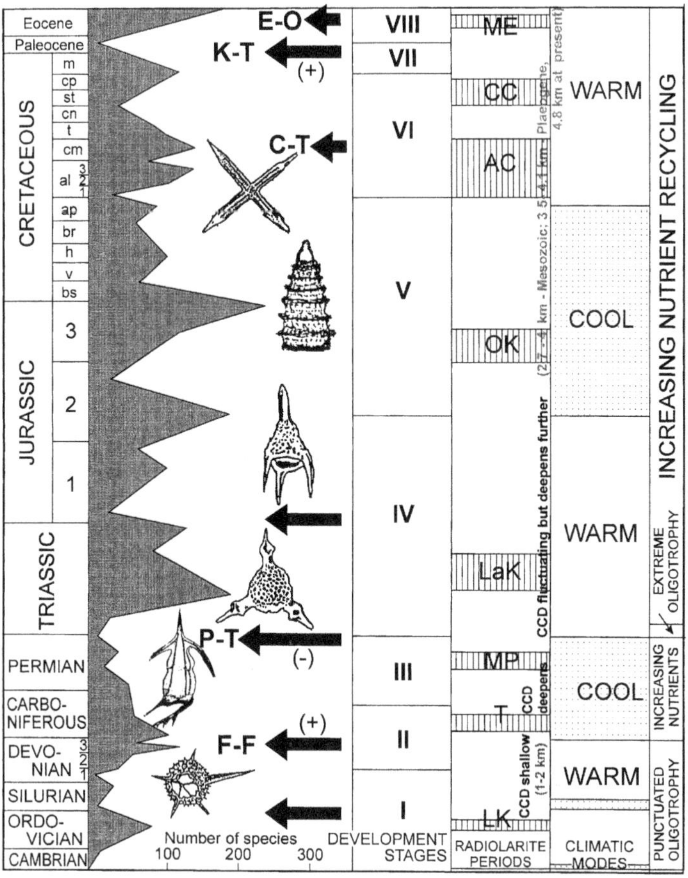

Figure 11.25 Variability in diversity of radiolarian species from the Cambrian to the Eocene, and main trends of global oceanic paleoenvironments (climate and availability of nutrients). Arrows indicate major global bioevents, with + and – indicating maximum and minimum abundances in radiolarians, respectively (F–F: Frasnian–Famennian boundary; P–T: Permian–Triassic boundary; C–T: Cenomanian–Turonian boundary; K–T: Cretaceous–Tertiary boundary; E–O: Eocene–Oligocene boundary). Development stages delimit major steps in the evolution of radiolarians. Major accumulation of radiolarian remains occurred during radiolarite periods (LK: Llandeilan–Caradocian; T: Tournaisian; MP: Mid-Permian; LaK: Ladinian–Carnian; OK: Oxfordian–Kimmeridgian; AC: Albian–Cenomanian; CC: Coniacian–Campanian; ME: Mid-Eocene). Reprinted from Racki, G., Cordey, F., 2000. Radiolarian paleoecology and radiolarites: is the present the key to the past? Earth-Science Reviews, 52, 83–120.

of marine organic matter at the base of black-shale intervals, whereas radiations follow black-shale episodes. The Middle Eocene peak in abundance follows an interval of extensive volcanism (Chron 24 plate reorganization) and warm climate, and could be related to elevated concentrations in silica and other nutrients at a time of evolving ocean circulation. Because of subsequent diagenetic alteration, these intervals are generally marked by compact radiolarites in ancient sediment series. Besides enhanced vertical circulation, nutrient supply and preservation of silica in the ocean, these intervals may principally reflect elevated silica concentration. It is probable that intense hydrothermalism and volcanism, during phases of accelerated seafloor spreading and plate tectonic activity, increased the concentration of silica (and other nutrients) in the ocean at a regional or global scale, facilitating the development of radiolarians.

Radiolarian skeletons have decreased in weight and thickness since the Cretaceous, following a decline in the concentration of dissolved silica in ocean waters. It is remarkable that the concentration in silica decreased as diatoms increased in diversity and abundance in the Cretaceous, and expanded their habitats in the Eocene (Figure 11.26). Therefore, the evolutive trend of radiolarians may record a competitive pressure for dissolved silica, and the ecology of modern radiolarians reflects an adaptation to Cenozoic, silica-depleted oceans. Besides, thin radiolarian skeletons are more sensitive to dissolution in silica-depleted oceans after death and oxidation of organic elements.

Sediment trap experiments in the equatorial Atlantic have demonstrated that fluxes of polycystine radiolarians below surface water masses reflect fairly well their abundance in the plankton. In those areas, the dissolution of biosiliceous elements seems to occur deeper in the water column. Although a diversity of polycystine taxa

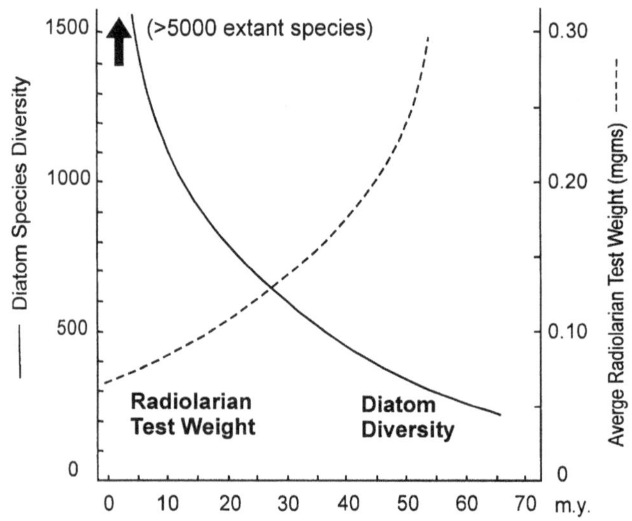

Figure 11.26 Evolution trends of radiolarian test weight and diatom diversity (which approximately follows diatom abundance) for the past 70 Myr. Note anticorrelation. Reprinted from Racki, G., Cordey, F., 2000. Radiolarian paleoecology and radiolarites: is the present the key to the past? Earth-Science Reviews, 52, 83–120.

were recorded (187), only six species accounted for more than 50% of the fluxes. Seasonal variations were limited, since maximum (summer) fluxes were nine times higher than minimum (winter) fluxes. For comparison, differences of two orders of magnitude are sometimes observed between highest and lowest fluxes of other plankton. Overall, clear similarities between living assemblages and surface sediments are limited to high-latitude areas, whereas significant dissimilarities occur in subtropical areas. However, distinct radiolarian assemblages coincide with hydrographic boundaries and concentrations of nutrients. For example, assemblages closely related to upwelling conditions have been described in surface sediments off Somalia, where radiolarian abundances are also five to ten times higher than in less productive areas of the Indian Ocean. Thus, polycystine radiolarians are valuable indicators of changes in fertility, and useful for reconstructing past hydrographic conditions.

11.2.2. Diatoms

Diatoms are unicellular algae which belong to the class Bacillariophyceae. The diatom cell consists in a protoplasm which contains the same organelles as other eukaryotic algae, i.e. a nucleus, a Golgi body, mitochondria, etc. The protoplasm also contains chromatophores (for photosynthesis), globules of fatty substance (for nutrition) and vacuoles (for buoyancy). The chromatophores vary in number, size, shape and position according to the species, and give color to the cell. The cell wall is made of hydrated amorphous silica (opal A) and separated in two valves connected by a girdle. The upper valve (epivalve) is slightly bigger than the lower valve (hypovalve) and fits on it, enclosing the protoplasm into a frustule. A thin organic coating protects the frustule. Diatom frustules may vary from 2 μm to 2 mm in size, most forms ranging from 10 to 100 μm. Diatoms are autotrophic, photosynthetic organisms which make their organic substance using energy from the sun, and their presence is therefore limited to the photic zone. However, there are some exceptions, i.e. a few species which are heterotrophic and need a source of organic carbon. Besides, they need a variety of nutrients: phosphorus, nitrate and silica are the most important, but sulfur, iron, manganese and some vitamins are essential for many species. They are therefore especially abundant in many highly productive regions of the modern ocean, from coastal upwellings and equatorial divergences to river plumes and areas of important atmospheric dust supply. Diatoms are principally planktonic but benthic, infaunal and epifaunal forms are also found in shallow areas. Planktonic diatoms are either solitary or colonial algae, which may form chains of considerable length.

Diatoms reproduce principally by cell division, the average cycle being about 8 h. They double in volume during the cell cycle, as they duplicate their organelles and chromosomes. Each daughter cell inherits a valve from the parent cell, which becomes an epivalve. Therefore, each daughter cell synthesizes a new, slightly smaller, hypovalve. As a result, the vegetative multiplication of diatoms usually leads to a progressive diminution in size of their frustules. This is compensated by episodic sexual reproduction and formation of an auxospore, which expands before generating a new frustule.

The valves consist in a network of silica bridges which connect systems of silica ribs, and delimit the pores (Figure 11.27). The rib system is organized around an initial silica ring or bar (sternum). New elements are formed inside vesicles and added to the wall. Differences in the organization of the rib system have been used in the classification of diatoms. Centric diatoms are organized around a silica ring. They have spherical, triangular or cylindrical frustules with concentric or radial sculpture. The majority of planktonic diatoms are centric forms (Figure 11.28). Pennate diatoms are organized around a silica bar. They have elongate frustules with bilateral symmetry around a median line (raphe) which generally coincide with the initial silica bar of each valve and strengthens its structure. Most benthic diatoms, but also many Antarctic planktonic diatoms, are pennate forms (Figure 11.29).

Diatoms are non-motile and have a density close to or slightly higher than that of seawater, which increases as the frustules grow and get thicker. They have

Figure 11.27 Details of the central area (left) and areolae (right) of the diatom *Asteromphalus hiltonianus* highlighting the general organization of silica bridges and pores. Note that the areolae looks like a finely perforated sieve plate. Modified from Haq, B.U., Boersma, A. (Editors), 1978. Introduction to marine micropaleontology. Elsevier, Amsterdam.

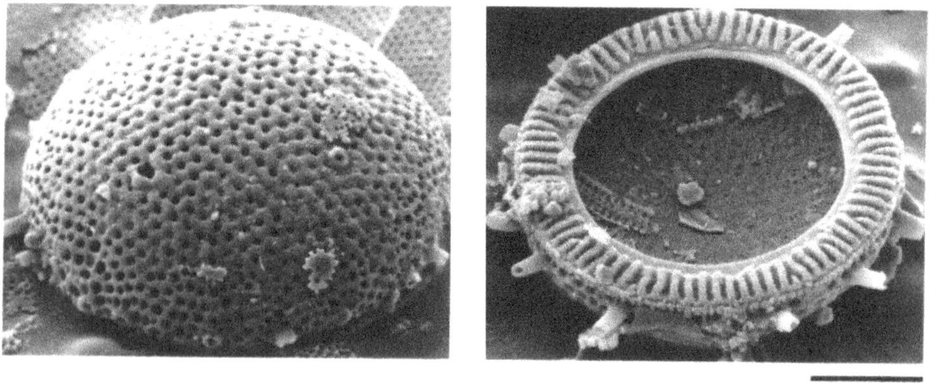

Figure 11.28 Example of centric diatom: the valve exterior (left) and interior (right) of the Neogene Antarctic species *Thalassiosira spumellaroides*. Scale bar: 10 μm. Reprinted from Barker, P.F., Kennett, J.P., O'Connell, S. (Editors), 1990. Proceedings of the Ocean Drilling Program, Scientific Results, volume 113. Ocean Drilling Program, College Station, TX.

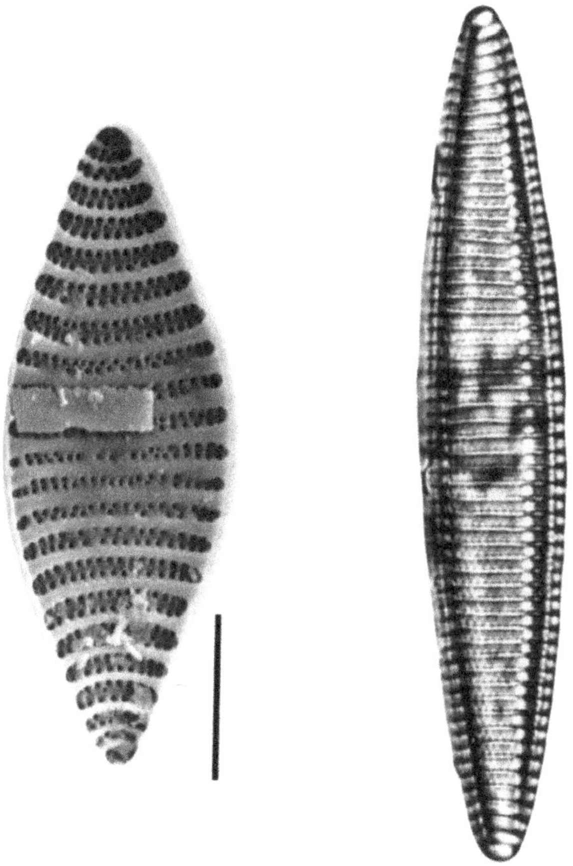

Figure 11.29 Examples of pennate diatoms: the pelagic Neogene Antarctic species *Nitzschia barronii* (left) and *Nitzschia reinholdii* (right). Scale bar: 10 μm. Reprinted from Barker, P.F., Kennett, J.P., O'Connell, S. (Editors), 1990. Proceedings of the Ocean Drilling Program, Scientific Results, volume 113. Ocean Drilling Program, College Station, TX.

accordingly a tendency to sink. Sinking rates are diminished by the shape of the frustule, the secretion of a low-density mucilage around the cell, and the replacement of high-density ions (such as calcium and magnesium) by low-density ions (such as sodium and potassium) within the vacuoles.

Although the oldest known diatoms have been found in Jurassic deposits, they principally developed during the Cretaceous and underwent major evolutionary radiation in the Middle Cretaceous, with the development of heavily silicified species. It is possible that diatoms already existed in the Early Mesozoic but were not fossilized. The centric diatoms first appeared and were the unique group to occur in the Cretaceous and until the Late Paleocene. Their diversity has remained essentially the same since the Eocene. Pennate diatoms radiated in the Late Paleocene and have shown significant radiation during the Cenozoic, especially since the Miocene. About 600 genera and 20,000 species have evolved since the Cretaceous.

Approximately 10,000 species are living on the modern Earth, where pennate forms largely dominate.

Diatoms tolerate a wide range of environments and are abundant in the phytoplankton and the phytobenthos of marine and freshwater ecosystems. For example, polar forms survive very low temperatures and absence of sunlight for part of the year. Also, diatoms are sometimes found in wet soils. However, they are absent from hypersaline environments. When environmental conditions are adverse, diatoms may form resting spores which may produce new frustules when conditions improve. For example, resting spores are part of the annual cycle of some freshwater diatoms, which survive seasonal desiccation. In the Arabian Sea, resting spores which slowly settle in the water column are brought to the surface late in spring when intermediate waters upwell in response to wind stress, facilitating a rapid phytoplanktonic bloom. Resting spores may survive for several decades in anoxic conditions.

Planktonic diatom populations are principally controlled by the availability of silica. In culture, diatom cells starved of silica stop dividing. By contrast, limitation of other nutrients such as nitrogen leads to minor changes, such as a lengthening of the cell cycle. In the ocean surface waters are largely undersaturated in silica, which is present in the form of orthosilicic acid $Si(OH)_4$. Soluble silica is transported to specific vesicles of the diatoms, where polymerization produces solid silica in the form of hydrated amorphous opal $SiO_2.nH_2O$. Silica uptake is especially rapid during a short interval of the diatom cycle, i.e. the formation of new valves which immediately follows cell division. Diatom populations are concentrated near the thermocline, which varies in depth regionally and/or seasonally. During intervals of shallow thermocline and stratification of the surface mixed-layer, i.e. stable hydrographic conditions, diatoms bloom in response to improved light conditions and increase their uptake of orthosilicic acid in proportion. Seasonal blooms frequently result in a significant decrease of the concentration in soluble silica of surface waters, and subsequent decline of diatom populations. For example in the Southern Ocean, a high diatom biomass develops during summer blooms in regions of high concentration in orthosilicic acid principally, removing more than 70% of the soluble silica initially present in the uppermost 50 m of the water column (Figure 11.30). The diatom bloom finishes by the end of the summer when the thermocline is destabilized and the mixed-layer deepens: most diatoms disappear from uppermost waters, where the concentration in orthosilicic acid increases.

Like other phytoplankton, diatoms are grazed by heterotrophic organisms. Grazing by zooplankton such as dinoflagellates and by microcrustaceans removes the organic matter, leaving the frustules, or frustule fragments, exposed to dissolution (see Section 11.2.3) as they sink in the water column. Grazing by larger organisms such as fish allows the frustules to sink rapidly within fecal pellets where they are protected from dissolution. Besides, diatoms have a tendency to assemble and form aggregates, especially when blooms terminate. The mechanism is very similar to flocculation, and allows a rapid transfer of diatoms to the seafloor. As a result, diatom dissolution principally occurs in undersaturated surface waters, and within the uppermost sediment close to the ocean/sediment interface (see Section 11.2.3). Nevertheless, only a small proportion of living diatoms is finally buried in oceanic

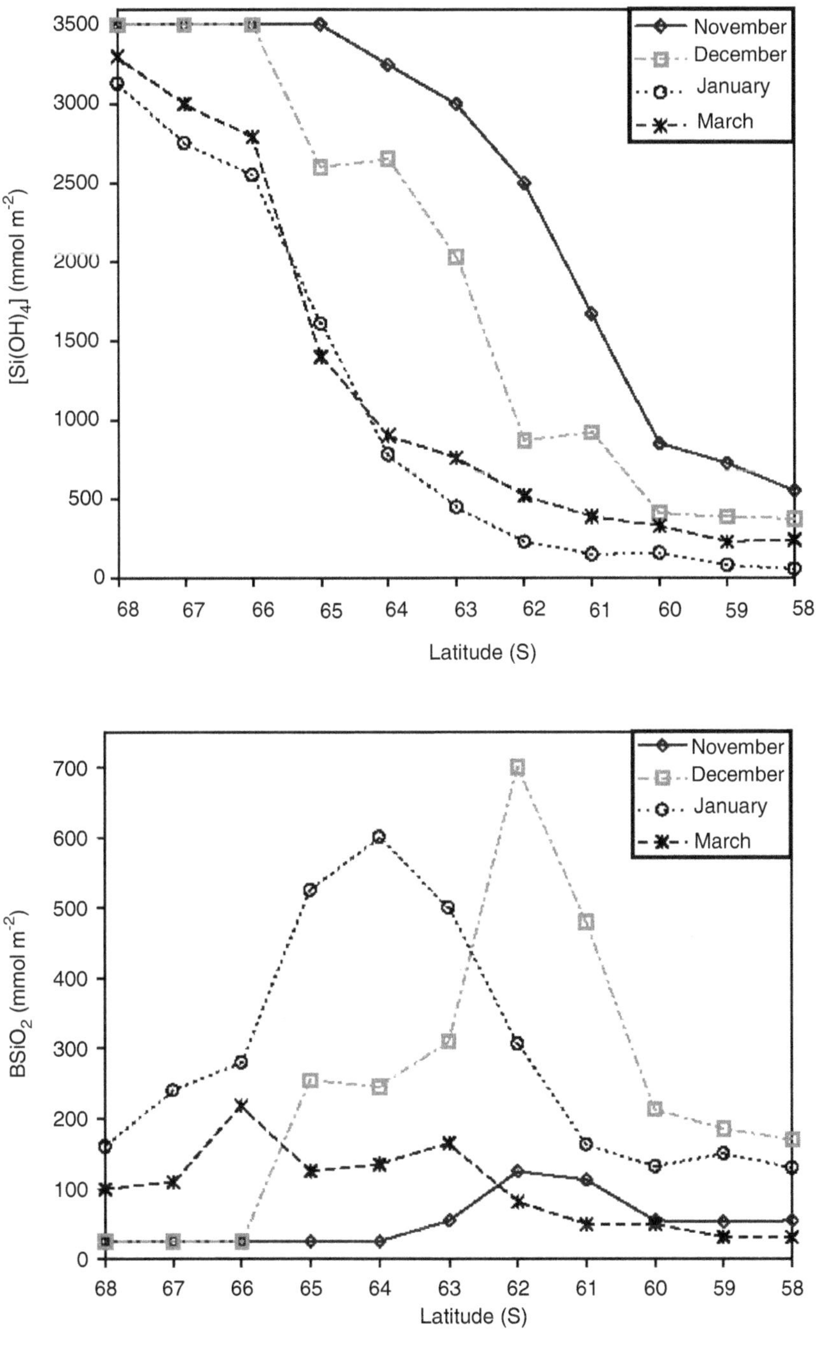

Figure 11.30 Seasonal variability of concentrations in silicic acid in the upper 50 m of the water column (top) and biogenic silica in the photic zone (bottom), in the Pacific sector of the Southern Ocean. Note significant silica depletion of surface waters during summer blooms in regions of maximum productivity. Modified from Sigmon, D.E., Nelson, D.M., Brzezinski, M.A., 2002. The Si cycle in the Pacific sector of the Southern Ocean: seasonal diatom production in the surface layer and export to the deep sea. Deep-Sea Research II, 49, 1747–1763.

sediments: estimates range frequently between 6% and 10% of the diatom biomass in areas where they dominate in oceanic sediments, e.g. in tropical upwelling areas off Somalia and in the Southern Ocean. Moreover, lightly silicified diatom species increase in importance in surface waters during intervals of maximum productivity, when the concentration of orthosilicic acid in seawater is minimum. Lightly silicified species dissolve easily and are generally not preserved in the sediment where the relative proportion of heavily silicified species increases. In fine, diatom assemblages in surface sediments are somewhat different from surface water communities. For example, the predominance of the heavily silicified *Fragiliariopsis kerguelensis* in many sediments of the Southern Ocean (where it represents up to 80% of the diatom assemblage) reflects its dominance in early spring blooms and late summer communities, but not the diversity of summer communities which account for the same proportion of export as the spring bloom (Figure 11.31).

Nevertheless, fossil diatom assemblages preserve information on surface water properties. First of all, in many areas including the highly productive Southern Ocean, the dominance of diatoms in surface sediments reflects elevated production rates in the photic zone and fluxes to the seafloor. Besides, the biogeographic distribution of diatom assemblages and relative abundances of diatom species show close relationships with environmental conditions in surface waters (Figure 11.31). In the Southern Ocean for example, the distribution of diatom species appears principally controlled by the pattern of sea surface temperature and sea ice cover. There, the southernmost extension of the tropical/subtropical assemblage coincides with average sea surface temperatures of 11°C and the subtropical front.

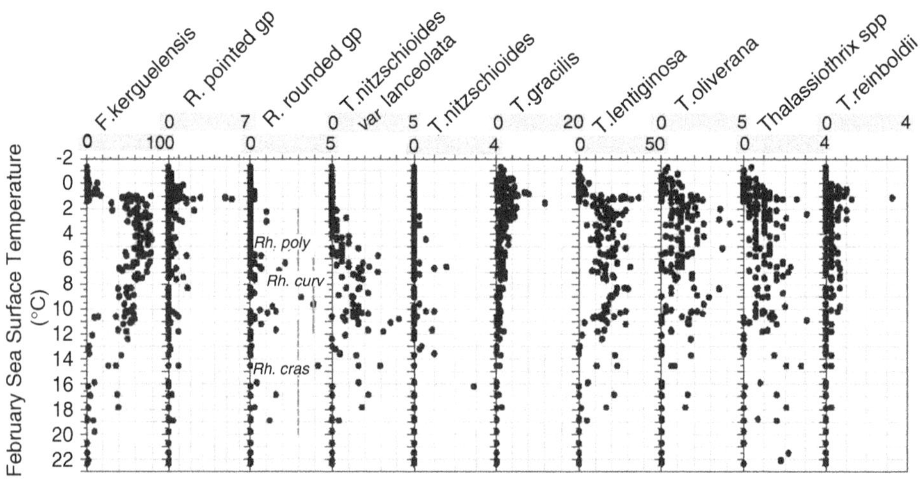

Figure 11.31 Relative abundances of selected diatom species in surface sediments of the Southern Ocean, plotted against summer sea surface temperatures. Note the large dominance of the heavily silicified *Fragiliaropsis kerguelensis* (left), and the responsiveness of some species to changes in sea surface temperature. Reprinted from Crosta, X., Romero, O., Armand, L.K., Pichon, J.-J., 2005. The biogeography of major diatom taxa in Southern Ocean sediments: 2. Open Ocean related species. Palaeogeography, Palaeoclimatology, Palaeoecology, 223, 66–92.

Beyond the subtropical front, three diatom assemblages are identified from north to south:

- A warm Antarctic assemblage is characterized by the *Rhizosolenia* rounded group, the *Thalassionema nitzchioides* group and its variety *lanceolata*.
- A pelagic open ocean assemblage is composed of *Fragiliaropsis kerguelensis* (Figure 11.32), *Thalassiosira lengitinosa, Thalassiosira oliverana* and the *Thalassiothrix* spp. Group.
- A cool Antarctic assemblage is characterized by the *Rhizosolenia* pointed group, the *Thalassiosira gracilis* group and *Trichotoxon reinholdii*.

These open ocean assemblages are replaced further south by cryophilic diatoms, which vary in abundance with the annual duration of sea ice cover. Among them, several species such as *Thalassiosira tumida* strongly increase in abundance in areas where sea ice is present more than seven months per year, whereas other species such as the *Thalassiosira antarctica* group (Figure 11.32) show a steady increasing response to sea ice duration.

11.2.3. The Dissolution and Preservation of Biogenic Silica

Despite continuous supply of silica by runoff and hydrothermalism principally, the ocean remains significantly undersaturated and dissolution of biogenic silica occurs at all depths of all oceanic regions. Undersaturation is maximum in the uppermost waters of the photic zone, where microorganisms extract high quantities of silica to build their mineral elements (Figure 11.33). This is especially the case during phytoplankton blooms, and diatoms currently play a major role in the burial and recycling of biogenic silica.

The dissolution rate of biogenic opal in seawater is expressed by the equation

$$V_{dis} = K([Si(OH)_4]_{sat} - [Si(OH)_4]) \times A_{sp}$$

where K is the velocity of the reaction, $[Si(OH)_4]_{sat}$ the concentration in orthosilicic acid for saturation, $[Si(OH)_4]$ the actual concentration in orthosilicic acid and A_{sp} the specific surface of the particles. However, dissolution rates are influenced by a variety of parameters:

- First, the velocity of reaction and concentration for saturation increase with temperature and in theory the dissolution rate should increase by an order of magnitude for a 15°C increase in seawater temperature. In the ocean, 15–25 time increases in dissolution rates with temperature have been observed: minimum dissolution rates occur in areas of the Southern Ocean where surface water temperatures vary from −1.5°C to 1.5°C, whereas maximum dissolution rates occur in tropical areas where sea surface temperatures of 14–22°C have been recorded. However, dissolution rates weaken with decreasing seawater temperatures in the deep ocean.
- Biogenic opal is rarely pure, and includes trace elements extracted from seawater. Among them, several metals including Al and Fe may decrease dissolution rates. The role of Al seems especially important, and it has been demonstrated that both

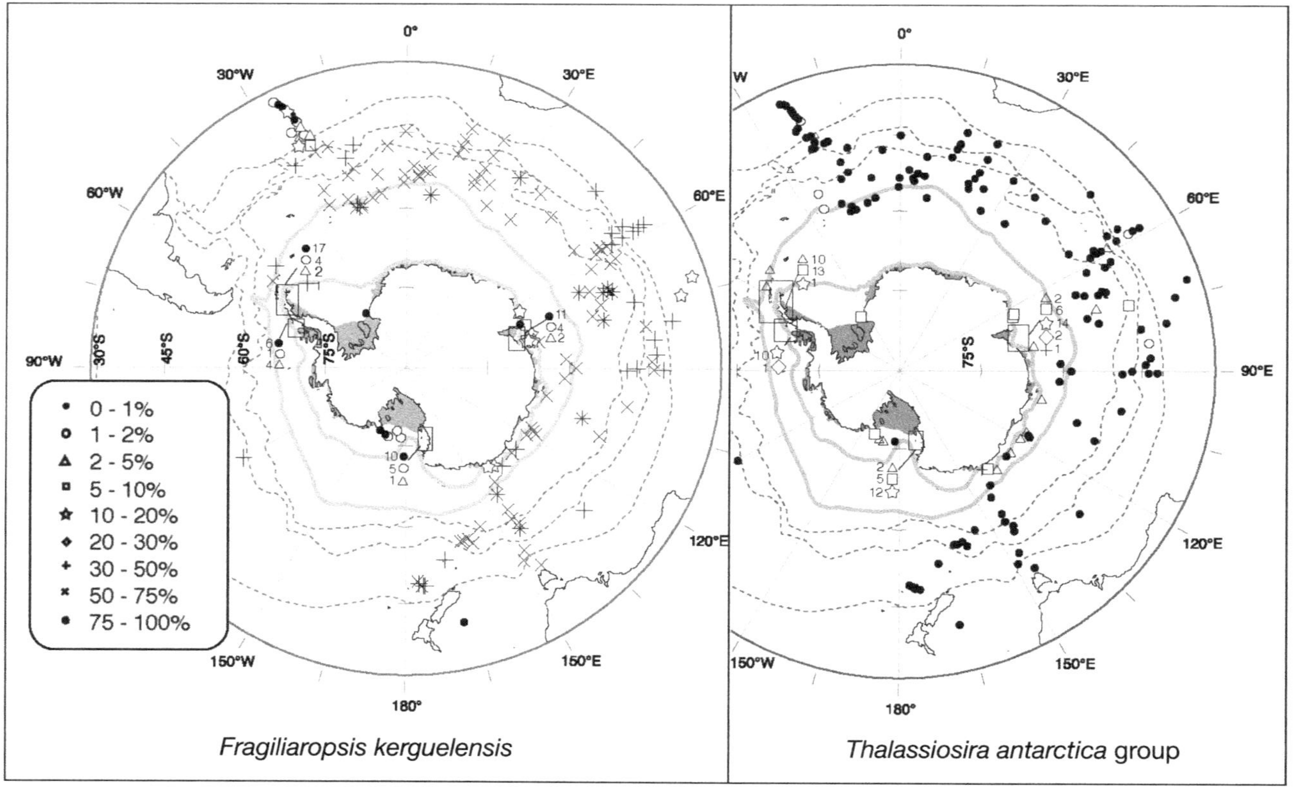

Figure 11.32 Distribution of the pelagic *Fragiliaropsis kerguelensis* and cryophilic *Thalassiosira antarctica* group relative abundances in modern sediments of the Southern Ocean. Dotted lines represent oceanic fronts, gray lines the location of winter and summer maximum sea ice extent and shaded areas the regions of permanent sea ice cover. Note the coincidence of maximum relative abundances with open, cold antarctic waters and durable sea ice cover, respectively. Modified from Crosta, X., Romero, O., Armand, L.K., Pichon, J.-J., 2005. The biogeography of major diatom taxa in Southern Ocean sediments: 2. Open ocean related species. Palaeogeography, Palaeoclimatology, Palaeoecology, 223, 66–92 and Armand, L.K., Crosta, X., Romero, O., Pichon, J.-J., 2005. The biogeography of major diatom taxa in Southern Ocean sediments: 1. Sea ice related species. Palaeogeography, Palaeoclimatology, Palaeoecology, 223, 93–126.

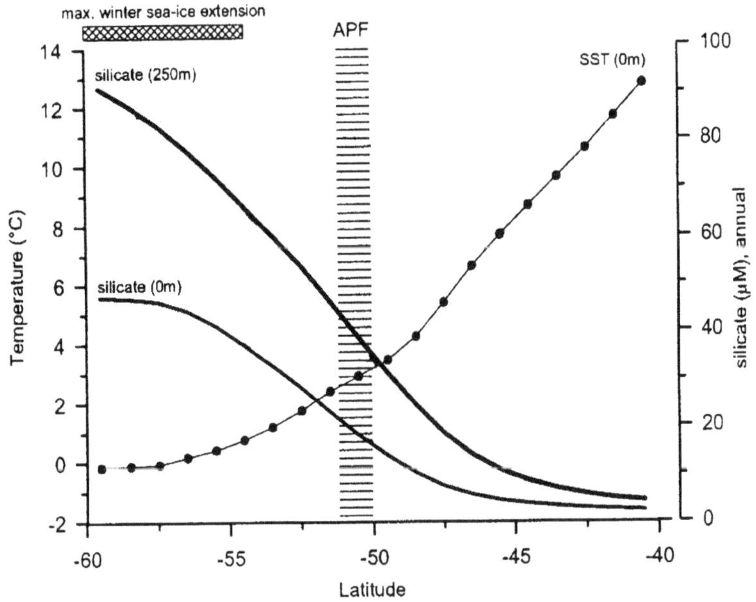

Figure 11.33 Annual averages of sea surface temperature (SST) and silicate concentrations (surface and 250 m) along a meridional transect at 0° across the southernmost South Atlantic and adjacent Southern Ocean. APF: Antarctic Polar Front. Note lower concentrations of silicate at the surface and in regions of higher SST, where silica dissolution increases. Modified from Fischer, G., Gersonde, R., Wefer, G., 2002. Organic carbon, biogenic silicia and diatom fluxes in the marginal winter sea-ice zone and in the Polar Front Region: interannual variations and differences in composition. Deep-Sea Research II, 49, 1721–1745.

the solubility and dissolution rates of biogenic silica decrease with increasing Al/Si ratios.

- The dissolution of opaline remains also varies according to the species. The thickness and degree of silicification of the skeletons and frustules, the number and arrangement of the pores which influence the specific surface, are important factors. As a consequence, the radiolarians are generally more resistant to dissolution than the diatoms, especially the spumellarians.

Maximum dissolution of biogenic silica is recorded in the photic zone, where concentrations in orthosilicic acid are the lowest, and seawater temperatures are the highest of the water column (Figure 11.33). In the Ross Sea area where surface temperatures are low all year round, most dissolution occurs in the upper 50 m and it is estimated that about 65% of the biogenic silica (principally produced by phytoplankton) dissolves. Further dissolution in surface waters results in downward silica fluxes at 250 m which are about 70% of those at 50 m. Fluxes at 250 m and on the seafloor (here at water depths of 500–850 m) are remarkably similar, illustrating the rapid sinking of the organisms as aggregates or within fecal pellets which minimizes the exposure of biosiliceous elements to undersaturated waters. Transport via fecal pellets may increase significantly during bloom periods, associating higher surface productivity to enhanced biosiliceous fluxes to the seafloor.

However, only a small fraction of the biogenic silica that reach the seafloor is finally incorporated to surface sediments, indicating that significant dissolution also occurs on the seafloor. This is probably because opal particles remain in contact with aggressive seawater for a long time before burial. Opal preservation increases in areas of higher sedimentation rates, where the exposure to undersaturated deep waters is limited. It has been demonstrated in the Ross Sea that as much as 60% of the biosiliceous remains deposited on the seafloor are preserved in sediments accumulating at sedimentation rates above 16 cm/kyr, decreasing to near zero where sedimentation rates are below 2 cm/kyr. Higher sedimentation rates may occur in regions where significant proportions of siliciclastic elements accumulate, but also in areas of high productivity and efficient transfer of biogenic elements to the seafloor.

On the average, only 6% of the initial siliceous biomass is preserved in Ross Sea sediments. Roughly similar values are found in other areas of the Southern Ocean. However, this proportion increases to 20% in some areas of the productive Antarctic Circumpolar Current, where the water depth of the Southern Ocean is of 4,000–5,000 m. Although some lateral advection may occur, the most likely mechanism for this relatively high degree of preservation involves the nature and abundance of plankton grazers. The krill are major grazers in the open waters of the Southern Ocean and produce fecal pellets that sink rapidly: settling rates vary from 100–800 m/day for some species, to 1,500–2,700 m/day for others. Via the rapid transfer of biogenic material from the upper photic zone to the deep ocean, the krill ensure a better preservation of biogenic silica in surface sediments. In fine, opal preservation in many areas of the Southern Ocean is only slightly better than in other oceans where the average preservation is approximately 3%. Higher accumulation rates of biogenic silica in surface sediments look principally associated to enhanced production rates in the photic zone and transfer in the water column, at least in the Southern Ocean.

The limited preservation of biosiliceous remains in surface sediments is associated to a selective dissolution of slightly silicified organisms which in some cases may significantly alter the composition of microfossil assemblages. In the equatorial Atlantic, the plankton sinking below the photic zone is largely dominated by diatoms and silicoflagellates during the productive upwelling season. Sinking diatom assemblages are dominated by the small and slightly silicified species of the *Nitzschia bicapitata* group which account for more than 30% of the total on the average, increasing as to exceed 50% of the diatom flux during intervals of maximum productivity. By comparison, the more robust *Azpeitia* group accounts for less than 10% of the assemblage on the average, higher relative abundances being recorded during the less productive season (Figure 11.34). As a result of the preferential dissolution of slightly silicified organisms, the proportion of *Nitzschia bicapitata* group species decreases below 10% of the diatom assemblage in surface sediments, whereas the proportion of the more robust species of the *Azpeitia* group increases to more than 20%. There, dissolution obviously erased the upwelling signal from the sedimentary diatom assemblage. In the Southern Ocean in contrast,

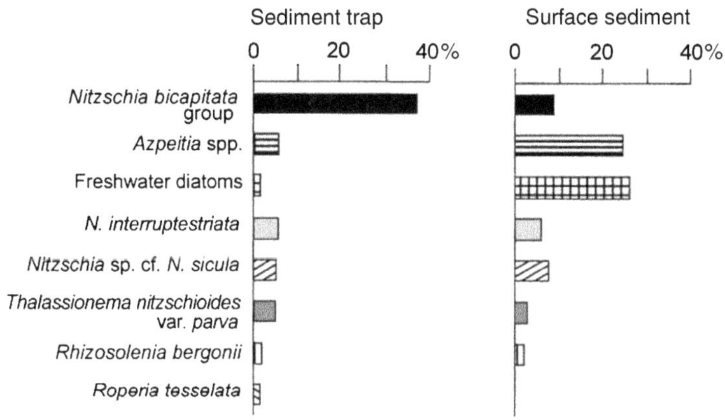

Figure 11.34 Relative abundances of the most common diatom species in sediment traps and surface sediments of the Atlantic Ocean near the Equator. Note percentage variations of the lightly silicified *Nitzschia bicapitata* and heavily silicified *Azpeitia*, as a result of preferential dissolution of lightly silicified remains in the water column. Modified from Romero, O., Fischer, G., Lange, C., Wefer, G., 2000. Siliceous phytoplankton of the western equatorial Atlantic: sediment traps and surface sediments. Deep-Sea Research II, 47, 1939–1959.

the presence of heavily silicified diatom species such as *Fragiliariopsis kerguelensis* which dominate during bloom seasons, and *Thalassiosira lengitinosa* which is typical of open ocean assemblages (see Section 11.2.2) allows the preservation of the productivity signal and pattern of sea surface conditions in surface sediments.

11.2.4. The Diagenesis of Biogenic Silica

Silica dissolution continues within a few centimeter of the uppermost sedimentary column, decreasing in importance as the concentration of pore waters in silicic acid increases. In 5–10 cm silicic acid concentrations reach asymptotic values and silica dissolution decreases accordingly (Figure 11.35). Asymptotic concentrations in silicic acid show a strong regional variability, from 50 µm/l to about 900 µm/l (Figure 11.36). Therefore, opal is preserved although pore water concentrations in silicic acid do not reach saturation level, which is of about 1,100 µm/l at 4°C. In fact, the dissolution rate of opal in sediments is controlled by a variety of parameters including (but not limited to) the reactivity of the silica surface, the amount of exposed surface area, the pH and the degree of undersaturation of pore waters. For example, preferential dissolution of the most soluble elements on the seafloor (i.e. those with abundant pores, suture surfaces separating silica spherules and surface and compositional defects) tend to the accumulation of less soluble elements. Also, the adsorption of alkali cations such as Na^{+} onto surface silica molecules may facilitate the dissolution of silica, whereas the adsorption of other cations such as Al^{3+} may inhibit its dissolution.

Pore water concentrations in Al increase significantly in the sedimentary column, where pH values above seven may initiate its solution (see Section 10.1.1). Al in oceanic sediments is principally derived from siliciclastic and volcanic

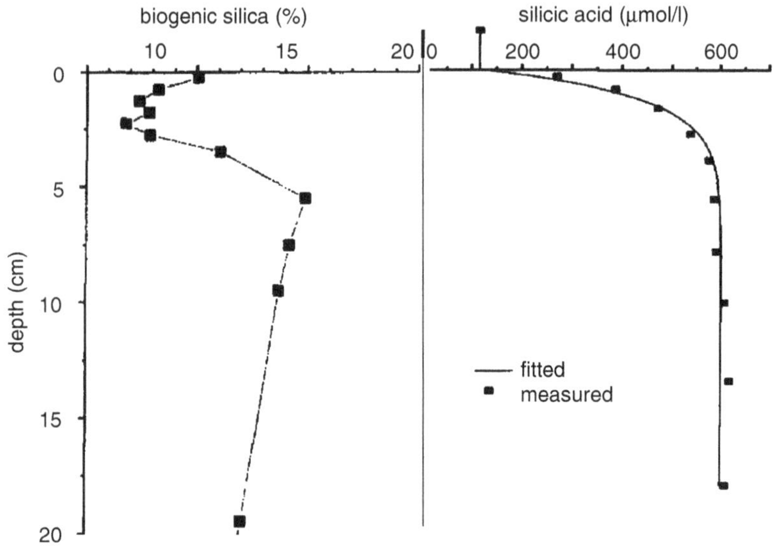

Figure 11.35 Percentages of biogenic silica (left) and pore water concentration in silicic acid (right) in surface sediments sampled at 1580 m water depth of the Indian Ocean off Somalia. The profiles suggest that dissolution of biogenic silica is active in the topmost sediment, but ceases as concentrations in silicic acid reach asymptotic values at about 5 cm below seafloor. Modified from Koning, E., Brummer, G.-J., Van Raaphorst, W., Van Bennekom. J., Helder, W., Van Iperen, J., 1997. Settling, dissolution and burial of biogenic silica in the sediments off Somalia (northwestern Indian Ocean). Deep-Sea Research II, 44, 1341–1360.

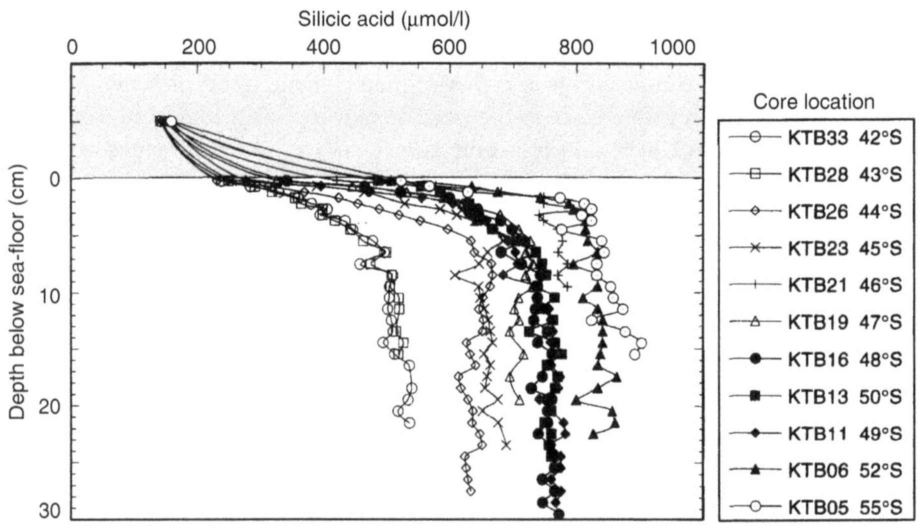

Figure 11.36 Pore water concentrations in silicic acid of surface sediments along a N–S transect of the Crozet Basin, Southern Ocean. The transect extends from the subtropical convergence to beyond the polar front. Note the regional variability of concentrations in dissolved silica, highest concentrations being recorded in areas of coldest water. At all sites, asymptotic values are reached within 2–10 cm below seafloor. Modified from Rabouille, C., Gaillard, J.-F., Tréguer, P., Vincendeau, M.-A., 1997. Biogenic silica recycling in surficial sediments across the Polar Front of the Southern Ocean (Indian Sector). Deep-Sea Research II, 44, 1151–1176.

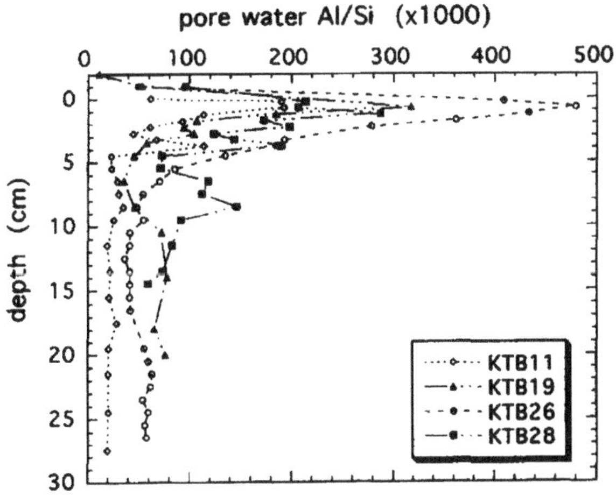

Figure 11.37 Atomic ratios of aluminum and silica in pore waters of surface sediments of the Crozet Basin, Southern Ocean. Sampling sites are those of Figure 11.36. Besides regional variability, note peak abundances of Al in the uppermost 5–10 cm of the sediment. Concentrations in Al reach peak values as dissolved silica attains asymptotic values (see Figure 11.36). Reprinted from Van Cappellen, P., Qiu, L., 1997. Biogenic silica dissolution in sediments of the Southern Ocean. I. Solubility. Deep-Sea Research II, 44, 1109–1128.

elements. Then, siliceous microfossils may incorporate Al cations within their surface lattice by substitution for Si, reducing their solubility. In topmost sediments of the Southern Ocean, highest pore water concentrations in dissolved Al often coincide with lowest silica solubilities (Figures 11.36 and 11.37). In fine, the main parameter that influences the concentration of pore waters in dissolved silica is the availability of dissolved Al. Therefore, the diagenesis of biogenic silica is principally controlled by its degree of solubility, which results from a complex interplay between silica dissolution and reprecipitation, and the uptake or release of Al by silica surfaces. At relatively high concentrations of Al in pore waters, part of the silicic acid may combine with Al to form amorphous Al–Si coatings (a protection against dissolution) or facilitate the formation of authigenic minerals. This is a probable explanation for the presence of zeolites such as clinoptilolite in the upper sedimentary column of many oceanic areas.

The porosity of biosiliceous oozes near the sediment/ocean interface is the highest of marine sediments (80–90%) and is only slightly affected by compaction in the upper part of the sedimentary column. Porosity values of 75–80% are still common at burial depths of 200–300 m. Below, the solubility of siliceous microfossils made of amorphous opal (opal A) increases with temperature and pressure providing that particles are not protected from dissolution, for example by Al–Si coatings or authigenic overgrowths. The solubility of amorphous biogenic silica is especially responsive to temperature, and dissolution is more important in areas of elevated heat flow. The dissolution of amorphous opal is followed by the recrystallization of disordered cristobalite and trydimite (opal CT) within the adjacent pore space when concentrations in silicic acid reach a threshold. Opal CT offers less reactive surface sites to pore waters and is more resistant to dissolution than biosiliceous remains.

As pressure and temperature further increase deeper in the sedimentary column, opal CT may dissolve and recrystallize into microcrystalline quartz. The diagenetic alteration of biogenic silica through dissolution and precipitation may start for temperatures as low as 40–50°C for opal CT formation, and approximately 60°C for quartz formation. The recrystallization of biosiliceous remains into opal CT and of opal CT into quartz are both associated to significant decreases in the silicic acid content of pore waters, sediment porosity and sediment volume (compaction). These changes in the mineralogy and physical properties of the sediment occur within a few meters and often characterize diagenetic fronts of regional importance. This is for example the case in Paleogene sediments of ODP Site 1172 in the Tasmanian sector of the Southern Ocean (Figure 11.38), where the diagenetic fronts also correspond to increased P wave velocities and related seismic horizons.

In biosiliceous as in other sediments, preferential precipitation of opal CT and quartz may occur within intervals of higher silica content or higher porosity

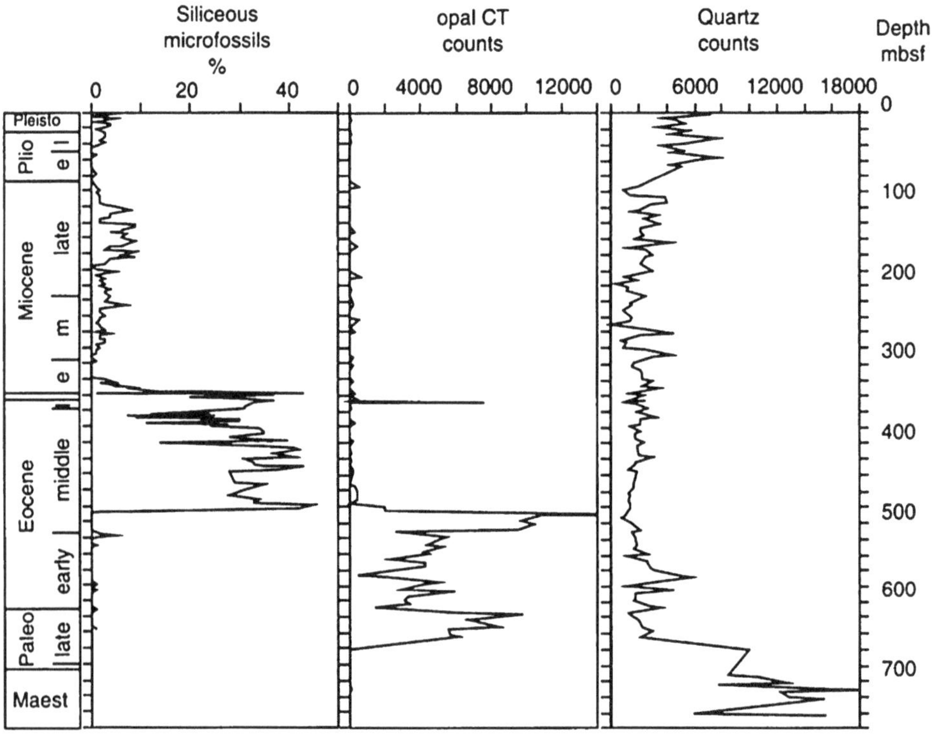

Figure 11.38 Transition from siliceous microfossils (from smear-slides) to opal CT and quartz (from X-ray diffraction data) as a result of diagenetic alteration at ODP Site 1172 on the East Tasman Plateau, Southern Ocean. Note that changes in mineralogy occur within a few meters and form distinct diagenetic fronts, which also coincide with decreased porosities and decreased pore water contents in dissolved silica. Reprinted from Robert, C. in Exon, N., Kennett, J.P., Malone, M. (Editors), 2004. The Cenozoic Southern Ocean: Tectonics, sedimentation, and climate change between Australia and Antarctica. Geophysical Monograph, volume 151. American Geophysical Union, Washington, DC.

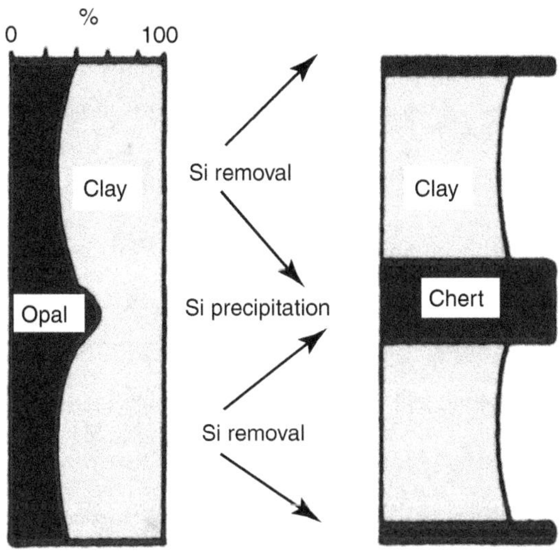

Figure 11.39 Formation of chert beds in hemipelagic sediments. Opal dissolution is especially active in intervals of lower abundances (left). Dissolved silica is transported by circulating pore waters, and precipitates in the form of microcrystalline quartz within silica-rich intervals where chert beds progressively develop (right). Modified from De Wever, P., Azéma, J., Fourcade, E., 1994. Radiolaires et radiolarites: production primaire, diagenèse et paléogéographie. Bulletin des Centres de Recherches Exploration-Production Elf Aquitaine, 18, 315–379.

(Figure 11.39). This is the origin of many chert beds, whereas chert nodules rather precipitate onto the surface of nuclei. Chert beds and nodules are characterized by very low porosities below 10% and increase in size as circulating pore fluids bring in more dissolved silica. Their growth ceases when the sediment becomes impermeable because of cementation. In some cases, minor amounts of biosiliceous remains initially dispersed within biogenic carbonate oozes or hemipelagic muds maybe entirely concentrated within chert beds or nodules as a result of diagenetic alteration. In other cases, volcanic ash or glass layers may provide silica for chert formation.

FURTHER READING

De Wever, P., Azéma, J., Fourcade, E., 1994. Radiolaires et radiolarites: production primaire, diagenèse et paléogéographie. Bulletin des Centres de Recherches Exploration-Production Elf Aquitaine, 18: 315–379.

Einsele, G., 1992. Sedimentary basins, evolution, facies, and sediment budget. Springer, Berlin.

Fischer, G., Wefer, G., 1999. Use of proxies in paleoceanography. Springer, Berlin.

Haq, B.U., Boersma, A., 1978. Introduction to marine micropaleontology. Elsevier, Amsterdam.

Kennett, J.P., 1982. Marine geology. Prentice-Hall, Englewood Cliffs, NJ.

Lee, J.J., Anderson, O.R., 1991. Biology of Foraminifera. Academic Press, London.

Nichols, G., 1999. Sedimentology and stratigraphy. Blackwell, London.

Racki, G., Cordey, F., 2000. Radiolarian paleoecology and radiolarites: is the present the key to the past?. Earth-Science Reviews, 52: 83–120.

Ragueneau, O., Tréguer, P., Leynaert, A., Anderson, R.F., Brzezinski, M.A., De Master, D.J., Dugdale, R.C., Dymond, J., Fischer, G., François, R., Heinze, C., Maier-Reimer, E.,

Martin-Jézéquel, V., Nelson, D.M., Quéguiner, B., 2000. A review of the Si cycle in the modern ocean: recent progress and missing gaps in the application of biogenic opal as a paleoproductivity proxy. Global and Planetary Change, 26: 317–365.

Round, F.E., Crawford, R.M., Mann, D.G., 1990. The diatoms: Biology and morphology of the genera. Cambridge University Press, Cambridge.

Other references used in this chapter

Abelmann, A., Gowing, M.M., 1997. Spatial distribution pattern of living polycystine radiolarian taxa — baseline study for paleoenvironmental reconstructions in the Southern Ocean (Atlantic sector). Marine Micropaleontology, 30: 3–28.

Andruleit, H., Stäger, S., Rogalla, U., Cepek, P., 2003. Living coccolithophores in the northern Arabian Sea: ecological tolerances and environmental control. Marine Micropaleontology, 49: 157–181.

Armand, L.K., Crosta, X., Romero, O., Pichon, J.-J., 2005. The biogeography of major diatom taxa in Southern Ocean sediments: 1. Sea ice related species. Palaeogeography, Palaeoclimatology, Palaeoecology, 223: 93–126.

Beavington-Penney, S.J., Racey, A., 2004. Ecology of extant nummulitids and other larger benthic Foraminifera: applications in paleoenvironmental analysis. Earth-Science Reviews, 67: 219–265.

Bijma, J., Faber, W.W., Hemleben, C., 1990. Temperature and salinity limits for growth and survival of some planktonic foraminifers in laboratory cultures. Journal of Foraminiferal Research, 20: 95–116.

Bijma, J., Hemleben, C., 1994. Population dynamics of the planktic foraminifer *Globigerinoides sacculifer* (Brady) from the central Red Sea. Deep-Sea Research I, 41: 485–510.

Bijma, J., Hemleben, C., Oberhänsli, H., Spindler, M., 1992. The effects of increased water fertility on tropical spinose planktonic foraminifers in laboratory cultures. Journal of Foraminiferal Research, 22: 242–256.

Bollmann, J., 1997. Morphology and biogeography of *Gephyrocapsa* coccoliths in Holocene sediments. Marine Micropaleontology, 29: 319–350.

Boltovskoy, D., Alder, V.A., Abelmann, A., 1993. Annual flux of radiolaria and other shell plankters in the eastern equatorial Atlantic at 853 m: seasonal variations and polycystine species-specific responses. Deep-Sea Research I, 40: 1863–1895.

Caulet, J.-P., Vénec-Peyré, M.-T., Vergnaud-Grazzini, C., Nigrini, C., 1992. Variation of South Somalian upwelling during the last 160 ka: radiolarian and foraminifera records in Core MD85674. In: C.P. Summerhayes, W.L. Prell, K.C. Emeis (Editors), *Upwelling systems: Evolution since the early Miocene* Geological Society. London.

Conan, S.M.-H., Ivanova, E.M., Brummer, G.-J.A., 2002. Quantifying carbonate dissolution and calibration of foraminiferal dissolution indices in the Somali Basin. Marine Geology, 182: 325–349.

Crosta, X., Romero, O., Armand, L.K., Pichon, J.-J., 2005. The biogeography of major diatom taxa in Southern Ocean sediments: 2. Open Ocean related species. Palaeogeography, Palaeoclimatology, Palaeoecology, 223: 66–92.

Exon, N., Kennett, J.P., Malone, M. (Editors), 2004. The Cenozoic Southern Ocean: Tectonics, sedimentation, and climate change between Australia and Antarctica. Geophysical Monograph, volume 151. American Geophysical Union, Washington, DC.

Ganssen, G., Wefer, G. (Editors), 2000. Particle flux and its preservation in deep-sea sediments. Deep-Sea Research II, 47: 1679–2279.

Giraud, F., Pittet, B., Mattioli, E., Audouin, V., 2006. Paleoenvironmental controls on the morphology and abundance of the coccolith *Watznaueria britannica* (Late Jurassic, southern Germany). Marine Micropaleontology, 60: 205–225.

Hagino, K., Okada, H., Matsuoka, H., 2005. Coccolithophore assemblages and morphotypes of *Emiliana huxleyi* in the boundary zone between the cold Oyashio and warm Kuroshio currents off the coast of Japan. Marine Micropaleontology, 55: 19–47.

Jorissen, F.J., de Stigter, H.C., Widmark, J.G.V., 1995. A conceptual model explaining benthic foraminiferal microhabitats. Marine Micropaleontology, 26: 3–15.

Kuroyanagi, A., Kawahata, H., 2004. Vertical distribution of living planktonic foraminifera in the seas around Japan. Marine Micropaleontology, 53: 173–196.

Nelson, D.M., DeMaster, D.J., Dunbar, R.B., Smith, W.O., 1996. Cycling of organic carbon and biogenic silica in the Southern Ocean: estimates of water column and sedimentary fluxes on the Ross Sea continental shelf. Journal of Geophysical Research, 101: 18519–18532.

Pondaven, P., Ragueneau, O., Tréguer, P., Hauvespre, A., Dezileau, L., Reyss, J.-L., 2000. Resolving the "opal paradox" in the Southern Ocean. Nature, 405: 168–172.

Romero, O., Armand, L.K., Crosta, X., Pichon, J.-J., 2005. The biogeography of major diatom taxa in Southern Ocean sediments: 3. Tropical/subtropical species. Palaeogeography, Palaeoclimatology, Palaeoecology, 223: 49–65.

Romero, O., Fischer, G., Lange, C., Wefer, G., 2000. Siliceous phytoplankton of the western equatorial Atlantic: sediment traps and surface sediments. Deep-Sea Research II, 47: 1939–1959.

Sato, T., Yuguchi, S., Takayama, T., Kameo, K., 2004. Drastic change in the geographical distribution of the cold-water nannofossil *Coccolithus pelagicus* (Wallich) Schiller at 2.74 Ma in the late Pliocene, with special reference to glaciation in the Arctic Ocean. Marine Micropaleontology, 52: 181–193.

Schiebel, R., Hemleben, C., 2000. Interannual variability of planktic foraminiferal populations and test flux in the eastern North Atlantic Ocean (JGOFS). Deep-Sea Research II, 47: 1809–1852.

Schmidt, D.N., Renaud, S., Bollmann, J., Schiebel, R., Thierstein, H.R., 2004. Size distribution of Holocene planktic foraminifer assemblages: biogeography, ecology and adaptation. Marine Micropaleontology, 50: 319–338.

Schmiedl, G., de Bovée, F., Buscail, R., Charrière, B., Hemleben, C., Medernach, L., Picon, P., 2000. Trophic control of benthic foraminiferal abundance and microhabitat in the bathyal Gulf of Lions, western Mediterranean Sea. Marine Micropaleontology, 40: 167–188.

Sigmon, D.E., Nelson, D.M., Brzezinski, M.A., 2002. The Si cycle in the Pacific sector of the Southern Ocean: seasonal diatom production in the surface layer and export to the deep sea. Deep-Sea Research II, 49: 1747–1763.

Van Cappellen, P., Qiu, L., 1997. Biogenic silica dissolution in sediments of the Southern Ocean. I. Solubility. Deep-Sea Research II, 44: 1109–1128.

Van Cappellen, P., Qiu, L., 1997. Biogenic silica dissolution in sediments of the Southern Ocean. II. Kinetics. Deep-Sea Research II, 44: 1129–1149.

Van der Zwaan, G.J., Duijnstee, I.A.P., den Dulk, M., Ernst, S.R., Jannink, N.T., Kouwenhoven, T.J., 1999. Benthic foraminifers: Proxies or problems? A review of paleoecological concepts. Earth-Science Reviews, 46: 213–236.

Villa, G., Palandri, S., Wise, S.W., 2005. Quaternary calcareous nannofossils from Periantarctic basins: paleoecological and paleoclimatic implications. Marine Micropaleontology, 56: 103–121.

Zaric, S., Donner, B., Fischer, G., Mulitza, S., Wefer, G., 2005. Sensitivity of planktic foraminifera to sea surface temperature and export production as derived from sediment trap data. Marine Micropaleontology, 55: 75–105.

Zielinski, U., Gersonde, R., 1997. Diatom distribution in Southern Ocean surface sediments (Atlantic sector): implications for paleoenvironmental reconstructions. Palaeogeography, Palaeoclimatology, Palaeoecology, 129: 213–250.

Ziveri, P., Thunell, R.C., Rio, D., 1995. Export production of coccolithophores in an upwelling region: results from San Pedro Basin, Southern California Borderlands. Marine Micropaleontology, 24: 335–358.

CHAPTER TWELVE

Organic Sediments

Organic sediments include a significant proportion of organic compounds of marine and/or continental origin. In the ocean, organic compounds are for the most part produced by living organisms which are principally concentrated in the photic zone and the contribution of phytoplankton dominates over that of zooplankton, necton and benthos. Organic compounds of continental origin are for the most part derived from vegetation and soils. Organic matter is therefore more abundant in oceanic regions of important biological activity such as upwelling and divergence areas. Organic matter is also an important component of marine sediments in coastal and continental margin environments where benthic activity increases and river-borne terrigenous elements accumulate.

12.1. Organic Elements in the Water Column

12.1.1. Sources of Organic Compounds

Organic matter is synthesized by living organisms. The major elements found in organic compounds are by far carbon and hydrogen, which are associated to oxygen and nitrogen principally. These elements are all abundant in the atmosphere and hydrosphere. The production of organic matter starts with photosynthesis, which combines carbon and hydrogen from carbon dioxide and water to generate initial organic compounds and release oxygen. The simplest organic compound produced by photosynthesis is glucose, from which living organisms can synthesize complex carbohydrates, proteins, lipids and lignin principally. Photosynthesis is carried out by primary producers of organic matter, that is phytoplankton and algae in the ocean, and higher plants on the continents.

Using solar energy, photosynthesis takes place in those areas where sunlight is available. In the ocean where dissolved carbon dioxide is available at all depths but where the penetration of sunlight decreases rapidly with depth, photosynthesis is limited to the upper 100–200 m of the water column (the photic zone). On the continents, photosynthesis concentrates on surfaces where sunlight and water from precipitation are concurrently available. In all cases, the intensity of photosynthesis varies with the availability of solar energy at all time scales, from daily and seasonal to orbital variabilities. In addition, the availability of water is a limiting factor on the continents, where photosynthesis strongly decreases in extreme environments such as high latitude and tropical deserts.

In the ocean, diatoms, haptophyte algae (among them coccolithophores) and dinoflagellates are major primary producers of organic matter (autotrophic organisms). Their distribution and abundance are largely controlled by environmental parameters such as food availability and sea surface temperature (see Sections

11.1.2 and 11.2.2). Among them, the availability of specific nutrients such as Si, P, K, SO_4, NO_3 and Fe is especially important. Therefore, the distribution of organic matter in oceanic surface waters closely follows that of phytoplanktonic organisms. Major source areas of phytoplanktonic organic matter include regions of equatorial divergence and coastal upwelling, the Antarctic Circumpolar Current system and river plumes. Besides, the production of phytoplanktonic organic matter is enhanced during spring blooms and transient events such as wind mixing episodes, decay of eddies and windblown dust deposition. Overall, diatoms tend to dominate each time conditions become favorable to the development of phytoplankton, and a direct coupling between opal and organic matter production is frequently observed in productive surface waters. Autotrophic organisms are grazed by heterotrophic organisms, which derive their energy from oxidizing organic compounds and are secondary producers of organic matter. Heterotrophic organisms principally include zooplankton (herbivorous and carnivorous) and necton (which live on plankton and/or fish). Foraminifers and radiolarians are secondary producers which are especially abundant in fertile waters (see Sections 11.1.1 and 11.2.1). Therefore, the energy stored by autotrophic organisms during photosynthesis in the form of carbohydrates and lipids principally is transferred to higher trophic levels. This energy is extracted during respiration and used by heterotrophic organisms for their vital processes which include, for example, growth, reproduction, and locomotion. This energy is also partly used to build new organic compounds, principally proteins and lipids which however are of lower energy content than primary organic compounds.

Autotrophic and heterotrophic organisms contribute different organic compounds which vary in composition and abundance with environmental parameters, principally insolation, food availability and water temperature. Overall, the composition of the organic matter released in surface waters is a mélange of their respective contribution. In the upper photic zone of the equatorial Pacific divergence for example, the production of organic carbon varies seasonally with the intensity of the upwelling and abundances are symmetric about the equator (Figure 12.1). Organic compounds are largely dominated by amino acids (derived from proteins), carbohydrates and lipids which together contribute about 82% of the Total Organic Carbon (TOC). Lipids for the most part consist of fatty acids, associated to sterols and minor amounts of alcohols, hydrocarbons, alkenones and other compounds. By contrast chlorophyll, an essential pigment of phytoplankton, only accounts for about 0.2% of the TOC. Uncharacterized organic compounds account for the remaining 18% of the TOC.

On the continents, trees and grass are primary producers of organic compounds, which are therefore largely dominated by cellulose and lignin, the latter being especially concentrated in trees. By comparison with marine organic compounds, lignin and cellulose are relatively poor in hydrogen. Besides, continental organic matter also contains a variety of components including pollen, waxes and humic acids. Overall, tropical rain forests are currently a major contributor of continental organic matter. Organic compounds derived from the vegetation, together with those derived from the soils, are transported to the ocean by running waters principally. This contribution peaks during flooding events, when the erosion of

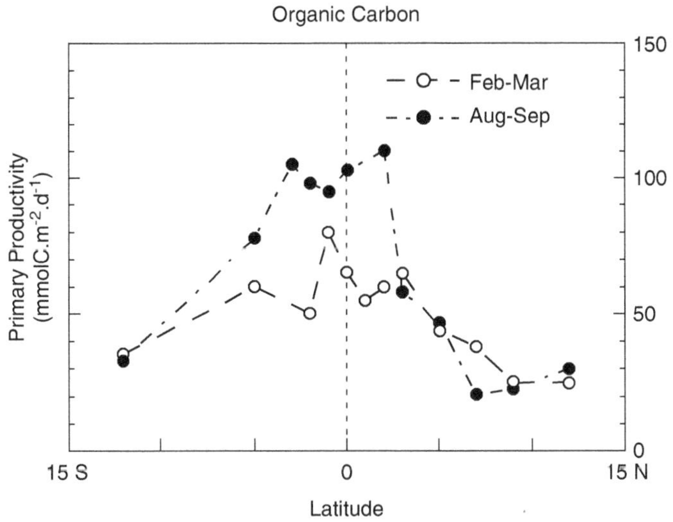

Figure 12.1 Primary productivity across the equatorial divergence zone of the Pacific Ocean. Note symmetry about the equator and seasonal variability of productivity. Modified from Berelson, W.M., Anderson, R.F., Dymond, J., De Master, D., Hammond, D.E., Collier, R., Honjo, S., Leinen, M., Mc Manus, J., Pope, R., Smith, C., Stephens, M., , 1997. Biogenic budgets of particle rain, benthic remineralization and sediment accumulation in the equatorial Pacific. Deep-Sea Research II, 44, 2251–2282.

soils and their vegetation cover increases due to intense precipitation. Organic elements are dispersed within river plumes and therefore are concentrated in continental margin areas. In addition, organic molecules are often sorbed onto mineral surfaces, especially fine-grained mineral particles such as clays which have large surface areas. Therefore, at least part of the organic elements of continental origin settle together with terrigenous river loads and remain in continental margin areas where they are likely to be further reworked together with terrigenous sediments (see Sections 10.2 and 10.5).

12.1.2. Fluxes of Organic Matter in the Water Column

A significant part of the photosynthetic organic matter of marine origin is recycled within the photic zone. Organic compounds are used as food or oxidized in the photic zone, with strong regional variations. Where primary production is low, most organic matter is consumed and downward fluxes are of minor importance. In the oligotrophic subtropical gyre of the Atlantic Ocean, only 0.6% of the particulate organic carbon is exported downward, out of the photic zone. In highly productive areas, a significant proportion of organic matter contributes to the downward fluxes of particles. In the equatorial Pacific divergence area for example, 30–60% of the photosynthetic carbon is respired back to inorganic carbon or released in a dissolved form within a day. At 105 m water depth in the lower photic zone, floating particles principally consist of phytoplankton (principally diatoms), foraminifers, small copepods and fecal pellets. The proportion of particulate organic carbon within floating particles varies seasonally, from 7–24% to 9–38%. When averaged over a full year, the fluxes of organic carbon in the lower photic zone are symmetric about the equator, just like the primary production. Besides, organic compounds at 105 m water depth are very similar to those in the upper photic zone (see Section 12.1.1), with about 21% of the TOC remaining uncharacterized. In addition, the organization of the trophic web modulates the production of particulate organic carbon. The development of heterotrophic organisms results in a global decrease of the gross mass of organic matter, as primary organic compounds are either spent for energy or transformed into compounds of lower energy content. In the Southern Ocean for instance, diatoms account for about 90% of the organic carbon released in the Polar Front area and 75% in the Antarctic Circumpolar Current where the grazing pressure is higher. In such highly productive areas, there is a clear relationship between the production of biogenic opal and organic carbon (Figure 12.2). The organization of the trophic web also modulates the relative proportion of organic compounds. For example, the ingestion of phytoplankton by herbivores results in the metabolism of specific lipids. Other lipids are rejected in fecal pellets, which are also enriched in zooplankton-derived compounds: among them cholesterol and wax esters, which are absent from phytoplanktonic lipids.

Export fluxes of particulate organic carbon below the photic zone of highly productive areas are extremely variable in intensity. In the Southern Ocean, 30–50% of the particulate organic carbon produced annually is exported. In some areas of

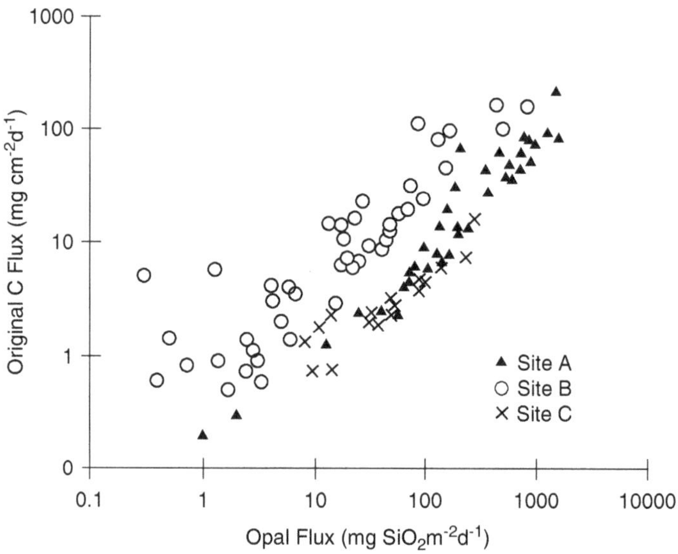

Figure 12.2 Fluxes of organic carbon and biogenic opal below the photic zone, at three mooring sites of the Ross Sea. Sampling encompasses a two-year interval. Note relationship between both fluxes. Reprinted from Nelson, D., DeMaster, D., Dunbar, R., Smith, W., 1996. Cycling of organic carbon and biogenic silica in the Southern Ocean: Estimates of water-column and sedimentary fluxes on the Ross Sea continental shelf. Journal of Geophysical Research, 101, 18519–18532.

the equatorial Pacific divergence by contrast, the proportion decreases below 10%. Higher export fluxes are often associated with food webs that are dominated by larger plankton such as diatoms, which aggregate and are rapidly removed downward during blooms. In addition, wind-blown dust and fine-grained terrigenous particles transported in surface waters may stimulate the formation of such aggregates. Higher export fluxes also occur in areas where blooms of phytoplankton are associated to rapid grazing by large heterotrophic organisms (krill, for instance) and related release of fecal pellets.

Fluxes of particulate organic carbon decrease with depth in the water column because of remineralization principally, the recycling rates being lower for rapidly sinking particles. This is illustrated by the evolution of the Si/C ratio with depth in the water column (Figure 12.3). In the northeast Atlantic, the ratio strongly increases between 1,000 and 3,000 m, by a factor of two. The ratio is about one order of magnitude higher in the productive Southern Ocean where values of 0.05–1 in the photic zone grade to values of 1–11 at 1,200 m and 4–16 at 4,000 m. The relative increase of the ratio with depth is lower in the Southern Ocean, where high diatom production and grazing ensure a rapid downward transfer of biogenic and organic particles. The recycling rate of organic carbon in the water column therefore looks highly variable. In most cases however, recycling principally occurs within and immediately below the photic zone. It is for example estimated that about 90% of the particulate organic carbon exported from the photic zone is remineralized before reaching a depth of 1,000 m in the Antarctic Circumpolar Current area of the Southern Ocean (Figure 12.4). In the upwelling area of the Arabian Sea off Oman, about 92% of the carbon fixed by primary producers is recycled in the uppermost 100 m of the water column, increasing to 99% at 1,000 m water depth. The relative proportion of organic compounds also varies strongly with depth in the water column, as illustrated in the equatorial Pacific divergence area. Overall, the proportion of uncharacterized organic carbon increases from 18% in the photic zone to 68% in the lower part of the water column below 3,500 m. Amino acids and lipids account for most of the loss in characterized organic carbon, whereas the proportion of carbohydrates remains relatively stable by comparison (Figure 12.5). In detail, the degradation of lipids is highly selective, with decreasing abundances of compounds attributable to phytoplankton to the advantage of compounds derived from zooplankton with depth, a regression of the more labile storage lipids, and a significant progression of compounds indicative of bacterial reprocessing in deep waters.

Despite severe recycling of organic compounds in the upper part of the water column, fluxes of particulate organic carbon to the deep ocean frequently reflect changes in the productivity of surface waters (Figure 12.6). Changes at seasonal to interannual scale are recorded. They often coincide with variations in surface water hydrography and climate. For example, the Benguela current system of the South Atlantic, which flows along the coast of Africa, is associated to an important upwelling and is a highly productive region. Off Namibia at 30°S, the primary production of organic matter (Figure 12.7) is lowest in winter (June and July) and increases to a summer maximum lasting from November to February, and this pattern is reflected in the annual distribution of the fluxes of particulate organic

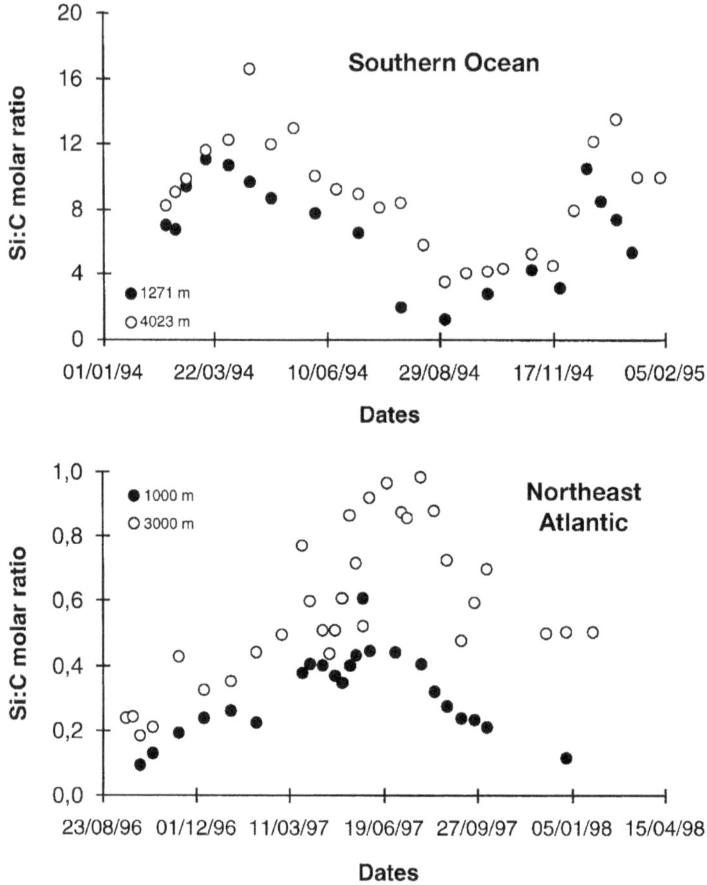

Figure 12.3 Seasonal variations of the Si/C ratio at intermediate and deep-water depths of the Southern Ocean and northwest Atlantic Ocean, from sediment trap measurements. Note elevated ratios in the more productive Southern Ocean. Higher ratios at deep-water depth illustrate the rapid decay of organic matter in the water column. Modified from Ragueneau, O., Tréguer, P., Leynaert, A., Anderson, R.F., Brzezinski, M.A., De Master, D.J., Dugdale, R.C., Dymond, J., Fischer, G., François, R., Heinze, C., Maier-Reimer, E., Martin-Jézéquel, V., Nelson, D.M., Quéguiner, B., 2000. A review of the Si cycle in the modern ocean: recent progress and missing gaps in the application of biogenic opal as a paleoproductivity proxy. Global and Planetary Change, 26, 317–365.

carbon at 1,000 m water depth. Further north at 20°S in the Walvis Ridge area, carbon fluxes show distinct fall (May and June) and spring (September–November) maxima. The spring maximum coincides with the drift of large upwelling filaments above the study area. In contrast, there is no local increase in productivity during the fall maximum. However, two maxima in productivity are clearly visible in spring and fall further east near the shelf break where they coincide with increased trade-wind activity. It is assumed that organic elements of low density are transported downward and westward to the study area, over a distance of a few hundred of kilometers. The lateral transport of particles is supported by a time shift

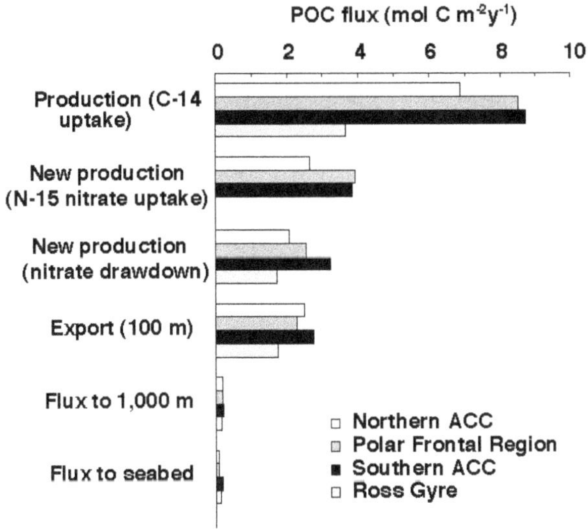

Figure 12.4 Annual fluxes of particulate organic carbon in four areas of the Southern Ocean. POC, particulate organic carbon and ACC, Antarctic circumpolar current. Estimates from two different methods are given for new production. Note that fluxes strongly decrease in the upper part of the water column. Modified from Nelson, D.M., Anderson, R.F., Barber, R.T., Brzezinski, M.A., Buesseler, K.O., Chase, Z., Collier, R.W., Dickson, M.-L., François, R., Hiscock, M.R., Honjo, S., Marra, J., Martin, W.R., Sambrotto, R.N., Sayles, F.L., Sigmon, D.E., 2002. Vertical budgets for organic carbon and biogenic silica in the Pacific sector of the Southern Ocean, 1996–1998.

of about two or three months between the maxima of productivity in coastal areas (March–May) and carbon fluxes in the study area (May and June). Besides, changes of annual fluxes of organic carbon by a factor of two are attributed to the changing intensity of the Benguela upwelling induced by a variable trade-wind activity.

Carbon fluxes below upwelling areas are the most sensible to changes in hydrography and climate. However, the nature and availability of nutrients may interact with those parameters for modulating the export production of organic carbon. This is the case in the Arabian Sea where upwelling conditions are principally controlled by the trade winds. There, a reversal of the surface current system is triggered by the onset of the southwestern monsoon winds late in the spring. The surface winds are focused within a narrow corridor off the Arabian coast and the resulting Ekman transport induces an offshore flow in surface waters. This is associated with the development of strong upwelling systems along the coasts and in the open ocean. Upwelled waters enriched in nutrients support an important productivity for the entire duration of the southwestern monsoon, from June to October. Upwelling conditions start in the open ocean where silica depleted sub-surface waters enriched in nitrate are brought into the photic zone (Figure 12.8). As a result, non-siliceous organisms including calcareous microorganisms expand rapidly, but their development is limited by the availability of nitrate. Downward fluxes of particulate organic carbon are predominantly controlled by these organisms in May and June (Figure 12.9). Upwelling conditions extend with a delay of about

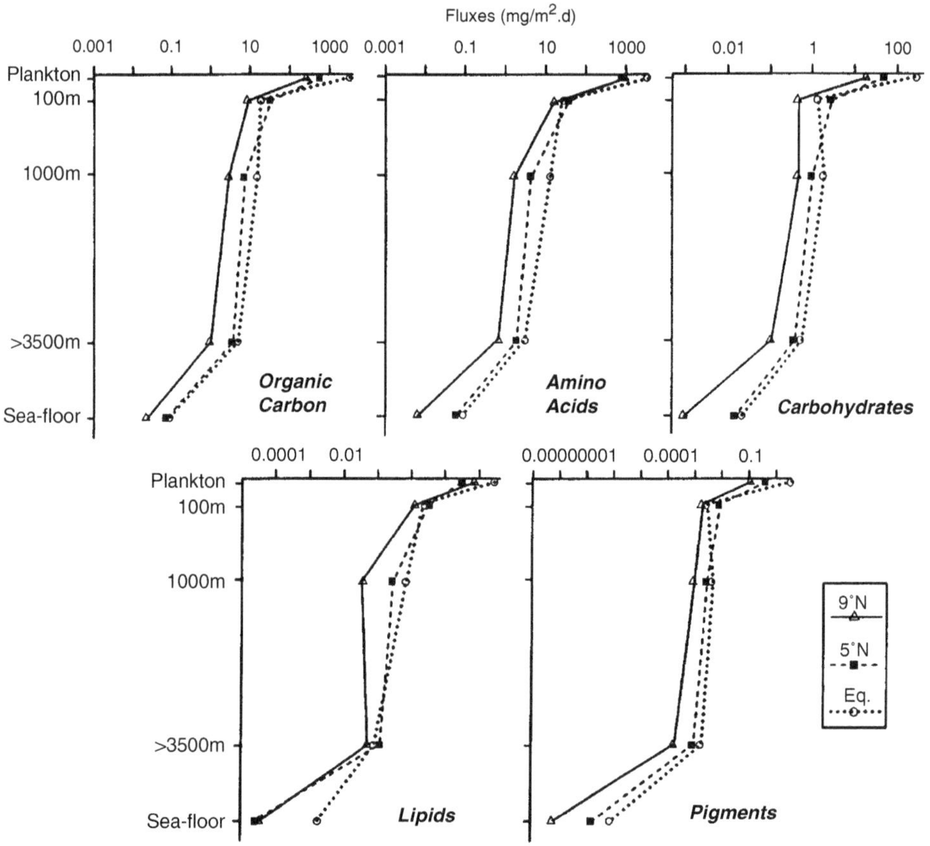

Figure 12.5 Fluxes of major organic compounds in the equatorial divergence area of the Pacific Ocean. Note enhanced degradation in the upper water column and near the water/sediment interface. Note also the important decay of amino acids and lipids. Reprinted from Wakeham, S.G., Lee, C., Hedges, J.I., Hernes, P.J., Peterson, M.L., 1997. Molecular indicators of diagenetic status in marine organic matter. Geochimica et Cosmochimica Acta, 61, 5363–5369.

two or three weeks to coastal areas where deeper waters enriched in silica are brought to the surface. This marks the onset of a diatom bloom and further development of other siliceous microorganisms. Diatoms follow the path of surface waters toward the open ocean, consuming the dissolved silica available and facilitating the succession of non-siliceous organisms. Related fluxes of organic carbon at 3,000 m water depth respond to the onset of upwelling conditions with a delay of two weeks approximately. In the transition area between coastal and open ocean upwellings, organic carbon fluxes at 3,000 m water depth first increase in response to the onset of the open ocean upwelling. About six weeks later, a decreasing trend of the carbonate to opal ratio indicates a progressively increased influence of the coastal upwelling. Variabilities in open ocean and coastal upwelling velocities are reflected in the fluxes of organic carbon to the deep ocean with a delay of approximately two and eight weeks respectively. At the end of the southwestern

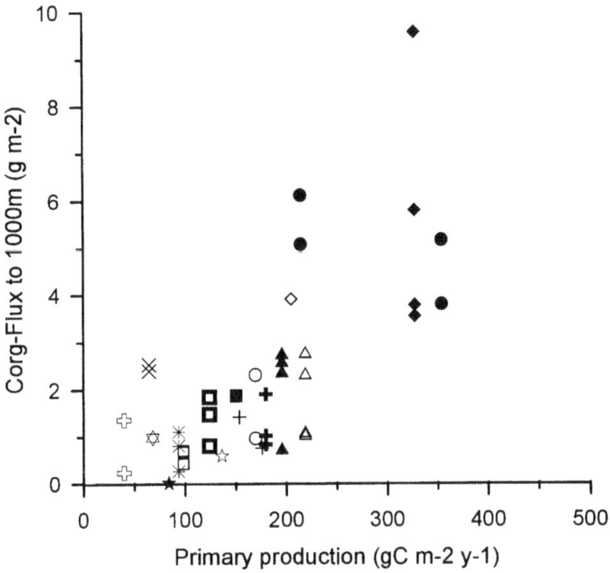

Figure 12.6 Relationship between primary production and fluxes of particulate organic carbon to 1,000 m water depth, as deduced from measurements at a variety of sites in productive areas of the equatorial and southeast Atlantic Ocean and Weddell Sea. Reprinted from Fischer, G., Ratmeyer, V., Wefer, G., 2000. Organic carbon fluxes in the Atlantic and the Southern Ocean: Relationship to primary production compiled from satellite radiometer data. Deep-Sea Research II, 47, 1961–1997.

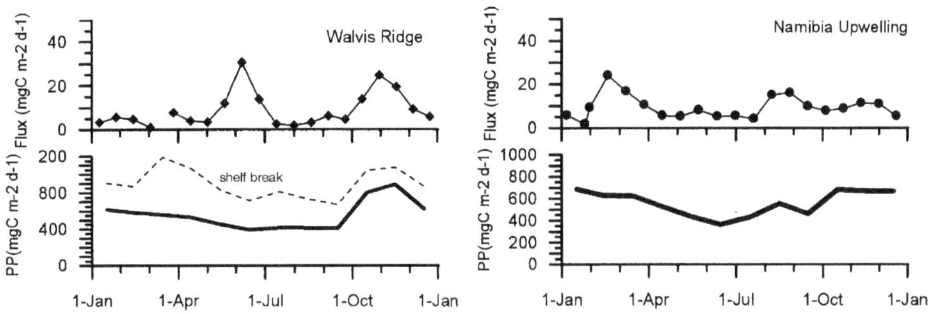

Figure 12.7 Primary production and fluxes of particulate organic carbon to 1000 m water depth within the Benguela current and upwelling system, southeast Atlantic Ocean. Note seasonal variations in primary production and carbon fluxes at both locations. Modified from Fischer, G., Ratmeyer, V., Wefer, G., 2000. Organic carbon fluxes in the Atlantic and the Southern Ocean: relationship to primary production compiled from satellite radiometer data. Deep-Sea Research II, 47, 1961–1997.

monsoon, an increasing trend of the carbonate to opal ratio indicates a decreasing influence of diatoms on carbon fluxes. A comparison of the variations in upwelling velocities and carbon fluxes suggests that carbon fluxes are principally controlled by the coastal upwelling. However, upwelling velocities show interannual variability in

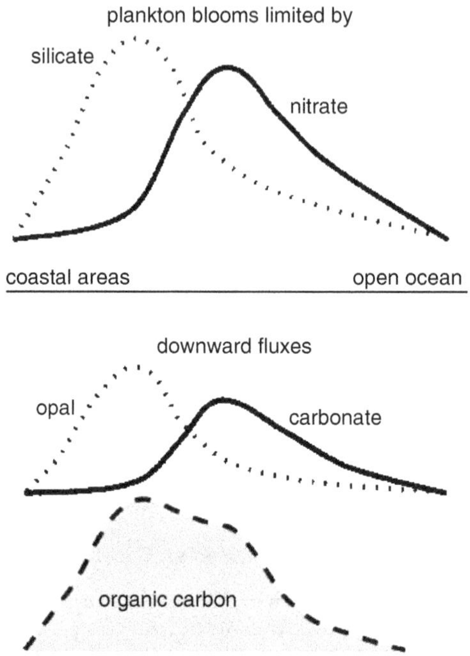

Figure 12.8 Geographical development of plankton blooms in the western Arabian Sea and related fluxes of particles to the deep ocean. Note the larger contribution of siliceous organisms to fluxes of particulate organic carbon. Modified from Rixen, T., Haake, B., Ittekot, V., 2000. Sedimantation in the western Arabian Sea: The role of coastal and open-ocean upwelling. Deep-Sea Research II, 47, 2155–2178.

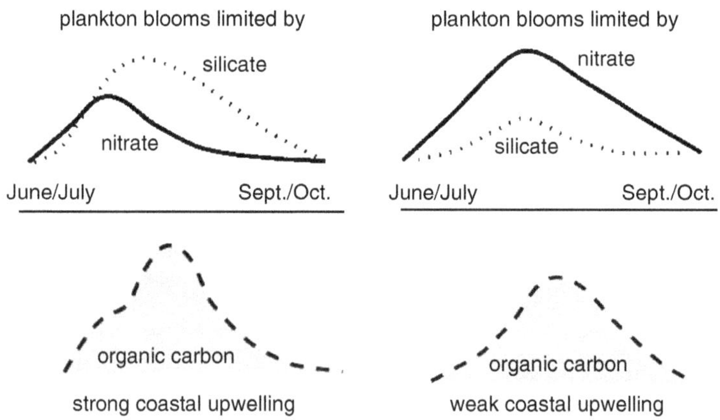

Figure 12.9 Seasonal development of plankton blooms in the western Arabian Sea and related downward fluxes of particulate organic carbon in open ocean areas. Note differences between years of strong coastal upwelling (left) and years of weak coastal upwelling (right). Modified from Rixen, T., Haake, B., Ittekot, V., 2000. Sedimantation in the western Arabian Sea: The role of coastal and open-ocean upwelling. Deep-Sea Research II, 47, 2155–2178.

the open ocean as in coastal areas. This is globally reflected in the carbon fluxes which are lowest during years of weak coastal upwelling, although no simple relationship has been established yet (Figure 12.9).

12.2. Organic Compounds in Sediments

12.2.1. Remineralization and Preservation of Organic Matter in Bottom Waters and Surface Sediments

A further, major decrease in fluxes of particulate organic carbon is observed near the water/sediment interface. In the Arabian Sea for example, only 2–18% of the particulate organic carbon that reach bottom water environments are finally incorporated to the sediment. In addition, the proportion of uncharacterized organic carbon further increases, from 68% in the bottom waters to 80% in the surface sediments of the equatorial Pacific divergence. Again, amino acids and lipids (especially polyunsaturated fatty acids) decrease the most in abundance. To summarize, remineralization processes are associated to a transition from highly labile and predominantly well-characterized organic compounds in the photic zone to predominantly uncharacterized and resilient organic material in surface sediments. In fine, only a minor proportion of the organic matter produced in the photic zone is buried in surface sediments. Overall, it is estimated that a mere 0.3–0.4% of the organic carbon from the photic zone is currently buried in surface sediments. Most sediments in oligotrophic areas of the oceanic gyres are free of particulate organic carbon. In productive open ocean areas such as the Antarctic Circumpolar Current region of the Southern Ocean, burial efficiency varies between 0.08% and 0.02% of the annual production of particulate organic carbon. In fact, about 90% of the organic carbon burial occurs in continental margin areas. There, elevated production of organic matter in coastal upwelling areas, river plumes and other near-shore areas interferes with the input of terrigenous organic matter. Burial efficiencies are however highly variable.

In the deep ocean, benthic organisms live on the organic particles that reach the seafloor. Benthic communities include a variety of organisms, ranging from bacteria to foraminifers and worms, which play a significant role in recycling organic compounds via their vital processes. At the water/sediment interface bacterial abundances can be 10^3–10^4 times higher than in overlying waters, and bacterial activity significantly higher than in underlying sediments. The recycling of organic matter near the water/sediment interface therefore involves both biological and chemical processes (see Section 2.3.10). Depending on environmental conditions in bottom waters and surface sediments, different electron acceptors are used for organic matter oxidation. Oxygen is used first, as its reaction with organic compounds yields the most energy. When oxygen is used up, nitrogen (from nitrates), manganese and iron (from oxides) and sulfur (from sulfates) are successively used as electron acceptors. In most pelagic sediments, where the input of organic compounds is low, oxygen is by far the dominant electron acceptor and accounts for up to 99% of organic matter oxidation. Electron acceptors other than oxygen play a critical role in continental margin areas of high organic input. There, the

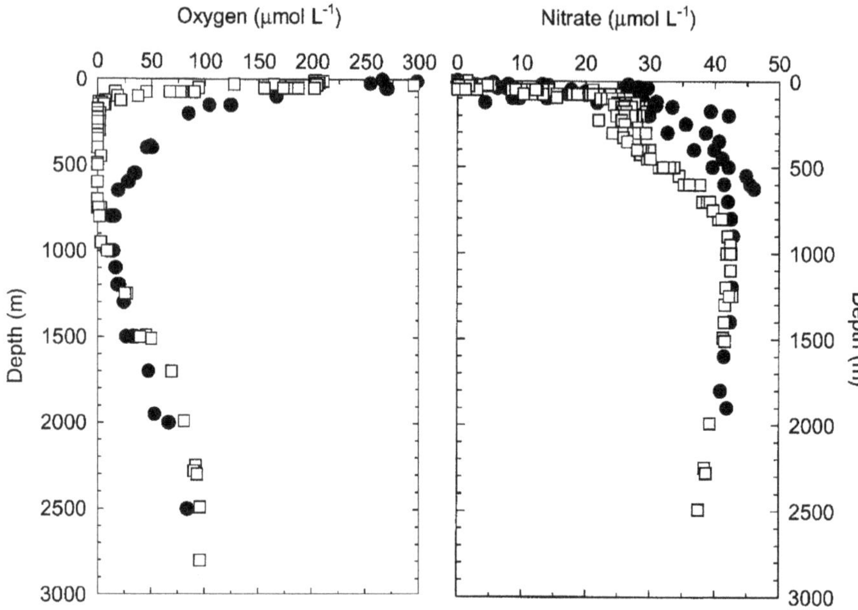

Figure 12.10 Seawater concentrations in oxygen and nitrate at two locations of the northeast Pacific margin: Mazatlan in Mexico (open squares) and Washington State (filled circles). Note the stronger oxygen-minimum zone between 100 and 1,000 m off Mazatlan, which also partly coincides with depleted concentrations in nitrate. Reprinted from Hartnett, H.E., Devol, A.H., 2003. Role of a strong oxygen-deficient zone in the preservation and degradation of organic matter: A carbon budget for the continental margins of nortwest Mexico and Washington State. Geochimica et Cosmochimica Acta, 67, 247–264.

oxygen of pore waters is frequently used up before all reactive organic compounds are consumed, especially where sediments are bathed by poorly oxygenated waters (oxygen-minimum zone of coastal upwellings, for instance). Nitrate reduction is therefore a common process in continental margin sediments (Figure 12.10), whereas sulfate reduction principally occurs in estuarine environments where organic contents are the highest. In well-oxygenated areas of the North American margin of the Pacific Ocean, oxygen consumption accounts for about 70% of organic matter oxidation. Nitrate and sulfate reduction account for 10–20% and 5–20% of the total, respectively. Within the oxygen-minimum zone in contrast the role of oxygen is highly variable, accounting for 5–45% of organic matter oxidation. There, denitrification is responsible for 40–70% of organic matter oxidation, and sulfate reduction for 5–25%. In both areas the role of Mn and Fe is very minor, because the sediments are depleted in manganese and iron oxides. The oxidation of organic matter releases carbon dioxide and H^+ in pore waters, which partly diffuse to bottom waters where they interfere with carbonate/bicarbonate reactions (see Section 11.1.3). In some cases however, the quantity of carbon dioxide and H^+ remaining in the sediment is sufficient to decrease the pH of pore waters which become more aggressive regarding calcite. The loss in alkalinity is then compensated by dissolution of sediment carbonates. This metabolic dissolution of carbonates may occur at all depths, including above the lysocline.

In some cases, the source of organic matter seems to influence the rates of decomposition. Algal fatty acids and hydrocarbons, which are in the C_{14}–C_{18} range, are considered as more reactive than those from vascular plants which are in the C_{22}–C_{32} range. This is however not confirmed for other lipids such as sterols and *n*-alkanes. Among lipids of marine origin, molecules containing functional groups such as fatty acids and alcohols are more susceptible to degradation than other compounds, such as hydrocarbons. Also, large organic compounds must be first hydrolyzed to smaller molecules by extracellular enzymes, so they can be transported through cellular membranes. Then, bacteria consume the hydrolyzed compounds for their vital processes. Sulfate-reducing and methanogenic bacteria which are at the end of the sedimentary food chain (see Section 2.3.10) ultimately consume simple organic compounds such as short-chain fatty acids and carbon dioxide, and release sulfide or methane. For example, carbohydrates of high molecular weight such as polysaccharides are hydrolyzed to monosaccharides, which are likely to be remineralized rapidly. In fine, degradation processes in the water column and near the water/sediment interface leads to increased relative abundances of more resistant compounds. Among lipids, long and straight-chain fatty acids, alcohols, alkanes and alkenones derived from phytoplankton and higher plants are selectively preserved in sediments, together with compounds indicative of bacterial reprocessing. Among carbohydrates, polymeric carbohydrates such as cellulose, chitin, dextrin, alginate and lignin are more easily preserved.

The way organic elements are bundled together or with other particles also influences their vulnerability. For example, the aggregation of diatoms during blooms and the rapid sinking and incorporation of the aggregates to surface sediments facilitates the preservation of the associated organic compounds. Also, the association of organic compounds with clay minerals and colloidal materials in continental environments acts as a protection against degradation and facilitates their accumulation in continental margin areas. However, it has been frequently observed that organic matter in continental margin sediments is adsorbed onto mineral surfaces, and is especially concentrated in fine-grained sediments such as clays. On the continental shelf and slope off the Columbia River in the northeast Pacific Ocean after removal of discrete organic debris which account for about 10% of the total organic carbon, a clear correlation exists between organic carbon percentages and surface areas of sediment particles, which is most important for sheet silicates (Figure 12.11). Surface area therefore appears as an important parameter controlling the concentration of organic matter in shelf deposits and in fine terrigenous sediments such as black shales. Fresh organic matter being generally in the form of particles, some degree of solubilization is necessary before the organic compounds are sorbed onto minerals. Concerning organic elements of continental origin this is probably done in the source area, as minerals in soils and river suspension frequently carry organic coatings. However, it has been demonstrated that in shelf sediments of the Amazon organic coatings are less abundant than in river suspensions, and in part consist of organic compounds of marine origin. The mechanisms involved in the adsorption and desorption of organic compounds remain unclear, but the observation helps understanding the decreased preservation efficiency of continental organic matter in most modern marine environments. Nevertheless, organic

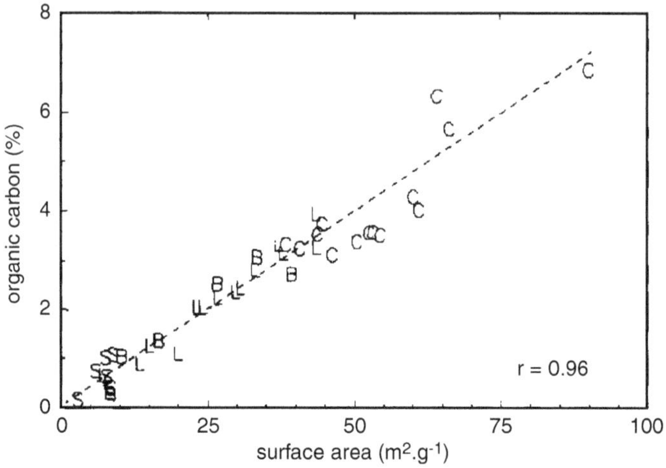

Figure 12.11 Correlation between organic carbon contents and mineral surface area in suspended sediments from the estuary of the Columbia River and surface sediments from the adjacent shelf and slope of the northeast Pacific margin. B, bulk sediment; S, sand fraction; L, silt fraction and C, clay fraction. Modified from Hedges, J.I., Keil, R.G., 1995. Sedimentary organic matter preservation: An assessment and speculative synthesis. Marine Chemistry, 49, 81–115.

compounds sorbed onto mineral surfaces are relatively protected against degradation. It has been suggested that preferential location of sorbed organic compounds in small mesopores at the mineral surfaces, and/or chemical bondings between organic molecules and minerals may limit the efficiency of enzymes and subsequent bacterial degradation. In addition, adsorption/desorption processes might favor the formation of resistant macromolecules (Figure 12.12).

Long-term exposure to oxygen and other electron acceptors such as Mn and Fe finally results in significant degradation of sorbed organic compounds. In the presence of oxygen, Mn and Fe can be oxidized either spontaneously or by bacteria. Metal-reducing bacteria are facultative anaerobes capable of growing on diverse organic elements. This is illustrated in turbidite sediments emplaced about 140 kyr ago in the Madeira abyssal plain of the North Atlantic, which were exposed to oxygenated bottom waters for about 10 kyr before emplacement of the next gravity flow. There, oxygen penetrated into the sediment and an oxic layer about 40 cm thick developed at the surface of the turbidite. Concentrations in organic carbon decreased from 0.93–1.02% in the core of the turbidite to 0.16–0.21% above the redox front, and pollen from 1,500 to 1,600 grains/g to near zero. In contrast, sorbed organic matter and pollen grains remained stable below the redox front: despite high sulfate concentrations in pore waters, acquired from sea water during the emplacement of the turbidite, sulfate reduction did not intervene in remineralization processes. This is probably because sulfate-reducing bacteria are strict anaerobes which can only use specific, fermentatively produced organic compounds, and many organic compounds of continental origin are resistant to fermentative break down. Long-term exposure to oxygen was therefore sufficient to cause considerable remineralization of sorbed organic compounds and complete disparition of pollen grains which on the contrary seems to resist degradation in

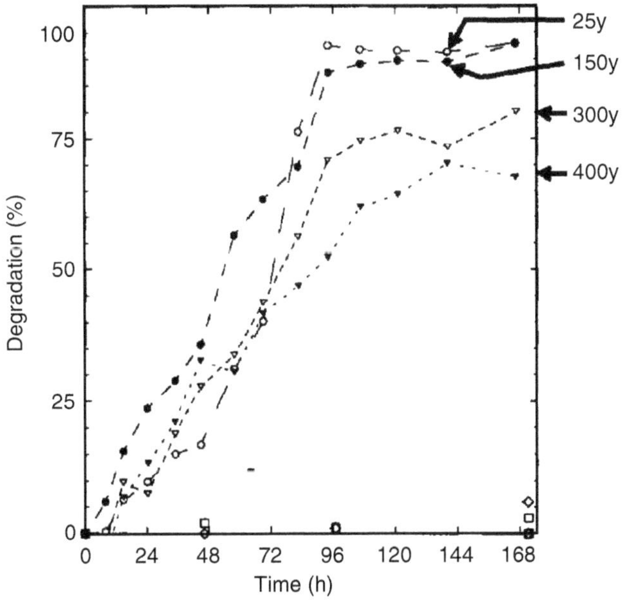

Figure 12.12 Time necessary for chemically degrading organic material in four samples from the Washington State area of the northeast Pacific margin. Note that the proportion of resistant compounds increase with the age of the sediment. Modified from Hedges, J.I., Keil, R.G., 1995. Sedimentary organic matter preservation: An assessment and speculative synthesis. Marine Chemistry, 49, 81–115.

anoxic environments. This is supported by exceptional abundances of coated grains and the remarkable preservation of pollen grains in anoxic coastal environments. Molecular oxygen could be necessary for extensive degradation of carbon-rich organic materials such as lignin, pollen, aliphatic polymers and microbial lipids and others by oxygen-requiring enzymes, leading to the selective preservation of these materials in anoxic environments.

The rate of organic matter oxidation decreases sharply with increasing depth below the water/sediment interface. Degradation rates are generally high enough to deplete pore waters in oxygen and/or nitrate within the top centimeter (Figure 12.13). In the northwest Atlantic Ocean, oxidation rates at the interface are 2.5–10 times those at 2 cm, depending on the location. In the Atlantic as in the Pacific oceans, the oxidation of organic matter seems unambiguously concentrated within the topmost 2 cm of the sediment. It is estimated that at least 70% of organic matter degradation in sediments occur within this narrow interval (Figure 12.14). Besides, the lifetime of organic compounds near the water/sediment interface is highly variable, from a few weeks to several hundred years, depending on the nature of the compounds. In the equatorial Pacific Ocean, two labile organic fractions are described. One accounts for 70–90% of the oxygen consumption, has a lifetime of a few months and likely consists of phytoplanktonic detritus. Temporal variations in degradation rates are likely caused by this fraction and coincide with changes in surface conditions such as temperature, upwelling activity and productivity. The

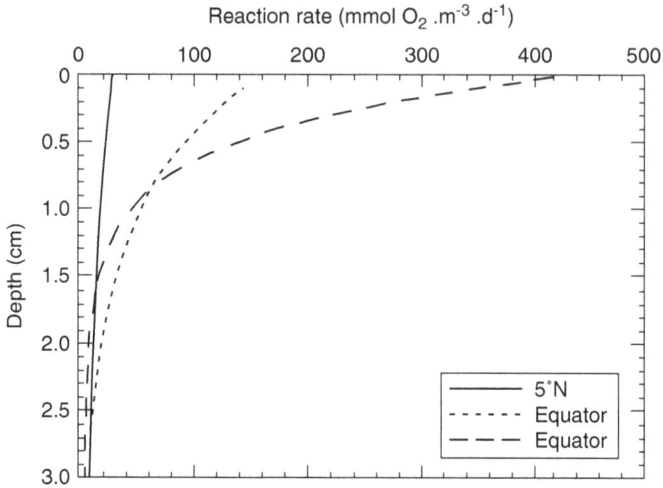

Figure 12.13 Variability of oxygen reaction kinetics with depth in surface sediments, at two sites of the equatorial Pacific Ocean. The site at the equator is within the productive divergence zone whereas the site at 5°N is within a less productive area. Different curves for the site at the equator correspond to different calculation methods. Note that the reaction is more important at the equator and that most of it occurs within the topmost 2 cm. Modified from Hammond, D.E., McManus, J., Berelson, W.M., Kilgore, T.E., Pope, R.H., 1996. Early diagenesis of organic material in equatorial Pacific sediments: Stoichiometry and kinetics. Deep-Sea Research II, 43, 1365–1412.

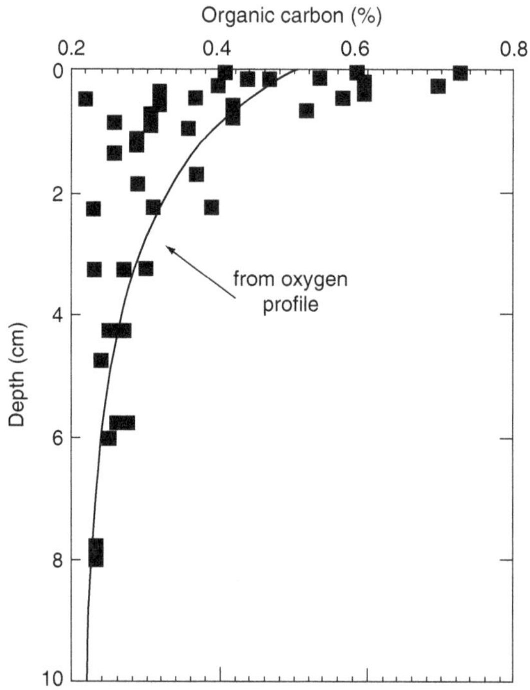

Figure 12.14 Organic carbon contents of surface sediments in the equatorial Pacific Ocean. Samples come from several cores retrieved from the same area. Note rapid decrease of organic carbon percentages within the topmost 2 cm. Modified from Hammond, D.E., McManus, J., Berelson, W.M., Kilgore, T.E., Pope, R.H., 1996. Early diagenesis of organic material in equatorial Pacific sediments: Stoichiometry and kinetics. Deep-Sea Research II, 43, 1365–1412.

second fraction accounts for 10–30% of the oxygen consumption and has a lifetime of a few decades. The labile components coexist with a third, much more resistant, organic fraction.

12.2.2. Factors Controlling the Accumulation of Organic Material in Sediments

Most organic compounds of marine origin are rapidly oxidized in the water column as well as in surface sediment, where degradation is concentrated within the topmost 2 cm of the sedimentary column. Besides, degradation occurs in oxic as well as in anoxic environments, and at very comparable rates. By comparison, many organic compounds of continental origin and/or sorbed onto mineral surfaces are globally more resistant to degradation (or better protected against degradation), except in oxic environments providing that exposure to molecular oxygen is sufficient. Therefore, a variety of parameters and sometimes complex processes control the preservation and accumulation of organic material.

A rapid burial of organic elements may preserve organic compounds from degradation. The abundance of carbon that is not remineralized in surface sediments often correlates with both the fluxes of carbon at the water/sediment interface and sedimentation rates. For that reason, sediments in regions of high accumulation rates may contain high proportions of organic matter. This is the case in the Guaymas basin of the Gulf of California where abundant organic material of predominantly planktonic origin accumulates at all water depths. There, the contents in organic carbon range from 1.05% to 5.35%. Similar proportions of particulate organic carbon in laminated sediments deposited under dysoxic to anoxic conditions within an oxygen-minimum zone and in homogeneous sediments deposited under oxic conditions illustrate the dominant control of burial over oxygen availability in the formation of organic-rich facies (Figure 12.15). Moreover, maximum abundances in organic carbon are recorded in homogeneous sediments below the core of the oxygen minimum. However, the pattern of preservation is sometimes rather complex. In the Arabian Sea for instance, preservation of organic material in surface sediments within and below the oxygen-minimum zone involves compounds resistant to oxidation, sorption onto mineral surfaces, and hydrodynamic control on sediment distribution.

Dysoxic and anoxic environments may be sites where organic material preferentially accumulates. The Santa Monica and other basins of the southern California margin are floored by laminated sediments of Holocene Age implying low oxygen concentrations in response to a strong upwelling and related productivity, and poor bottom water ventilation. A return to bioturbated conditions during the late seventies could have been caused either by increased oxygen supply to the basins bottom waters, or by reduced fluxes (and related oxidation rates) of organic carbon. No evidence for an increased oxygenation of bottom waters has been found in the records. Besides, accumulation rates of organic carbon and biogenic carbonates decreased concurrently in the Santa Monica basin (whereas a decrease in diatom fluxes was observed in the adjacent Santa Barbara basin), suggesting that carbon fluxes and productivity both decreased (Figure 12.16). In addition, detailed investigations of carbon isotopes in benthic foraminifers and

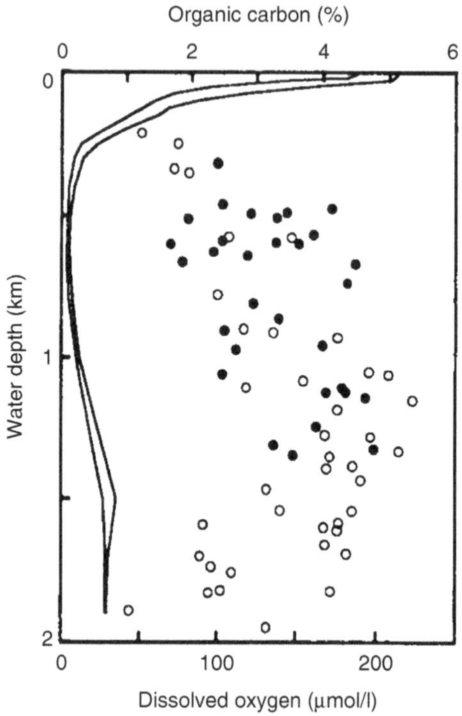

Figure 12.15 Depth distribution of organic carbon contents (circles) of surface sediments, and dissolved oxygen contents (solid lines) of sea water in the Guaymas Basin, Gulf of California. Filled circles represent laminated sediments, whereas empty circles represent homogeneous sediments. Note similar carbon contents in laminated and homogeneous sediments, and carbon maximum in homogeneous sediments below the oxygen-minimum zone. Modified from Calvert, S.E., Bustin, R.M., Pedersen, T.F., 1992. Lack of evidence for enhanced preservation of sedimentary organic matter in the oxygen minimum of the Gulf of California. Geology, 20, 757–760.

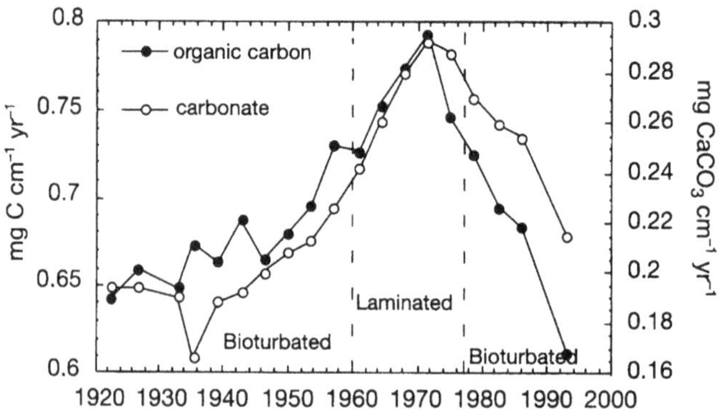

Figure 12.16 Accumulation rates of organic carbon and biogenic carbonate in the Santa Monica Basin of the northeast Pacific Ocean, from the 1920s to the 1990s. Note concomitant variations in organic carbon and productivity, as deduced from biogenic carbonate accumulation rates. Reprinted with permission from Macmillan Publishers LTD; Stott, L.D., Berelson, W., Douglas, R., Gorsline, D. Increased dissolved oxygen in Pacific intermediate waters due to lower rates of carbon oxidation in sediments. Nature, 407, 367–370, copyright 2000.

accumulation rates of organic carbon highlight a close correlation between changes in carbon oxidation and bottom water oxygenation that favored the return of benthic life. In such areas, upwelling activity and related productivity exert a major control on both the oxygen content of bottom waters and the organic carbon content of surface sediments. Additional support is provided by the Late Pleistocene record of organic carbon at ODP Site 1017 off Point Conception (further north along the California margin), where maxima in organic carbon concentration are recorded in sediments which accumulated in oxygen-depleted environments during warm interstadial events. Organic carbon maxima coincide with episodes of enhanced export of organic and other biogenic material from surface waters. They indicate increased productivity and upwelling as winds intensified in response to higher continent/ocean thermal gradients. This pattern of sedimentation changed some 13 kyr ago after the Alleröd, when advection of more oxygenated intermediate waters from the California current increased.

Dysoxic to anoxic benthic environments may be sites where organic material is preferentially preserved. This is generally the case in regions where organic compounds of continental origin and/or sorbed onto mineral surfaces are abundant. This is also the case in areas of the Northern California margin where labile organic compounds derived from phytoplankton dominate. There, diatomaceous sediments containing up to 3% organic carbon are principally derived from high productivity associated to summer upwelling conditions driven by the southward flowing California current. Despite the presence of an oxygen-minimum zone that extends from 600 to 1,200 m water depth, oxygen contents at all depths are currently sufficient to support a benthic fauna and recent sediments are bioturbated. During the Late Pleistocene in contrast, oxygen contents were too low to support a benthic fauna and laminated sediments were preserved within the oxygen-minimum zone. Lower oxygen concentrations in the Late Pleistocene were attributed to stronger upwelling activity and related organic productivity. The preserved organic material is much more abundant in Late Pleistocene laminated sediments than in Holocene bioturbated sediments. Besides, organic compounds in Late Pleistocene laminated sediments are richer in hydrogen than in Holocene bioturbated deposits, suggesting lower degradation. As organic material is largely of marine origin, it has been inferred that its preservation was improved in laminated sediments associated to the stronger, Late Pleistocene oxygen-minimum zone (Figure 12.17). However, it remains unclear if sorption onto mineral surfaces played a role in the preservation of otherwise relatively labile organic compounds or not. Additional evidence for preferential preservation of organic material in dysoxic to anoxic environments is provided by a comparison of the carbon budgets of the east Pacific margins of Washington State and Mazatlan (Mexico). In both regions, upwelling conditions and related productivity are associated to oxygen-deficient bottom waters at mid-slope. Productivity and accumulation rates are higher on the Washington margin where oxygen contents on the seafloor are nevertheless sufficient to support burrowing organisms, whereas the Mazatlan margin is characterized by a severe oxygen-minimum zone. In those areas, the preservation of organic material (Figure 12.18) is not primarily controlled by accumulation rates or productivity since both are higher on the Washington margin where burial efficiencies are lowest. Burial

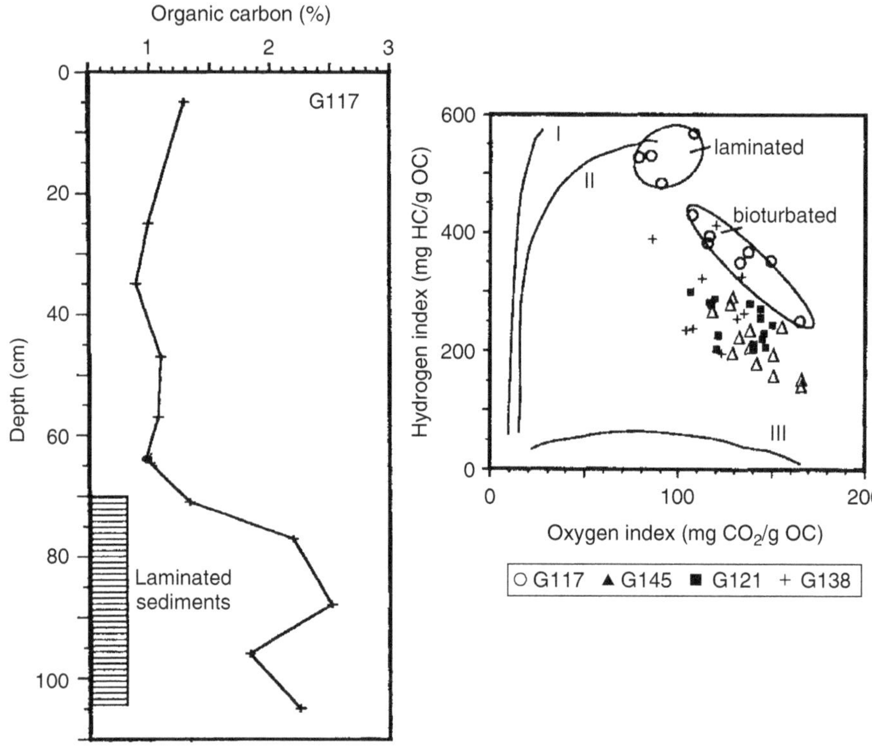

Figure 12.17 Organic carbon contents in a core (G117) taken from the oxygen-minimum zone of the Northern California margin, northeast Pacific Ocean (left). Note higher carbon abundances in laminated sediments associated to the stronger Late Pleistocene oxygen minimum. On the Van Krevelen diagram (right), sediments from cores G121 and G138 come from the deep, oxygenated part of the margin whereas sediments from cores G117 and G145 come from the oxygen-minimum zone. Lower values of the hydrogen index in the bioturbated sediments of core G117 indicate more advanced degradation of organic compounds. Modified from Dean, W.E., Gardner, J.V., Anderson, R.Y., 1994. Geochemical evidence for enhanced preservation of organic matter in the oxygen minimum zone of the continental margin of northern California during the late Pleistocene. Paleoceanography, 9, 47–61.

efficiencies are however relatively comparable on the Washington margin (12–17%) and in oxic sediments of the shelf and deep slope of the Mazatlan margin (19–23%), where they increase to 38% within the oxygen-minimum zone. The oxidation of organic material is highest in Washington margin sediments because of the availability of oxygen and nitrate as electron acceptors in overlying waters (Figure 12.10). In addition, the quantity of nitrate available for carbon oxidation is increased by coupled nitrification–denitrification processes, which require oxygen. Carbon oxidation is also facilitated by oscillating redox conditions associated with burrowing. In contrast, the oxidation of organic material is lowest in sediments within the oxygen-minimum zone of the Mazatlan margin where overlying waters are anoxic and poor in nitrate (Figure 12.10). In addition, nitrate contents are not reinforced by nitrification processes. Despite significant sulfate

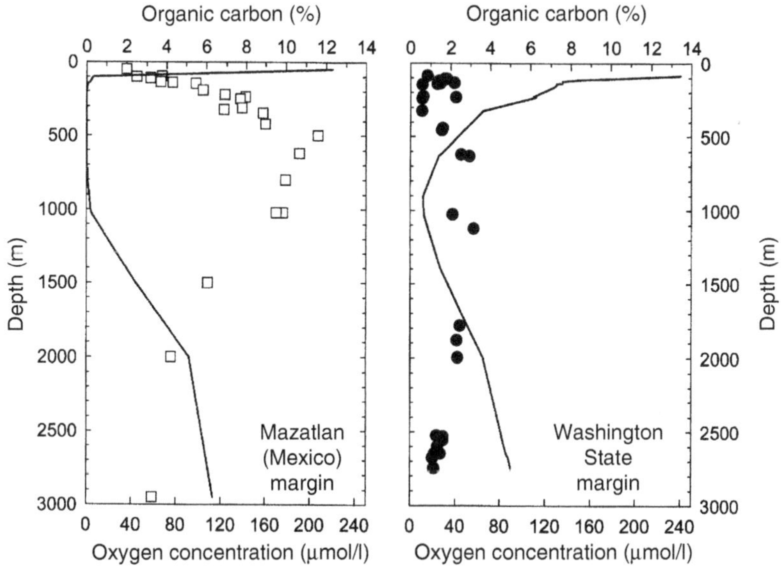

Figure 12.18 Organic carbon contents in surface sediments from the Mazatlan and Washington margins of the northeast Pacific Ocean. Oxygen profiles are those of Figure 12.10. Note maximum contents of organic material within the oxygen-minimum zone of the Mazatlan margin. Modified from Hartnett, H.E., Devol, A.H., 2003. Role of a strong oxygen-deficient zone in the preservation and degradation of organic matter: A carbon budget for the continental margins of northwest Mexico and Washington State. Geochimica et Cosmochimica Acta, 67, 247–264.

reduction, the availability of electron acceptors is limited and oxidation rates are low. As a consequence, oxygen content exerts a major control on the preservation of organic carbon in Mazatlan margin sediments.

12.3. The Diagenesis of Organic Material: Formation and Migration of Fossil Fuels

12.3.1. The Transformation of Kerogens

The anaerobic oxidation of organic compounds slows down rapidly with depth in surface sediments and ceases when all sulfate is consumed, at depths which generally range between 1 m and 15 m below seafloor. In regions where significant proportions of organic matter accumulates (i.e., in continental margin sediments) the organic material the most resistant to oxidation, which is preserved and/or reprocessed through bacterial activity, is known as kerogen. Kerogens are insoluble physical mixtures of selectively preserved, resistant biopolymers of high molecular weight (> 10,000) and diverse origins, which may include specific macromolecules such as lipids in their matrix, and can be to some extent altered. The composition

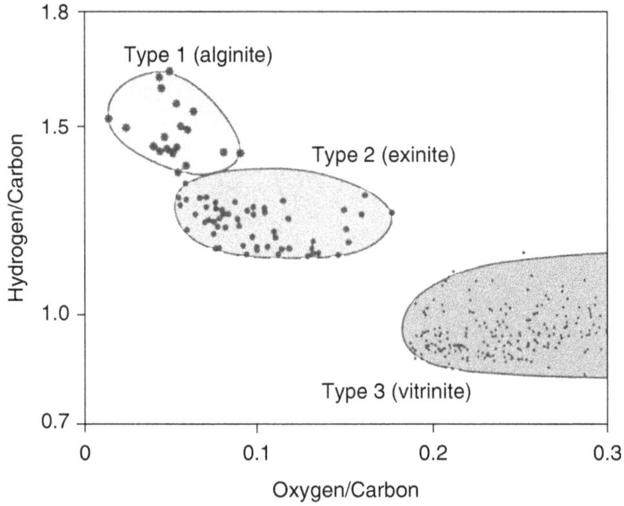

Figure 12.19 Distinction between the three major types of kerogens, based on their hydrogen/carbon and oxygen/carbon ratios. Kerogen of type 4 (inertinite) is not represented. Modified from Biju-Duval, B., 1999. Géologie sédimentaire, Technip, Paris.

of kerogens varies with their initial biochemical composition and their degree of diagenetic alteration and thermal evolution.

Based on the origin and nature of the initial biomass, four main types of kerogens are defined. They are distinguished from their maceral contents. Macerals are discrete particles of insoluble organic material which can be identified and represent residual detritus from various sources: for example, alginite is derived from algae, exinite from vegetal membranes, sporinite from spores, vitrinite from higher plant debris, resinite from resins and huminite from soils. Kerogens are also distinguished from their contents in carbon, hydrogen and oxygen, as determined using pyrolysis methods and expressed through hydrogen/carbon (H/C) and oxygen/carbon (O/C) ratios (Figure 12.19):

- Kerogen of type 1 is also named alginite, after its principal maceral component. This kerogen is rich in hydrogen and relatively poor in oxygen, and is therefore characterized by high H/C and low O/C ratios. It is principally derived from algal lipids, and includes abundant compounds typical of bacterial reprocessing. Type 1 kerogen principally produces paraffinic chains, with minor aromatic compounds, and has a very good potential for the production of oil.
- Kerogen of type 2 is also named exinite, after its principal maceral component. This kerogen has lower hydrogen and higher oxygen contents than alginite and is characterized by intermediate H/C and O/C ratios. It is principally derived from leaf cuticles, pollen and spores, plant waxes, fats and resins, and includes products derived from marine phytoplankton. Type 2 kerogen produces a variety of hydrocarbons, with aromatic and naphtenic compounds being more abundant than for type 1 kerogen. It has a good potential for the production of oil and gas.

- Kerogen of type 3 is also named vitrinite, after its principal maceral component. This kerogen is poor in hydrogen and relatively rich in oxygen, and is characterized by low H/C and high O/C ratios. It is principally derived from lignin and other higher plant components. Type 3 kerogen produces polyaromatic compounds principally. Its potential for oil is lower than that of other kerogen types, but it has a high potential for gas and vitrinite is also a major constituent of coals.
- Kerogen of type 4 is also named inertinite. It includes a variety of constituents of vegetal origin, which have been frequently reworked and strongly oxidized. For that reason, inertinite is very poor in hydrogen and has a very low H/C ratio. Its potential for oil and gas is extremely weak.

The kerogen of type 1 is typically found in dark, carbon-rich, finely laminated or massive sediments which accumulated in lacustrine or marine environments. It is however relatively rare in the ocean, where algal products are among the most labile and are frequently mixed with significant proportions of organic material derived from the continent. In contrast, the kerogen of type 2 is much more frequent and is typically found in continental margin environments where organic material derived from upwelling related high productivity is associated to river-borne organic material derived from soils and higher plants. This is for example the case along the Pacific margin of North America, where kerogen of type 2 is found in oxygen-deficient as well as in oxic sediments. The kerogen of type 3 is typically found in deltaic environments, from the subaerial delta plain to the submarine delta fan. Characteristic examples include the Niger delta, and the delta of the Mahakam River of Indonesia where continental relief and humid tropical climates allowed the accumulation of type 3 kerogen within terrigenous sediments during the entire Neogene. The composition of kerogens and their distribution in marine sediments highlight the better preservation of organic material derived from vegetation, and their concentration in continental margin sediments.

The initial diagenesis of kerogens starts below the sulfate-reduction zone in recently deposited sediments and comprises microbial and chemical alteration processes at low temperature: bacterial fermentation increases in importance with depth, and bacteria of the Archaea Group produce increasing quantities of biogenic gas, which is about 99% methane. Methane production generally occurs through two distinct pathways where hydrogen and acetate are the principal electron donors:

– carbon dioxide reduction

$$CO_2 + 4H_2 \rightarrow CH_4 + 2H_2O;$$

– acetate fermentation

$$CH_3COOH \rightarrow CH_4 + CO_2$$

The boundary between the sulfate-reduction and methane-generation zones is marked by a specific horizon, the sulfate–methane interface, where sulfate is consumed for the anaerobic oxidation of methane (Figure 12.20). Up to 50% of the organic carbon initially contained in kerogens can be transformed into methane,

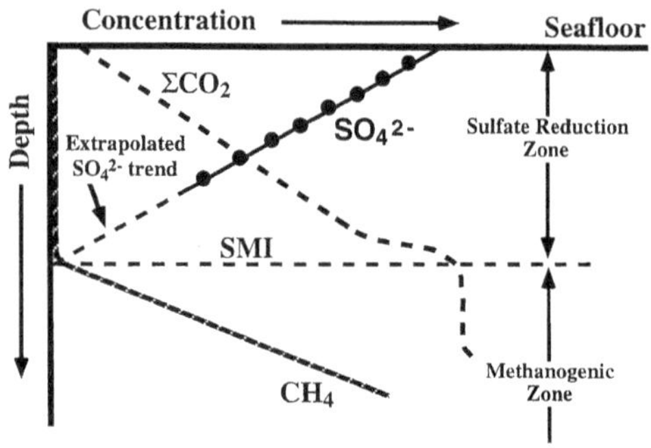

Figure 12.20 Summary highlighting the succession of major geochemical processes in anoxic sediments. SMI, sulfate–methane interface. Reprinted from Borowski, W.S., Paull, C.K., Ussler W., 1999. Global and local variations of interstitial sulfate gradients in deep-water, continental margin sediments: Sensitivity to underlying methane and gas hydrates. Marine Geology, 159, 131–154.

generally within the upper 1,000 m of the sedimentary column. Microbially mediated methane production is enhanced in anoxic and sulfate-poor sediments of moderate sedimentation rates, and for temperatures of 35–45°C. In fact, bacterial fermentation and the related production of methane are principally controlled by temperature and cease around 70°C.

In the initial diagenesis zone, kerogens represent 80–95% of the organic material which is preserved in sediment series. They are associated to small quantities of bitumen, an organic fraction which is extractable with organic solvents. At this stage bitumen are mainly derived from organic compounds of low molecular weight, inherited from the precursor biogenic material and slightly altered by diagenetic geochemical reactions.

The catagenesis corresponds to the thermal alteration of kerogens. Although pressure and temperature both take part in the alteration of kerogens in sediment series, temperature principally controls the diagenetic evolution. The proportion of extractable compounds (bitumen) augments during the catagenesis. At this stage, bitumen are principally released from the kerogen via a rupture of weak bondings, which intensifies as catagenesis progresses. They are often considered as precursors for oil and gas formation. Water and carbon dioxide are first released from the kerogens, as temperatures raise above 50°C. This is illustrated by a strong decrease of the O/C ratio of the kerogens (Figure 12.21). The thermal breakdown of kerogens increases in intensity as temperatures reach 60–80°C, releasing heavy oils (C_{15} and more compounds) principally, together with smaller amounts of lighter hydrocarbons of paraffinic and aromatic type (C_{8-15} hydrocarbons). The production of lighter compounds increases with temperature. Light oils (C_{2-7} hydrocarbons) are released together with methane (wet gas) when temperatures reach 120–150°C. The kerogens lose important quantities of hydrogen during the formation of hydrocarbons (oil window). This is illustrated by a drastic decrease of the H/C ratio of the kerogens (Figure 12.21). Besides temperature, the generation of hydrocarbons is controlled by the degree of maturity of the kerogen. For instance, the threshold for oil production varies with the type of kerogen and its composition: a kerogen of type 2 releases oil at lower temperatures when enriched in sulfur.

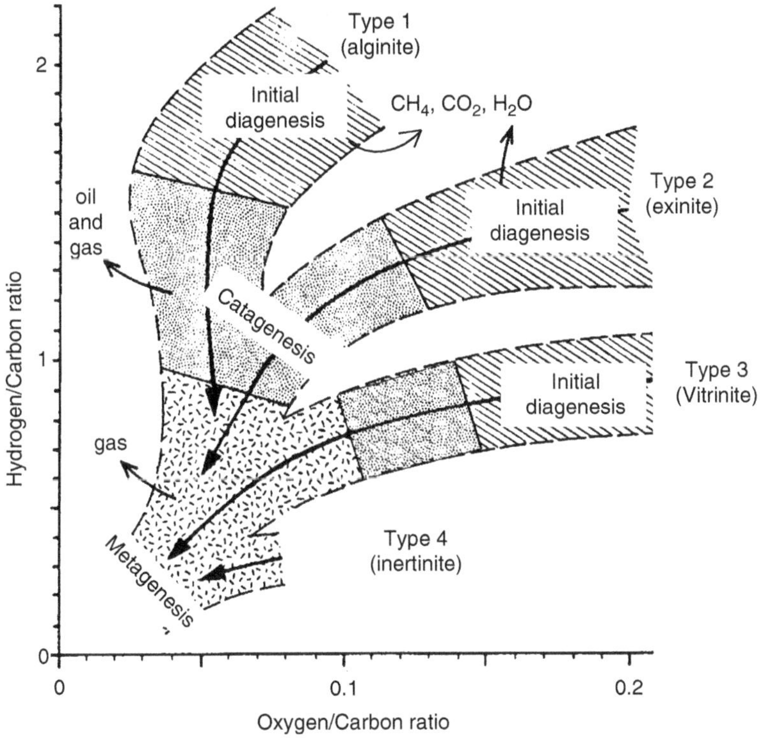

Figure 12.21 Diagenetic evolution of major kerogen types. Note significant loss in hydrogen as hydrocarbons are released, the convergence of evolution paths and the formation of carbon residue. Modified from Einsele, G., 1992. Sedimentary basins, Springer, Berlin.

Under average heat flow conditions, oil formation occurs at burial depths of 2,000 m and more. This is illustrated by the Liassic black shales of the Paris Basin (see Section 4.3.2) which contain a kerogen of type 2. The kerogen is poorly altered and immature at the periphery of the basin where the black shales outcrop, as well as in most areas of the basin where they have been buried to shallow depths. The black shales reached maturity for oil formation in the central part of the basin only, at burial depths of 2,000–2,500 m. The formation of oil may occur at shallower burial depths in regions of higher heat flow. In such areas the kerogen evolves more rapidly and the steps of oil formation are shortened. As a result the kerogen releases a variety of hydrocarbons where lighter compounds predominate. This is the case in many rift areas, such as the Rhine Graben of Western Europe and the Guaymas Basin of the Gulf of California.

The metagenesis is the last stage of the diagenetic alteration of kerogens, which starts at temperatures above 150°C. Dry gas only is stable at such temperatures and methane (thermogenic methane) principally is released. In addition, oil and bitumen which may reach such temperatures because of increasing subsidence or changing heat flow are further cracked to gas. This is especially the case for oil derived from the less productive type 3 kerogen which is commonly retained within the remaining kerogen, as observed in the sediment series of the Mahakam River delta of Indonesia. At this stage the hydrogen content of the remaining kerogen is very low. All kerogens show very similar H/C and O/C ratios and are strongly enriched in carbon (Figure 12.21). The carbon residue derived from kerogens of type 1 (which has a potential for oil and gas of about 90%) and type 2 (which has a potential for oil and gas of about 60%) becomes a minor constituent of the parent rock. In contrast, the carbon residue derived from type 3 kerogen, which potential for oil and gas may hardly reach 25%, still forms a major constituent of the parent rock and may further evolve to coal. The quality of coals varies with the detailed composition of the original kerogen, and the carbon content which varies from 70–75% in lignite to 90–95% in anthracite as metagenesis progresses.

12.3.2. The Migration of Biogenic Gas and the Dynamics of Gas Hydrates

As biogenic gas form during early stages of the diagenetic evolution of organic material and at relatively shallow depths of the sedimentary column, they are easily expelled upward to the seafloor, together with pore fluids. In conditions of relatively high pressure and low temperature that prevail at shallow burial depths of sediment series bathed by cold waters, biogenic gas may combine to pore waters to form gas hydrates, providing that gas concentration exceeds saturation. Gas hydrates are solid phases composed of water and low molecular weight gases where methane may account for as much as 99% of the gas (Figure 12.22). Other compounds include hydrogen sufide, ethane and carbon dioxide principally. In gas hydrates, hydrogen-bonded water molecules form ice cages where molecules of methane are trapped. Inclusion of gases causes water to solidify in a cubic system rather than the usual hexagonal system, the inclusion of gas molecules strengthening the hydrate structure. However, the gas and ice are linked through weak van der Waals forces only. This makes the hydrate structure rather unstable, for example under changing conditions of pressure and temperature.

Pure methane hydrates form at pressures of 48–50 atm and temperatures of 4–6°C. The presence of gases of higher molecular weight causes the hydrates to form at either higher temperature or lower pressure. Also, concentrations of sodium chloride in pore waters may lower the temperature of hydrate formation by as much as 2°C, whereas the presence of nitrogen may increase the pressure of hydrate formation. Yet, conditions favorable to the formation of gas hydrates are commonly found in the upper few 100 m of rapidly accumulating marine sediments, such as prograding continental shelves and accretionary wedges (see Section 8.1.3), as soon as water depth exceeds 300–500 m (Figure 12.23). Relatively high porosities and permeabilities are also important factors facilitating the transport of methane to the upper sediment, and its accumulation. For example on the Blake Ridge of the northwest Atlantic Ocean, the sediments below the hydrate zone contain about 25% of carbonates, decreasing to 8% within the hydrate zone where the quantity of biosiliceous remains and related porosity increase. It is assumed that abundant biosiliceous remains reduce the capillary forces between grains and change the size and shape of the pore space, facilitating the formation of gas hydrates. However, gas hydrates do not develop within the original sediment pore space, although hydrate grains are sometimes disseminated within the sediment. Gas hydrates generally

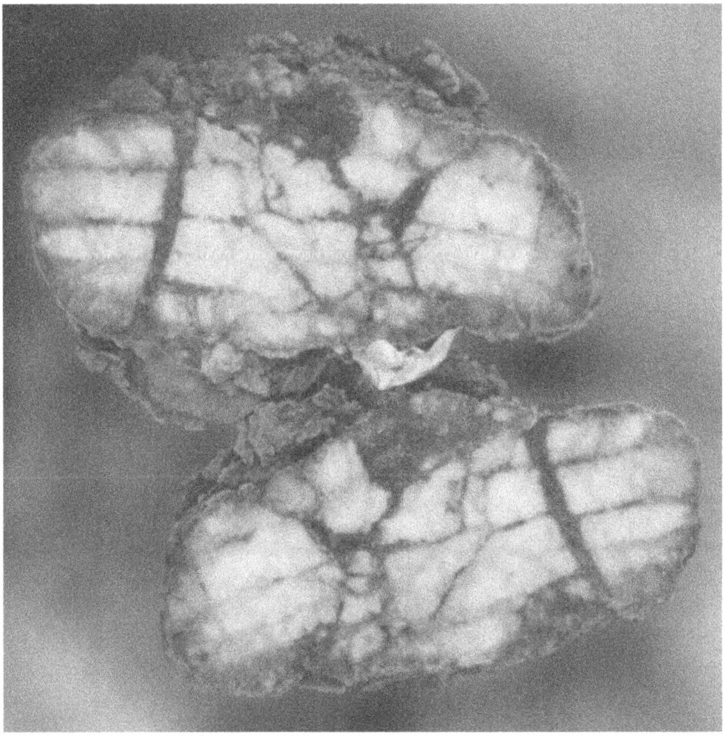

Figure 12.22 Fragments of gas hydrates, drilled on the Hydrate Ridge of the Cascadia margin, northeast Pacific Ocean. Reprinted from Trehu, A.M., Bohrmann, G., Rack, F.R., Torrres, M.E. et al., 2003. Proceedings of the Ocean Drilling Program, Initial Reports, volume 204. Ocean Drilling Program, College-Station, TX.

occur as layers, lenses, nodules and veins several millimeters to decimeters thick, filling fractures and joints, with common evidence that sediment was displaced or fractured during hydrate growth. Hydrates are generally oriented parallel to bedding, but sometimes cut the bedding planes obliquely. On average, gas hydrate represents about 2% of the sediment pore space, but this may locally increase up to 40% of the pore space. As a consequence, the formation of gas hydrates may significantly alter the physical properties of the sediment, especially its porosity, permeability and related fluid migration pattern. Also, the mechanical properties and consolidation processes are altered, as well as the composition of the pore fluids which increase in salinity.

Although gas hydrates are rarely found within the upper 40 m of the sedimentary column, they may locally occur in the uppermost sediment or even outcrop on the seafloor in areas of high methane fluxes as is the case on the Cascadia margin and the Santa Barbara and Guaymas basins off northwest America. While gas hydrates are present from tens to hundreds of meters below seafloor throughout their entire zone of stability, they are principally concentrated in specific intervals such as faults and more favorable lithologies. They are also concentrated near the lower boundary of the gas hydrate stability zone, where the contrast in acoustic impedance between

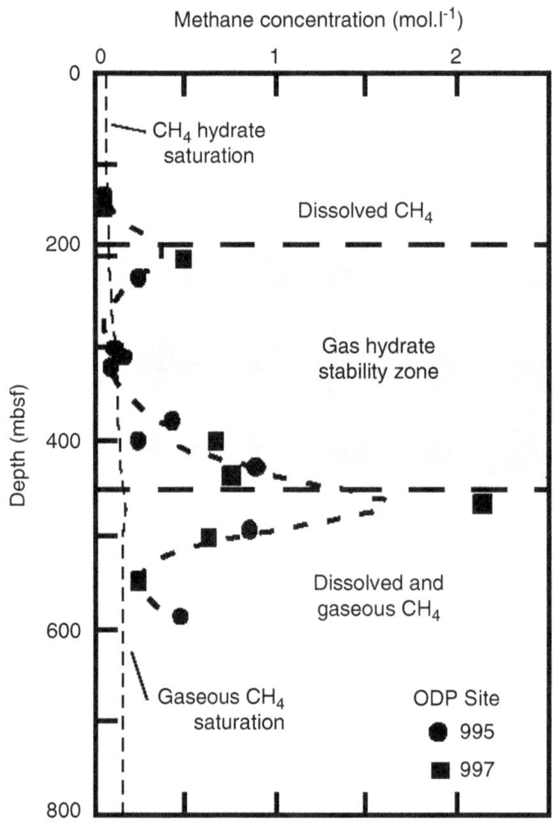

Figure 12.23 Methane concentrations at ODP Sites 995 and 997 drilled on Blake Ridge, northwestern Atlantic Ocean. The shaded area represents the domain where hydrate formation may occur. The dashed line represents the threshold for gas hydrate formation (within and above the gas hydrate stability zone) and for methane gas bubble formation (below the gas hydrate stability zone). Note maximum concentrations in methane near the lower boundary of the gas hydrate stability zone. Modified from Borowski, W.S., 2004. A review of methane and gas hydrates in the dynamic, stratified system of the Blake Ridge region, offshore southeastern North America. Chemical Geology, 205, 311–346.

sediments containing solid hydrates above and free gas below (negative polarity) generates a strong seismic reflector (Figure 12.24). This reflector mimics the morphology of the seafloor, as a consequence of the role of pressure in hydrate formation and is designated as bottom simulating reflector (BSR). BSRs are commonly used to evaluate the presence of gas hydrates in sediment series. Gas hydrates may represent significant reservoirs of fossil fuel, because they contain as much as 164 times the saturation concentration of methane at standard pressure and temperature. In addition, there may be significant quantities of methane trapped beneath the hydrate horizons in the form of free gas or dissolved in pore fluids.

The gas hydrate stability zone is not an impermeable barrier to fluid migration, and regions of hydrate formation are often characterized by active venting associated to authigenic carbonate structures, and/or a pockmarked seafloor. Active venting

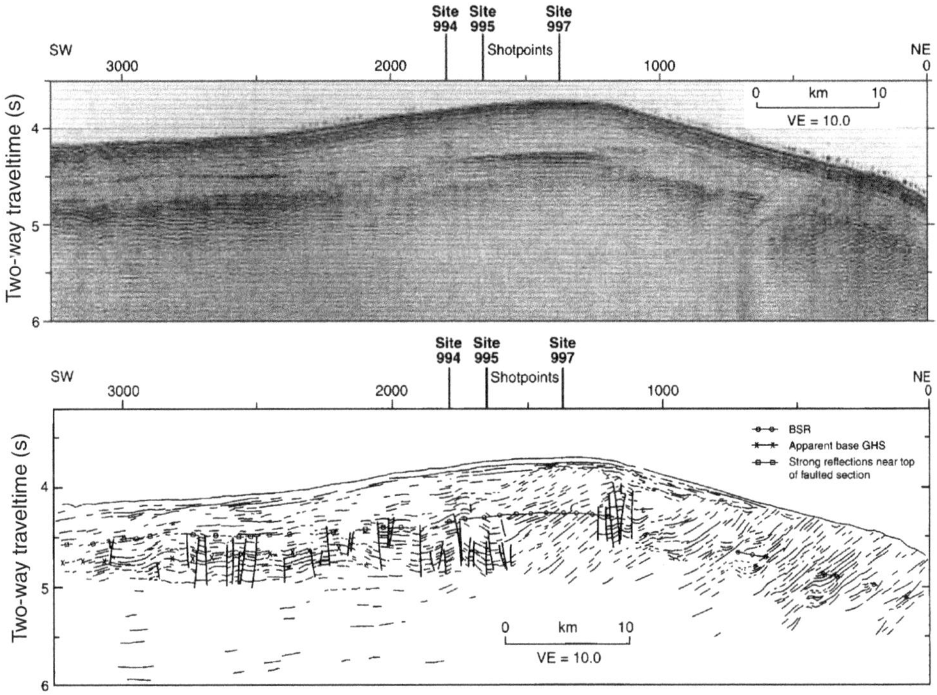

Figure 12.24 Seismic profile across Blake Ridge, northwestern Atlantic Ocean (top) and its interpretation (bottom). BSR, bottom simulating reflector; GHS, gas hydrate stability zone. ODP sites drilled on Blake Ridge are projected on profile. Modified from Paull, C.K., Matsumoto, R., Wallace, P.J. et al., 1996. Proceedings of the Ocean Drilling Program, Initial Reports, volume 164. Ocean Drilling Program, College-Station, TX.

generally coincides with fault systems, on passive margins of high accumulation rates as in convergent areas of accretionary wedges, where fluid release is important. In accretionary wedge areas, the BSR is either pushed upwards or weak, suggesting destabilization of gas hydrates at depth. The fault systems act as conduits which rapidly channel warm pore fluids from beneath the BSR up to the seafloor, where methane is released. On Hydrate Ridge on the Cascadia Margin of northwest America, the composition of the vent water indicates a discharge of freshwater (from hydrate destabilization) together with deep pore water. There, the release of methane-charged fluids from active vents generates plumes in the lower water column, which are several hundreds of meters high and several kilometers wide (Figure 12.25). Methane concentration within the plumes may reach as much as 7,4000 nl/l, whereas the average concentration in the deep ocean is below 20 nl/l. In other areas, methane bubbles rise from pockmarks or small chimneys. The dissociation of gas hydrates starts at their surface, where the formation of methane bubbles cause the formation of pores. This is an endothermic reaction which consumes energy from the adjacent environment which decreases in temperature. This in turn favors ice formation, and the preservation of the remaining hydrates at the edge of their conditions of stability.

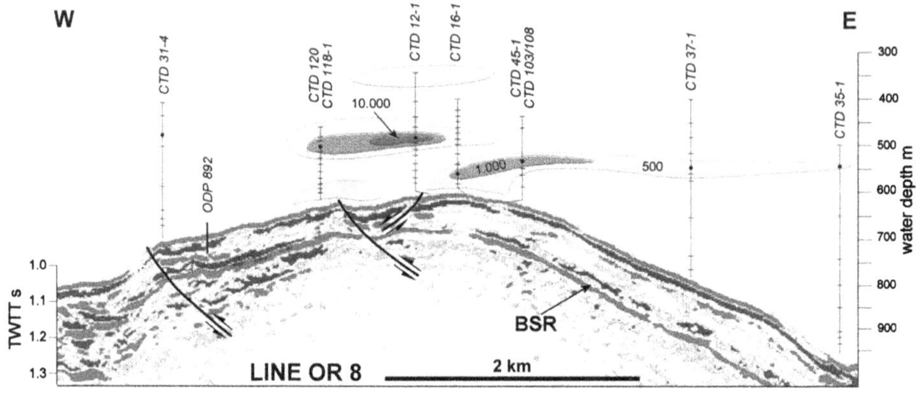

Figure 12.25 Methane plume over Hydrate Ridge, Cascadia margin of the northeast Pacific Ocean. Methane concentrations were measured from water samples and replaced on seismic profile. BSR, bottom simulating reflector; TWT, two-way traveltime. Reprinted from Suess, E., Torres, M.E., Bohrmann, G., Collier, R.W., Greinert, J., Linke, P., Rehder, G., Trehu, A., Wallmann, K., Winckler, G., Zuleger, E., 1999. Gas hydrate destabilization: enhanced dewatering, benthic material turnover and large methane pluimes at the Cascadia convergent margin. Earth and Planetary Science Letters, 170, 1–15.

The methane which currently escapes from the seafloor to the ocean is aerobically oxidized into carbon dioxide. However, a significant part of the methane currently released from hydrate fields is anaerobically oxided in shallow sediments and associated to the formation of authigenic carbonates. Siderite may form within the gas hydrate stability zone of high alkalinity. The oxidation of methane migrating above the sulfate–methane interface produces bicarbonate

$$CH_4 + SO_4{}^{2-} \Longleftrightarrow HCO^{3-} + HS^- + H_2O$$

which may lead to oversaturation of pore fluids with respect to carbonate. Carbonate then combines to Ca but also to other cations such as Mg, Sr and Ba, as to produce authigenic minerals. Besides, sulfur may combine to Fe to form framboidal pyrite. The most commonly found authigenic carbonates are magnesian calcite, dolomite, and aragonite, in the form of nodules, crusts or cements (Figure 12.26) as observed on Blake Ridge in the Atlantic Ocean. Aragonite preferentially forms close to the seafloor. The formation of carbonate crusts and cements is sometimes followed by brecciation or exhumation due to seepage-induced disturbances.

Although currently of minor importance, hydrate destabilization is continuous and pockmark fields such as those observed on the California margin indicate that intervals of violent methane release occurred in the past. Pockmarks are crater-like structures which develop as the dissociation of gas hydrates creates overpressured conditions and failure of the sediment fabric. As a consequence, the sediment is fluidized within the rising gas plume. Mechanical sediment failure caused by gas hydrate dissociation may also initiate slope instability and trigger gravity flows and submarine landslides, intervals of high hydrate concentration (i.e., BSRs) acting as decollement zones. It is probable that episodes of intense dissociation of gas hydrates, pockmark activity and

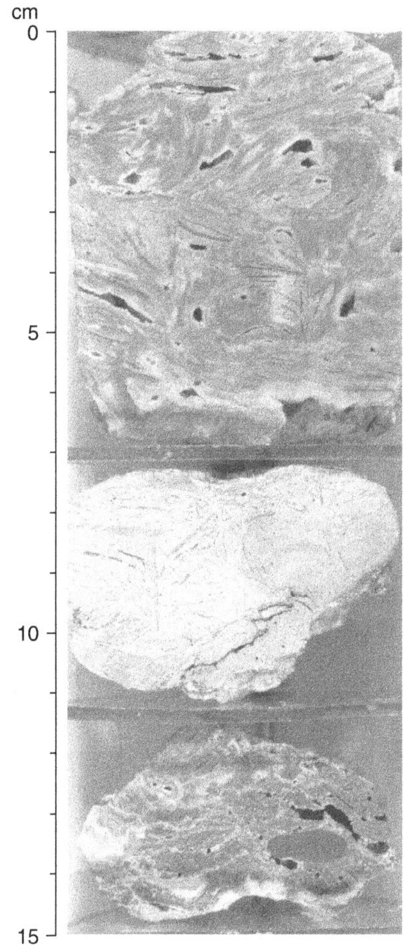

Figure 12.26 Authigenic carbonates at ODP Site 996 on Blake Ridge, northwestern Atlantic Ocean (interval 164-996E-3X-CC, 0–15 cm). (Top) Biocalcirudites mainly composed of carbonate-cemented shell fragments. (Bottom) Calcirudites with intraclasts and bioclasts. Reprinted from Paull, C.K., Matsumoto, R., Wallace, P.J. et al., 1996. Proceedings of the Ocean Drilling Program, Initial Reports, volume 164. Ocean Drilling Program, College-Station, TX.

related slope instabilities have released considerable quantities of methane in the environment. In convergent margin areas, sediment deformation and earthquake activity may trigger local destabilization of gas hydrates. Since they are principally found at shallow burial depth, gas hydrates should also respond rapidly to changes in sea level and sea water temperature. Reduced hydrostatic pressure associated to sea level fall during glacials could destabilize gas hydrates and release methane in the ocean and atmosphere, mitigating the impact of glaciation. However, this is not supported in detail by ice records which show increased methane contents at glacial terminations only. It is probable that the decrease of sea water temperature during glacials was sufficient to offset the impact of sea level change and that glacials were rather periods of gas hydrate build-up at ocean scale. It is likely that gas hydrate destabilization preferentially occurred at stadial and glacial terminations as a consequence of increased production of warmer intermediate water. As warm conditions progressed, gas hydrates became gradually more stable because of higher hydrostatic pressure associated to sea level rise and steady bottom water temperatures. Major episodes of gas hydrate dissociation may also coincide with brief intervals of global warming that occurred throughout geologic times, such as the latest Paleocene thermal maximum. It is for example speculated that either a switch of deep water formation from southern to northern high latitudes caused a deep ocean warming of 3–5°C, or that increased tectonism at ocean scale accelerated the deformation and uplift of accretionary wedges, both triggering catastrophic destabilization of gas hydrates in the latest Paleocene.

12.3.3. The Migration and Accumulation of Thermogenic Hydrocarbons

Hydrocarbons generated during the thermal alteration of kerogens sometimes remain within the parent rock to form bituminous sediments or sedimentary rocks. This is the case when the porosity and permeability of the sediment series are very low (bituminous schists). In most cases however, hydrocarbons are expelled from the parent rock. This migration of hydrocarbons is principally a consequence of high pressure and related compaction, which increase with burial depth. When kerogens reach the catagenesis zone, a significant proportion of pore fluids have already been expelled from the sediment series. The newly formed hydrocarbons saturate the porous space where the hydrostatic pressure increases. The degree of saturation and related hydrostatic pressure vary with the quantity of hydrocarbons released, as well as with the nature and physical properties of the parent rock: pore pressure is higher in parent rocks of lower porosity. Progressively, a pressure gradient develops between the parent rock and unproductive sediments acting as drains. The expulsion of hydrocarbons from the parent rock (primary migration) is controlled by both the pressure gradient and the capacity of the drain to allow further migration, which is a function of the porosity and permeability of the sediment (Figure 12.27). The presence of low molecular weight compounds such as methane also facilitates the primary migration of hydrocarbons.

The transfer of hydrocarbons through the draining sediments and sedimentary rocks corresponds to the secondary migration. Because of differences in molecular weight, the different types of hydrocarbons migrate at different pace and are progressively separated into distinct phases, the draining sediments also acting as filters. As for pore fluids (see Section 2.3.10) the flow of hydrocarbons through sediment series can be estimated, using the multiphasic Darcy's law. The most efficient drainage is ensured through rocks and sediments of high porosity and permeability such as sands and sandstones, grainstone carbonates, but also fault systems and discontinuities. Thermogenic hydrocarbons migrate together with the residual pore fluids and are sometimes released to the surface or to the seafloor. Thermogenic hydrocarbons being principally generated at significant burial depth of continental margins where sediment series are complex, their upward migration is in most cases rapidly limited by sediment layers of low permeability such as clays and claystones. In contrast, thermogenic hydrocarbons may migrate laterally over a

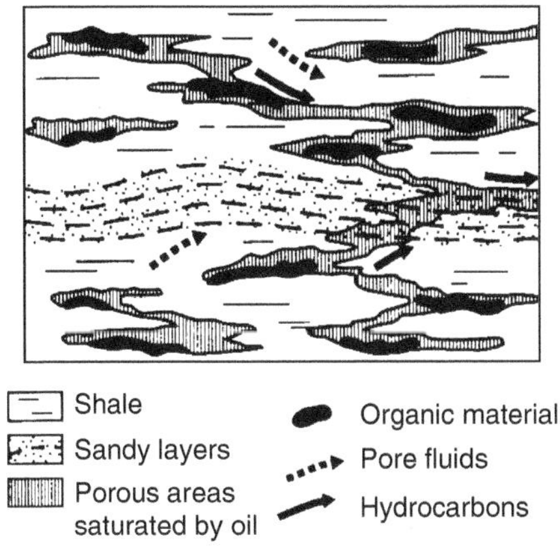

Figure 12.27 Primary migration of hydrocarbons. Oil and gas saturate the porous space and are expelled together with pore fluids toward a porous and permeable drain. Modified from Biju-Duval, B., 1999. Géologie sédimentaire, Technip, Paris.

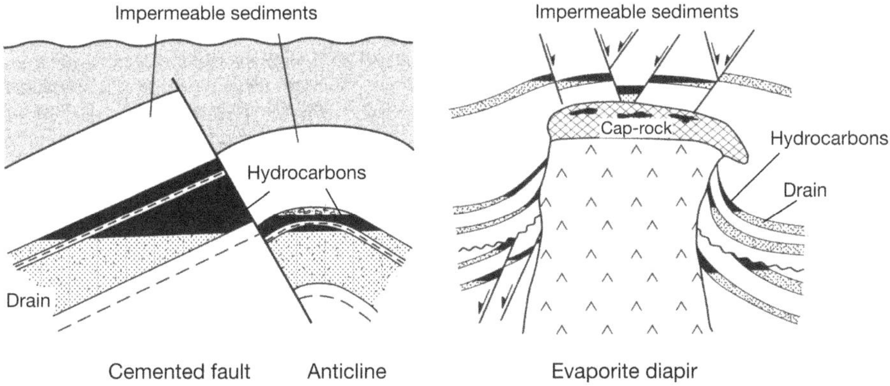

Figure 12.28 Example of hydrocarbon traps associated to cemented faults, sediment deformation (anticline) and salt tectonics (evaporite diapir). Modified from Biju-Duval, B., 1999. Géologie sédimentaire, Technip, Paris.

few tens of kilometers, because sediment facies commonly extend over large areas at basin scale. Hydrocarbons are trapped when they can no longer progress through the draining facies and structures. This is the case for example when the sediment facies changes, sediment series are deformed, fault systems are cemented, or impermeable evaporite diapirs cut sediment series (Figure 12.28). Hydrocarbons progressively accumulate within the porous and permeable sediment which turns into a reservoir. The quality of a reservoir is a function of its porosity, permeability and extension of facies. Depending on burial depth and local conditions of pressure and temperature a further thermal alteration of hydrocarbons may occur within the reservoirs, leading to increasing proportions of light compounds.

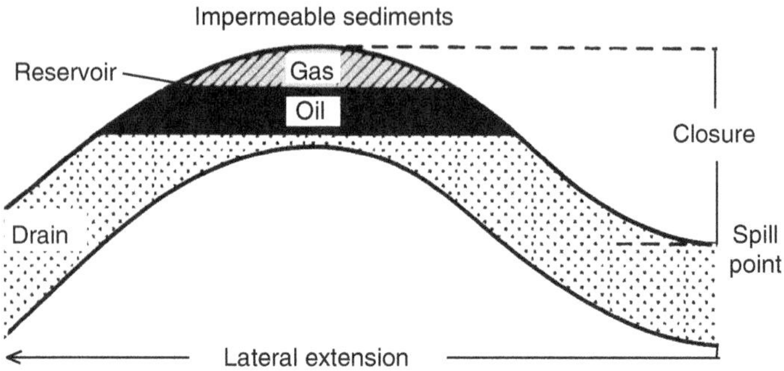

Figure 12.29 Main characteristics of an anticline hydrocarbon trap. The closure and lateral extension control the potential extent of the reservoir. Modified from Stoneley, R., 1995. An introduction of petroleum exploration for non-geologists, Oxford University Press, Oxford.

There is a variety of situations where hydrocarbons can be trapped within reservoirs. In delta fan environments, sandy channel facies (transformed into sand or sandstone lenses through burial) sealed by fine siliciclastic facies represent potential reservoirs. In a different continental shelf setting, this is also the case of porous carbonate reef facies capped by fine siliciclastics. In deeper areas of passive continental margins, the evaporites which accumulated during the early stages of the margin development and were deeply buried and deformed by salt tectonics as the continental shelf progressed, commonly trap hydrocarbons within the porous and permeable siliciclastic sediments they breach. In deformed sedimentary basins, hydrocarbons are more frequently trapped within anticline structures or against cemented fault systems. In all cases, the vertical (closure) and lateral extension of the trap exert a major control on the quantity of hydrocarbons stored in the reservoir (Figure 12.29). Hydrocarbon traps are not totally impermeable and may allow further migration (dysmigration). For example, the vertical amplitude of anticline structures frequently allows fluids to escape from beneath (spill point) and continue their migration within the draining facies. In other cases, the trap allows a limited upward migration which provides indices for potential accumulation of hydrocarbons below.

FURTHER READING

Biju-Duval, B., 1999. Géologie sédimentaire. Technip, Paris.

Einsele, G., 1992. Sedimentary basins, evolution, facies, and sediment budget. Springer, Berlin.

Engel, M.H., Macko, S.A., 1993. Organic geochemistry: Principles and applications. Plenum Press, New York.

Kennett, J.P., Cannariato, K.G., Hendy, I.L., Behl, R.J., 2003. Methane hydrates in Quaternary climate change. American Geophysical Union, Washington, DC.

Mao, W., Koh, C.A., Sloan, E.D., 2007. Clathrate hydrates under pressure. Physics Today, 60: 42–47.

Philp, R.P., 2003. Formation and geochemistry of oil and gas. Treatise on Geochemistry, 7: 223–256.

Other references used in this chapter

Arnosti, C., Holmer, M., 1999. Carbohydrate dynamics and contributions to the carbon budget of an organic-rich coastal sediment. Geochimica et Cosmochimica Acta, 63: 393–403.

Berelson, W.M., Anderson, R.F., Dymond, J., De Master, D., Hammond, D.E., Collier, R., Honjo, S., Leinen, M., Mc Manus, J., Pope, R., Smith, C., Stephens, M., 1997. Biogenic budgets of particle rain, benthic remineralization and sediment accumulation in the equatorial Pacific. Deep-Sea Research II, 44: 2251–2282.

Bice, K.L., Marotzke, J., 2002. Could changing ocean circulation have destabilized methane hydrate at the Paleocene/Eocene boundary? Paleoceanography, 17. doi: 10.1029/2001PA000678.

Bohrmann, G., Kuhs, W.F., Klapp, S.A., Techmer, K.S., Klein, H., Murshed, M.M., Abegg, F., 2007. Appearance and preservation of natural gas hydrate from hydrate ridge sampled during ODP Leg 204 drilling. Marine Geology, 244: 1–14.

Borowski, W.S., Paull, C.K., Ussler, W., 1999. Global and local variations of interstitial sulfate gradients in deep-water, continental margin sediments: Sensitivity to underlying methane and gas hydrates. Marine Geology, 159: 131–154.

Calvert, S.E., Bustin, R.M., Pedersen, T.F., 1992. Lack of evidence for enhanced preservation of sedimentary organic matter in the oxygen minimum of the Gulf of California. Geology, 20: 757–760.

Canuel, E.A., Martens, C.S., 1996. Reactivity of recently deposited organic matter: Degradation of lipid compounds near the sediment-water interface. Geochimica eet Cosmochimica Acta, 60: 1793–1806.

Cowie, G.L., Calvert, S.E., Pedersen, T.F., Schulz, H., Von Rad, U., 1999. Organic content and preservational controls in surficial shelf and slope sediments from the Arabian Sea (Pakistan margin). Marine Geology, 161: 23–38.

Dean, W.E., Gardner, J.V., Anderson, R.Y., 1994. Geochemical evidence for enhanced preservation of organic matter in the oxygen minimum zone of the continental margin of Northern California during the late Pleistocene. Paleoceanography, 9: 47–61.

De Master, D.J., Ragueneau, O., Nittrouer, C.A., 1996. Preservation efficiencies and accumulation rates for biogenic silica and organic C, N, and P in high-latitude sediments: The Ross Sea. Journal of Geophysical Research, 101: 18501–18518.

Fischer, G., Ratmeyer, V., Wefer, G., 2000. Organic carbon fluxes in the Atlantic and the Southern Ocean: Relationship to primary production compiled from satellite radiometer data. Deep-Sea Research II, 47: 1961–1997.

Ganeshram, R.S., Calvert, S.E., Pedersen, T.F., Cowie, G.L., 1999. Factors controlling the burial of organic carbon in laminated and bioturbated sediments off NW Mexico: Implications for hydrocarbon preservation. Geochimica et Cosmochimica Acta, 63: 1723–1734.

Hammond, D.E., McManus, J., Berelson, W.M., Kilgore, T.E., Pope, R.H., 1996. Early diagenesis of organic material in equatorial Pacific sediments: Stoichiometry and kinetics. Deep-Sea Research II, 43: 1365–1412.

Hartnett, H.E., Devol, A.H., 2003. Role of a strong oxygen-deficient zone in the preservation and degradation of organic matter: A carbon budget for the continental margins of northwest Mexico and Washington State. Geochimica et Cosmochimica Acta, 67: 247–264.

Hedges, J.I., Keil, R.G., 1995. Sedimentary organic matter preservation: An assessment and speculative synthesis. Marine Chemistry, 49: 81–115.

Hendy, I.L., Pedersen, T.F., Kennett, J.P., Tada, R., 2004. Intermittent existence of a southern California upwelling cell during submillenial climate change of the last 60 kyr. Paleoceanography, 19: PA3007. doi: 10.1029/2003PA000965.

Hendy, I.L., Pedersen, T.F., 2005. Is pore water oxygen content decoupled from productivity on the California Margin? Trace element results from ODP Hole 1017E, Santa Lucia slope, California. Paleoceanography, 20: PA4026. doi: 10.1029/2004PA001123.

Lee, C., Murray, D.W., Barber, R.T., Buesseler, K.O., Dymond, J., Hedges, J.I., Honjo, S., Manganini, S.J., Marra, J., Moser, C., Peterson, M.L., Prell, W.L., Wakeham, S.G., 1998. Particulate organic carbon fluxes: Compilation of results from the 1995 US JGOFS Arabian Sea process study. Deep-Sea Research II, 45: 2489–2501.

Martin, W.R., Sayles, F.L., 2003. The recycling of biogenic material at the sea floor. Treatise on Geochemistry, 7: 37–65.

Mayer, L.M., 1994. Surface area control of organic carbon accumulation in continental shelf sediments. Geochimica et Cosmochimica Acta, 58: 1271–1284.

Naehr, T.H., Eichhubl, P., Orphan, V.J., Hovland, M., Paull, C.K., Ussler, W., Lorenson, T.D., Greene, H.G., 2007. Authigenic carbonate formation at hydrocarbon seeps in continental margin sediments: A comparative study. Deep-Sea Research II, 54: 1268–1291.

Nelson, D.M., Anderson, R.F., Barber, R.T., Brzezinski, M.A., Buesseler, K.O., Chase, Z., Collier, R.W., Dickson, M.-L., François, R., Hiscock, M.R., Honjo, S., Marra, J., Martin, W.R., Sambrotto, R.N., Sayles, F.L., Sigmon, D.E., 2002. Vertical budgets for organic carbon and biogenic silica in the Pacific sector of the Southern Ocean, 1996–1998. Deep-Sea Research II, 49: 1645–1674.

Paull, C.K., Matsumoto, R., Wallace, P.J. et al., 1996. Proceedings of the Ocean Drilling Program, Initial Reports, volume 164. Ocean Drilling Program, College-Station, TX.

Ragueneau, O., Tréguer, P., Leynaert, A., Anderson, R.F., Brzezinski, M.A., De Master, D.J., Dugdale, R.C., Dymond, J., Fischer, G., François, R., Heinze, C., Maier-Reimer, E., Martin-Jézéquel, V., Nelson, D.M., Quéguiner, B., 2000. A review of the Si cycle in the modern ocean: Recent progress and missing gaps in the application of biogenic opal as a paleoproductivity proxy. Global and Planetary Change, 26: 317–365.

Rixen, T., Haake, B., Ittekot, V., 2000. Sedimantation in the western Arabian Sea: The role of coastal and open-ocean upwelling. Deep-Sea Research II, 47: 2155–2178.

Snyder, G.T., Hiruta, A., Matsumoto, R., Dickens, G.R., Tomaru, H., Takeuchi, R., Komatsubara, J., Ishida, Y., Yu, H., 2007. Pore water profiles and authigenic mineralization in shallow marine sediments above the methane-charged system on Umitaka Spur, Japan Sea. Deep-Sea Research II, 54: 1216–1239.

Stott, L.D., Berelson, W., Douglas, R., Gorsline, D., 2000. Increased dissolved oxygen in Pacific intermediate waters due to lower rates of carbon oxidation in sediments. Nature, 407: 367–370.

Suess, E., Torres, M.E., Bohrmann, G., Collier, R.W., Greinert, J., Linke, P., Rehder, G., Trehu, A., Wallmann, K., Winckler, G., Zuleger, E., 1999. Gas hydrate destabilization: Enhanced dewatering, benthic material turnover and large methane plumes at the Cascadia convergent margin. Earth and Planetary Science Letters, 170: 1–15.

Trehu, A.M., Bohrmann, G., Rack, F.R., Torrres, M.E. et al., 2003. Proceedings of the Ocean Drilling Program, Initial Reports, volume 204. Ocean Drilling Program, College-Station, TX.

Trehu, A.M., Long, P.E., Torres, M.E., Bohrmann, G., Rack, F.R., Collett, T.S., Goldberg, D.S., Milkov, A.V., Riedel, M., Schultheiss, P., Bangs, N.L., Barr, S.R., Borowski, W.S., Claypool, G.E., Delwiche, M.E., Dickens, G.R., Gracia, E., Guerin, G., Holland, M., Johnson, J.E., Lee, Y.-J., Liu, C.-S., Su, X., Teichert, B., Tomaru, H., Vanneste, M., Watanabe, M., Weinberger, J.L., 2004. Three-dimensional distribution of gas hydrate beneath Hydrate Ridge: Constraints from ODP Leg 204. Earth and Planetary Science Letters, 222: 845–862.

Wakeham, S.G., Hedges, J.I., Lee, C., Peterson, M.L., Hernes, P.J., 1997. Compositions and transport of lipid biomarkers through the water column and surficial sediments of the equatorial Pacific Ocean. Deep-Sea Research II, 44: 2131–2162.

Wakeham, S.G., Lee, C., Hedges, J.I., Hernes, P.J., Peterson, M.L., 1997. Molecular indicators of diagenetic status in marine organic matter. Geochimica et Cosmochimica Acta, 61: 5363–5369.

CHAPTER THIRTEEN

Hydrogenous Sediments

Initially, inorganic precipitation from seawater was considered as the key process at the origin of the hydrogenous fraction of marine sediments. In most cases however, biological and particularly bacterial activity, hydrothermalism, alteration of substrates and early diagenetic processes interact in the formation of hydrogenous sediments. As a consequence, hydrogenous sediments include a variety of components. Besides evaporites which mainly form during the earliest stages of ocean evolution, hydrogenous components include hydrothermal deposits, metalliferous elements such as polymetallic nodules and crusts, and a variety of authigenic minerals: glauconite and phosphate which form in continental shelf and slope environments, barite which is principally found in pelagic sediments, zeolites such as phillipsite and clay minerals such as Fe-smectite and celadonite which may derive from the submarine alteration of fresh basaltic substrates. Among them, the context for evaporite formation is given in chapters 3 (Rift Systems) and 5 (Crustal Fissure Systems), whereas hydrothermal deposits are detailed in Section 6.4.2 (Hydrothermal Deposits of Active Mid-Oceanic Ridges). This chapter focuses on metalliferous and authigenic deposits which have not been described before and are either widespread in the ocean or characteristic of specific environmental conditions. They include polymetallic nodules and crusts, glauconite and phosphates.

13.1. Polymetallic Nodules and Crusts

First recovered from the deep ocean during the *Challenger* Expedition, polymetallic nodules (Figure 13.1) are mainly concentrated on the seafloor of the deepest basins whereas polymetallic crusts are principally associated to submarine volcanic systems and seamounts.

13.1.1. Formation and Composition of Metalliferous Deposits

Manganese is the principal chemical element found in metalliferous deposits, where it is generally associated to iron. Manganese concentrations of 10–25% are frequent, but may locally increase up to 50%. Iron concentrations are generally lower, below 20–25%. Associated elements include copper, zinc, nickel, chromium and lead, but in much smaller proportions generally ranging from a few parts per million to a few percents. Manganese is abundant in seawater, and is predominantly of hydrothermal origin like most associated elements. The predominance of manganese in metalliferous sediments of the deep ocean is a consequence of the rapid incorporation of ore elements such as nickel and zinc into sulfide minerals of hydrothermal deposits (see Section 6.4.2). Then, hydrothermal fluids mix with

Figure 13.1 Polymetallic crust (left) and nodule (right). Modified from Murray, J., Renard, A.F., 1891. Report on Deep-Sea Deposits, Based on Specimens Collected During the Voyage of *H.M.S. Challenger* in the Years 1872–1876. Eyre and Spottiswoode, London.

seawater. Metalliferous elements precipitate in higher proportions in those oceanic regions adjacent to areas of active volcanism and hydrothermalism, i.e., active spreading ridges and regions of intraplate, hot-spot volcanism where seamounts are abundant. Metalliferous elements are also widely dispersed beneath hydrothermal plumes and can participate in the formation of hydrogenous minerals in oceanic regions remote from the source area. However, their accumulation so as to form metalliferous deposits requires very low sedimentation rates of other sediment particles. These conditions are fulfilled in deep oceanic basins of weak terrigenous input, low surface productivity and/or significant dissolution of biogenic and organic elements. As a consequence, metalliferous deposits are more abundant in the deep basins of the Indian Ocean and especially the Pacific Ocean. The relationship between spreading ridge and seamount activity, and the occurrence of metalliferous deposits, suggests that intervals of increased spreading rates (which are also intervals of enhanced intraplate volcanism and seamount activity) may coincide with increased accumulation rates of metalliferous deposits. It is generally admitted that metalliferous oxides and hydroxides accrete at growth rates that range from 0.5–10 mm/Myr, but some estimates may include intervals of non-deposition.

Overall, there is a strong relationship between seawater and the geochemical composition of many metalliferous deposits. Although chemical elements of hydrothermal origin largely dominate, local environmental conditions may also play a role in the elemental content of metalliferous deposits. For example, polymetallic crusts and nodules may include elements of biogenic origin (Ba, Ca, P, etc.) or detrital origin (Si, Al, etc.).

Manganese occurs in three oxidation states (Mn^{2+}, Mn^{3+}, Mn^{4+}), is easily oxidized and chemically behaves like a number of elements including magnesium, iron, nickel and cobalt, giving rise to a variety of oxide and hydroxide minerals. Metalliferous oxides and hydroxides typically have a poor crystallinity and form by catalytic oxidation and adsorption onto substrates such as pre-existing oxides, minerals or rock fragments. Microorganisms such as bacteria may act as catalysts, and can significantly increase the rate of metal oxidation. The most common minerals found in manganese-rich oxides and hydroxides include:

- Vernadite, which is the dominant manganese-bearing phase found in both polymetallic crusts and nodules. It is a fine-grained phase of poor crystallinity. Its structure is unclear, but chemical analyses show that vernadite commonly includes minor percentages of Fe, K, Mg, Ca and Ba, as well as water.
- Birnessite, which is an important component of polymetallic crusts and nodules, and is also found in hydrothermal deposits. Birnessite is generally fine grained and has a relatively poorly organized sheet structure. The sheet units are made of MnO_6 octahedral layers, free valences being filled by exchangeable cations and water in an interlayer position. Exchangeable cations principally include Ni and Ba, but also Co, Na, Ca and Mg. Dehydrated birnessite sometimes occurs in hydrothermal environments, where the presence of exchangeable, interlayered cations protect to a certain extent the mineral structure from collapse.
- Todorokite, which is an iron-free hydrous manganese oxide and a major component of polymetallic nodules. Like zeolites, todorokite has a tunnel-like structure. This structure is made of triple chains of MnO_6 octahedra and hosts a variety of cations as well as water. Typically, lower valence cations such as Mn^{3+} and Mn^{2+}, but also Ni^{2+}, Mg^{2+} and others substitute for Mn^{4+} to compensate for the tunnel cations. Because of important substitution, metalliferous deposits dominated by todorokite may contain up to several percent of Ni, Co and Cu. Todorokite is principally an early diagenetic mineral which resists dissolution because of significant substitution of cations.
- Manganite, where about one-half of the oxygen atoms in the MnO_6 octahedra is replaced by hydroxyl anions. Manganite is found in hydrothermal vein deposits, but also as an alteration product of pre-existing Mn-bearing minerals.

Manganese-rich minerals are often associated to iron oxides and hydroxides, such as feroxyhyte and goethite. Metalliferous deposits also include a variety of mineral phases such as clay minerals, apatite, quartz and feldspars, but generally in limited amounts. Metalliferous deposits therefore mainly include oxide and hydroxide mineral phases that precipitate at the sediment–water interface, but also others that are of terrigenous or biogenic origin, or form via early diagenetic processes.

13.1.2. Polymetallic Crusts

Polymetallic crusts consist of oxide and hydroxide layers where manganese and iron dominate. They form on rocky substrates of seamounts, active ridges and submarine plateaus, where active bottom circulation maintains the seafloor free of sediment. Polymetallic crusts accrete at very slow rates of 0.5–5 mm/Myr on the average, but higher accretion rates (as high as 120 mm/Myr) have been recorded in some cases. In hydrothermally active areas, polymetallic crusts often include volcaniclastic material cemented by manganese oxides where birnessite dominates. Those hydrothermal crusts have elevated Mn/Fe ratios and low contents of other elements such as Zn, Ni and others because of the preferential incorporation of certain elements into sulfide minerals. Where freshly emitted basalts interfere with seawater, their alteration into palagonite is associated to the development of zeolites (phillipsite) and clay minerals (Fe-smectite) and to the loss and gain of chemical elements. Especially, concentrations in Mn, Fe, Ti and K increase. This is followed in some cases by the precipitation of metalliferous oxides and hydroxides which may progressively form a protective veneer against further alteration.

The nature and composition of polymetallic crusts vary strongly within hydrothermally active areas. In the Kurile Island arc of the north Pacific Ocean, three types of coexisting polymetallic crusts have been described:

- Black massive crusts (type 1) composed of Mn oxides with smooth to botryoidal surface and a complex internal structure are typical submarine manganese crusts. Overall they have high Mn/Fe ratios, with dominant birnessite and rare vernadite and goethite in the basal layers which consist of debris cemented by metalliferous oxides. The body of the crust consists of radial mammillae (aggregates of birnessite and vernadite) cemented by thin layers of dominant birnessite, whereas the upper layer is made of thin dendrites of birnessite and todorokite, cemented with massive material identical in composition to the cement of the basal layer.
- Variegated (black, brown and green) crusts (type 2) made of clay, amorphous Fe–Si phases and Mn oxides are found on sediment substrates. Overall, these crusts have low Mn/Fe ratios. They principally consist of green clay cementing debris or semi-lithified stratified deposits. The green clay is composed of nontronite, a Fe-smectite which is here associated to feroxyhyte. The crusts also include brown veinlets of amorphous Fe–Si phases and fissures which are filled with black Mn oxides, principally birnessite and todorokite. In some cases, variegated crusts show alternating black, brown and green layers which illustrate changing physico-chemical conditions within the hydrothermal plume.
- Composite crusts (type 3) are black and massive at their base, but brown to light colored at their top. The black and massive lower part is very similar in texture and composition to type 1 crusts (see above). The upper part of these crusts consists of brown layers of vernadite, feroxyhyte and goethite, alternating with light clay layers. The Mn/Fe ratio decreases in the upper part of the crust, which is also enriched in Co, Cu and Ni. Hydrothermal influence is significant at the base of the crust but decreases in the upper crust which is more typical of hydrogenous conditions. This evolution most probably reflects a declining hydrothermal contribution to the site. In addition, the clear succession of Mn-rich hydrothermal and Mn-poor hydrogenous crusts suggests a negligible diagenetic contribution to their formation.

The texture and mineralogy of polymetallic crusts clearly show a great variability, which most probably reflects the influence and chemical characteristics of hydrothermal plumes, as well as their evolution through time (Figure 13.2).

The influence of hydrothermal plumes is highlighted through a comparison of the fine chemical composition of polymetallic crusts. In the central Pacific Ocean for instance, the metalliferous elements found in polymetallic crusts from the Pitcairn Island hot-spot area show significant differences in nature and especially in

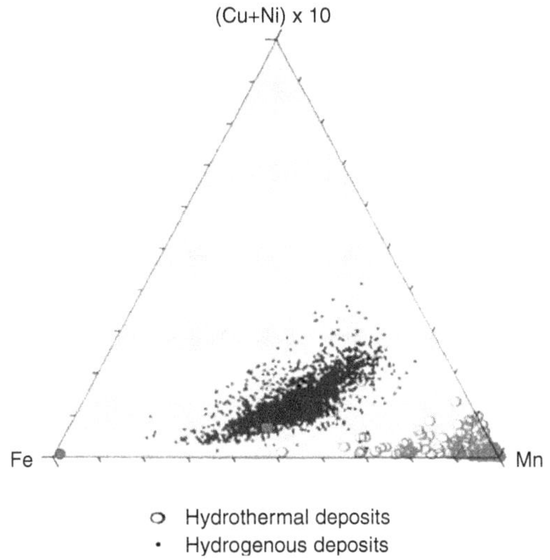

Figure 13.2 Chemical composition of hydrothermal and hydrogenous polymetallic deposits from the northwest Pacific Ocean. Note the lower contents of Mn and higher contents in Fe, Cu, and Ni in hydrogenous relative to hydrothermal deposits, i.e., with decreased influence of hydrothermal sources. Modified from Glasby, G.P., Cherkashov, G.A., Gavrilenko, G.M., Rashidov, V.A., Slovtsov, I.B., 2006. Submarine hydrothermal activity and mineralization on the Kurile and Western Aleutian Island arcs, N.W. Pacific. Marine Geology, 231, 163–180.

abundance (Figure 13.3). A crust retrieved from the active hot-spot area is made of coarse-grained, porous layers of sediment including abundant microfossils, impregnated by metalliferous oxides, and capped by finely laminated metalliferous oxides (hydrothermal crust). This crust is compared to a typical hydrogenous crust of high porosity especially in the upper layers, made of black to red-brown material (with biogenic carbonates in the lower layers), which has been retrieved from an adjacent, inactive seamount. The most remarkable difference between both crusts concerns their Mn/Fe ratios which attain maximum values close to 2,440 in the hydrothermal crust but vary between 0.54 and 1.68 in the hydrogenous crust, illustrating the predominance of Mn in hydrothermal plumes close to their source area. This is also expressed by percentage abundances of Mn and Fe: the hydrothermal crust is characterized by abundances of Mn up to 50% and Fe below 0.6%, whereas the hydrogenous crust contains up to 20–25% of each element (Figure 13.3). In addition, the hydrogenous crust contains much more Ni, Zn and Cu than the hydrothermal crust, and also contains Co and Pb which are absent from the hydrothermal crust. Besides, the layers which correspond to different stages of the crust formation contain significantly different contents in metalliferous elements, indicative of differences in bottom water chemistry. In the hydrothermal crust for example, Mn abundances are minimum (and Fe abundances maximum) in the lower part made of sediment impregnated by metalliferous oxides. In contrast, Mn abundances are maximum (and Fe abundances minimum) in the upper part of the crust made of finely laminated metalliferous oxides, illustrating an increased influence of the hydrothermal plume. In the hydrogenous crust the different layers, which have

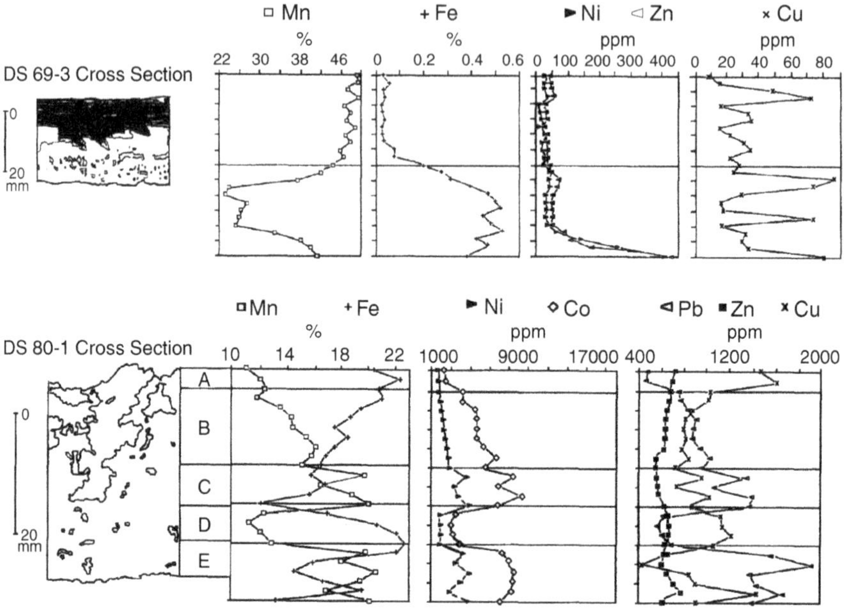

Figure 13.3 Cross-sections of a hydrothermal crust (top) and hydrogenous crust (bottom) highlighting compositional variations. Letters (A, B and A–E) represent crust layers of different texture and porosity (see description in Section 13.1.2). Note higher contents of Fe, Ni, Co, Cu and Zn in the hydrogenous crust, anticorrelation between contents of Mn and Fe, and the dominance of Fe in the highly porous, upper layers of the hydrogenous crust. Reprinted from Glasby, G.P., Stüben, D., Jeschke, G., Garbe-Schönberg, C.-D., 1997. A model for the formation of hydrothermal manganese crusts from the Pitcairn Island hotspot. Geochimica et Cosmochimica Acta, 61, 4583–4597.

been recognized from variations in porosity and abundance of detrital material, are characterized by significant differences in abundances of all metalliferous elements. It is remarkable that abundances of Mn (and Ni, Co, Pb) decrease in the most porous upper layers, where Fe (and Zn, Cu) contents increase. Such differences may result either from changes in hydrothermal activity or bottom water conditions.

Hydrogenous crusts rapidly increase in importance with distance from the hydrothermal source area. In the central Pacific Ocean, for example, they represent by far the most widespread polymetallic deposits on seamounts and plateaus.

13.1.3. Polymetallic Nodules

Polymetallic nodules are commonly found in the deepest oceanic basins of more than 4,500 m water depth, where very low sedimentation rates of a few millimeters per million years are associated to an efficient bottom circulation. They are however more abundant on abyssal hills and seamounts than in abyssal lows. Polymetallic nodules are concretions of irregular, spheroidal to ellipsoidal shape, which commonly are 20–80 mm in size. Smaller nodules are generally sub-spheroidal, the degree of sphericity decreasing as the nodule size increases. This suggests that nodules grow more rapidly along their horizontal axes, progressively taking a sub-ellipsoidal shape. However, larger nodules sometimes show irregular shapes. The upper surface of the nodules (which is in contact with seawater) may be smooth to finely granulated. In contrast, the lower surface of the nodules (which is in contact with sediments) may be rough and coarsely granulated. The surface texture of polymetallic nodules also varies at regional scale, and nodules may have overall smoother surfaces in some areas compared to others. In the central Indian Ocean, for example, polymetallic nodules in the deepest abyssal plain have a coarser surface texture than those in elevated areas.

In cross-section, polymetallic nodules generally show a succession of concentric layers of varied thicknesses enclosing a nucleus, which acts as an initial adsorption surface for metalliferous oxides and hydroxides (Figure 13.4). Nuclei generally consist of rock fragments such as basalt or palagonite, biogenic fragments such as shark teeth or ear bones, and ancient, smaller nodules. The concentric layers consist of alternating bands of lighter and darker color and sometimes include biogenic remains. In detail, the microstructure of polymetallic nodules is of laminar, dendritic and globular types (Figure 13.5), which may eventually interfere. Discrete laminar microstructures generally lack lateral continuity and seem preferentially associated to finely granulated surfaces, whereas dendritic microstructures rather characterize coarsely granulated surfaces. Globular microstructures commonly develop around biogenic or terrigenous detritus incorporated into the nodules. These inclusions are commonly altered by intranodule diagenesis, the microfossils being also filled by metalliferous oxides. Occasionally, micro-unconformities reflect depositional hiatuses

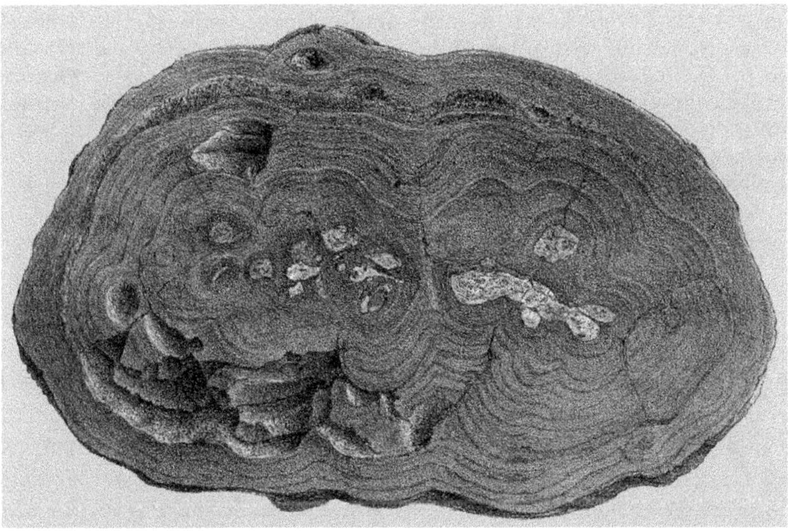

Figure 13.4 Cross-section of a polymetallic nodule. Note the presence of several nuclei and the number of concentric layers. Modified from Murray J., Renard, A.F., 1891. Report on Deep-Sea Deposits, Based on Specimens Collected During the Voyage of *H.M.S. Challenger* in the Years 1872–1876. Eyre and Spottiswoode, London.

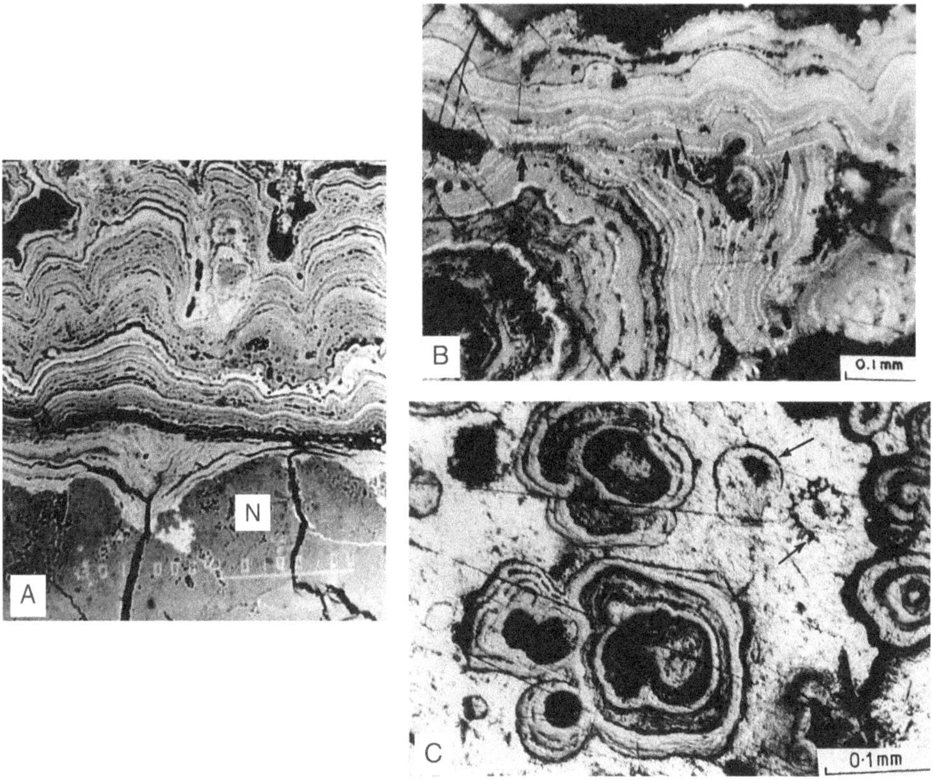

Figure 13.5 Microstructure of a polymetallic nodule. (A) Laminar microstructures resting onto a nucleus (N) and grading to dendritic microstructures outwards. (B) Unconformity (black arrows) truncating pre-existing laminae of todorokite (light gray) and vernadite (dark gray), and capped by further laminae of the same minerals. (C) Globular microstructures of todorokite (light gray) and vernadite (dark gray) enclosing microfossils. Other microfossils are filled with todorokite (black arrows). Reprinted from Banerjee, R., Roy, S., Dasgupta, S., Mukhopadhyay, S., Miura, H., 1999. Petrogenesis of ferromanganese nodules from east of the Chagos Archipelago, Central Indian Basin, Indian Ocean. Marine Geology, 157, 145–158.

or erosion, suggesting that the growth of polymetallic nodules is not continuous. Many nodules also show radial cracks which may either concern specific depth zones of the nodule, or the entire nodule. Radial cracks frequently displace pre-existing microstructures, and are generally filled with secondary material. The most abundant minerals found in polymetallic nodules are vernadite and todorokite, associated to other metalliferous oxides and hydroxides such as feroxyhyte.

The mineralogical and chemical composition of polymetallic nodules suggest that they form via either hydrogenetic precipitation from seawater or diagenetic remobilization of metalliferous elements principally. In fact, individual nodules often represent mixtures of material derived from different accretionary processes. Overall, nodules or nodule layers where vernadite dominates may also contain minerals of various origins such as phillipsite, quartz and feldspars. They are enriched in Fe and Co and their Mn/Fe ratio is around one, which is close to the values found in seawater. This material is considered as representative of hydrogenous processes, i.e., direct precipitation from seawater. In contrast, nodules or nodule layers where todorokite dominates show enrichment in Mn, Ni, Cu, and Zn. They have Mn/Fe ratios significantly greater than two and are considered as indicative of diagenetic accretionary processes (Figure 13.6). In this case, metalliferous oxides and hydroxides are made of chemical elements released into pore waters by the diagenesis of underlying sediments. However, polymetallic nodules strongly influenced by hydrothermal processes and containing abundant birnessite are also observed in areas adjacent to hydrothermal sources.

The hydrogenetic or diagenetic character of polymetallic nodules seem principally controlled by sedimentary environments. In some cases, the upper and lower surfaces of polymetallic nodules show significant differences in composition: the smooth to finely granulated upper surface largely reflects hydrogenetic processes, whereas the rough and coarsely granulated lower surface indicates a dominance of diagenetic processes. This is interpreted as the result of an active circulation of oxygenated bottom waters, combined to oxic diagenesis within the underlying pelagic sediment. In other cases, polymetallic nodules are alternately composed of microlayers with diagenetic and hydrogenetic characteristics. It has been suggested that the lowering of pH during the formation of Mn-rich diagenetic minerals such as todorokite might in turn favor the precipitation of other elements such as Fe and Si. Therefore, these nodules primarily formed by diagenetic

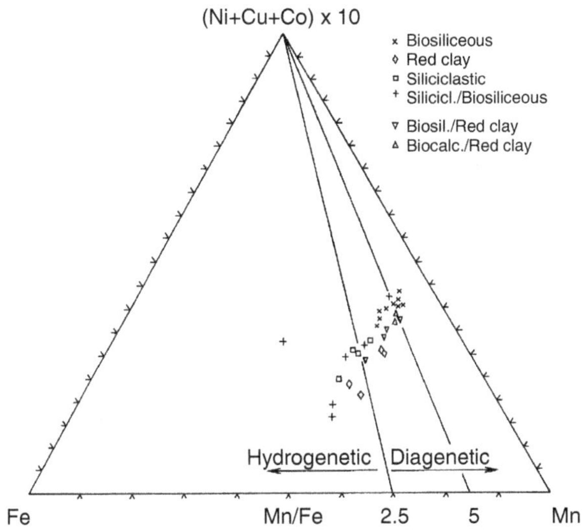

Figure 13.6 Chemical composition and dominant accretion processes of polymetallic nodules. Note that the diagenetic nodules are characterized by high Mn/Fe ratios and contain more Co, Cu and Ni than hydrogenetic nodules. Also, hydrogenetic accretion dominates on siliciclastic and red clay substrates, whereas diagenetic accretion dominates on biogenic substrates. Reprinted from Banerjee, R., Roy, S., Dasgupta, S., Mukhopadhyay, S., and Miura, H., 1999. Petrogenesis of ferromanganese nodules from east of the Chagos Archipelago, Central Indian Basin, Indian Ocean. Marine Geology, 157, 145–158.

accretion during some periods, and subordinately by hydrogenetic accretion. In many oxic environments however, diffusive fluxes of remobilized metals to the water–sediment interface are currently insufficient to account for their accumulation in diagenetic nodules. In such environments, metalliferous oxides often precipitate within oxygenated sediments at the periphery of oxygen-depleted microenvironments, i.e., burrows and organic remains (Figure 13.7). For example, micronodules are considered as derived from the precipitation of metal ions remobilized in oxygen-depleted microenvironments nearby. Also, higher quantities of metal ions could be transferred to the water–sediment interface during intermittent mixing of the uppermost sediment layer by bioturbation or bottom current activity. In this case, metal ions remobilized from the sediment accrete as diagenetic microlayers onto nodule surfaces during intervals of active bioturbation and/or bottom water activity, whereas hydrogenetic microlayers accrete during intervals of lower bioturbation and/or bottom water activity (Figure 13.8). Therefore, the nature of surface sediments and the hydrology of the ocean, as well as their control processes such as productivity and climate, may exert a major influence on the formation, composition and distribution of deep-sea polymetallic nodules.

Polymetallic nodules frequently show a succession of laminar, dendritic and/or globular microstructures, and hiatuses, arranged as to form an apparent pattern of cyclic growth. In the central Pacific Ocean, some nodules expose four orders of cyclic growth pattern: laminae bands, laminae zones, laminae groups and laminae pairs (Figure 13.9). The detailed chemical composition of such a nodule, along a 2.6-mm thick succession of six laminae bands, shows that abundances of Mn and Fe

Figure 13.7 Burrow tube coated with metalliferous oxides and connected to a polymetallic nodule. Reprinted from Jung, H.-S., Lee, C.-B., 1999. Growth of diagenetic ferromanganese nodules in an oxic deep-sea sedimentary environment, northeast equatorial Pacific. Marine Geology, 157, 127–144.

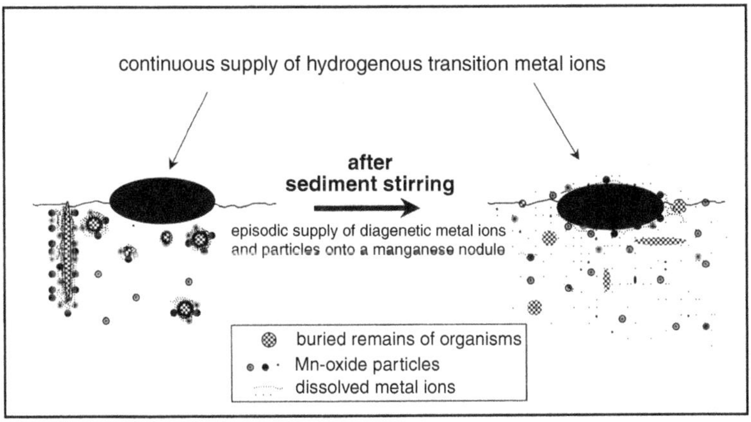

Figure 13.8 Schematic model summarizing the diagenetic and hydrogenetic accretion of metal ions and particles onto a nodule surface in oxygenated sedimentary environments of the deep ocean. Note continuous supply of hydrogenous metal ions, concentration of metalliferous oxide particles around oxygen-depleted microenvironments, and their transfer onto nodule surfaces via mixing of surface sediments. Reprinted from Jung, H.-S., Lee, C.-B., 1999. Growth of diagenetic ferromanganese nodules in an oxic deep-sea sedimentary environment, northeast equatorial Pacific. Marine Geology, 157, 127–144.

are generally anticorrelated, reflecting the variability of diagenetic and hydrogenetic processes. A few intervals of low Mn and Fe contents coincide with maximum abundances of Si and Al, indicating the presence of detrital particles, and a discontinuity is visible between Laminae Bands III and IV. Spectral analyses conducted on the geochemical signals for Laminae Bands IV–VI, where no discontinuity is observed, reveal significant spectral peaks which are close to the average thicknesses of laminae zones and groups. When spectral data are tuned to Milankovitch orbital series, the growth cycles of laminae zones (41.5 kyr) and groups (24 kyr) are close to the periodicities of obliquity and precession, respectively. In addition, they suggest a growth cycle of about 91 kyr for laminae bands, and an average growth rate of 4.5 mm/Myr for this section of the nodule. Besides, radiometric ages obtained from Laminae Bands IV–VI suggest an average growth rate of 4.6 mm/Myr, the entire succession of laminae bands extending from 50–570 ka in age. This strongly supports the hypothesis that the nodule may record Milankovitch orbital signatures. With some exceptions like isotope stage 6, Mn contents are on the average somewhat higher, and Fe contents lower, during glacial than during interglacial intervals (Figure 13.10). This may reflect an increased importance of diagenetic processes during glacials when the circulation of Antarctic Bottom Water is more active, and hydrogenetic processes during interglacials. In addition, concomitant maxima in Si and Al, indicating the presence of terrigenous particles, increase in frequency during interglacials, suggesting enhanced contribution of eolian dust and/or reduced bottom current activity.

The formation of polymetallic nodules is controlled by a variety of parameters directly related to local and regional oceanographic environment such as marine productivity and hydrology and characteristics of surface sediments (which evolve

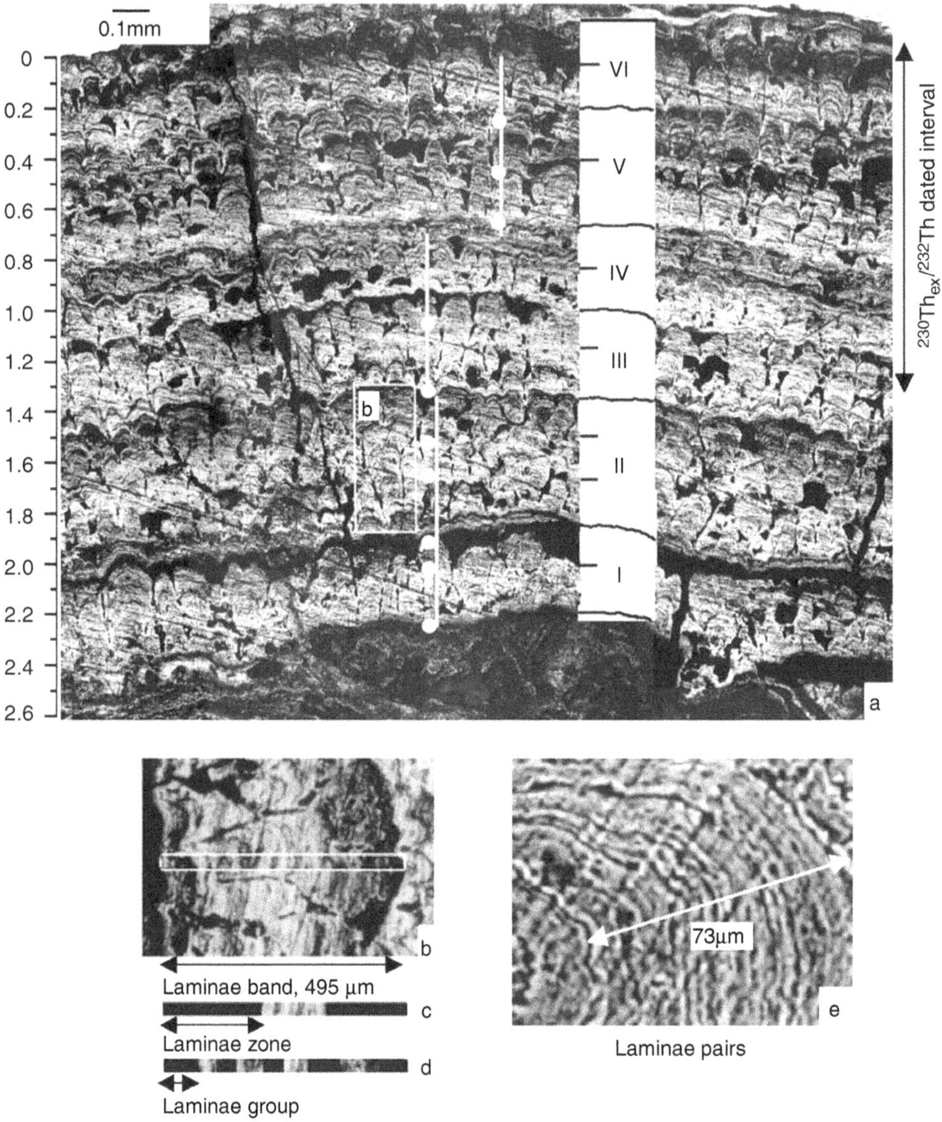

Figure 13.9 Cyclic growth pattern of the outer coating of a central Pacific polymetallic nodule. Roman numbers correspond to Laminae Bands I–VI (see text). Geochemical series shown on Figure 13.10 were obtained along the white lines. Note the hierarchy of laminae: Laminae Band II (b) includes three Laminae Zones (c) made of two to four Laminae groups (d) which consist of a number of laminae pairs (e). Reprinted from Han, X., Jin, X., Yang, S., Fietzke, J., Eisenhauer, A., 2003. Rhythmic growth of Pacific ferromanganese nodules and their Milankovitch climatic origin. Earth and Planetary Science Letters, 211, 143–157.

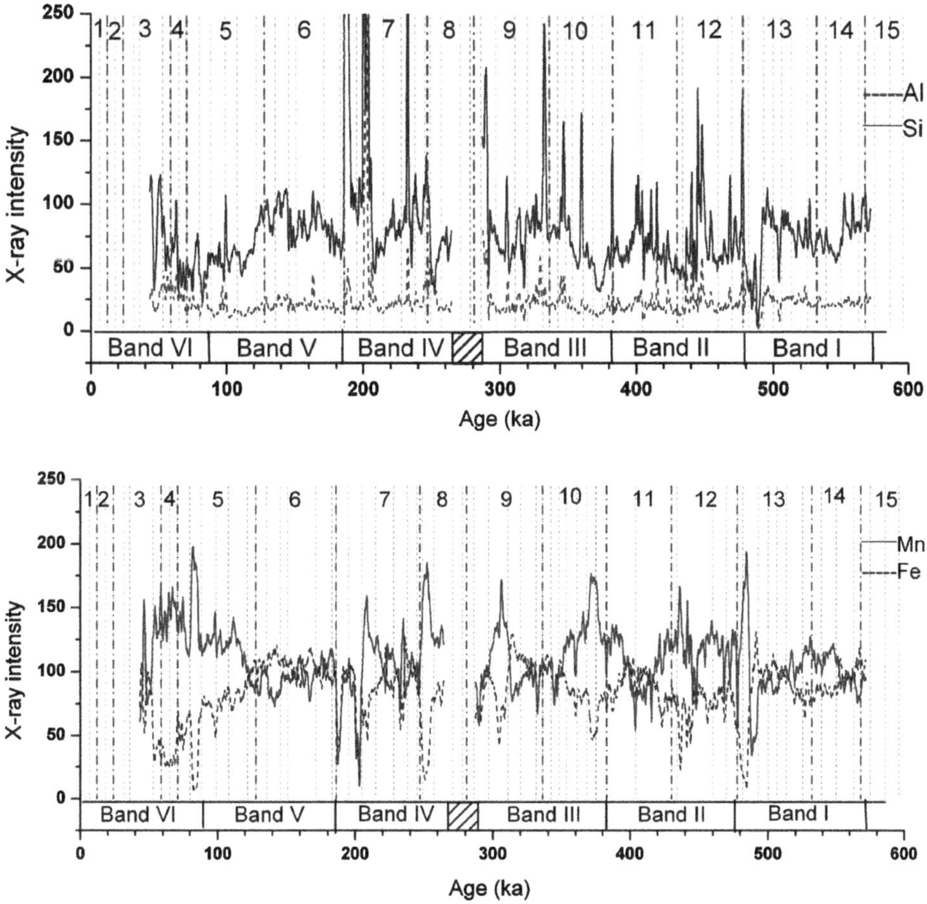

Figure 13.10 High-resolution elemental profile of the outer coating of a central Pacific polymetallic nodule. The profile corresponds to Laminae Bands I–VI on Figure 13.9. Ages (see Section 13.1.3 for dating method) and Pleistocene isotope stages are shown. Note apparent cyclicities on records, anticorrelation between Mn and Fe, and high contents of Si and Al during many intervals of low Mn and Fe. Also, average Mn contents are commonly higher during glacials. Reprinted from Han, X., Jin, X., Yang, S., Fietzke, J., Eisenhauer, A., 2003. Rhythmic growth of Pacific ferromanganese nodules and their Milankovitch climatic origin. Earth and Planetary Science Letters, 211, 143–157.

with global climate), as well as to the availability of metal elements (which are of hydrothermal origin principally). As a consequence, the distribution, nature and composition of polymetallic nodules may change significantly at ocean scale.

- Polymetallic nodules from areas of active intraplate volcanism and hydrothermalism of the central Indian Ocean are characterized by higher contents in metal elements such as Co, Cu, Ni and Zn. This is independent of the diagenetic or hydrogenetic nature of the nodules, suggesting that bottom water and surface sediment both benefited from metal enrichment related to volcanic and hydrothermal activity. Besides, the nodules from elevated areas such as inactive seamounts contain higher concentrations in Fe and Co whereas the nodules from the abyssal plain have higher values of Mn, Ni and Cu, suggesting that diagenetic processes dominate in the deeper abyssal plain whereas hydrogenetic processes increase in importance in elevated areas.

- When the chemical composition of metal elements (especially the Mn/Fe ratio and the contents of Mn, Ni and Cu relative to those of Fe and Co) are compared to the composition of the surface sediment, the degree of diagenetic accretion in the nodules increases with the percentage of biogenic remains (especially siliceous microfossils) in the sediment (Figure 13.6). In other words, hydrogenetic accretion processes dominate on terrigenous/red clay substrates, whereas diagenetic accretion processes dominate on biosiliceous substrates. This is confirmed by the dominance of todorokite in nodules resting on biogenic sediments.
- In areas of high primary productivity, a sufficiently large proportion of organic matter may reach the seafloor, where degradation of organic compounds results in the formation of oxygen-depleted sedimentary environments. The related remobilization and transfer of metal ions to the water–sediment interface facilitates the diagenetic accretion of nodules. This is, for example, the case in the highly productive equatorial Pacific Ocean, where polymetallic nodules of diagenetic origin dominate. There, polymetallic nodules are found below the carbonate compensation depth where carbonate dissolution limits the particle rain to the seafloor. The resulting low sedimentation rates provide conditions favorable to the development of nodules. The supply of organic matter to the seafloor decreases with distance from the area of maximum productivity, as does the importance of diagenetic accretion. In contrast, hydrogenetic accretion processes increase in importance with distance from the productive area, and progressively dominate nodule formation. As the composition of seawater is relatively constant throughout the region, this spatial trend most probably reflects a dilution by diagenetic accretion processes.

13.1.4. Other Types of Metalliferous Accumulations

Accumulations of polymetallic crusts and nodules are generally found near the sediment–water interface of deep oceanic environments. However, metalliferous deposits may also occur in shallow, nearshore environments, as well as within hemipelagic sequences which accumulated at higher rates than the deep sediments of the central oceanic basins.

13.1.4.1. Baltic Sea polymetallic nodules

Polymetallic nodules are for example found at shallow water depth (15–20 m) of the Mecklenburg Bight, western Baltic Sea. This region is characterized by a stratification of the water column in summer, associated to oxygen depletion in bottom waters as the degradation of organic matter progresses. The ventilation and oxygenation of the water column improve during the fall due to frequent stormy weather, whereas the sediment remains anoxic all the year round. In winter, Mn and other metals slowly dissolve within the anoxic sediment, migrate upwards with oxygen-depleted pore waters and precipitate onto suspended particles of the nepheloid layer as they drift into oxygenated bottom waters. When water stratification develops in summer and anoxia extends to bottom waters, fluxes of metals from the sediment and dissolution within an anoxic nepheloid layer result in significant enrichment of the anoxic bottom waters in dissolved metals (Figure 13.11). Locally, concentrations in Mn may increase up to 200 times higher than the average value in oxygenated waters. As mixing and oxygenation of the entire water column progress in autumn, the metals precipitate around particles or form a new layer around pre-existing nodules. Nodule formation occurs in elevated areas, which are relatively more oxygenated. These nodules increase very rapidly in size, their average growth rate being about 2 mm/kyr.

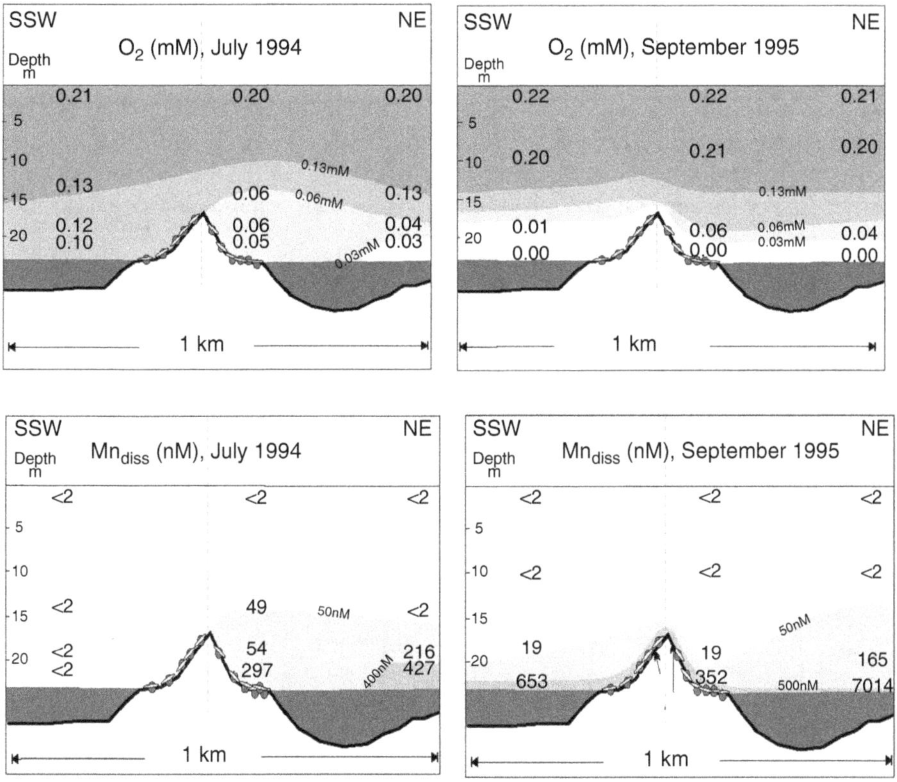

Figure 13.11 Oxygen and manganese concentrations in the water column of the Mecklenburg Bight of the western Baltic Sea at the beginning (July) and the end (September) of the stratification period. Note significant increases in Mn near the seafloor as water stratification and bottom anoxia progress, and the presence of nodules in elevated areas which are more sensible to winter ventilation and related precipitation of metalliferous oxides. Modified from Hlawatsch, S., Neumann, T., Van den Berg, C.M.G., Kersten, M., Harff, J., Suess, E., 2002. Fast-growing, shallow-water ferro-manganese nodules from the western Baltic Sea: origin and modes of trace element incorporation. Marine Geology, 182, 373–387.

Whereas the assemblages of chemical elements in Baltic Sea and deep-sea nodules are rather similar, concentrations in Co and especially Zn are significantly higher in Baltic Sea nodules. This is a probable consequence of anthropogenic metal input into the Baltic Sea.

13.1.4.2. East Mediterranean metalliferous layers

East Mediterranean sapropels, and especially the most recent Holocene sapropel (S1), are usually overlain by two discrete layers enriched in metalliferous oxides. Mediterranean sapropels are recognized with their dark color, due to high contents in organic material. However, sapropels are generally thinner than initially, because post-depositional oxidation removed part of the original organic material in the upper section of the sapropels (Figure 13.12). The lower metalliferous layer marks the limit of the residual sapropel and almost certainly corresponds to metal oxide precipitation at the boundary between anoxic and oxic sedimentary environments. Metal oxide precipitation probably followed the return of oxygenated bottom waters, penetration of oxygenated waters into the sediment and subsequent oxidation of the uppermost initial sapropel. The lower metalliferous layer clearly results from diagenetic alteration.

In contrast, the upper metalliferous layer marks the limit of the initial sapropel, as defined by geochemical and productivity indicators such as Ba concentrations

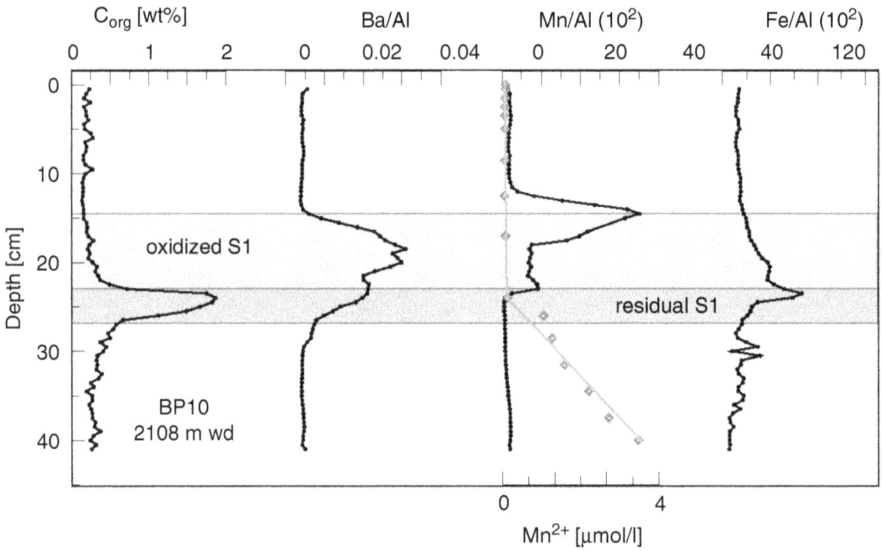

Figure 13.12 Variability of organic carbon and metal elements in Holocene Mediterranean sapropel S1 and adjacent sediment, as recorded off Lybia in the Ionian Sea. Mn^{2+} values are from pore waters. Initial extension of Sapropel S1 is estimated from organic carbon and Ba concentrations. Note increased contents in Mn (from Mn/Al ratios) at the current and initial upper boundaries of the sapropel. Reprinted from Reitz, A., Thomson, J., De Lange, G.J., Hensen, C., 2006. Source and development of large manganese enrichments above eastern Mediterranean sapropel S1. Paleoceanography, 21, doi:10.1029/2005PA001169.

(Figure 13.12). The upper metalliferous layer seems related to the return of oxygenated waters at the end of sapropel times. It is probable that dissolved Mn and other metals were transferred from anoxic sediment to oxygen-depleted bottom waters while sapropel sediment accumulated, and that metal oxides precipitated onto sediment particles as bottom water ventilation and related oxygen contents increased at the end of sapropel formation.

By far, the largest concentrations in metalliferous elements (especially Mn) are found at localities between 1,100 and 2,000 m water depth (Figure 13.13). This does not reflect the degree of anoxia and related remobilization of metals since intermediate-depth sapropels are relatively poor in organic material, highest concentrations being recorded at maximum depths of the basins. It is likely that higher quantities of dissolved metals were in fact released from the deepest sediment into bottom waters during sapropel formation. Dissolved metals migrated within deep, oxygen-depleted waters, where concentrations increased. Precipitation of metal

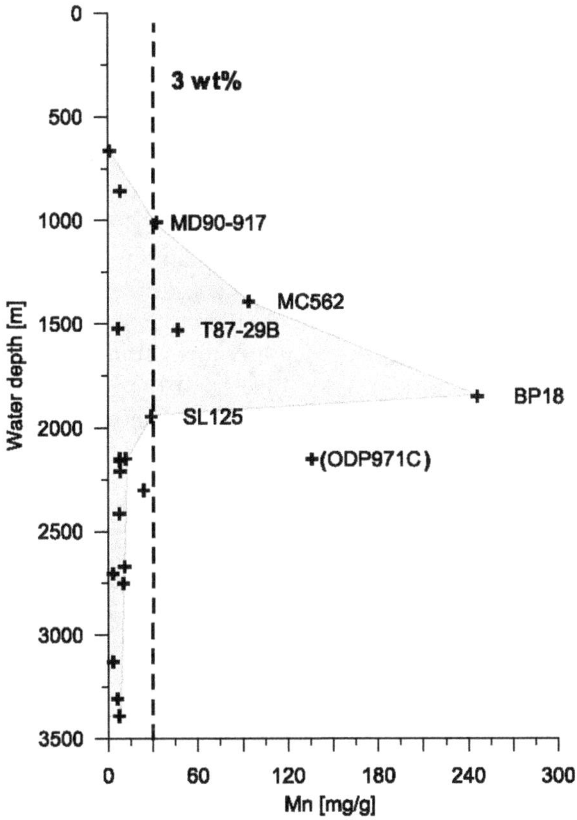

Figure 13.13 Mn contents in the upper metalliferous layer of Sapropel S1, at different water depths of the East Mediterranean Sea. Note maximum concentration between 1,100 and 2,000 m water depth. Reprinted from Reitz, A., Thomson, J., De Lange, G.J., Hensen, C., 2006. Source and development of large manganese enrichments above eastern Mediterranean sapropel S1. Paleoceanography, 21, doi:10.1029/2005PA001169.

oxides probably occurred preferentially at the transition depth between deep, oxygen-depleted waters and more oxygenated, intermediate waters, the particulate oxides settling onto higher topography being better preserved from further dissolution.

13.2. Glauconite and Other Green Clays

Green clays include a variety of clay minerals which contain a relatively high proportion of iron and potassium. They occur as sand-sized grains, principally found in shelf sediments bathed by hydrodynamically active, oxygenated waters (Figure 13.14). Most of the time green clays fill microfossil shells, bioclasts and fecal pellets, or occur as spherules within rock fragment and sediment pore space. Besides, green clays may occur on a variety of substrates such as micas, lithoclasts, and hardgrounds. Green clays look closely associated to relatively confined microenvironments.

Occurrences of green clay grains of diverse chemical and mineralogical compositions within a single area suggest that they evolved from mineral assemblages similar to those found in adjacent sediment. In surface sediments from the Gulf of Guinea for instance, gray fecal pellets contain the same clay minerals as adjacent sediments, whereas higher contents of iron-rich clay minerals such as Fe-smectite are recorded in pellets of lighter (ochre to light-green) color, lower organic carbon contents, and better dissolution of biogenic carbonate remains. It is likely that authigenic iron-rich clays develop within the pore space created via the dissolution of particulate organic matter and biogenic carbonate. In detail, green clays develop progressively, providing that they remain for long periods of time near an interface between oxidizing and reducing sedimentary environments. It is estimated that minimum time spans of 1–10 kyr are necessary for green clay formation. During the early stages of green clay formation, minor amounts of iron-rich minerals (Fe-smectite principally) evolve from the initial clay assemblage, Fe being remobilized from oxygen-depleted sediments below. By the same time, the

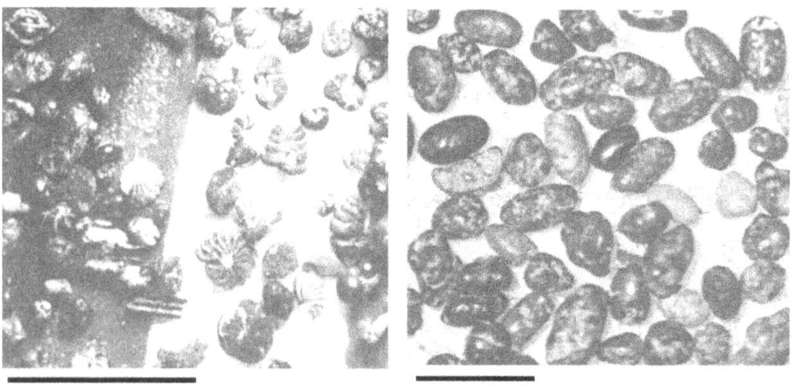

Figure 13.14 Green clay grains from the Senegal Shelf of the central Atlantic Ocean. Scale bar: 1 mm. Left: sample with dominant microfossil infillings; right: fecal pellets. Modified from Odin, G.S. (Editor), 1988. Green Marine Clays, Elsevier, Amsterdam.

proportion of potassium slightly increases. Authigenic iron-rich minerals progressively fill the pore space of the grain, the minerals being still poorly crystallized. At this stage, potassium contents generally increase to 4–6%. This is followed by a series of recrystallizations which progressively erase the original structure of the grain. This is associated to an increase in volume, formation of fractures, and further incorporation of potassium which may increase to 6–8%. Fractured dark-green grains from the Gulf of Guinea contain clay minerals enriched in both iron and potassium only, with no trace of the original terrigenous clay assemblage. Ultimately, authigenic minerals may fill the fractures and further, diagenetic recrystallization of the clays may occur after burial of the green clay grains.

Two major types of green clays are recognized in modern, shallow marine environments, based on differences in mineralogy, conditions of formation and distribution:

- Verdine is generally found in low latitude environments (Figure 13.15), from nearshore areas to about 200 m water depth, maximum concentrations being recorded between 20 and 60 m water depth. In such areas terrigenous clay assemblages generally contain significant amounts of kaolinite, commonly associated to smectite and other clay minerals. The main constituent of verdine is phyllite V, a poorly crystallized 7-Å thick clay mineral which contains abundant Fe (18–25%). This mineral is associated to other clays such as phyllite C, a 14.5-Å

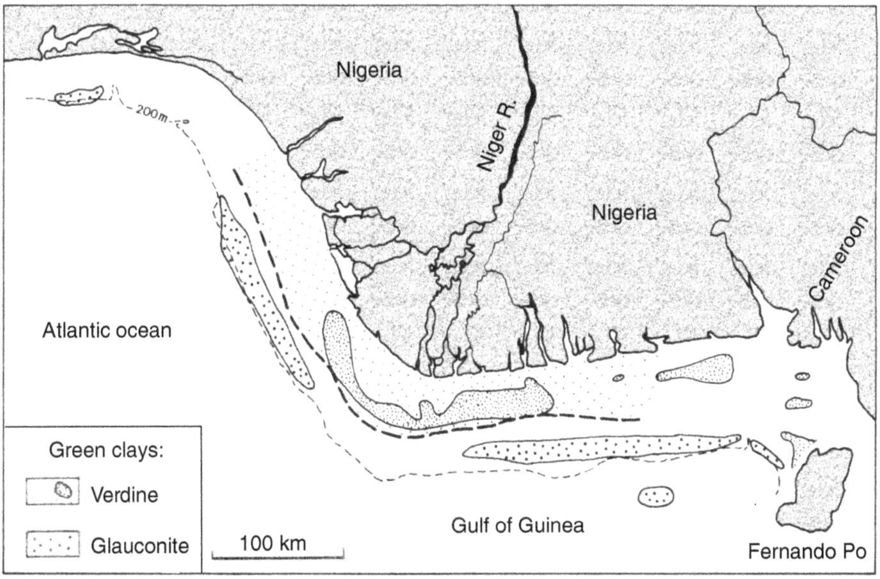

Figure 13.15 Distribution of verdine and glauconite grains on the continental shelf off the Niger Delta, Atlantic Ocean. Note occurrences of verdine at shallower depths of the shelf, in areas most influenced by freshwater discharges and terrigenous sources (areas of higher dot densities represent verdine concentrations above 2% of the bulk sediment), and glauconite at greater water depths, in areas less influenced by river discharges. Modified from Odin, G. S. (Editor), 1988. Green Marine Clays, Elsevier, Amsterdam.

thick clay mineral which is very close to phyllite V in composition. Verdine usually develops in areas strongly influenced by freshwater discharges and/or of terrigenous substrates, for example, the periphery of river mouths and mangroves. It is likely that abundant particulate iron brought to the ocean by run-off and remobilized within oxygen-depleted sedimentary environments facilitates the formation of verdine, providing that sedimentation rates are sufficiently low.

- Glauconite grains can be found on continental shelves and other elevated areas of most latitudes (Figure 13.15), maximum concentrations being recorded between 150 and 300 m water depth. Glauconite formation may also sometimes occur at relatively deep water depth. For example, glauconite formation took place at several occasions on the Ivory Coast–Ghana Ridge (2,100 m water depth) of the equatorial Atlantic Ocean during the Pleistocene. Glauconite may therefore evolve from a variety of clay assemblages. The main constituent of glauconite grains is the glauconite mineral, a 10-Å thick clay mineral which contains more than 20% Fe and more than 4% K. Poorly evolved glauconite grains may also contain Fe-smectite and/or a variety of Fe-rich random mixed-layered clays. Glauconite is less dependent on fluvial input than verdine, and principally occurs in environments of active bottom water circulation.

Verdine and glauconite are characteristic of distinct and specific environments. Because of its association with low sedimentation rates and active bottom circulation, glauconite is considered a reliable indicator of current-induced hiatuses, providing that the grains and adjacent sediment do not show traces of reworking.

Glauconite being an indicator of specific environmental conditions, its conditions of formation have been investigated in detail. First of all, K and Rb contents progressively increase during glauconitization, and are good indicators of the degree of maturity of glauconite grains (Figure 13.16). However, the chemical composition of glauconitic material seems significantly influenced by the nature of the initial substrate. For instance, the progressive increase in K of glauconitic material that forms on carbonate-rich substrates is often associated to an intake in Mg, Si and Fe, and release of Ca. In contrast, the progressive increase in K of glauconitic material found on siliciclastic substrates is associated to an intake in Fe, significant release of Al and slight release of Ca and Mg, but does not show significant correlation with contents in Si (Figure 13.16). In some cases, occurrences of relict minerals such as quartz may affect these relationships. Yet, it is remarkable that in all cases highest contents in potassium (above 7%) are associated to decreased contents in iron.

The chemical evolution of glauconite grains is associated to a comparably significant mineralogical evolution. On the Ivory Coast–Ghana Ridge of the equatorial Atlantic Ocean, for example, the clay fraction of Pleistocene sediments is dominated by kaolinite and smectite, associated to illite, quartz and calcite. There, only minor amounts of kaolinite occur in light-green grains where Fe-smectite dominates, whereas smectite–glauconite mixed-layered clays are major minerals in mature dark-green grains. Also, sediments from the Gulf of Lions in the western Mediterranean Sea contain a terrigenous clay fraction dominated by illite, associated to a variety of minerals including chlorite, smectite, kaolinite, quartz, feldspars and calcite. Similar assemblages are found in beige fecal pellets from areas of high

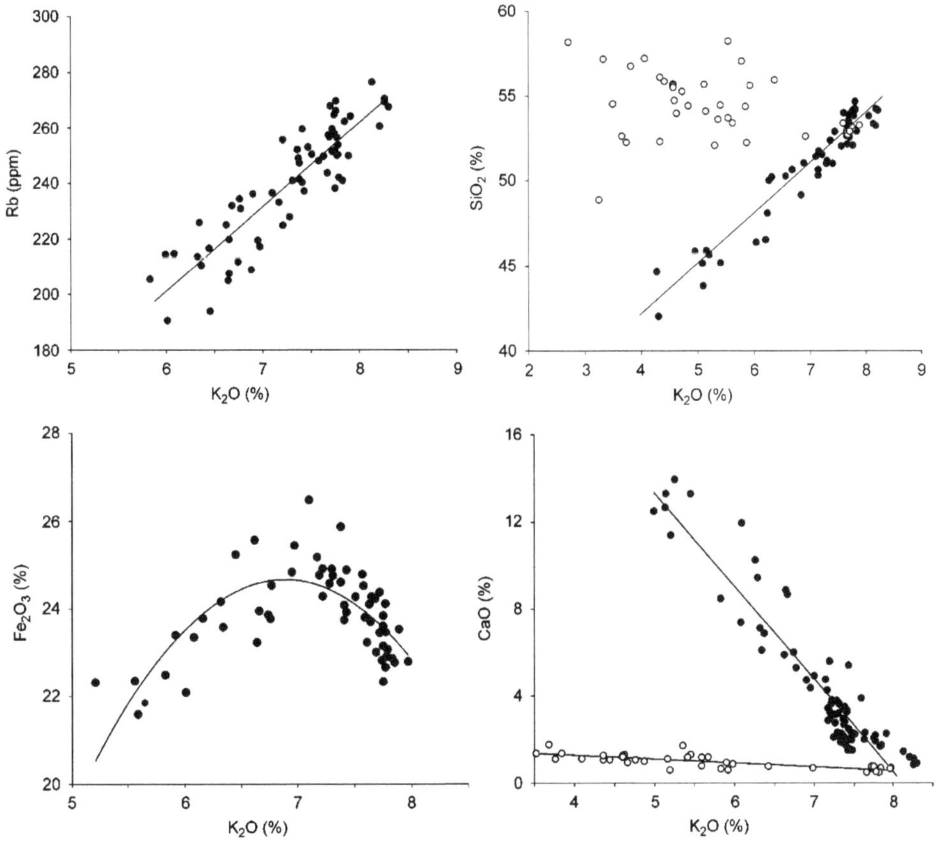

Figure 13.16 Variations of Rb, Si, Fe, and Ca relative to K in glauconite grains of various maturities. Black dots: glauconite grains associated to carbonate-rich substrates; white dots: glauconite grains associated with siliciclastic substrates. Note differences in Si uptake and Ca release for glauconite grains from carbonate-rich and siliciclastic substrates, and release of Fe for mature glauconite grains of high K content. Modified from Amorosi, A., Sammartino, I., Tateo, F., 2007. Evolution patterns of glaucony maturity: a mineralogical and geochemical approach. Deep-Sea Research II, 54, 1364–1374.

sedimentation rates off the Rhône Delta, but green fecal pellets in areas and intervals of low sedimentation rates contain significantly different minerals: Fe-smectite (nontronite) dominates in light-green grains, whereas nontronite–glauconite mixed-layered clays dominate in more mature dark-green grains. It is probable that increased concentrations of chemical elements within the pellets due to partial dissolution of terrigenous minerals and intake from the sediment was followed by the development of new minerals in the form of tiny crystals. Also, the intake of Fe (associated to Fe-smectite) probably preceded the intake of K, which is specifically associated to glauconite clay layers.

Glauconite (and other green clay) grains are generally concentrated on the seafloor or within specific intervals of the sedimentary column, whereas the substrates from which they evolve (microfossils, fecal pellets, etc.) are generally

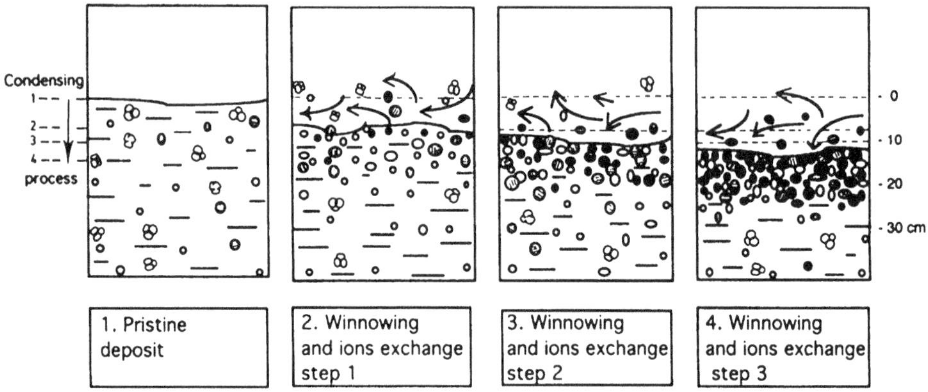

Figure 13.17 Simplified model for the formation and accumulation of glauconite grains. The grains increase in darkness with maturity. Note that persistent winnowing and erosion by bottom currents facilitate both the maturation and concentration of glauconite grains. Reprinted from Giresse, P., Wiewióra, A., 2001. Stratigraphic condensed deposition and diagenetic evolution of green clay minerals in deep water sediments on the Ivory Coast–Ghana Ridge. Marine Geology, 179, 51–70.

scattered within the sediment. In addition, accumulations of glauconite grains commonly include samples that show distinct degrees of maturity. It is likely that continuing bottom current activity preferentially removes the fine sediment fraction over great distances. In contrast fecal pellets, sandy microfossils filled with sediment and glauconite grains, which increase in density with maturity, remain near the seafloor where glauconitization proceeds. Bottom conditions such as current velocity and seafloor morphology may provide local and/or episodic conditions favorable to the accumulation of glauconite grains (Figure 13.17). Therefore, the mechanisms involved in the development of glauconite grains and minerals are also involved in their concentration and accumulation. In passive margin sediments, accumulations of mature glauconite grains frequently occur above unconformities, within condensed sections at the base of transgressive sequences of prograding continental shelves. There, glauconitization is favored by low accumulation rates during intervals of increasing relative sea level and subsequent intervals of maximum sea level (stillstand), when terrigenous loads bypass the shelf to accumulate on the continental slope principally (see Section 6.1.2). This explains the preferential occurrence of glauconite grains in shallow shelf environments of the modern ocean. In contrast, regressive sequences may contain small quantities of reworked glauconite grains only.

13.3. Phosphates and Phosphorites

Phosphates are disseminated within marine sediments where their average concentration is about 0.1%. However, phosphate concentrations may locally increase as to form phosphorites, which are sediments containing more than 20% phosphates. Phosphates are principally made of apatite, which in fact is a diversified family of phosphatic minerals: $Ca_5(PO_4)_3 \cdot (F, OH)$. Apatites may also contain other elements, such as Fe and Al. Among apatites, carbonate fluorapatite (francolite) dominates in marine deposits, hydroxyapatite is principally found in bones, and collophane are sedimentary apatites of uncertain structure and composition.

The most important (pre-anthropogenic) source of phosphorus is igneous and related sedimentary rocks that outcrop on continents. Phosphorus is a soluble anion (see Section 10.1) which is associated to oxygen (and partly made bioavailable) in the form of phosphate (PO_4) through chemical weathering reactions. Phosphate concentrations in soils commonly range from 0.02% to 1%. Organic and inorganic phosphate compounds are predominantly particulate, and transported to the ocean by running waters principally, but also by winds. Although a significant proportion of phosphate is rapidly buried together with terrigenous sediments off river mouths, about 25% are transferred to ocean water in a bioavailable form due to desorption reactions. It is estimated that at ocean scale, almost all bioavailable phosphate is used in the photic zone where it is incorporated into organic compounds. This is especially the case in areas of high productivity such as coastal upwellings and oceanic divergences (Figure 13.18), but also deltaic environments. Most phosphate is released in the water column as organic compounds are oxidized (see Section 12.1), and re-used by marine organisms. A minor proportion of about 5% phosphate reaches the seafloor where it interacts with the early diagenetic alteration of organic material. Phosphate being closely associated to organic compounds, its transfer to surface sediments is facilitated in areas of higher organic fluxes to the sediment–water interface (see Section 12.2).

Most phosphorite occurs as nodules of coarse sand to pebble size, but also as peloids and mollusk molds. Among them, pale-brown friable nodules are interpreted to be younger than hard, black nodules: francolite largely dominates

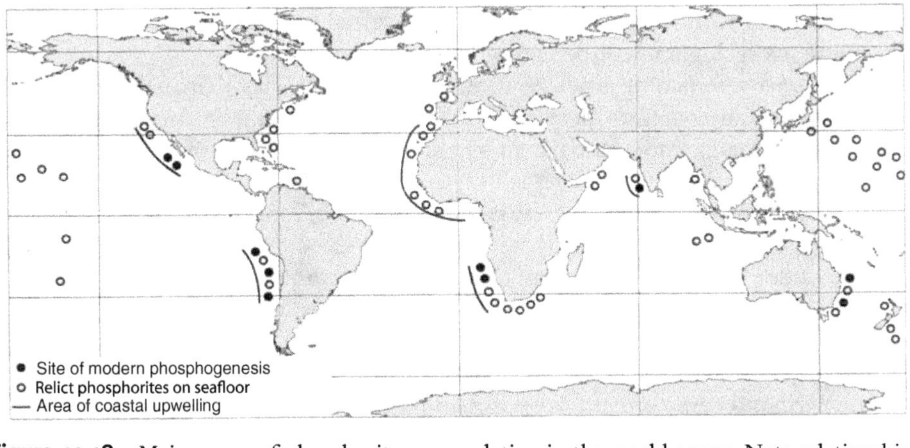

Figure 13.18 Major areas of phosphorite accumulation in the world ocean. Note relationships with areas of coastal upwelling and other regions of high productivity. Note also the importance of ancient phosphorites on the sea floor. Modified from Föllmi, K.B., 1996. The phosphorus cycle, phosphogenesis and marine phosphate-rich deposits. Earth-Science Reviews, 40, 55–124.

in hard nodules but is commonly associated to an amorphous collophane phase in friable nodules, suggesting better lithification of phosphorites through time. This is a direct consequence of the slow kinetics of francolite formation in oceanic environments. Phosphorite nodules may also contain varying amounts of siliciclastics such as quartz, feldspars and clay minerals, authigenic and diagenetic elements such as pyrite and glauconite grains, organic material and/or biogenic carbonates such as foraminifers. Francolite forms by precipitation near the water–sediment interface. Elevated pore-water concentrations in phosphate, that are necessary for francolite formation, are obtained through the degradation of organic material (see Section 12.2). In oxygen-depleted sedimentary environments, the transfer of phosphate from organic compounds to pore waters is facilitated by adsorption/desorption processes onto Mn and Fe particles, which are redox dependent. In some cases, francolite precipitates directly by nucleation and crystallization onto the surface of siliciclastic and/or biogenic elements. In other cases, francolite replaces pre-existing minerals, principally biogenic carbonates. Once formed, initial francolite crystals provide reactive surfaces for further francolite accretion, which may continue as long as environmental conditions in the ocean and on the seafloor remain favorable.

Phosphorite formation near the water–sediment interface being a relatively slow process associated to significant organic fluxes to the seafloor and remineralization of organic material, modern phosphorites are found in areas of high productivity, low oxygen contents on the seafloor, and low sedimentation rates. These conditions are locally met for example on the Atlantic margin of South Africa, where active upwelling cells and an oxygen-minimum zone in the ocean ensure significant organic fluxes to the seafloor, whereas arid conditions on the continent limit the distribution of siliciclastics and help maintaining low sedimentation rates. There, phosphorites occur on the modern shelf and within condensed Pleistocene sediment series that form during intervals of transgression and highstands of sea level, and related migration of active upwelling cells onto the inner shelf. Nevertheless, small nodules of francolite are sometimes scattered within organic-rich sediments of higher sedimentation rates.

In fact, many francolite nodules show complex histories that include multiple episodes of phosphogenesis. On the continental shelf of southeast Africa close to the Cape Canyon for example, a dark and hard nodule provides evidence for several generations of francolite (Figure 13.19). The inner part of the nodule contains sand-sized quartz grains and probably grew rapidly, around 5.5–6.3 Ma. The outer part of the nodule contains abundant pyrite grains and developed around 1.0–1.8 Ma, probably under reducing conditions. The inner and outer part of the nodule are separated by an orange layer containing iron oxides that suggest a period of weathering, probably in subaerial conditions. Therefore, individual francolite nodules may show multiple episodes of accretion sometimes separated by several million years, and can survive reworking and weathering once they are sufficiently lithified. Because of their high density and resistance, well-lithified francolite nodules may concentrate in areas of active bottom circulation and reworking, within condensed series characterized by major changes in lithology, unconformities and hiatuses.

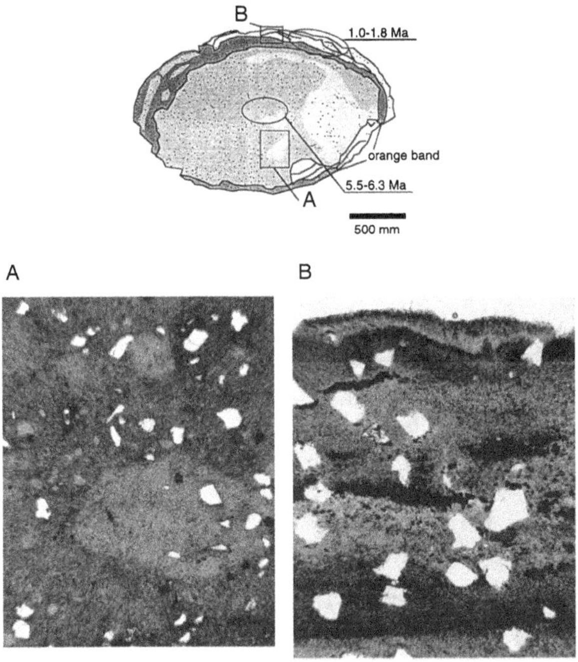

Figure 13.19 Schematic structure of a phosphorite nodule from Pleistocene shelly sands off southeast Africa. Note distinct intervals of phosphogenesis several million years apart, with an interval of probable subaerial weathering (orange band) in between. (A) Detail of the inner part of the nodule, which includes subangular peloids and (B) detail of the outer part of the nodule, with opaque pyrite-rich layers. Note the presence of light, angular to subangular, sand-sized quartz grains scattered throughout. Modified from Compton, J.S., Mulabisana, J., McMillan, I.K., 2002. Origin and age of phosphorite from the Last Glacial Maximum to Holocene transgressive succession off the Orange River, South Africa. Marine Geology, 186, 243–261.

This illustrates complex relationships between phosphogenesis, marine environments and climate. Phosphogenesis started off southeast Africa in the Late Oligocene during a major marine transgression, triggered by globally warmer climate conditions and related decreased volumes of Antarctic ice. This initial interval of phosphogenesis persisted until the earliest Miocene and led to the accumulation of lithified francolite nodules. However, an important interval of phosphogenesis coincides with a major marine regression of Middle and Late Miocene Age caused by the expansion of Antarctic ice sheets. Although a regression would shift the area of higher productivity to the outer shelf and slope, concomitant increases in circulation and upwelling activity probably resulted in an overall enhanced productivity and higher fluxes of organic material to the seafloor. It is remarkable that Late Miocene phosphogenesis occurred during short, third-order highstands of sea level superimposed to the global regression, and that francolite replaced pre-existing biogenic carbonates principally. Since the Pliocene, fluctuations of sea level may have been too rapid to facilitate the accumulation of abundant, lithified phosphorite nodules during transgressions, friable phosphorite nodules being easily fragmented and removed during regressions.

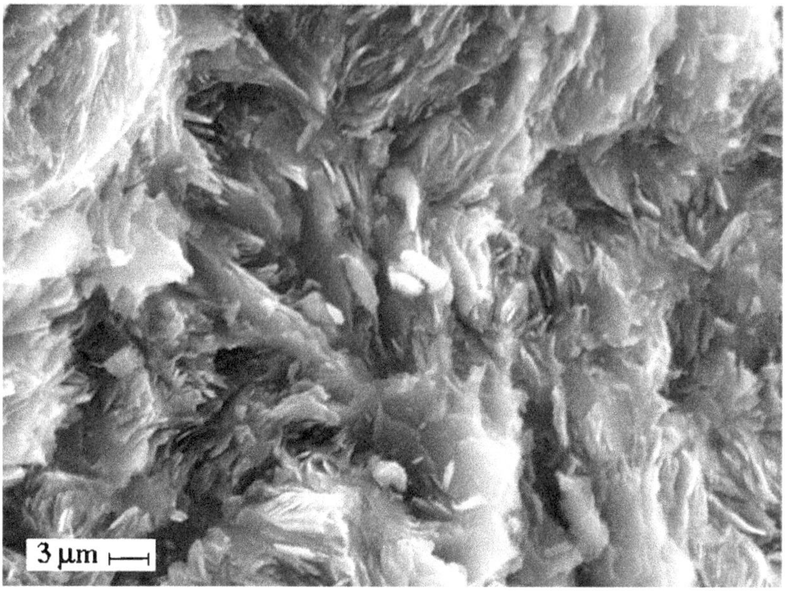

Figure 13.20 Scanning electron microscope image of a francolite–glauconite grain from the southeast African shelf. Note the concomitant presence of globules and bars of francolite, and lamellar glauconite. Reprinted from Wigley, R., Compton, J.S., 2007. Oligocene to Holocene glauconite–phosphorite grains from the head of the Cape Canyon on the western margin of South Africa. Deep-Sea Research II, 54, 1375–1395.

Francolite cements frequently include a variety of particles, among them pyrite and glauconite grains. Pyrite is considered indicative of formation within the sulfate-reduction zone of the surface sediment. The association of francolite with glauconite grains suggests a formation within rather similar sedimentary environments. On the outer continental shelf of southeast Africa in the Cape Canyon area, glauconite–phosphorite grains represent foraminifer tests where carbonate has been replaced by francolite, and infilled by variable proportions of glauconite and francolite (Figure 13.20). Francolite dominates over glauconite in Late Oligocene to Early Miocene biogenic carbonates, whereas glauconite dominates in principally siliciclastic Late Miocene to Pleistocene deposits. This highlights how these authigenic processes are sensitive to a delicate balance between productivity and terrigenous delivery.

FURTHER READING

Chamley, H., 1999. Clay sedimentology. Springer, Berlin.

Einsele, G., 1992. Sedimentary basins, evolution, facies, and sediment budget. Springer, Berlin.

Föllmi, K.B., 1996. The phosphorus cycle, phosphogenesis and marine phosphate-rich deposits. Earth-Science Reviews, 40: 55–124.

Kennett, J.P., 1982. Marine geology. Prentice-Hall, Englewood Cliffs, N.J.

Odin, G.S. (Editor), 1988. Green marine clays. Elsevier, Amsterdam.

Slansky, M., 1986. Geology of sedimentary phosphates. North Oxford Academic, London.

Other references used in this chapter

Amorosi, A., Sammartino, I., Tateo, F., 2007. Evolution patterns of glaucony maturity: a mineralogical and geochemical approach. Deep-Sea Research II, 54: 1364–1374.

Banerjee, R., Roy, S., Dasgupta, S., Mukhopadhyay, S., Miura, H., 1999. Petrogenesis of ferromanganese nodules from east of the Chagos Archipelago, Central Indian Basin, Indian Ocean. Marine Geology, 157: 145–158.

Compton, J.S., Mulabisana, J., McMillan, I.K., 2002. Origin and age of phosphorite from the last Glacial maximum to Holocene transgressive succession off the Orange River, South Africa. Marine Geology, 186: 243–261.

Compton, J.S., Wigley, R., McMillan, I.K., 2004. Late Cenozoic phosphogenesis on the western shelf of South Africa in the vicinity of the Cape Canyon. Marine Geology, 206: 19–40.

Giresse, P., Wiewióra, A., 2001. Stratigraphic condensed deposition and diagenetic evolution of green clay minerals in deep water sediments on the Ivory Coast–Ghana Ridge. Marine Geology, 179: 51–70.

Giresse, P., Wiewióra, A., Grabska, D., 2004. Glauconitization processes in the northwestern Mediterranean (Gulf of Lions). Clay Minerals, 39: 57–73.

Glasby, G.P., Cherkashov, G.A., Gavrilenko, G.M., Rashidov, V.A., Slovtsov, I.B., 2006. Submarine hydrothermal activity and mineralization on the Kurile and western Aleutian Island arcs, N.W. Pacific. Marine Geology, 231: 163–180.

Glasby, G.P., Stüben, D., Jeschke, G., Garbe-Schönberg, C.-D., 1997. A model for the formation of hydrothermal manganese crusts from the Pitcairn Island hotspot. Geochimica et Cosmochimica Acta, 61: 4583–4597.

Han, X., Jin, X., Yang, S., Fietzke, J., Eisenhauer, A., 2003. Rhythmic growth of Pacific ferromanganese nodules and their Milankovitch climatic origin. Earth and Planetary Science Letters, 211: 143–157.

Harris, L.C., Whiting, B.M., 2000. Sequence-stratigraphic significance of Miocene to Pliocene glauconite-rich layers, on- and offshore of the US Mid-Atlantic margin. Sedimentary Geology, 134: 129–147.

Hlawatsch, S., Neumann, T., Van den Berg, C.M.G., Kersten, M., Harff, J., Suess, E., 2002. Fast-growing, shallow-water ferro-manganese nodules from the western Baltic Sea: origin and modes of trace element incorporation. Marine Geology, 182: 373–387.

Jung, H.-S., Lee, C.-B., 1999. Growth of diagenetic ferromanganese nodules in an oxic deep-sea sedimentary environment, northeast equatorial Pacific. Marine Geology, 157: 127–144.

Kasten, S., Glasby, G.P., Schulz, H.D., Friedrich, G., Andreev, S.I., 1998. Rare earth elements in manganese nodules from the South Atlantic Ocean as indicators of oceanic bottom water flow. Marine Geology, 146: 33–52.

Knoop, P.A., Owen, R.M., Morgan, C.L., 1998. Regional variability in ferromanganese nodule composition: northeast tropical Pacific Ocean. Marine Geology, 147: 1–12.

Li, Y.H., Schoonmaker, J.E., 2003. Chemical composition and mineralogy of marine sediments. Treatise on Geochemistry, 7: 1–35.

Mukhopadhyay, R., Iyer, S.D., Ghosh, A.K., 2002. The Indian Ocean nodule field: petrotectonic evolution and ferromanganese deposits. Earth-Science Reviews, 60: 67–130.

Reitz, A., Thomson, J., De Lange, G.J., Hensen, C., 2006. Source and development of large manganese enrichments above eastern Mediterranean sapropel S1. Paleoceanography, 21. doi: 10.1029/2005PA001169

Wen, X., De Carlo, E.H., Li, Y.H., 1997. Interelement relationships in ferromanganese crusts from the central Pacific Ocean: their implications for crust genesis. Marine Geology, 136: 277–297.

Wigley, R., Compton, J.S., 2007. Oligocene to Holocene glauconite–phosphorite grains from the head of the Cape Canyon on the western margin of South Africa. Deep-Sea Research II, 54: 1375–1395.

Subject Index

www.ingramcontent.com/pod-product-compliance
Lightning Source LLC
LaVergne TN
LVHW011858200926
848484LV00002B/482
* 9 7 8 0 4 4 4 5 1 8 1 7 0 *